INDUSTRIAL WATER RECLAMATION AND REUSE TO MINIMIZE LIQUID DISCHARGE

WEF Special Publication

2021

Water Environment Federation
601 Wythe Street
Alexandria, VA 22314-1994 USA
https://www.wef.org

ISBN 978-1-57278-360-7

About WEF

The Water Environment Federation (WEF) is a not-for-profit technical and educational organization of 35,000 individual members and 75 affiliated Member Associations representing water quality professionals around the world. Since 1928, WEF and its members have protected public health and the environment. As a global water sector leader, our mission is to connect water professionals; enrich the expertise of water professionals; increase the awareness of the impact and value of water; and provide a platform for water sector innovation. To learn more, visit www.wef.org.

Prepared by the Industrial Water Reclamation and Reuse to Minimize Liquid Discharge Task Force of the Water Environment Federation

Brian Moore, PhD, *Chair*
David Riedel, PE, *Vice-Chair*

Jorge Aguinaldo
Ali Safarzadeh-Amiri
Sara Arabi, PhD, PE, BCEE
Tina Arrowood
Hossain M. Azam
Somnath Basu
Asher Benedict, PE
Todd Boykin, PE
Javier Casas
Christina Casler
Dylan Christenson
Joseph Cleary
Ian Cooper
Erik Desormeaux
Gennaro Dicataldo, PhD, PE
Karen Marie Dietze
Elyse Dumas
Val S. Frenkel
Mikel E. Goldblatt, PE
Al Goodman
Richard T. Grubbs
Edward Horai
Graham Juby, PhD, PE
Raj Kailasam, PhD
Dustin Klempel
Walt Kozlowski
Christopher Lewis
Yanjin Liu
Alison Ling, PhD, PE
George Patrick, PE
Peter Petersen
Patrick Regan
Eric Rosenblum
Kushal Seth
Marcia Sherony
Rajesh Shenoy
Mark Strahota
Derek Stevens
Paul Sun
Brent Sutter
Leila Tootchi
Jimmy Yu, PhD
Usama Zaher

Under the Direction of the Industrial Wastewater Subcommittee of the Technical Practice Committee

2021

Water Environment Federation
601 Wythe Street
Alexandria, VA 22314-1994 USA
http://www.wef.org

Special Publications of the Water Environment Federation

The WEF Technical Practice Committee (formerly the Committee on Sewage and Industrial Wastes Practice of the Federation of Sewage and Industrial Wastes Associations) was created by the Federation Board of Control on October 11, 1941. The primary function of the Committee is to originate and produce, through appropriate subcommittees, special publications dealing with technical aspects of the broad interests of the Federation. These publications are intended to provide background information through a review of technical practices and detailed procedures that research and experience have shown to be functional and practical.

Contents

List of Figures

List of Tables

Preface

This book serves as a comprehensive summary of water reclamation technologies and practices for minimizing liquid discharge for a broad range of industries. It will be of use to practitioners, consultants, management, and research and development professionals who want a timely update on the technologies in use and on those technologies on the horizon to provide the highest rates of water recovery in industrial practice. The industrial sectors covered include oil and gas (unconventional oil and gas and produced and petroleum refining water), pharmaceutical, textile, automotive, manufacturing, energy, and food and beverage. The book covers a much broader perspective than North America, including technology development and sustainable applications within the water-scarce yet growing regions of Central and South America, Asia, and the Middle East.

This first edition of this manual was produced under the direction of Brian Moore.

Authors' and reviewers' efforts were supported by the following organizations:

Amiri Clean Water
Arcadis
Barr Engineering
Black & Veatch
Brown and Caldwell
Carollo Engineers
CDM Smith
Envirospectives
Evoqua
Fluor
Geosyntec Consultants
Gradiant
Greeley and Hansen
Headworks International
PepsiCo
Phillips 66
Solenis
Stantec
Trojan Technologies
Woodard and Curran

Malcolm Mann of Saltworks Technology, Richmond, BC, prepared the section on Electrodialysis Reversal in Chapter 5.

Saurabh Deshpande of Headworks International, Houston, TX, conducted a thorough literature search and detailed editorial work in compiling Chapter 5.

1

Introduction

Sara Arabi, PhD, PE, BCEE; Elyse Dumas;
Alison Ling, PhD, PE; & Asher Benedict, PE

1.0 OVERVIEW AND PURPOSE

It is typically recognized that water is at the core of sustainable development and a key determinant in all aspects of social, economic, and environmental development. More efficient use of our water resources in the future will be vital to meeting all these needs. As a group, industrial processes use a significant amount of water, so industrial water reclamation, reuse, and recycling represent a major opportunity area for improving water efficiency. Industrial water reuse differs from municipal reuse in two key ways: the range of applications are more varied and the geographic locations are typically more dispersed.

This publication serves as a comprehensive summary of water reclamation technologies and practices for minimizing liquid discharge in a broad range of industries. It includes detailed technological information as well as case studies, with input from academia, equipment manufacturers, consultants, and industry. The intent of this publication is to update the state of literature on the topic and serve as a reference for industry practitioners and management, engineering and energy consultants, academia, and research and development firms. This publication addresses technologies that provide the highest rates of water recovery, including currently available and upcoming technologies. The industrial sectors to be covered are diverse; some highlight sectors include the oil and gas sector (unconventional oil and gas, produced and petroleum refining water); pharmaceutical; textiles; automotive; manufacturing; energy; and food and beverage.

2.0 ORGANIZATION AND STRUCTURE OF THE BOOK

The remainder of Chapter 1 defines water reuse and summarizes factors driving the industry to seek to minimize liquid discharges, followed by key (US-focused) regulatory considerations, opportunities, water quality implications, and a summary of available technologies.

Chapter 2 identifies opportunities to assess and execute industrial water reuse projects. This chapter introduces additional opportunities and challenges for selected industrial sectors and provides an implementation path for typical industrial wastewater reuse projects.

Chapter 3 outlines fundamentals of wastewater treatment, water reclamation, and reuse, with sections on biological treatment, precipitation, solid/liquid separation, membrane filtration, adsorption and ion exchange, reverse osmosis, and forward osmosis.

Chapter 4 focuses specifically on advanced oxidation processes (AOPs). Advanced oxidation process technologies include ultraviolet (UV)/peroxide, ozone/peroxide, UV/ozone/peroxide, UV with titanium oxide, and Fenton's reaction.

Chapter 5 addresses desalination and management of brines associated with industrial water reclamation and reuse. Desalination includes the discussions on electrodialysis reversal and freeze crystallization technologies. Brine management consists of brine concentration and concentrate disposal. Evaporation and dynamic vapor recovery are discussed as the technologies for brine concentration and thermal crystallization. Deep-well injection and landfilling are presented as the technologies for ultimate disposal of concentrate brine.

Chapter 6 discusses technologies for minimization of liquid discharge across a wide range of applications. Specifically called out are food and beverage, power, oil exploration and production, refining, and textile industries.

Chapter 7 provides information on the applicability of using pilot treatment systems for determining treatment process efficiencies and potential scale-up factors for full-scale treatment systems. Case studies for wastewater reuse in industries are also provided.

3.0 WATER USE AND REUSE

Water conservation efforts within industry can include the following categories: reduce, reuse and recycle (sometimes termed "R3"). Reduce projects entail using less water within a given process (i.e., reducing flowrate or duration of water use); reuse projects entail taking the waste effluent from one area and directly reusing it (as a water supply) within another area or process; recycle consists of treating a waste effluent to improve the quality such that it can be reused again. Frequently, these strict definitions for reuse and recycle tend to be relaxed, such that any project that involves using water again (whether with or without treatment) is termed as water "reuse". This book focuses upon recycle, as well as reuse of water within industry; however, for simplicity throughout the text we will often refer to these (reuse and recycle) projects as "reuse".

Water and energy use are inherently connected because of the large amounts of energy required to treat and transport water. Maintaining water in an engineering setting and treating it for a specified purpose through water reuse can reduce overall energy consumption associated with water needs. The U.S. Environmental Protection Agency (U.S. EPA) provides specific guidelines on improving energy efficiency and sustainability through water reuse (U.S. EPA, 2012).

While reuse of treated municipal wastewater for agricultural, commercial, or indirect or direct potable reuse has received significant attention, industries account for half of water use in the United States (Dieter et al., 2018). As a result, industrial water reuse has the potential to provide significant conservation of water resources as well as operational cost savings. According to a 2015 study, water withdrawals for thermoelectric power accounted for 41% of national use in 2015, with withdrawals for livestock and aquaculture at 3%, mining at 1%, and other self-supplying industries at 5% (Dieter et al., 2018). Some industries draw water from public water supply, which accounted for 15% of water use (Dieter et al., 2018). Industrial water use is a strong candidate for water reuse because water typically remains in a central location from withdrawal to use to discharge. Industrial facilities often have diverse water quality needs across operations, creating opportunity for reuse of effluent from high-quality users in lower quality applications. Water reuse also presents an opportunity for industries to save costs on water appropriation, water treatment, and wastewater treatment.

4.0 WATER REUSE DRIVERS

Water scarcity and sustainability are the primary drivers for water reuse. Until recently, water demands for domestic and commercial uses were met directly from naturally existing surface or groundwater via human-made storage and distribution structures such as dams, canals, and reservoirs. The rapid rise in the world's population combined with economic development, droughts, climate change, and mismanagement of water resources in some parts of the world have placed additional stress on the water supply. The resulting global water supply shortage has led countries, municipalities, and other governing bodies to seek alternate sources of water and improve upon existing water management practices.

The drivers for industrial water reuse may be broadly classified in three categories: costs associated with water supply and treatment, specific industry or process demands, and regulatory requirements. These primary categories, with effects from either physical, social, economic, or political factors, have led to many water conservation and reuse efforts. For example, in coastal areas with a high tourism influx, the social driver to protect marine ecosystems has produced water management practices to control negative effects of water resource recovery facility (WRRF) discharges. As another example, transporting water to locations without a local source, such as agricultural irrigation and onshore drilling activities, incurs high costs and environmental effects.

Table 1.1 presents these three key water conservation drivers. Water conservation measures may be undertaken to avoid high costs associated with freshwater purchase and/or wastewater treatment in the form of surcharges for exceedances. Permitting clarity is one item that affects decision-making for water reuse. An example of key performance indicator (KPI) metrics is water conservation or reduction in potable water usage, which are drivers for implementation of water reuse strategies. Site constraints include infrastructure limitations and the distance between source, treatment, and end use. Chapter 2 elaborates upon water efficiency drivers. An example of a regulatory driver would be requiring the permittee to improve processes to reduce the raw water demand, which is further discussed in the next section.

TABLE 1.1 Industrial water conservation drivers.

Water Supply Drivers	Industry/Process Drivers	Discharge/Regulatory Drivers
Supply costs	KPI water metrics used	Discharge costs
Water scarcity	Site constraints	Regulations parity
	Business branding/ stewardship	Discharge permit
	Financial incentives/ funding availability	Regulations clarity

5.0 OVERVIEW OF REGULATIONS FOR INDUSTRIAL WATER REUSE

There are no federal regulations or standards that govern the reuse or recycle of water in the United States. In the absence of federal regulations or standards, each state is allowed to develop programs for water reuse to meet specific needs while ensuring the reuse projects are designed, constructed, and operated in a manner that protects human health and represent a beneficial reuse of water. Existing state water reuse regulations typically focus more on municipal or agricultural reuse than on industrial reuse. Industrial reuse may reflect more of an internal cycling process with occasional blowdown to external waters, and thus is more subject to National Pollutant Discharge Elimination System regulations than to water reuse regulations focused on municipal reuse. When water reuse regulations do apply, they often depend on the specific application and final use of the waters and whether or not they are expected to come into contact with people enough to affect public health (U.S. EPA, 2012).

Only 25 out of the 50 U.S. states provide regulations or guidelines with the intent of overseeing industrial water reuse, with the most developed regulations being in Arizona, California, Florida, Hawaii, North Carolina, and Virginia (U.S. EPA, 2012). These guidelines are specified for cooling water blowdown, and typically include requirements for specific unit processes as well as water quality requirements associated with organic matters, suspended solids, and human pathogen indicators.

In addition to these state regulations, industrial users are also held to a production standard when their product is made for human consumption. In this case, the more stringent of the two standards typically applies. In addition, some regulating bodies also specify the need to have backup systems for reliability in the event of failure of a unit process.

6.0 WATER REUSE OPPORTUNITIES

Figure 1.1 presents a graphic of industrial water reuse opportunities.

In the industrial sector, reused water has been most broadly applied as makeup water for cooling towers in pulp and paper mills, textile mills, and other facilities. More recently, industries have embraced using treated wastewater as boiler feed, for equipment washdown, and in power generation. Another example currently being implemented is the reuse of water for hydraulic fracturing in the oil and gas industry. There is little evidence of applications for direct use of reuse water within products for human consumption because this typically requires the greatest amount of treatment to alleviate product manufacturing risks or unacceptable public health and safety concerns (WRF, 2016). Water reuse potential across different industries is rated in Table 1.2 (Visvanathan & Asano, 2009). As discussed in Chapter 2, water reuse potential is industry specific; some of the opportunities for each industry are also discussed in Chapter 2.

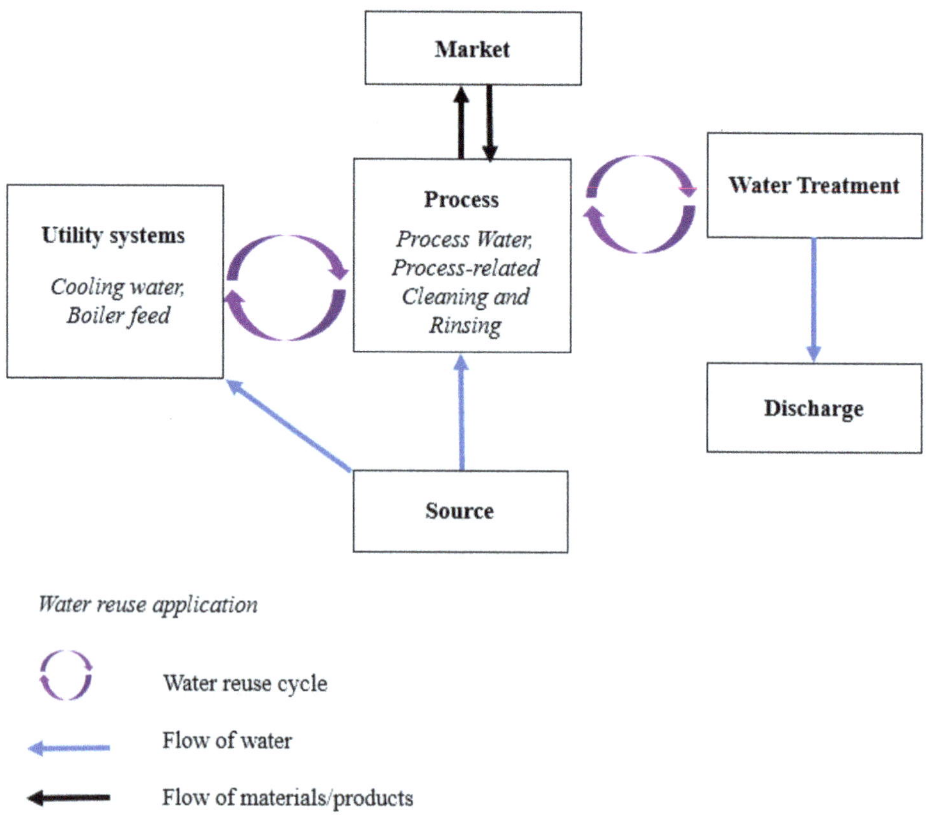

FIGURE 1.1 Industrial water reuse opportunities.

TABLE 1.2 Water reuse potential by industries (Visvanathan & Asano, 2009).

High Potential	Medium Potential	Low Potential
Pulp and paper	Slaughterhouse	Tanneries and leather finishing
Cotton textile	Dairy	Pesticide
Glass and steel	Canning and food processing	Rubber
	Distillery	Aluminum
	Wood	Explosives manufacturing
	Textile	Paint manufacturing
	Photographic processing	
	Chemical	
	Fertilizer	
	Oil refining	
	Petroleum	
	Electroplating	
	Metal processing	
	Power generation	
	Pharmaceuticals	

Figure 1.2 shows the water volume by use (%) across industry verticals for Canada (Kozar, 2018). It indicates medium to high reuse potential for the food, pulp and paper, downstream energy, metals and mining, and transportation equipment industries. While Figure 1.2 presents the reuse volume (%) for industries in Canada, this information can be easily used in reference to industries in the United States given the strong similarities in the industry operations and technologies in the two countries. In Figure 1.2, the upstream industry includes the exploration and production sector, and the downstream industry includes oil refineries, petrochemical plants, petroleum products distributors, retail outlets, and natural gas distribution companies.

7.0 WATER REUSE OPTIONS AND WATER QUALITY IMPLICATIONS

As technology advances and regulations or requirements surrounding industrial water reuse become more prevalent, new industries continue to explore expanded opportunities for water reuse, either in traditional water reuse applications or in emerging applications discussed throughout this book.

Many of the traditional industrial water reuse applications focus on utility demands such as heating and cooling. These applications are ideal

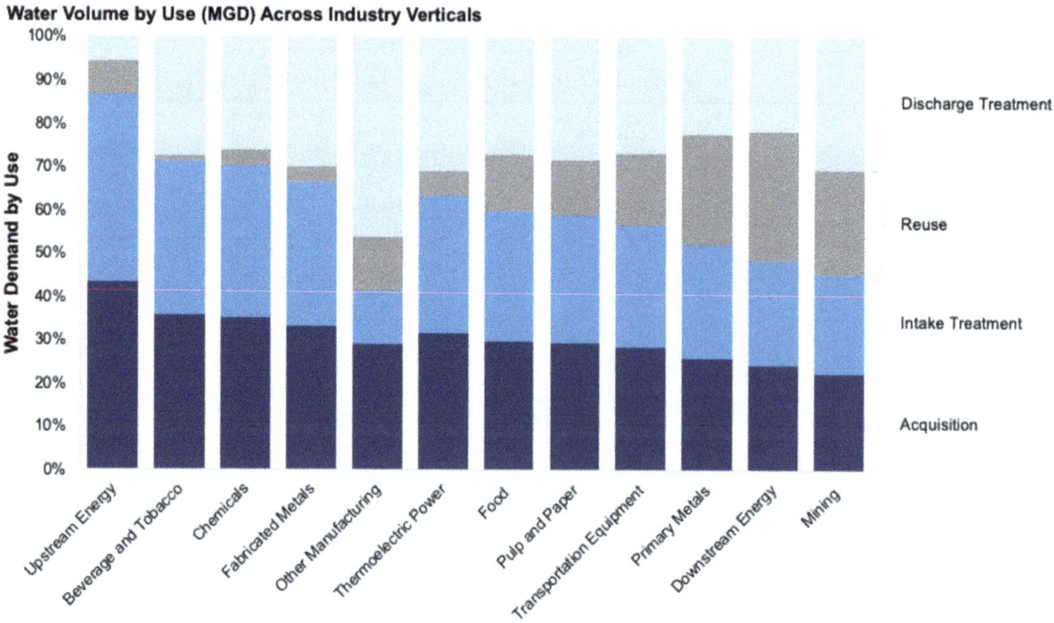

FIGURE 1.2 Water volume by use (%) across industry verticals (Kozar, 2018).

candidates for industrial water reuse because of the large amounts of water they use, but also the relatively low water quality they require. Reuse of water for cooling tower makeup allows for much higher concentrations of dissolved solids and other constituents when compared to drinking water or typical wastewater pretreatment requirements. Treatment systems that generate industrial cooling water target key contaminants, such as chlorides and multivalent ions, to optimize cooling tower performance and maximize cycles of concentration, whereas other contaminants that do not directly affect health and safety or cooling tower performance can remain untreated.

Another opportunity for industrial water reuse includes low-quality applications such as using recycled water for general washdown or for cleaning select pieces of manufacturing equipment. These applications benefit from the low risk of contamination in the unforeseen event of water treatment equipment failure. However, facilities must take care when deciding if, when, or how this lower-quality reuse water will be used. For additional information on how select industries are navigating industrial water reuse opportunities such as these, see Chapter 6.

Although the water quality required for these low-risk applications is often readily achievable, industrial facilities will not typically implement treatment systems solely for small volume water reuse applications. For this reason, these smaller, low-risk applications are often coupled with larger volume systems with multiple points of use within a single facility, such as cooling tower makeup if the facility has sufficiently large cooling needs.

Traditional water reuse applications in nontraditional industries (food and beverage, chemicals, pharmaceuticals, etc.) represent a growing opportunity for industrial water reuse. Although recycled water is used for the same applications as early industrial water reuse adapters (pulp and paper or textile mills, general manufacturing, etc.), these emerging applications may require reuse water to be of a quality similar to, or exceeding, potable water quality because of the environment in which they are used.

Depending on the industry and potential risk at the point of use, industrial water reuse systems may fall under the third category: future industrial water reuse applications. The use of recycled water for product washdown is not new, for example; but, if this washdown water is used on surfaces that directly contact certain food or beverage products (or the food and beverage product themselves), the reuse water must now meet or exceed primary drinking water quality standards set forth by U.S. EPA. Similarly, applying recycled water to generate steam is not new, but when this steam is used in clean rooms or areas of pharmaceutical production, it must meet a different, much more stringent level of quality than has historically been required, rendering such a project to be deemed high risk with minimal appetite for implementation.

Other emerging opportunities for water reuse do not require water quality to be as high as the aforementioned industries, but they face their own set of logistical challenges, which includes storage, delivery or transport, and proximity of use. Chapter 7 dives into the unique opportunities and challenges for water reuse in the power, oil and gas, and refining industries and a case study is presented in Chapter 7 for a pilot study of petroleum process produced-water.

Industrial water treatment for reuse must be seen as a fit-for-purpose exercise. If a facility cannot match the reuse water quality available with the water quality required for a certain process within the same facility, a partnership between two or more companies may provide a beneficial opportunity for reuse. A manufacturing company that generates a large quantity of wastewater may not be able to treat their wastewater to a high enough quality for reuse within internal process manufacturing, but a nearby power plant may be able to use that recycled water for cooling and power production. Partnerships such as these not only alleviate some of the water withdrawals that the purchasing partner would have otherwise needed, but could also generate revenue for the facility selling newly treated recycled water. Other successful examples of industrial water reuse partnerships include partnerships between refining sites, pulp and paper mills, and nonpotable municipal recycled water.

As direct and indirect potable water reuse become more commonplace in the municipal sector, regulations for municipal reuse water quality may set a precedent for future high-risk industrial reuse applications. Either through industrial quality requirements or municipal regulations, opportunities to couple resource recovery may help offset the increasing complexity of treatment required to produce higher quality reuse effluent and the systems necessary to provide the appropriate level of risk management.

It is important to note that recycled water quality for all of these industrial applications, especially high-quality emerging applications, is often doubted or scrutinized for its perceived danger. Companies must uphold their water quality requirements not only for the health and safety of their facility, employees, and the quality of their product, but also for their public image. Implications of failed water reuse quality can have crippling effects on a company's reputation, sales, and production. Chapters 2 through 5 contain guidance to develop a reuse project, including specific unit processes, on how to design and operate safe industrial water reuse treatment systems.

8.0 WATER RECYCLE TECHNOLOGIES

A variety of unit treatment processes have proven successful in industrial recycle projects. Some of the most common treatment processes include biological treatment, physical/chemical treatment, adsorption using granular activated carbon (GAC), AOPs, and membrane-based treatment processes such as microfiltration or ultrafiltration and reverse osmosis. The treatment approach for a specific reuse project is selected based on effluent discharge requirements and constraints specific to the users of the recycle water. Table 1.3 presents the advantages and disadvantages of common treatment technologies for water recycle.

8.1 Biological Treatment

Biological treatment is used to treat organics in industrial water and wastewater applications. Biological treatment is typically an end-of-pipe/facility-wide treatment approach rather than a point of use treatment. Industrial applications typically use aerobic biological treatment for low-strength wastewaters and anaerobic biological treatment for high-strength wastewater. Biological treatment is typically coupled with inorganics removal systems such as coagulation–flocculation as a pretreatment step or membrane filtration as a posttreatment or as part of the biological process itself (membrane bioreactor). This process produces biological sludge, which

TABLE 1.3 Application, advantages, and disadvantages of water recycle technologies.

Treatment Technology	Best at Removing	Advantages	Disadvantages
Biological treatment	Organics and nutrients	Proven effectiveness for organics removal	Produces biological sludge, significant energy consumption, prone to process upsets and influent fluctuations
Precipitation	Metals and suspended solids	High degree of metals and suspended solids removal, comparatively low capital cost	Significant sludge generated, disposal/landfilling costs, safety issues with some chemicals, can have high O&M cost
Solid–liquid separation	Particles and oils	Low capital cost	Some sludge does not settle well and may be difficult to handle, may generate odors
Membrane filtration (ultrafiltration or microfiltration)	Particles	Low suspended solids effluent achieved	Relatively high capital cost, no removal of dissolved constituents
Adsorption (GAC)	Organics	Less wastewater volume (backwash) to manage than reverse osmosis	Disposal required at bed exhaustion, media not regenerated on-site
Sodium zeolite softener	Hardness, multivalent cations	Less wastewater volume (brine) to manage than reverse osmosis	Highly saline waste brine can create disposal challenges in some cases
Ion exchange	Metals and salts	Lower capital cost than reverse osmosis	Requires strong acids/bases for regeneration, causing some facilities to opt for offsite (third-party) regeneration of media

(*continued*)

TABLE 1.3 Application, advantages, and disadvantages of water recycle technologies. (*Continued*)

Treatment Technology	Best at Removing	Advantages	Disadvantages
Reverse osmosis	Nearly all contaminants	High degree of removal for all constituents, reliable treatment	High capital and O&M cost, reject water may require treatment and disposal, requires pretreatment to limit fouling, fouling issues for waters with high hardness or silica, sensitivity to presence of solvents and other organics, highly sensitive to oils and greases
Forward osmosis	Nearly all contaminants	High degree of removal, lower operating costs than reverse osmosis	Relatively new technology, reject water may require treatment and disposal. Limited installed-base leading to reluctance for adoption in many cases
Advanced oxidation	Organics, disinfection byproducts, pharmaceutical and personal care products	High degree of degradation for target organic constituents	High capital and O&M costs, including high energy cost, requires pretreatment to limit competition for oxidant, extremely expensive to treat to full mineralization; generation of organic oxidation byproducts in typical application

O&M = operations and maintenance

requires disposal. In the past decade membrane bioreactors (MBRs) have become a favorable biological treatment process because of footprint savings compared to conventional activated sludge systems, improvements in energy efficiency, decreasing capital costs, and production of a high-quality effluent suitable for many reuse applications with little or no additional treatment.

8.2 Precipitation

Coagulation and flocculation processes have been widely used to remove inorganics and suspended solids by precipitating them from solution. The precipitation process using coagulation and flocculation is used as a pre-treatment step followed by solids separation. It enables the reduction of suspended solids, color and turbidity, colloidal particulate, and some dissolved metals. The most commonly used coagulants are trivalent aluminum and iron salts, and poly aluminum chloride. Polymer flocculants are often dosed to bind coagulated particles into larger particles for downstream removal. This process produces chemical sludge that requires disposal.

8.3 Solid–Liquid Separation

Solid–liquid separation is typically achieved using sedimentation, dissolved air flotation, or filtration processes for removal of suspended solids. Dual-media filtration is a filtration method that uses two different types of filter media, typically sand and finely granulated anthracite. Sedimentation tanks function based on gravity and addition of chemicals (i.e., coagulation and/or flocculation) may be required to enhance settling.

8.4 Membrane Filtration

Membrane filtration using microfiltration/ultrafiltration is frequently used for water reuse projects to remove particles. Membrane (microfiltration and ultrafiltration) filtration are physical filtration processes where water is passed through a membrane to separate microorganisms and suspended particles larger than the filter's pore size from the water. Membranes are divided into microfiltration, ultrafiltration, and nanofiltration based on their nominal pore size range. The typical nominal pore size of a microfiltration membrane is approximately 0.1 to 10 µm; for ultrafiltration, approximately 0.01 to 0.1 µm; and for nanofiltration, approximately less than 0.001 µm. The microfiltration/ultrafiltration processes can be used as the main water treatment process for reuse or as a pretreatment step for reverse osmosis. Microfiltration/ultrafiltration membranes replace clarification as the solid–liquid separation process in MBR systems. Nanofiltration is used for treatment of color and dissolved solids for industrial reuse applications. Characteristics of nanofiltration fall between ultrafiltration and reverse osmosis, and nanofiltration functions by both pore-size flow (convective) and the solution-diffusion mechanisms.

8.5 Adsorption and Ion Exchange

Activated carbon can be used to remove dissolved chemicals through adsorption. Adsorption of organics is commonly used in applications without a membrane process or can be used as a pretreatment step for removal of organics (and/or chlorine) before reverse osmosis. This is typically achieved using pressure vessels filled with GAC, which is replaced or regenerated regularly to maintain the adsorption capacity. Frequency of carbon replacement or regeneration dictates the viability of using GAC. Powdered activated carbon can alternately be used to achieve similar organics adsorption. Powdered activated carbon is typically added to a solids removal process, so it produces additional sludge and requires more volumetric use of carbon.

In an ion exchange process, a resin media binds and then exchanges certain ions (e.g., sodium or hydrogen) for other ions with a similar electrostatic charge (e.g., metals dissolved in wastewater). Ion exchange can remove both cations (e.g., metals) and anions (e.g., nitrates and sulfates) from wastewater, and produces a regeneration waste that must be managed.

8.6 Reverse Osmosis

Reverse osmosis is a method of removing dissolved salts and other constituents from water. Pressure is used to force the water through a semipermeable membrane that transmits the water but stops most dissolved materials from passing through the membrane. This is one of the only processes capable of reducing both total dissolved solids and total organic carbon. Pretreatment for reverse osmosis typically consists of cartridge or bag filtration as well as addition of special chemicals to prevent scale from fouling the reverse osmosis surface. Posttreatment could include alkali addition to make the water less corrosive and readjustment of the pH to acceptable levels for reuse.

Concentrate (i.e., reject, brine) management is a critical consideration in evaluating the viability of reverse osmosis for water reuse. In some cases/areas, the dissolved solids content of this brine is too high if a facility chooses direct surface discharge or discharge to a local municipal WRRF as brine management options. At coastal facilities, a brine waste stream can be discharged to a highly saline environment (i.e., ocean or sea). At inland facilities where discharge of brine to a surface waterbody or public wastewater facility is not permitted, solutions such as deep well injection would need to be considered. Other brine management options include evaporation ponds, brine concentration, mechanical evaporation, or off-site disposal. Significant research is currently being conducted in the water industry related to concentrate treatment and volume reduction strategies to reduce the associated costs. Chapter 5 further describes desalination and brine management options.

8.7 Forward Osmosis

Forward osmosis is the natural diffusion of water through a semipermeable membrane from a solution of a lower concentration to a solution with a higher concentration. The semipermeable membrane acts as a barrier that allows small molecules such as water to pass through while blocking larger molecules like salts, sugars, starches, proteins, viruses, bacteria, and parasites. Forward osmosis is considered a low-resource water treatment technology compared to energy-intensive reverse osmosis systems. Forward osmosis, in combination with a reverse osmosis reconcentration system, is a relatively new technology used to recover high-quality water.

8.8 Advanced Oxidation

Advanced oxidation is one of the processes that can be used as a treatment process in advanced water treatment systems. An AOP that generates the hydroxyl radical is typically used to provide destruction of recalcitrant and trace organics, converting them to organic byproducts. Hydroxyl radicals are typically formed through a reaction of some combination of ozone, UV light, hydrogen peroxide, chlorine, or reduced iron (Fenton's reagent). The UV-based AOP process combines advanced oxidation and photolysis. Hydrogen peroxide is an excellent source for generating hydroxyl radicals; however, it is an expensive and potentially hazardous chemical. Advantages have been demonstrated for UV/chlorine AOPs including effective trace organic oxidation, lower energy consumption required for the UV system, and lower chemical costs.

9.0 CHALLENGES OF WATER REUSE

One challenge of reusing treated wastewater is public perception of the water being dirty because its source is a "wastewater" treatment facility. The absence of dedicated distribution systems for nonpotable water and the potential for cross-contamination are other major challenges. While some states do have regulations and standards that protect water reuse projects from improper use, most states do not. The challenges to reuse are industry specific; Chapter 2 further elaborates upon these challenges for each industry.

As an example, a barrier to water reuse projects in the food and beverage industry is getting the buy-in from all regulatory stakeholders on the applicability of reusing treated wastewater. It is important to consider challenges for each industry because a driver for one industry may be a challenge for the other. For example, a regulatory framework for one industry may be a driver, while presenting itself as a challenge for another.

In general, high treatment cost associated with membrane treatment, which is a foundation for many water recycle processes, is a challenge for water reuse implementation. Pretreatment requirements are driven by the need to prevent unacceptable membrane fouling, and posttreatment is used to remove additional contaminants that are not removed in the membrane treatment systems. In addition, concentrate management can affect large costs and regulatory burden.

Long return on investment (ROI) timeframe is often considered a key market inhibitor for industrial water reuse. In many industries achieving 2- to 3-year ROI on reuse/recycling opportunities will encourage project implementation (WRF, 2016). Readily implementable reuse opportunities requiring lower capital investment are typically installed first (e.g., cooling tower makeup water), and studies for additional reuse opportunities are delayed or not implemented at full scale until the cost of treatment modifications will provide a reasonable ROI (WRF, 2016). To the same degree that an ROI analysis includes cost factors that are not reflected in a simple payback calculation, a triple bottom line (TBL) analysis captures factors a traditional ROI might ignore. Chapter 2 further elaborates upon TBL analysis and social ROI.

10.0 REFERENCES

Dieter, C., Maupin, M., Caldwell, R., Harris, M., Ivahnenko, T., Lovelace, J., Barber, N., & Linsey, K. (2018). Estimated use of water in the United States in 2015. *Geological Survey Circular,* 1441.

Kozar, M. (2018). *Opportunities in U.S. industrial water: Market size, trends, and forecast 2018–2022.* Bluefield Research.

U.S. Environmental Protection Agency. (2012). *Guidelines for water reuse* (EPA/600/R-12/618). U.S. Environmental Protection Agency.

Visvanathan, C., & Asano, T. (2009). Wastewater recycle, reuse, and reclamation. Vol I: The potential for industrial wastewater reuse. In *UNESCO and Encyclopedia of Life Support Systems.* Eoloss Publishers Co. Ltd.

WateReuse Foundation. (2016). *Drivers, successes, challenges and opportunities for onsite industrial water reuse: A path forward for collaboration and growth.* WateReuse Foundation.

2

Planning for Water Recycling and Reuse

Eric Rosenblum; Brian Moore, PhD; Christina Casler; Rajesh Shenoy; & Todd Boykin, PE

1.0 INTRODUCTION AND SUMMARY

Industry runs on water. As much as they require raw materials and energy, all industries to some degree depend on a steady supply of water for a wide range of uses. Indeed, without water no other resources (including energy) would be available. As noted in a U.S. Environmental Protection Agency (U.S. EPA) report, "Direct use of water [in] agriculture, forestry, mining, energy resource extraction, manufacturing, electric power production, and public water supply . . . supports activity elsewhere in the economy, creating a ripple effect as goods and services are produced and transferred through supply chains until they reach the final consumer" (U.S. EPA, 2012).

In the global marketplace, factory inputs can come from far away. In response, industry, as a whole, has invested significant time and money into enhancing supply chains for key resources, including raw materials, and energy. Recognizing water as a critical component, companies have begun looking for ways to bolster the security of their water "supply chain" as well. (Kammeyer, 2018; Morrison & Schulte, 2010). Even facilities located in areas with established utilities have invested in treatment facilities to enhance their water supply reliability through on-site wastewater reuse (Rosenblum et al., 2012).

Each of the treatment methods discussed in this book can produce water suitable for reuse. However, it takes more than equipment to overcome the challenges water reuse projects face. It takes careful planning and thoughtful engagement of facility staff and management, along with a thorough knowledge of facility operations.

This chapter provides guidance on how to plan and implement projects that reduce, reuse, and recycle water to minimize liquid discharge; the organization of the chapter roughly follows that process.

The first step is to understand how water is used in a given industry and facility (Section 1.1) and to establish the goals of a wastewater minimization program (Section 1.2). Motivations for water conservation include water supply shortages, discharge regulations and restrictions, cost-cutting initiatives, and corporate sustainability policies. Once a company's goals are clear, internal stakeholders at all levels must commit to them and accept responsibility for their parts of the planning process (Section 1.3).

The facility can then investigate opportunities to "reduce, reuse, and recycle" water and wastewater (Section 2.2). Because water use reduction is often the most economical way to reduce discharges, early assessment of conservation potential not only finds reductions requiring the least amount of effort, it also ensures a more accurate estimate of wastewater flows eventually available for reuse. For the same reason, connection to a municipal recycled water supply, if available, should be considered before investing in more costly on-site treatment options. (Although it does not minimize facility discharge, per se, use of municipally supplied recycled water reduces overall discharge into local receiving waters and reduces demand on local water supplies.) The process of identifying water reduction potential is facilitated by creating an inventory of the quantity and quality of water demands and wastewater discharges (Section 2.3). By comparing the quality of wastewater produced to the water quality required for reuse in specific applications (Section 3.2 and Section 3.3), a suite of reuse and recycle alternatives can be identified and analyzed (Section 3.3). Recycling options can be evaluated based on a range of quantitative (Section 4.1) and qualitative (Section 4.2) criteria to determine which are most appropriate for the facility involved. In addition to capital and operating costs, these criteria include residuals management, permitting complexity, and effect on current operations. Taken together, they serve as the basis for calculating the return on investment (Section 4.3), which can, in turn, be used to evaluate treatment alternatives (Section 4.4). Finally, the recommended water minimization plan must be discussed with facility staff and corporate decision-makers alike in a manner that provides them with the information they need to select and support the most appropriate project (Section 5.0).

1.1 Water Use by Sector

Industrial water use can be categorized into several distinct types, which are common across most industries. Table 2.1 provides a general perspective of how different industrial sectors use water according to their production process, and Table 2.2 highlights the distinct challenges to on-site water treatment and reuse for each sector. Because of the common nature of end uses, reuse concepts can often be copied within a sector. The following section highlights a few examples of water reuse in key sectors. Chapter 6, "Minimum Liquid Discharge Technology Application," also presents detailed descriptions and case studies of specific discharge minimization technologies by sector.

1.1.1 Food and Beverage

Aside from agricultural irrigation, water is used in the food and beverage industry for cooking, processing, and transporting products as well as cooling, cleaning, and disinfecting equipment, as well as being used directly within the product (California Department of Water Resources, 2013; Moore & Buzby, 2017). The food and beverage processing industry requires plentiful amounts of water, making this sector the third largest industrial user of water (Barbera & Gurnari, 2018).

Table 2.3 shows how each food and beverage manufacturing subsector uses water. Water reuse in the food industry must be carefully coordinated because of strict regulations regarding the quality of water used in production of food and food packaging. However, because water use varies with the quantity and the quality of the products, facility capacity, processes, equipment used, automation levels, and cleaning operations systems, the International Life Sciences Institute concluded the following (International Life Sciences Institute, 2008): "In food processing a broad range of possibilities exist with regard to water management, including increased efficiency of water use and the promotion of water reuse."

Wastewater from food processing is often high in suspended and dissolved solids, with high concentrations of organic matter—typically measured as chemical oxygen demand (COD); biochemical oxygen demand (BOD); and fats, oils, and grease (FOG)—as well as nutrients such as ammonia and phosphorus. As a result, recycled food processing wastewater is rarely used as an ingredient in food or beverage products. For example, Coca-Cola only reuses water at their bottling plants for nonproduct purposes such as

TABLE 2.1 Characteristic water uses in specific industrial sectors (adapted from Moore & Buzby, 2017).

	Facility-Wide Water Uses			Product-Specific Water Uses					
Industry Sector	Heating and Cooling	Advanced Water Treatment	Equipment Cleaning	Casting and Machining	Cleaning	Conveyance	Formulation	Testing	Painting and Coating
Aerospace	x			x				x	x
Automotive	x	x		x	x		x		x
Chemical and Refinery	x	x	x		x		x		
Flat Glass Manufacture	x				x				
Food and Beverage	x	x	x		x	x	x		
Oil and Gas Exploration						x	x		
Paint and Coatings		x	x				x		
Power Generation	x	x	x						
Pharmaceuticals	x	x	x		x		x		
Textiles	x	x	x		x	x			x

TABLE 2.2 Water reuse challenges (by industrial sector).

Industry Sector	Challenges to Reuse	Comments
Auto and Aerospace	Solids, oils, and other contaminants in wastewater Large area plants require extended distribution of recycled water	Need to account for changing production lines, introduction of new technology
Electronics, Data Processing	24/7 operation requires careful monitoring of cooling water use	Multiple benefits for increased reliability, brand differentiation
Food and Beverage	Federal and state health regulations for indirect and direct product contact Brand sensitivity Some uses require U.S. Food and Drug Administration (FDA) approval	Water-intensive sector; many reuse opportunities. Note that enhanced care must be taken to ensure that all water in products is suitable for human consumption.
Oil and Gas Exploration	Presence of proprietary chemicals to facilitate hydrofracturing complicates reuse	Application of appropriate treatment demonstrated
Pharmaceuticals	High-strength organic waste from biological processes Federal regulations for indirect and direct product contact Very high requirements for ultrapure water for high revenue, high cost products Low appetite for reuse where recycled water is perceived as presenting too much risk Some uses require FDA approval	Reuse opportunities because of research and testing as well as manufacture. Need to account for changing production lines. Many pharmaceutical products go into the body; extremely high barrier for the use of recycle water within or in direct contact with products.
Power Generation	Additional water sources needed where on-site wastewater sources do not meet cooling demands	Treat cooling water to increase cycles of concentration.
Refining and Chemical	Treatment for use in high-pressure heating and cooling Admiralty metals sensitive to ammonia in treated wastewater	Ready availability of existing low-cost water supplies might affect project return on investment. High pressure boilers have high water quality requirements.
Textile Manufacturing	Concerns about effect of variable total dissolved solids on dye transfer properties	Water-intensive sector often sited in water-scarce areas

TABLE 2.3 Water use in food and beverage processing and potential water reuse opportunities.

Types of Food and Beverage Processing	Types of Water Reuse Applications
Animal Processing	**Environmental Control**
Animal Food	Air pollution reduction
Bakeries	Area cleaning/dust control
Beverages	Wastewater treatment/reuse
Dairy Products	**Process Water Use**
Fruits and Vegetables	Inclusion in product
Grains and Oil	Fluming/transport
Seafood	Product washing
Seeds	Cooking/autoclaving
Sugars and Confectionaries	Blanching/precook
	Peeling and preparation
	Processing animal parts
	Canning and bottling
	Can/bottle cooling/warming
	Conveyor lubrication
	Pump seal water and other uses
	Cleaning
	Clean in/out or place systems
	Can/bottle/package cleaning
	Transport vehicle cleaning
	Crate and pallet washing
	Other cleaning
	Domestic Uses
	Sanitation
	Irrigation
	Thermodynamic Processes
	Cooling towers
	Boilers
	Refrigeration
	Cogeneration and thermal recovery
	Air-conditioning
	Humidification
	Laboratory Operations
	Water Treatment

landscape irrigation, truck washing, cooling towers, and warehouse floor washing. (Moore & Buzby, 2017). Nevertheless, a significant amount of water is generated by bottle washing, heating and cooling, sanitizing floors, and so on; thus, there are many reuse applications that are both nonpotable and "nonproduct."

Two examples from South Africa illustrate how food manufacturers have been motivated to reuse water in response to local shortages. In 2012, Unilever Corporation in Durban achieved water security by installing a membrane bioreactor (MBR)/reverse osmosis facility that supplied 80% of the facility's needs with treated effluent, with the remainder of appropriate nonproduct uses supplied by rainwater and air-conditioning condensate. In the Western Cape, Nestle, Inc., reused condensate from its milk evaporation process for washing tanks and in boilers and cooling towers, cutting their water use per ton of product in half (2030 Water Resources Group, 2013). Chapter 6, Section 6.1, contains detailed case studies of minimum liquid discharge in the food and beverage sector.

1.1.2 Power Generation

Power generation represents the single largest use of water in the United States, accounting for nearly one-half of all withdrawals (U.S. EPA, 2012). Water in this sector is mostly used for cooling steam-driven generators, and the predominant mode is "once-through" cooling, in which water is withdrawn from local supplies, circulated through heat exchangers, then returned to the ocean, river, or stream. A nonconsumptive use, once-through cooling is typically practiced where water supplies are plentiful; in California, nearly all once-through cooling withdrawals are limited to ocean or estuarine saline supplies (California Department of Water Resources, 2013; National Academy of Sciences, 2012). By contrast, recirculating cooling systems, which account for approximately 8% of power plant water withdrawals, typically use fresh water from local supplies that is treated to prevent scaling, fouling, and corrosion, and discharged as blowdown to water resource recovery facilities (WRRFs). These are the facilities where water recycling projects are most likely to provide substantial benefits in terms of reduced water use and minimized discharge.

Power plants typically go through a rigorous permitting process, especially when they are located near metropolitan areas where they will share municipal water supplies. For this reason, power plants with recirculating cooling systems are frequently required to conserve water and to use regional recycled water supplies, where available. If located near other industrial facilities, they may use treated effluent from other processes as a source for their makeup cooling water. Whereas geothermal plants may still require significant volumes of water for steam production, other renewable energy sources such as solar photovoltaic and wind may require only small amounts of water for periodic cleaning (California Department of Water Resources, 2013). Chapter 6, Section 6.2, provides additional details related to the application of water treatment technology in power production.

1.1.3 Oil and Gas Exploration and Mining

Most extractive industries, including oil and gas exploration and mining for metals and minerals, require significant quantities of fresh water. Because these activities largely take place outside of urban areas, extensive effort is expended to acquire a sufficient supply of suitable-quality water. Similarly, disposal can be equally challenging because wastewater must often be confined to tailings ponds or injected to disposal wells. For these reasons, many extraction enterprises incorporate water treatment and recycling.

As noted in Chapter 6, gas exploration through hydrofracturing (also known as hydraulic fracturing and "fracking") consists of injection of water at high pressure and high flowrates to fracture rock formation overlying gas pockets. Typical hydrofracturing operations may use from 15 000 to 23 000 m^3/well (4 to 6 mil. gal/well). A variety of organic and inorganic chemicals are added to the water before injection to facilitate the hydrofracturing process, enhance hydrocarbon recovery/extraction, prevent corrosion, provide lubrication, and to prevent clogging and control biological growth. As much as one-third of that chemically treated water returns as "flowback" when gas is released. Additional water is also returned to the surface throughout the life of the well as an admixture of oil and water ("produced water"), which is typically disposed of through injection in dedicated disposal wells.

Recycling both flowback and produced water can potentially reduce both water transport and wastewater disposal costs. In some instances, reuse of hydrofracturing water can assist in the compliance with state and federal environmental regulations, and zero discharge has been accomplished by drying the residuals for solids disposal.

1.1.4 Refining and Chemical Manufacturing

Refineries and chemical plants are significant water users because water plays an integral part in cooling and heating. Cooling is required to control the product or process temperature, to facilitate chemical product phase change, and to extract heat input during steam use. The use of steam is critical in refining and chemical manufacturing, providing heat to drive chemical reactions, and enhance separations. Recovering and reusing steam condensate preserves some heat and helps to reduce the fresh water needed for steam production.

Because of the large water volumes needed for their operations, refineries and chemical plants are often located along or near rivers (or other large waterbodies) where large volumes of water can be accessed at minimal cost. The low price, ready access, and relative reliability of such a water supply can make it difficult for plant managers to justify investing in water recycling infrastructure unless there are other incentives or pressures that encourage water reuse. However, this does not mean that water reuse is never practiced in these sectors; even large chemical facilities have, in some cases, been induced to conserve and recycle water in response to supply risks, such as severe drought (Rosenblum et al., 2012). In some instances, internal water recycling or the use of municipal effluent have been required as a condition of permitting the construction of a refinery or chemical plant. In one such example, with monsoons becoming scarcer on India's east coast, the city of Chennai's water supply was threatened by seawater intrusion to its shrinking groundwater basin. In 1991, two local industries, Chennai Petroleum Corporation Limited and Chennai Fertilizers Limited, began purchasing 12 ML/d (3.2 mgd) and 16 ML/d (4.2 mgd), respectively, of secondary treated wastewater from the Chennai Metro Water Board that they treated to tertiary standards via biological treatment, microfiltration, pressure sand filtration, ammonia stripping, and reverse osmosis. The cost of water from the tertiary facility was reported to be little more than one-half of the utility's industrial potable water price, 35 INR/m^3 versus 60 INR/m^3 ($1.86 in U.S. dollars/gal versus $3.11 in U.S. dollars/gal) (Arceivala & Asolekar, 2006). Similar examples of water reuse exist for refineries. Specifically, Chapter 6 includes a detailed discussion of internal water conservation, recycling, and reuse strategies in refineries, including steam leak elimination and cogeneration.

1.1.5 Textiles

Textile production typically comprises the following four steps:

1. yarn formation,
2. fabric formation,
3. wet processing, and
4. fabrication.

Wet processing is the most water-intensive use, and includes sizing, desizing, scouring, bleaching, dyeing, and finishing. The dyeing process has the highest water consumption and generates 15% to 20% of the total textile wastewater flow. Taken together, dyeing, desizing, and scouring generate nearly 70% of the total textile wastewater flow; washing and bleaching also use significant amounts of water.

Textile wastewater typically has high concentrations of COD, BOD, alkalinity, color, refractory organics, and toxic chemicals such as heavy metals (Patel & Vashi, 2015). Color is the most difficult to treat component in dyeing effluent. At the same time, the textile industry has relatively strict water quality requirements for color, turbidity, pH, and hardness specific to the processes used. Dyes have to react strongly with the cloth such that discoloration or staining does not occur. Hardness may cause precipitation of some dyes and might increase the breakage of silk during reeling and throwing operations, while high turbidity, iron, and manganese content can cause staining during the production stage (Treweek, 1982).

Like all industrial sectors, the use of water in textile manufacturing has a sociological dimension as well. Many textile operations are located in developing countries like India and Bangladesh, where they provide important income opportunities to skilled and unskilled labor alike. Their demand for water, on the other hand, may contribute to problems of local water scarcity, which can result in the imposition of government constraints on freshwater withdrawals (Grönwall & Jonsson, 2017). This regulatory incentive may be sufficient to inspire companies to invest in advanced water treatment and reuse schemes.

For example, textile plants in Tiruppur, India, produce 80% of the country's knitwear. Their effluent was historically discharged (after minimum treatment) into the Noyyal River, eventually degrading its quality and threatening the livelihoods of downstream farmers. As a result, in 2015, the Indian Ministry of Environment, Forests, and Climate Change required all cotton and wool processors discharging more than 25 kL/d (660 gpd) to achieve zero liquid discharge (ZLD) (2030 Water Resources Group, 2013). Industries responded by reducing their water demand by 75% through on-site reuse, cutting freshwater withdrawals by 2400 m^3/d (634 000 gpd), and reducing pollution by retrofitting effluent treatment facilities with reverse osmosis and thermal evaporation processes. The ZLD approach improved the quality of local water supplies. Textile plants even reused the residual salt from thermal evaporation in the dyeing process. Chapter 6, Section 6.5, includes a detailed discussion of water recycling and reuse in the textile sector.

1.1.6 Pharmaceuticals

Pharmaceutical companies use water like most other industries—in boilers and cooling towers, and for cleaning, toilet flushing, and irrigation. In addition, like food and beverage producers, water is used as a primary ingredient in the product, and is produced as a byproduct. As a whole, the pharmaceutical sector has high water quality requirements relative to other industrial sectors. Different grades of water are used at different points in the manufacturing process, including several types of purified water, typically created through reverse osmosis or distillation.

Waste streams in this sector come from the synthesis of biological materials, from cleaning equipment (including test animal cages and reactors), washing of components and products, sanitation of water loops and equipment, and from byproducts of water treatment to create the purified water needed for drug manufacturing. When reverse osmosis treatment is used to purify local potable water, the relatively clean reverse osmosis reject water can be suitable for other uses, sometimes without further treatment. By contrast, the effluent from the production of proteins and other materials through biological synthesis often has a wide variety of organic constituents that can present unique challenges (Costello, 2010; Markardan, 2017).

In one instance, a facility proposed treating a combined waste stream of cleaning waste, biological process waste, and reverse osmosis reject water (from purification of fresh water) to be reused in cooling towers, boilers, and other areas in the facility. Even with the relatively clean reverse osmosis reject, the combined waste stream still had high levels of organics so additional treatment (GAC, membrane filtration) was required to reliably render this mixed wastewater suitable for reuse. As a consequence, managers decided to supplement the cooling tower demand with reverse osmosis reject alone and meet the balance of their goals through reduced water use. Also, because new drugs are continuously being developed and manufactured in this sector, an assessment of the potential for reuse in a biopharmaceutical facility must consider which existing processes will be maintained and which will be discontinued. Changes in water use could affect both the water demand of potential candidates for reuse water and supply of wastewater for future needs.

1.1.7 Automotive, Aerospace, and Locomotive Manufacturing

From a high-level perspective, automotive, aviation, and locomotive manufacturing operations use water in relatively similar ways. Machining and cutting processes are commonplace within these manufacturing settings. Metal working fluids (sometimes referred to as "coolant") are critical in these processes, directly contacting the metal being machined and providing lubrication, cooling, chip removal, and corrosion resistance. These fluids are actually 85% to 95% water, containing only 5% to 15% of coolant. For cost purposes, metal working fluids are reused as many times as possible.

Water also plays an important role during the surface treatment, washing, and rinsing of metal. Significant volumes of water are used to prepare the metal surface of automobile and locomotives so that paint adheres strongly to the metal. This entails washing the vehicle to remove any oils or dirt on the surface, followed by subsequent chemical treatment steps to condition the metal surface so that it will strongly hold the paint primer coat. Extensive rinsing is required between the various steps, often needing high-purity water for such rinsing. This paint prep process will typically represent 50% to 70% of the overall facility water use at an automobile assembly plant, with cooling towers representing another 10% to 30% of a site's overall water use.

The automotive industry tends to be progressive relative to sustainability and making efforts to minimize their environmental footprint (Moore & Buzby, 2017; Sarni, 2011). Although water efficiency is often the main driver within the automotive sector, company sustainability goals will often also play a role. Above all else, the most impressive water use and discharge minimization in these manufacturing facilities is found where facilities are located in highly water-stressed regions. In one of Toyota's Indian manufacturing facilities, for instance, more than 94% of industrial water demand is met by a combination of rainwater harvesting and recycled water. In addition to an MBR/reverse osmosis treatment facility, a 25 000-m^3 rainwater harvesting pond was installed as part of the corporate sustainability initiatives (Toyota, 2018). Water reuse within the automotive sector is discussed at greater length in Chapter 6, Section 6.4, including a detailed discussion of a large water recycle project.

1.1.8 Electronics and Data Processing

The use of water in the high-tech sector has much in common with the pharmaceutical industry. For example, manufacturing silicon chips and other electronic components often requires significant quantities of highly treated, ultrapurified water. The wastewater generated often contains a variety of organic constituents (in this case, solvents) that must be removed before reuse in many applications.

High-tech manufacturers have used a number of innovative strategies for treating and reusing wastewater. Whereas wastewater is typically treated for on-site reuse or treated effluent is supplied by a public agency, in at least one case, effluent from the chip manufacturing process is treated to such a degree that it is used to replenish the local groundwater supply. Intel Corporation designs and manufactures silicon chips at its Ocotillo facility in Chandler, Arizona, where it uses rinse-water collection systems with separate drains for ultrapure water waste. This "lightly used" rinse water is treated to drinking water standards and returned to the underground aquifer to replenish the municipal groundwater supply, while water from the manufacturing processes is reused in cooling towers and scrubbers. In addition, the Ocotillo campus takes recycled water from the City of Chandler for use in mechanical systems, landscape watering, and farm irrigation. In 2010,

Intel recycled approximately 25 000 m^3/d of water (5.5 mgd), equivalent to 25% of its total water withdrawals (U.S. EPA, 2017).

Another aspect of the high-tech sector is the operation of data centers. These facilities are, in some sense, the inverse of electrical power plants, in that multiple rows of servers consume electricity on a constant basis, generating significant quantities of heat in the process. Because the single largest use of water at a data center is for cooling, ZLD can be approached through sidestream softening, acid injection, and other treatment methods that increase cycles of concentration in cooling towers. And while data centers are not themselves water producers, they are among the largest customers of municipally treated recycled water and, in some instances, have been colocated with industrial wastewater generators to form a symbiotic industrial ecosystem.

1.2 Incentives for Water Conservation, Recycling, and Reuse

Despite these critical uses in industry, company managers are not always aware of the advantages of water reuse. Water comprises a relatively small percentage of the facility budget, and in some cases water management is outsourced. For this reason, the importance of reducing, reusing, and recycling water must be clearly articulated to, and all stakeholders must be included in, the process of implementing discharge minimization projects. This requires a shared understanding of the motivations to reduce water use and minimize wastewater discharge; motivations are likely to differ between stakeholders.

Common drivers to minimize liquid discharge include the following:

- limitations on water availability,
- limitations on wastewater discharge,
- risk of future curtailments including shortages because of climate change,
- desire to reduce or eliminate discharges of specific constituents of concern,
- need to reduce water supply or wastewater discharge costs, and
- desire to present the enterprise as socially responsible ("green branding").

Because water shortages and discharge limitations are typically related to geographic factors, the incentive to reuse industrial wastewater is often less closely tied to a type of industry rather than to its location. In water-stressed areas, for instance, all industries will be pressed to find ways to minimize water use; in areas that discharge into sensitive waterways, they may have to reduce the amount of wastewater they discharge, or be forced to produce effluent clean enough for reuse. Another factor that may influence a company's incentive to invest in on-site wastewater treatment and reuse is the availability of nonpotable industrial water. Where local utilities already treat municipal effluent for reuse, it may be more beneficial to the facility (and the community) to connect to the existing recycled water distribution system.

1.3 Confirming the Commitment of Management and Staff

Armed with an understanding of the extent of water use within the facility and the company's goals for minimizing both water demand and wastewater discharge, one can begin the process of developing and designing a project to meet these goals. For such a project to be implemented, however, that process must include the active participation of employees at many levels and divisions within the company.

Industrial reuse projects typically need to compete with other proposed process improvements for limited funding. As a result, project sponsors must not only demonstrate the cost-effectiveness of their proposals, they must also gain the support of the internal stakeholders affected by the project. The commitment of those responsible for the operation and management of the facility is as fundamental to the success of a wastewater minimization project as the selection of an appropriate treatment method. Without their support, it is unlikely that the company will construct a recycled water project, or even that sufficient information will be collected to allow a suitable project to be designed in the first place.

The importance of this step was recently confirmed by researchers who developed a charter to identify key issues and engage stakeholders, confirming their roles and responsibilities (Layton & Bowden, 2015). Another approach is the "Water Kaizen Blitz™" (WKB) process, based on "lean manufacturing" principles. Section 5.0 describes both of these techniques for obtaining the necessary support and involvement of company stakeholders in detail.

2.0 MAPPING WATER USE AND WASTEWATER DISCHARGES

2.1 Introduction

Once water reduction goals have been established and a team has been assembled, the next step is to trace water demands and wastewater discharges throughout the facility. Although water is used in diverse ways throughout an industry, a comprehensive map of water demand and quality requirements and wastewater discharges at every point of use within the facility is neither necessary nor possible. On the contrary, an adequate survey can be made by measuring the major water uses and discharges by type and location, with an eye toward identifying opportunities to *reduce* use, to redirect wastewater for *reuse* where possible, and to *recycle* treated wastewater throughout the facility.

2.2 Overview of Water Reuse Opportunities

Water discharge reduction projects can be classified into three general categories corresponding to the classic rubric of environmental management: reduce, reuse, and recycle (sometimes referred to as "the three Rs" or "R3"). Even before water demand and discharge volumes and concentrations have been identified, a simple list of the largest water demands and types of use can often suggest areas where water savings and reuse opportunities are likely to occur. Although reuse and recycling are the primary focus of this publication, as explained in the following subsections, it is equally important to consider water conservation when developing minimum discharge projects.

2.2.1 Reduce: Cutting Back on Water Use

Reduce refers to projects that decrease the demand for water by cutting back on either flowrate or duration of flow to water-using processes. "Reduce" projects typically are the most cost-effective projects that offer the greatest return on investment (ROI). (Determination of ROI is described more fully in Section 4.3.) These projects can often be quite simple, such as a small change to an operating procedure or the addition of a control valve. In other instances, it may be necessary to look more closely at a major process modification, such as substituting dry deposition for wet painting or replacing a cooling tower with an air-cooled system. Moore and Buzby (2017) developed the following list of questions to assist facility operations staff in identifying opportunities to reduce the use of water in industrial processes:

- Is the water flowing at a higher rate than needed to achieve the desired purpose for water use?
- Is water flowing to this process at a higher flowrate than is required by the manufacturer of the equipment or as noted in the standard operating procedure (SOP)?
- Is the line pressure higher than required for the process/equipment?
- Is water flowing when it does not need to be flowing?
- Is flowrate controlled in the process?

- Is water use manually controlled (i.e., controlled by operators)?
- Is water being discharged from the process before it needs to be (i.e., could the water be used longer, reducing the amount of influent water needed)?
- Does the operational procedure result in wastage of water?
- Does the SOP represent a water-efficient operation?
- Are there opportunities to automate the use of water in a manually controlled process?
- Are there alternative manufacturing techniques that do not require water or that require less water?

2.2.2 Reuse: "Direct" Use of Untreated Wastewater

Reuse refers to those projects where wastewater from one process is used as the water supply for another with little or no intervening treatment. Because industrial facilities typically have a diverse array of water-using operations, each producing effluent of a specific quality and each with distinct water quality needs, sometimes the effluent from one process may meet the quality requirements of another, creating an opportunity for reuse. In that case, plumbing typically represents the greatest capital costs, and the distance between the processes may determine the feasibility of the project.

2.2.3 Recycle: Treatment of Wastewater So It Can Be Used Again

Recycle projects (which are the primary focus of this book as a whole) involve the treatment of wastewater generated by one or more processes, sufficiently improving the water quality to enable its reuse back within the same processes or elsewhere in the facility. Because of the need for water treatment/purification equipment, recycle projects tend to be the most expensive category of conservation projects. Recycle projects will typically also produce a concentrated waste stream that must be taken into consideration in the planning stage.

2.2.4 Reduce First!

It is worth reiterating that an effective discharge minimization program must begin with minimizing a facility's water use to protect the value of future capital investments. On the demand side, if improvements are designed and constructed to supply a process with a certain recycled water flow and that demand subsequently decreases resulting from conservation measures implemented at a later date, the result is a stranded asset and a decrease in the value of the reuse project. Similarly, if water demand is cut back on a process whose discharge has been identified for reuse, the amount of wastewater available for reuse will be reduced in turn. Early evaluation and recommendation of water use reduction opportunities is essential to ensure that water reuse and recycling projects will achieve their projected goals.

2.3 Quantifying Water Demands and Quality Requirements

A facility *water balance* can help identify opportunities to reduce water use and minimize discharges. When considering water reuse and recycling projects, the water balance should include the *water quantity* (volume and flow) and *water quality* (salinity, pH, etc.) characteristics of both upstream water demands and downstream wastewater discharges, as well as the effluent from any treatment processes. This data can be used to identify opportunities to reuse water of varying quality in different areas of the facility. A water balance can provide critical information needed to identify major water consumers, reveal high-strength internal waste streams, and determine overall water and operational expenses (OpEx) savings to be expected for a given project as well as the financial return.

When a water balance has been developed from current and historical meter readings and laboratory data, it is important to "ground truth" the information by physically inspecting each of the major water-using processes in the facility, because field conditions may reveal important information not shown on the balance. Also, as previously noted, the effect of planned water efficiency improvements on water volumes should be considered to ensure that sufficient water will be available for reuse to meet the demand of identified applications.

A good water balance should describe the volume and flow of water used by processes throughout the facility, as well as the type of water used by each (e.g., city water, softened water, reverse osmosis water, etc.), but each individual meter need not be shown. As Figure 2.1 shows, water type can be depicted within a balance by color-coding the lines accordingly. Most sites will have minor water users that will not present much opportunity for water optimization (e.g., hand washing, safety showers, small process streams, etc.); these can be excluded or grouped together and labeled "minor uses."

The diversity of manufacturing processes can, indeed, make it feel overwhelming for an individual who is trying to identify opportunities to save water. However, considering the way in which water is being used (i.e., the purpose water serves) can greatly simplify the task of identifying/understanding all uses of water and develop water efficiency ideas for these processes (Moore & Buzby, 2017). Shifting one's focus from the "complexity of a piece of manufacturing equipment" to the "purpose served by water in

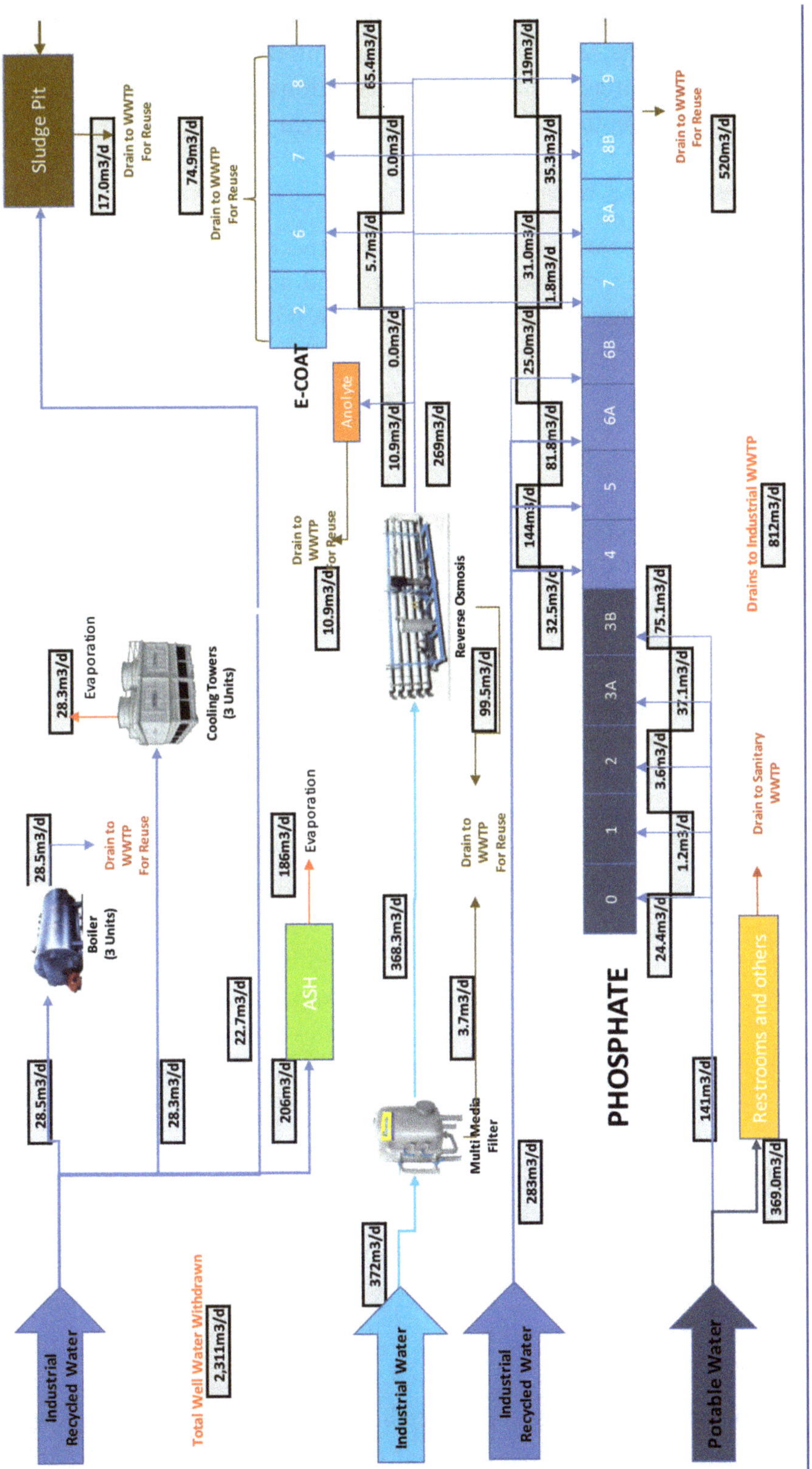

FIGURE 2.1 Sample water balance.

the process" can help one understand the rationale behind water use and quality requirements, which is helpful when attempting to identify water efficiency opportunities. The following subsections describe some of the key process areas that might be identified in a typical industrial water balance.

2.3.1 Washing and Rinsing

Product and process washing and rinsing activities (distinct from general cleaning) constitute a significant water demand at many facilities, and discharge from these processes can be clean and suitable for reuse. The specific purpose behind these activities varies by sector, but typically the water is used to clean or sanitize production areas or equipment associated with manufacturing, or to clean, condition, or sanitize raw materials or products before or during production. The following examples illustrate the range of uses:

- **Food and beverage/pharmaceuticals:** Washing and rinsing are strictly prescribed for quality control and to satisfy requirements by the U.S. Food and Drug Administration (FDA). Vessels that are used for the formulation of food products or pharmaceuticals that will enter the body (orally or via injection) are routinely washed and rinsed with extensively purified water to ensure cleanliness of the vessel; this is followed by a sanitization step where the vessel is exposed to high temperature water (often ~80 °C or ~175 °F in the pharmaceutical industry) or aggressively contacted with steam for an extended period to ensure no microorganisms or chemicals are present.
- **Automotive:** Washing and rinsing are critical steps in preparing a vehicle for painting. Some steps in this process can use city water, but many of the final rinse steps require highly purified water to ensure the vehicle surface is sufficiently clean to prevent potential paint coat defects. Counter-current rinsing practices (using the cleanest water at the end of the process and cascading it to reuse further upstream) are typically used within automotive assembly plant paint shops to reduce water use.
- **Refineries:** One washing application involves the use of water for washing the raw crude before processing to remove salts from the oil. This process does not have high water quality standards, typically reusing water from other processes.

2.3.2 Cooling

Many manufacturing processes produce heat, which needs to be removed for equipment to function properly. Because of its high heat capacity, abundant availability, and relatively low cost, water is typically used to extract heat from these processes. Many manufacturing processes will use heat exchangers, which are designed to enable efficient transfer of heat from a hot gas or liquid (e.g., oil, water, etc.) into a heat-receiving fluid, commonly water. Figure 2.2 depicts the fundamental function of a heat exchanger. After the cooling water exits the heat exchange process, it must be cooled before it can be used again for cooling purposes. In most cases, this is achieved through use of a cooling tower, as illustrated in Figure 2.3.

Water quality is critical to the cooling processes. As water evaporates in a cooling tower, dissolved constituents concentrate in the recirculating stream. Because heat exchangers typically have high skin temperatures, minerals in the cooling water can form scale that reduces the ability for heat to be transferred into the water, so a fraction of the water must be discharged ("blowdown") and supplemented with fresh water ("makeup") when operating setpoints are reached. Water hardness from calcium and magnesium, alkalinity, and silica can lead to the formation of scale upon heat exchange surfaces, and interfere with heat transfer. Chloride, sulfide, and aggregate dissolved minerals (measured as electrical conductivity) can all contribute to metal corrosion in heat exchangers, cooling towers, and piping. Coupons (small strips of metal) are inserted in the cooling water to detect problems, and water quality is routinely evaluated (e.g., Langelier Saturation Index ([LSI]) to estimate the potential for scaling and corrosion.

2.3.3 Manufacturing

Water quality requirements for manufacturing use vary widely between industries and processes. In general, the pharmaceutical, food and beverage, and electronics manufacturing sectors have higher water quality

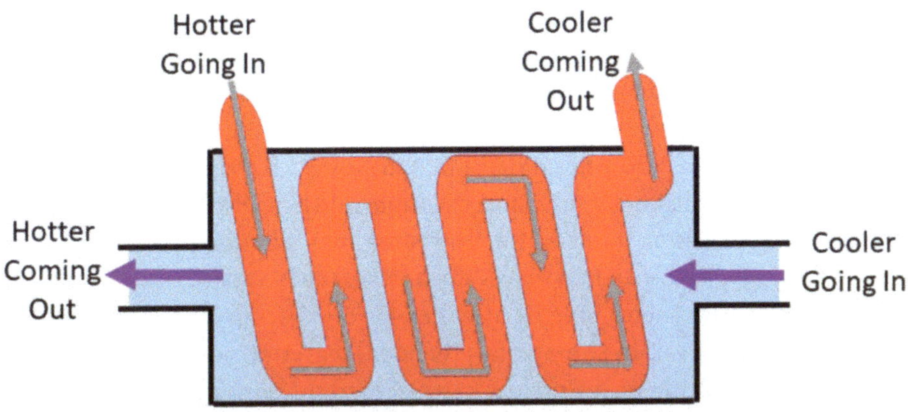

FIGURE 2.2 Diagram of a heat exchanger.

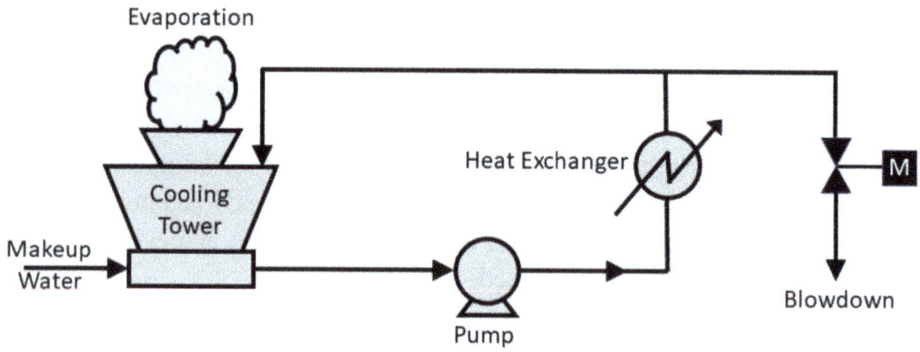

FIGURE 2.3 Cooling tower schematic.

requirements than other sectors, whereas oil and gas, mining, and pulp and paper industries tend to have lower water quality requirements. Manufacturing operations/processes with less stringent water quality requirements present a greater opportunity for reuse/recycle of water than those with high water quality needs, while those strictly regulated by the FDA or other agency may present and even greater challenge to water reuse.

2.4 Identifying Wastewater Discharges

A complete water picture must include the individual wastewater streams generated from major processes, including flowrates, major constituents, and relative contribution to the ultimate wastewater rate and composition. To match wastewater flows with reuse opportunities, it is essential to confirm wastewater quality, quantity and timing, how this varies, and what factors may affect it in the future.

2.4.1 Confirm Wastewater Flows

Unfortunately, water quality is rarely considered when developing a water balance, which hampers the analysis of reuse opportunities. Even identifying all pertinent effluent streams often requires referring to original facility design drawings or yard piping diagrams, followed by visual confirmation. (There is no substitute for walking through a facility following the water streams.) In addition, wastewater flowrates may change dramatically on an hourly, daily, and seasonal basis. If no permanent wastewater flowmeter exists, a temporary meter may be attached to key discharge line(s) to record hourly or daily flows. (In some areas, the local water or power utility makes such meters available at no or low cost.) Seasonal variability may require reference to facility-wide sewer discharge data. If the wastewater discharge is tied directly to facility production, it may be possible to estimate flows based on manufacturing records.

The effect of daily variability on the availability of wastewater as a supply is typically addressed through storage. Wastewater storage upstream of recycled water treatment may be needed to stabilize treatment processes and recycled water storage after treatment can help ensure consistent availability of treated water. Seasonal and market-driven changes in wastewater flows can be more difficult to manage. For example, changes in the type of item produced (or its rate of production) may limit the availability of recycled water. Just as designers must take into account the effects of water conservation, they must also anticipate the effect of potential changes in facility operation on wastewater supply availability. This consideration further reinforces the need to coordinate with all levels of facility management, including marketing and product development executives who may be more aware of planned changes to future facility operation.

2.4.2 Characterize Wastewater Constituents

The business case for an industrial water recycling project is influenced significantly by the cost of additional treatment required, which is dependent on the degree to which wastewater quality must be improved before it is reused. As a result, the success of an industrial water recycling process hinges on having an accurate wastewater quality characterization.

Some wastewater quality parameters (e.g., pH, conductivity) are easily measured with readily available data. Other important constituents, however, including BOD, nutrients, heavy metals, and volatile organics, typically need to be sent to a laboratory for analysis. Because these tests may be expensive, it is important to decide which parameters are important for the potential water uses identified and which wastewater discharge locations should be sampled. Sampling frequency should also be considered if water quality varies widely. Analysis of a composite sample (versus several grab samples) may be sufficient to characterize wastewater if storage is needed for the project.

For projects that do not involve additional treatment, the water quality needs of the processes that will reuse the wastewater determine the sampling parameters. It is important to assess the potential for product interference or the effect on process equipment. Table 2.4 lists some problems that may be encountered with the reuse of wastewater and the constituents with which they are associated.

In addition, wastewater recycling projects that require further treatment before reuse should include sampling for all constituents that might add complexity to the treatment process. For example, where membrane processes like reverse osmosis are used, the wastewater should be tested for oil and grease, COD, total suspended solids (TSS), turbidity, hardness, alkalinity, silica, iron, barium, strontium, and particulate density (silt density index). It may also be necessary to test the residual stream from the treatment process to ensure

TABLE 2.4 Sampling required to identify potential problems with wastewater reuse (Adapted from Victoria University Institute for Sustainability and Innovation, 2007. Reprinted with permission).

Potential Problem	Possible Cause	Parameters to Be Sampled
Corrosion	Pitting or thinning of metal because of acids, surface-active anions	pH; chloride and other halogens
Fouling	Biological growth, other surface interference with flow or heat exchange	BOD[a], TOC[b]; nutrients (phosphorus, total Kjeldahl nitrogen); fats, oils, and grease
Incomplete cleaning	Deposition of suspended or dissolved solids	TSS[c], TDS[d]
Overheating	Reuse of heated water	Temperature
Scaling	Precipitation and deposition of divalent cation oxides on metal	Alkalinity, hardness (calcium, magnesium), silica

[a]BOD = biochemical oxygen demand
[b]TOC = total organic carbon
[c]TSS = total suspended solids
[d]TDS = total dissolved solids

that wastewater from the facility continues to comply with local pretreatment discharge limitations and federal and state hazardous waste limits.

2.4.3 Locate Separated and Combined Waste Streams

Along with aggregate wastewater quality, it can be useful to sample individual upstream wastewater sources. Such sampling can pinpoint the origin of problematic constituents that may be isolated so that the remaining flow is easier to treat. Most industrial facilities combine wastewater from multiple processes to minimize plumbing costs, combining less contaminated streams that may be suitable for reuse with more heavily contaminated discharges. Segregation of these cleaner streams can simplify design and lower water reuse project costs.

Figure 2.4 depicts this situation. A manufacturing site required to reduce its discharge volume because of municipal water resource recovery facility (WRRF) capacity issues. The facility desired to reuse effluent within cooling towers and the scrubber, but the level of total dissolved solids (TDS) in the treated wastewater precluded direct reuse. To address this issue, it was planned to install a recycle system consisting of microfiltration and reverse osmosis at a cost in excess of $1 million (U.S. dollars). Fortunately, additional sampling of individual wastewater streams revealed that 90% of the

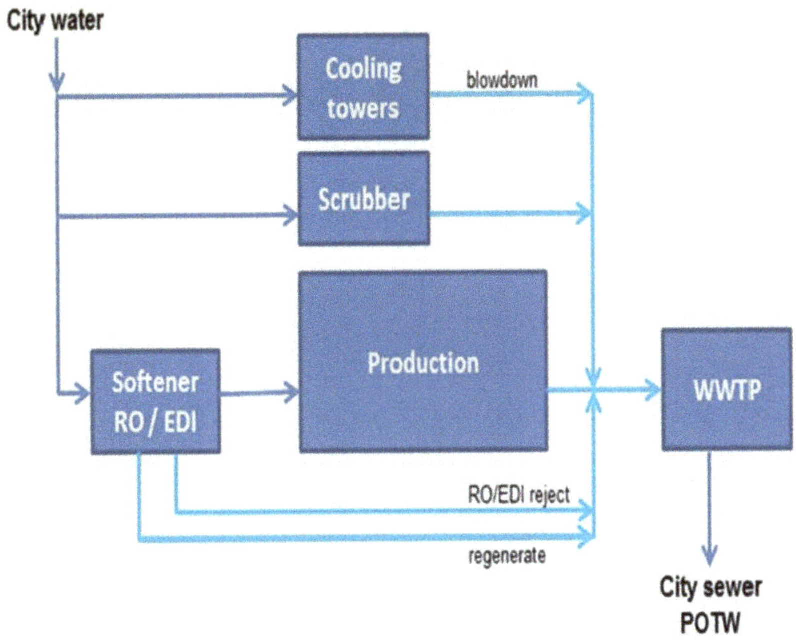

FIGURE 2.4A Block diagram of existing industrial facility without reuse.

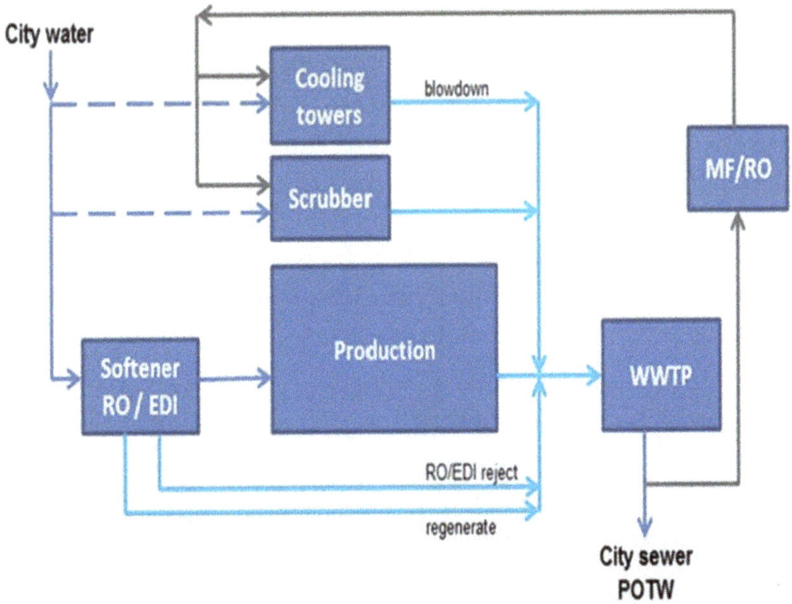

FIGURE 2.4B Block diagram of proposed reuse following microfiltration/reverse osmosis treatment.

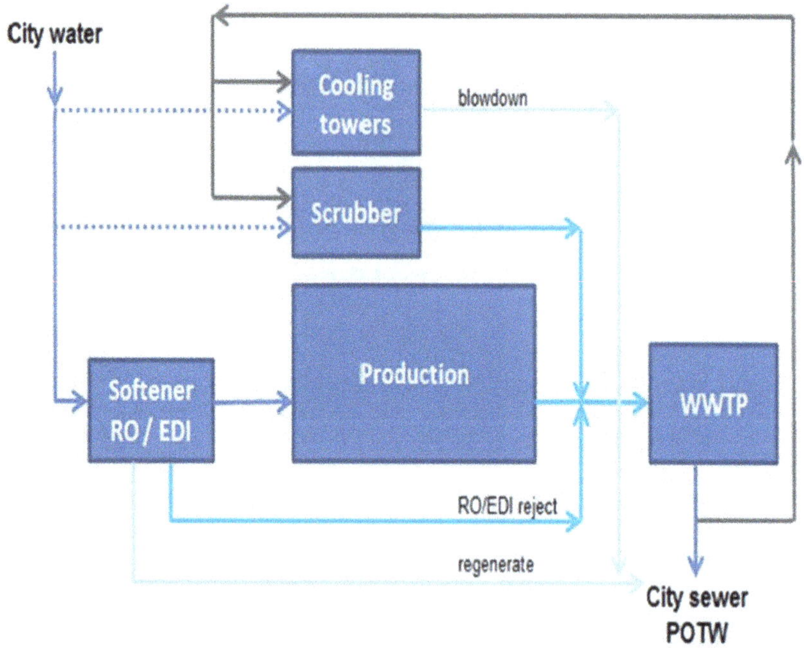

FIGURE 2.4C Block diagram of proposed reuse with isolation and diversion of softener reject and cooling tower blowdown to the municipal WRRF and no microfiltration/reverse osmosis treatment.

FIGURE 2.4 Block diagrams of water uses comparing alternate water reuse strategies.

TDS loading originated in waste from (a) water softener regeneration, (b) electrodialysis reject, and (c) cooling tower blowdown. Because these streams were also low in BOD, they were discharged directly to the municipal WRRF, thus making the industrial facility effluent sufficiently low in TDS to allow direct reuse in the cooling towers and scrubber without need for the recycling facility. As a result, the facility was able to reduce water use by 38 000 m^3/yr (10 mil. gal/yr) and reducing annual operating costs by more than \$100,000 while avoiding a \$1 million capital expense.

3.0 DEVELOPING TREATMENT ALTERNATIVES FOR WATER RECYCLING AND REUSE APPLICATIONS

3.1 Introduction

Recycled water can be manufactured to suit virtually any type of industrial use. The steps required to identify appropriate treatment alternatives to match recycled water demands are addressed in the following sections:

- Comparing the volume and flowrate of available flows with potential demands (Section 3.2)
- Comparing wastewater quality with water quality requirements of potential reuse applications (Section 3.3)
- Identifying treatment alternatives needed to produce water of appropriate quality and quantity for reuse (Section 3.4)

Once identified, treatment alternatives must be evaluated with respect to a wide range of criteria to find the project most suitable to accomplish management goals (Section 4.0).

3.2 Matching Wastewater Supply and Demand

Wastewater streams should be analyzed in parallel with the water demands, keeping in mind storage needs and loss of water through treatment trains (if applicable) to produce a water reuse project. In this phase, it is critical to recognize that the reliability of both wastewater discharges and water demands could be affected by water conservation measures or process changes.

As noted in Section 2.3, many wastewater flows are periodic in nature and vary with production types and schedules. In fact, large, continuous wastewater flows in industrial facilities are commonly those most readily reduced through conservation efforts. As a result, it is common for some type of storage to be required before wastewater treatment, even when batch treatment methods are used. By the same token, product water from the recycled water treatment process may also be stored to provide a steady supply of water for targeted reuse opportunities. By balancing influent and effluent storage, highly variable flows can be directed toward appropriate reuse, provided the overall volumes are sufficient to meet their intended demands.

3.3 Matching Wastewater Quality and Use

Once the quality of wastewater produced is assessed and high-strength streams have been identified and isolated, as appropriate, wastewater quality can be compared to the requirements of the water demands identified in the water balance.

3.3.1 Direct Reuse of Industrial Wastewater

As noted in Section 2.1, after water reduction, direct reuse of water without additional treatment is typically the most cost-effective method of reducing industrial discharges. There are many instances where this type of reuse can be identified. For instance, where products, containers, or facility equipment are subjected to multistage cleaning, the wastewater from the final rinse is frequently suitable for use in the primary or secondary wash cycles. (See Chapter 6 for a discussion of this type of reuse in food and beverage manufacturing.)

Similarly, when ultrapure water is provided for the manufacture of chemically sensitive products (pharmaceuticals, microchips), wastewater from advanced treatment is often appropriate for reuse. Although membrane treatment processes concentrate the constituents in potable water, their reject streams may even be suitable for cleaning or cooling, or for nonindustrial applications like toilet flushing.

Another approach is to redistribute wastewater without additional treatment after it has reached the quality required for disposal to the sewer system. In typical large manufacturing facilities, before discharge, all wastewater is treated at a minimum for oil and grease separation, flocculation, and precipitation to remove metals and pH stabilization. When sufficiently high levels of organics are present, biological treatment is common as well. The water produced by this treatment may be appropriate for a variety of in-facility uses such as makeup for cooling towers or scrubbers.

3.3.2 Treating Recycled Water for Reuse

When the opportunities for direct reuse have been exhausted, additional on-site demands can be satisfied by further treatment, tailored to provide water suitable for the greatest number of current and future demands. A wide range of technologies are available, singly or in combination, that can treat wastewater with virtually any degree of contamination to the degree that it can meet any industrial demand. However (and as further discussed in Section 3.4), the application of these technologies is always limited by local conditions, most often the budget allocated for the project and the location available for its construction.

With that in mind, a review of wastewater test results will indicate which constituents might interfere with potential applications, and to what level they must be removed to optimize water recycling. At that point, a preliminary assessment can be made of which treatment methods can produce water of acceptable quality. Table 2.5 presents a number of contaminants in treated wastewater and the potential problems they might cause in typical industrial applications. (Although this study refers to municipal wastewater, many contaminants are found in industrial wastewater as well.)

These contaminants can be removed by a variety of methods, each of which should be considered on a preliminary basis and screened for feasibility before being subjected to a more rigorous evaluation. Table 2.6 provides general guidance with respect to efficacy of various common treatment technologies for treatment of select contaminants. In conjunction with

TABLE 2.5 Summary of recycled water quality parameter and potential industry effects (Adapted from Victoria University Institute for Sustainability and Innovation, 2007. Reprinted with permission).

Parameter	Health	Corrosion	Scaling	Fouling	Process
Alkalinity		Y	Y		
Ammonia		Y		Y	
Calcium, Mg			Y		
Chloride		Y			
Color					Y
Dissolved solids		Y	Y		
Fluoride			Y		
Hardness		Y	Y		
Metals	Y				Y
Microbiological	Y	Y	Y	Y	PD[a]
Nitrate		Y			
Odor (BOD/DO)	Y				Y
pH		Y	Y		
Phosphate		Y	Y	Y	
Residual organics	Y			Y	
Salinity		Y			Y
Silica			Y		
Sulfate		Y	Y		
Suspended solids				Y	Y
Temperature	Y	Y	Y	Y	Y
Turbidity		Y	Y	Y	

[a]PD = process dependent

Table 1.3 (Chapter 1), this table can be used to identify technologies that may be appropriate for a given contaminant of interest.

Each of these technologies has different space requirements and operational characteristics as well as different capital and operating costs. For detailed information about the theory, design, and application of specific

TABLE 2.6 Treatment alternatives for removal of selected industrial waste contaminants (adapted from Moore & Buzby, 2017).

Treatment Method	Selected Contaminants							
	Suspended Solids	Oil and Grease		Mono-valent Ions	Dissolved Heavy Metals	Hardness	Dissolved Organics	Target Trace Organics
		High	Low					
SEPARATION AND FILTRATION								
Bag Filter	◑		○					
Sand Filter	◑		○					
Oil Water Separator		●	◑					
Dissolved Air Flotation	◑	◑	◑					
COAGULATION AND CLARIFICATION								
Chemical Precipitation	◑	◑	◑		●		○	
Chemical Softening	◑	◑	◑		●	●	○	
ADSORPION AND ABSORPTION								
Granular Activated Carbon	○		●				●	●
Ion Exchange Resin				●	●	●		

Organophilic Resin							●	●
Organoclay	○	◑	●					
OXIDATION								
Ozone			○				●	●
UV/H_2O_2							●	●
Fenton's Reagent			○				●	●
MEMBRANES								
Microfiltration	●		◑					
Ultrafiltration	●		●					
Nanofiltration					●	●	○	
Reverse osmosis				●	●	●	●	●
BIOLOGICAL								
Activated Sludge	○		○				●	●
MBR	●		○				●	●

Treatment Efficacy Symbol Definitions: ◑ = Poor to Fair; ○ = Fair to Good; ● = Good to Excellent
Source: Adapted from Tables 5.1 through 5.8, Moore & Buzby (2017).

treatment technologies, the reader is directed to Chapter 3, "Fundamentals"; Chapter 4, "Advanced Oxidation"; and Chapter 6, "Minimum Liquid Discharge Technology Applications."

3.4 Preliminary Screening and Development of Treatment Alternatives

Developing industrial wastewater treatment alternatives for reuse involves more than matching equipment performance to water quality specifications. A properly designed treatment alternative will also address other issues, including

- the water quality requirements of each of the potential recycle water customers,
- the compatibility of the treatment train with facility operations and maintenance (O&M), and
- the proximity of recycled water demands to the source and the need for a separate water distribution system.

Although these factors will be weighed again during the evaluation of a project, a preliminary screening exercise should be done to ensure that the treatment alternatives are among the most economical before committing them to more thorough—and more costly—project design.

Always an important aspect, project cost is largely influenced by the following three factors:

- The degree of treatment required
- The volume of water to be recycled
- The proximity of recycled water demands to the treatment system and to each other

Table 2.7 illustrates how these three factors can independently influence cost, with the uppermost cell (i.e., lightest color) in each column reflecting the lowest cost condition and the lowermost cell in each column representing the highest cost condition for that category. Projects will typically be the most cost-effective where lightly contaminated wastewater is available to serve a large volume of lower-quality water uses located near the treatment system. From this perspective, the development of treatment alternatives is frequently an iterative process, where many factors are repeatedly reviewed during the course of the project. The larger the project and the more variety of sources and end uses, the longer the analysis period.

TABLE 2.7 Correlation between project cost and selected industrial water reuse criteria.

Wastewater Treatment Requirements*	Volume of Reuse Demand	Proximity to Treatment System	Relative Cost of Alternative
Low	High volume	Adjacent	$
Moderate	Medium volume	Nearby	$$
High	Low volume	Remote	$$$

During the development of treatment alternatives, desktop analysis may not be enough with respect to the suitability of, or design criteria for, a treatment technology to a particular waste stream. Where applicable and available, information about similar technologies in similar applications can be used instead. However, if the use of the treatment technology is a unique application, if the waste stream is complex, or if stringent recycle water quality is required it may be necessary to construct a pilot treatment so that an appropriate treatment train can be designed. Piloting can range from bench-scale to skid-mounted systems.

Because the cost of the pilot facility is frequently a barrier to its adoption, the scope, budget, and need for such studies should be detailed in advance. An appropriate recommendation for a pilot test should include the specific design questions to be answered and a protocol for their determination. Chapter 7, "Technology Development Pilots," provides examples and recommendations for design and construction of pilot systems.

4.0 EVALUATING PROJECT ALTERNATIVES

After developing potential industrial water reuse project alternatives, the next step is to evaluate them to determine which will most effectively accomplish the company's goals and receive the approval and resources necessary for successful implementation. This section details an approach to evaluating recycled water projects, beginning with an identification of project costs and qualitative and quantitative benefits (Section 4.1 and Section 4.2). These factors must be accounted for in a way that will allow comparison of the various alternatives (Section 4.3). Commonly used assessment techniques include payback period, ROI, and cost-benefit analysis, while triple bottom line (TBL) accounting provides a more expanded view of costs and benefits, including environmental and social impacts not found in typical economic analysis. Although such factors may be beyond the concern of many managers, an increasing number of companies have adopted sustainability goals and corporate policies that require the consideration of "social return on investment" in making economic decisions.

Whichever method of accounting is used, projects are most often evaluated with the aid of a decision support matrix that compares the relative merit of the various options, sometimes facilitated by the use of decision support software (Section 4.4). Figure 2.5 presents a schematic in the form of a Gantt chart illustrating the various steps in the water reuse project. As indicated in the chart, some form of project evaluation should be conducted at each stage of the project development process based on the information available at the time—during project concept, preliminary and final design, and even after funding is secured. This entire evaluation process should be documented and the information made available for reference in the business case, which is discussed further in Section 5.0.

4.1 Quantitative Costs and Benefits

In most cases, quantitative effects can be expressed in monetary terms so that their relative costs (or savings) can be compared. The following criteria are regularly considered in evaluating recycled water projects: In addition

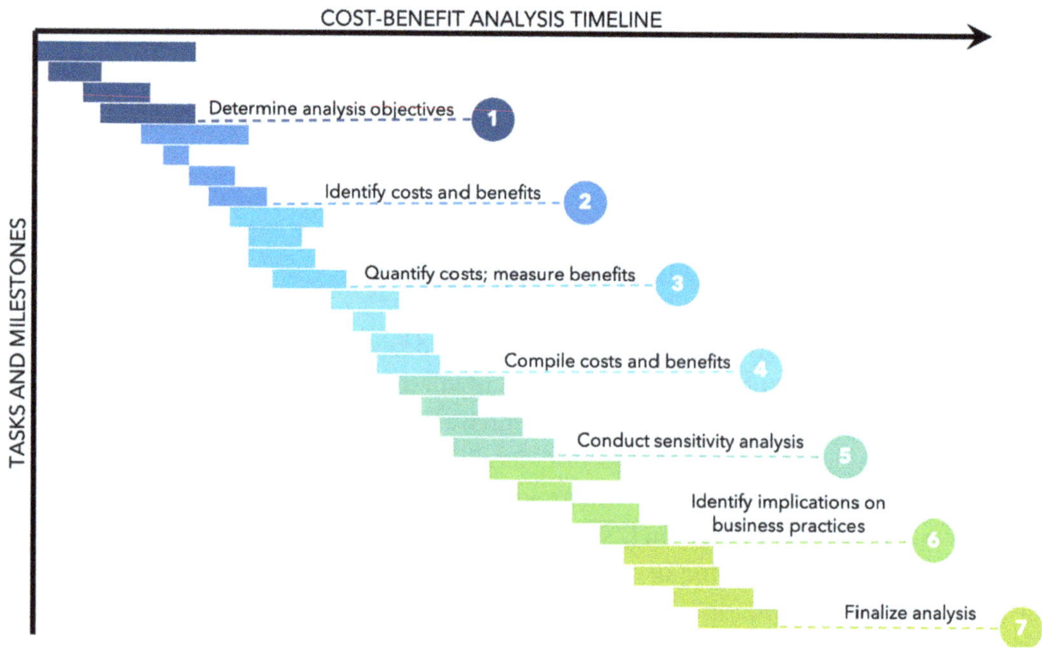

FIGURE 2.5 Gantt chart of alternatives analysis and project selection.

to project costs and the cost of residuals management, project energy use effects, utility savings, and costs related to operational complexity can all be considered in a quantitative manner.

4.1.1 Direct Costs and Savings

Capital expenditures (CapEx) and OpEx are the most straightforward portion of a cost-benefit analysis. However, it is important to ensure that all applicable expenses are accounted for throughout the project implementation. In addition to capital equipment and construction costs, capital expenditures also include all planning studies, permitting, preliminary design, site preparation, startup, and other one-time costs.

Operating expenses include staffing, chemicals, material replacements, maintenance, solids management, and power. The level of staffing depends on the complexity of the equipment, size of recycle system, as well as the permitting requirements and any specific labor requirements applicable at the site (e.g., unionized workforce). Before determining the cost of the staffing, it should first be determined whether the water recycling equipment will be run by existing staff, if new staff will be added, or if third-party operators will be contracted.

A net present worth analysis can be constructed from the combination of CapEx and OpEx. Table 2.8 provides an example of the various aspects of operating costs for common recycling facility unit processes that should be considered when estimating the OpEx of recycled water treatment facilities.

4.1.2 Residuals Management

Most water reuse projects will significantly reduce the quantity of wastewater discharged from the facility. Most recycle systems will make a clean stream for reuse, as well as producing a concentrated waste stream. Management of this waste stream must be considered.

Both reverse osmosis processes and physical separation processes create brines and concentrated waste streams that need to be evaluated if they violate local limits for discharge into the local wastewater treatment system. If so, they need to be neutralized or re-treated before discharge. If discharge is not possible, residuals may also be managed onsite via a solar pond or other evaporative method, or deepwell injection. Additional costs for waste stream treatment/management can add significant costs to projects and should not be ignored. These costs can be mitigated in many cases by initiating open discussions with the local wastewater agency and permitting agencies, negotiating for a mass-based discharge limit instead of concentration-based limits.

TABLE 2.8 Operating costs that can be incorporated when determining total purification costs for some common recycle technologies.

	Membranes	Sand Filters	Bag and Cartridge Filters	Carbon Organoclay	Softeners	Advance Oxidation
Water Losses	Membrane reject	Backwashing		Backwashing	Regeneration	
Chemicals	Membrane feed/reject/permeate chemicals (e.g., antiscalant, corrosion inhibitor, dechlorination, etc.)				Salt	Oxygen for O_3; H_2O_2, acid, iron, and base for Fenton's
Energy	Pumping through membrane					Ozone
Replacement/ Maintenance	Membrane replacement (only when performed on basis of membrane performance and not based on time in service)		Filter replacement	Adsorbent disposal and replacement		
Labor	Membrane replacement (only when performed on basis of membrane performance and not based on time in service)		Labor for charging filters	Labor for changing filters		

(*white cells = costs to incorporate; gray cells = minimal or no cost associated with this aspect*)

4.1.3 Utility Savings and the True Cost of Water

Although the operational cost of proposed recycle alternatives includes new expenses such as energy and chemicals, it is likely that a well-designed water recycling project will result in lower water and municipal WRRF charges. These savings may be calculated based on the estimated use of recycled water and the corresponding reduction in potable water charges, as well as reduced wastewater discharged volumes. In both instances, however, it will be necessary to confirm the estimated cost savings according to the utility billing formulas, which likely include both fixed and variable costs. Although mass loading charges may not be affected, sewer fees based on flow may be expected to decrease.

Another source of savings—and one frequently overlooked—is the reduction in the treatment provided to water after delivery by the utility. The operational cost of a proposed recycled water alternative should take into account not only reductions in the amount of treated water used, but also reductions in the level of treatment required to consider all those savings in the overall cost-benefit analysis.

4.1.4 Energy Demand Expressed in Terms of Greenhouse Gases

Power costs are already included in the OpEx calculation. However, the energy demand can also be quantified in terms of greenhouse gas (GHG) emissions. If the driver for the project is primarily to meet sustainability goals, this may be a relevant factor to consider.

For GHG emissions, calculations should include transportation of goods, chemical production, and material production emissions. Many online tools exist for calculating GHG emissions from a project, but most use methodology developed by the U.S. EPA and Intergovernmental Panel on Climate Change. Other resources include the Local Government Operations Protocol for more specific details. When full life-cycle GHG calculations are considered, more costly conservation projects may become favorable.

4.1.5 Operational Complexity and Effect on Existing Production

Operational complexity is related to the increased number of different processes involved in a project, the number of end uses, and the distance between source, treatment, and end use. Its effect can be estimated either quantitatively in terms of effect on production (positive or negative), in which case it can be considered as a cost or benefit of the project, or qualitatively for the purpose of comparing alternatives.

Calculating the effect of a reuse project on production requires open communication between project planners and operating staff. If information is insufficient to quantify the effect on operations, the relative complexity of alternative projects can be estimated qualitatively, instead. For example, alternatives could be ranked from 1 to 5, with 1 being "no additional effect on operations" and 5 being "significant additional complexity."

4.2 Qualitative Costs and Benefits

In addition to the quantitative criteria discussed in Section 4.1, there are many distinguishing factors that are not as readily expressed in monetary terms. As appropriate, the relative importance of these factors can be monetized through contingent valuation or other econometric methods. Alternatively, all factors (including costs and monetized benefits) can be assigned scalar values and compared in the final decision evaluation. In either case, these analyses should be preserved in tabular or narrative form for reference to aid in decision-making in the event that two or more projects appear equally suitable.

4.2.1 *Permitting Complexity*

Depending on the reuse strategy selected, local or state permits may be required before a project can be implemented (Figure 2.6). For example, if sanitary waste has been combined with the industrial effluent treated for reuse, state recycled water quality standards may be enforced along with restrictions governing recycled water distribution and use. Similarly, if treated wastewater will be reused for sanitary purposes (e.g., toilet flushing) or for irrigation in areas open to the public, local building officials may require that the facility follow the standards in the Uniform Plumbing Code (or the version adopted by a specific states). That code includes standards for water quality as well as for installation of nonpotable water piping and appurtenances (California Building Standards Commission, 2016; International Association of Plumbing and Mechanical Officials, 2015).

State regulations may also come into play where the use of nonpotable water crosses local jurisdictional boundaries. In many states, local utilities are given exclusive franchises to operate water systems or to sell water on a wholesale or retail level. In that case, transfer of recycled water from one facility to another—or use of wastewater from one facility in another—may trigger the need to negotiate with existing water supply or wastewater treatment agencies (Rosenblum, 2016).

The effect of permitting complexity on the suitability of a project alternative can be expressed in at least two ways. First, the feasibility of a project alternative may hinge on its ability to receive a permit from the approving body, resulting in a "go/no go" test. This might be the case, for example, where the ability to reuse wastewater is restricted by rules, as in some states, that delegate the right to wastewater to downstream water users (Strock, 2011). Second, the number of permits required for reuse, and the number of parties involved in obtaining those permits, directly affects the length of time required to implement a project (Rosenblum & Anderson, 2006). As a result, project alternatives with different permitting requirements might be distinguished by their projected start dates, which can be monetized by calculating the opportunity cost lost for the capital allocated to the projects.

4.2.2 Future Flexibility

Because industries are constantly changing, it is important to consider how a recycled water system will serve the facility in the future. If the source water to the facility and all manufacturing processes are stable, a water reuse or recycling program can be assured to have a consistent source and end use. However, if these elements change in the future, the demands on the treatment system will change with changing uses of water in the facility.

A robust treatment train should accommodate anticipated (current and future) changes in wastewater quality and swings in water demand, provided there is enough space for future treatment upgrades. In addition, if changes in flow are expected (e.g., expected facility growth) the facility may consider sizing infrastructure (pipelines and pump stations) for ultimate flows to avoid the cost of future updates. These decisions are best made in

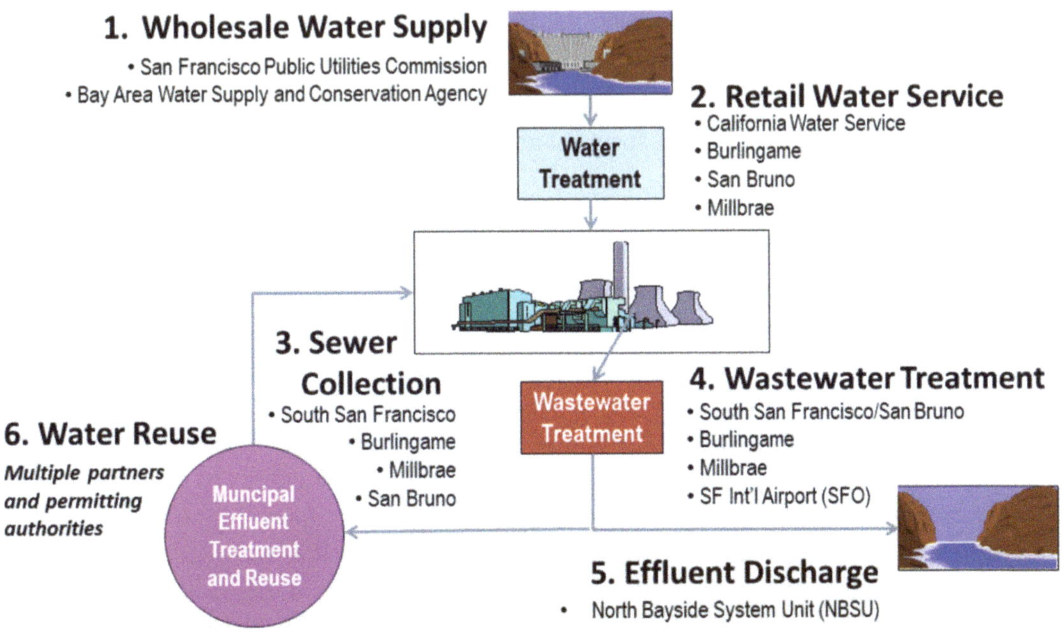

FIGURE 2.6 Multiple water and wastewater agencies involved in one northern California industrial reuse project (adapted from Rosenblum, 2016).

the planning stages to avoid more expensive future costs. They do carry a risk, however, in that any additional capital cost could become stranded if no expansion is required. As such, one needs to be practical and realistic in terms of expected future loading/flows when designing a system. If the life-cycle capital costs of project alternatives are compared as direct costs, then the ability of flexible alternatives to meet future needs should be at least counted as a qualitative benefit during project evaluation.

4.2.3 Reliability and Resiliency

Just as future flexibility is advantageous, the ability of proposed projects to ensure the availability of water in the future should also be considered, either qualitatively or in monetary terms. The benefits of having a reliable source of water can be estimated by considering the following questions (Moore & Buzby, 2017):

1. How vulnerable is the facility to a temporary water shortage?
    - What is the nearest water source outside of the local utility?
    - If that source is a waterbody, what sort of treatment would it require?
    - What permitting options exist for obtaining new water rights?
2. How much would a temporary water shortage affect production?
    - What is the value of a day's production to the facility?
    - What portion of the production would be impaired by a lack of water?
3. How likely are future water shortages to occur in the vicinity of the facility?

To gain an idea of the likelihood of future water shortages, it can be helpful to critically review the local utility's water management plans. Most water utilities prepare master planning documents periodically to assess the reliability of their water sources and to compare them to future demand projections. Although they may be unlikely to forecast shortages resulting in service curtailments, some sense of this possibility can be inferred from the size of the surplus they project as well as the extent to which drought and drought response is addressed in the report. One can also refer to water supply studies at the regional or state level, as well as statewide responses to drought such as in California, where a statewide mandate to reduce water consumption by 20% was implemented (CSWRCB, 2014). Many utilities passed on these restrictions to residential customers as well as industries.

The potential effect of drought can be quantified by considering both the likelihood that restrictions on water use may be imposed and the cost of adapting to them. For example, a facility might calculate the benefit of investing in a reliable water reuse system by considering:

1. The probability that at least 1 year in the next decade will experience a drought resulting in a water reduction of 20% or more; and
2. The cost of reducing productivity or redesigning the use of water and retraining operators in response to the restriction.

If it is estimated that there is a 1 in 5 chance of a severe drought occurring, and the cost of responding to water reductions is $1 million per year, the benefit of instituting a reuse program could be estimated (in a probability-weighted manner) as 0.2 × $1 million or $200,000 (corrected for present value, as appropriate). In regions where water shortages are not prevalent or predicted, other aspects of sustainable water use may be evaluated, including the rising cost of wastewater treatment or the future need to remove specific constituents (e.g., nutrients, heavy metals, trace organics) from the receiving stream.

4.2.4 Life-Cycle Assessment

One way of attempting to incorporate all significant quantitative and qualitative costs, benefits, and environmental effects is to perform a life-cycle assessment (LCA). An LCA is an approach that attempts to evaluate the entire effect of a product or process throughout its life, including "all energy and environmental implications . . . including design, planning, material extraction and production, manufacturing or construction, use, maintenance, and end-of-life fate of products (reuse, recycling, incineration, or landfilling)" (Stokes and Horvath, 2006). By looking at both the upstream and downstream costs and effects (e.g., environmental effects of manufacturing and disposing of reverse osmosis membranes), LCA provides a broader context for decision-making than traditional cost accounting, which is often limited to the capital and operating costs of treatment equipment without regard to its manufacture or disposal. The reader should be aware that LCA studies can be quite involved, and the expected usefulness/importance of the results that would be derived from conducting an LCA for evaluation of a recycle should first be carefully considered to ensure adequate value is provided.

4.2.5 Conservation, Regional Reuse Projects, and No-Project Alternative

Along with identified on-site treatment alternatives, project sponsors should include an assessment of "reduce" projects, regional reuse, and the "no project" alternatives (status quo). As noted in Section 2.2.4, water use reduction projects are often the most cost-effective ways to meet company water use and discharge minimization goals. In addition, some conservation strategies, like flow optimization projects, produce additional benefits in terms of energy conservation or reduced chemical use.

Use of municipally treated wastewater and/or participation in a regional water recycling program is another approach to reducing industrial water demand. Although using reclaimed water from a municipal WRRF may not minimize facility discharge, it does reduce demand on local drinking water supplies; additionally, many agencies are eager to connect industries because this year-round demand complements seasonal irrigation use. In some cases, water agencies have designed and built advanced municipal WRRFs adjacent to industrial customers specifically for their use.

The Chevron refinery in Richmond, California, is an example of a productive partnership with a local utility. Chevron uses up to 28 400 m^3/d (7.5 mgd) of East Bay Municipal Utility District (EBMUD) recycled water, including water treated to advanced standards. EBMUD's North Richmond Water Reclamation Plant produces up to 20 400 m^3/d (5.4 mgd) of tertiary disinfected water for use in Chevron's cooling towers, and another 7600 m^3/d (2 mgd) of advanced purified water (microfiltration and reverse osmosis) for use in Chevron boilers. In this partnership, EBMUD staff operate the facility and Chevron has contracted with EBMUD to pay for the water for a set timeframe (EBMUD, 2019).

In all cases, the "no project" alternative must also be considered. This analysis should include the effect of potential water shortages and the cost of compliance with future discharge regulations, as well as the cost of forgoing the many benefits of on-site treatment and reuse. Care should be taken, however, to avoid "double counting" such that the cost of coping with water shortage, for example, can be considered a future cost or lost opportunity, but not both.

4.3 Assessing Costs and Benefits

Quantitative and qualitative costs and benefits—once identified and accounted for—must be assessed in a manner that facilitates the comparison of project alternatives. Because industrial facilities are under continuous pressure to reduce costs and maximize returns, a financial cost/benefit ratio is often the first thing to be evaluated, typically expressed in terms of either payback period or ROI.

4.3.1 Payback Period

Payback period (Equation 2.1) simply reflects the time required for project earnings (or savings) to equal the net investment, as follows

$$\text{Payback Period} = \frac{\text{Investment Cost}}{\text{Annual Savings}} \tag{2.1}$$

When the cost of the project exceeds the annual savings, the payback period is expressed in years; however, when the annual payback exceeds the project cost, a payback period of months or even weeks can emphasize that value.

A simple payback period calculation can give sponsors a rough idea of the project's financial merit, to determine whether a project concept is worth moving to the design phase. A 2-year payback period is something of an industry benchmark, although the exact payback period threshold/requirement will vary from one company to another. Projects that take longer (than a company's target) to return capital cost are less likely to be funded, although sponsors can explore modifications or develop alternate solutions that are more likely to gain approval. Because the payback period of the "no project alternative" is typically assumed to be zero, facilities that rely exclusively on this method may have a default position to implement no new project proposals without a demonstrated ability to serve some critical need or an exceptionally short payback period.

4.3.2 Return on Investment

As a project is further developed, simple payback period may not adequately reflect its value or provide a useful comparison to other alternatives. In that case, the ROI may be used to evaluate its economic efficiency instead. Return on investment essentially represents the total net benefit of a project (rather than its annual return) divided by the total investment cost. A key benefit of ROI compared to simple payback period is that ROI is typically better able to account for the value of a project over time, especially when benefits accrue differently at different points in the life of the project.

For example, two projects requiring a capital investment of \$300,000 that provide annual savings of \$200,000 will each have a calculated payback period equal to 1.5 years. However, if one project takes 2 years longer to implement than the other, an ROI can discount those future savings based on the time value of the investment. Similarly, if the annual savings of a project increase from \$50,000 in the first year and \$100,000 in the second year to \$150,000 in the third year and \$200,000 per year in all years following, the ROI can also take these factors into account in calculating a net present worth of the total savings.

Both the payback period and ROI methods are limited to evaluation of quantitative aspects of proposed projects, either market-based factors like CapEx and OpEx costs, water rates and sewer fees, or qualitative factors that have been monetized like the risk of a water supply shortfall. Where project alternatives are close in cost and distinguished primarily by their qualitative benefits and/or costs (e.g., the ability to respond to future changes in production), simple payback and ROI may not be robust enough to reveal their differences. Although these qualitative factors can be reflected in a decision matrix (Section 4.4.1), they can also be appraised in expanded accounting methods like TBL accounting.

4.3.3 Triple Bottom Line and Social Return on Investment

To the same degree that an ROI analysis includes cost factors that are not reflected in a simple payback calculation, a TBL analysis captures factors a traditional ROI might ignore. The challenge, for many managers, is that many of these factors, like the environmental benefits of discharge minimization, are "externalities" that do not show up immediately on the facility's balance sheet. That does not mean, however, that those factors will not ultimately accrue to the company's benefit. Rather, it only means that they may not be reflected in the "2-year payback" window that commonly brackets industrial investments; they represent an uncertainty/risk making it impossible to precisely quantify timing and/or associated monetary effect. As noted by Obrien (2000), "The failure of cost-benefit analysis to include nonmonetary considerations is the main basis for rejecting monetary cost-benefit analysis as a sufficient basis for any environmental, public or private decision making."

One approach to expanding the scope of economic analysis is the use of TBL accounting. In addition to traditional calculation of ROI ("the bottom line"), TBL includes an account of environmental costs as identified in the LCA as well as an assessment of the social impacts of the project (Kenway et al., 2007). For example, if extensive industrial water use may reduce the amount of water available for public use, this effect would be noted, and the potential costs would be estimated or a scalar value would be assigned to this result. By the same token, if an on-site discharge minimization project lowers the cost of operating public WRRFs, those benefits would also be included in a TBL evaluation of the project. In many cases, these "collateral" costs and benefits can be monetized to reveal the full range of benefits associated with minimizing industrial discharges. Guidance on the practice of TBL analysis and social ROI assessment is beyond the scope of this chapter, but handbooks and analytical tools are widely available, some targeted to specific industrial sectors (Cashman & Mosley, 2016; Raucher, 2006; World Business Council for Sustainable Development, 2017).

Triple bottom line has become increasingly common in the evaluation of government-funded projects whose environmental and social benefits accrue to ratepayers as members of the general public. By contrast, in the private sector, where operating facilities are managed to reduce costs, maximize revenues, and increase shareholder value, the externalities reflected in a TBL analysis would seem to carry little weight. In truth, such broad considerations are rarely the most important to facility managers evaluating discharge minimization alternatives.

In response to pressure by shareholders, however, a significant portion of companies have set water conservation targets and other sustainability goals, which have the effect of encouraging managers to take efficient use of resources into account in their decision-making. These policies can influence production decisions up the supply chain as well (Nike, Inc., 2018). In addition to reinforcing the company's social "license to operate," such policies are often rewarded by the financial markets. In fact, many investment firms specifically consider a company's vulnerability to changing environmental and regulatory conditions when evaluating their investment potential (Bertolitti, 2019).

4.4 Comparing Alternative Treatment Schemes

Having identified and assessed costs, benefits, and other relevant factors, the project team can compare and evaluate alternatives to select the one most likely to meet their goals. Although the financial analysis (payback period, ROI, or TBL) will figure prominently in the analysis, some type of decision matrix is typically required to track both quantitative and qualitative criteria and to weight them according to the consensus of their importance.

4.4.1 Decision Support Matrix

In their review of decision support systems for domestic and industrial water and wastewater treatment process selection, Hamoudaet al. (2009) identified more than 20 factors that should be considered in treatment system design. They proposed that, in general, there are three approaches to decision-making: technical; technical and economic; and systems analysis, which considers "the bigger picture of cost, technical performance, as well as social, legal, and environmental interactions," and is more likely to bring about a sustainable result (Hamouda et al., 2009).

A basic decision support tool involves creating a matrix where the various factors identified earlier are converted into selection criteria by assigning scalar values that weight their importance. The translation of qualitative factors into quantitative criteria is a subjective exercise, and its validity depends on the consensus of the participants. In other words, so long as the project stakeholders agree on the relative importance of a qualitative factor, the assignment of that value is appropriate as far as influencing the final decision is concerned. In many cases, this may mean that projects will be selected on the basis of simple payback period or ROI; however, it is important to ensure that all factors are fairly considered, if only to identify in advance any "fatal flaws" that might make the project ultimately infeasible.

By subjecting all projects to this multicriteria analysis, each alternative can be ranked relative to the others, either through a single cumulative score or by its position in several independent categories. There are a number of software packages now available that support multicriteria decision analysis of water and wastewater treatment projects (e.g., Harris-Lovett et al. 2019). Although many of these tools are designed for use by municipal projects with significant environmental and social impacts affecting multiple diverse public and private stakeholder groups, some can be adapted to smaller projects, including industrial water treatment.

4.4.2 Sensitivity Analysis

The end result of the decision support program is a set scores for the various projects evaluated. Although the individual project score will be relatively uninformative, by comparing the scores of various projects, their relative value can be determined. An equally valuable use of a decision support scoring system, however, is that it can reveal which factors contribute most to the relative ranking of a project and how those rankings might be changed. It can be useful to understand what circumstances lead to a project being poorly ranked, and to consider how these conditions/circumstances might be mitigated. Similarly, a preferred alternative may rank poorly if conditions change, and the likelihood of such a change should also be evaluated. In many cases, the sensitivity analysis is as beneficial as the quantitative assessment itself.

5.0 MAKING THE BUSINESS CASE FOR REUSE

5.1 How Business Decisions Are Made

Whereas governments and public utilities may fund their enterprises through a combination of taxes and fees, private sector companies have fewer options when it comes to paying bills. They must purchase everything—including the design, construction, and operation of water recycling equipment—with money they make from the products they sell, or what they can borrow, sometimes at a high cost.

Given these financial constraints, a company will typically have a limited (and often fixed) budget for capital projects each year, and that budget is rarely sufficient to fund all projects requested across a facility. To compete for funding, projects must emphasize the financial value they provide to the company. This underscores the importance of making a business case for water conservation and reuse projects. An effective business case highlights the project's ROI, explains how it can be implemented without endangering the facility or its operation, and illustrates how its value compares favorably with other investment opportunities. This section will discuss, in general terms, how to prepare a business case that communicates the benefits of a recycled water program in an effective manner.

5.2 Making the Business Case

The most difficult part of implementing a recycled water project may not be selecting, designing, and building an effective treatment system, but rather convincing a small group of decision-makers to fund it. All the feasibility studies, conceptual designs, risk analyses, and cost-benefit exercises often culminate in a single 1-hour meeting. To walk out of that meeting with the necessary funding to implement the project, one must be prepared to convincingly answer the following questions:

- What goals does this project achieve?
- Why is this project chosen above the others?
- What is the consequence of not choosing a project?

Ultimately, the answers to those questions—and the essence of a successful business case—center on the true value of water and its long-term importance to facility operation, as measured during the project evaluation process. However, to communicate this information effectively to the decision-makers, the project sponsor must understand their roles and responsibilities within the organization.

5.3 Who Are the Decision-Makers?

Industrial facilities are highly structured institutions, aligning the efforts of many people with different areas of expertise—managers, engineers, operators, mechanics and electricians, accountants, human resource specialists, and product marketing experts, to name a few. In making a business case for industrial water reuse, project sponsors should clearly identify the parties in the organization who will review and approve the proposal. Some of the key decision-making roles include the following positions, based on their responsibility in the organization:

- The facility manager is in charge of the overall operation and success of the facility, and will likely have the greatest say in deciding which projects move forward. This position is also frequently responsible for coordinating the work of different divisions and sections, so the individual in that role may be particularly sensitive to how a project that benefits one group might affect the work of another. The facility manager may also be accountable for the financial performance of the facility, both in terms of operating within its budget and meeting its production targets; thus, projects that threaten to jeopardize either of these goals may not be adopted.

- The finance officer or facility accounting manager is responsible for the company balance sheet, and is typically asked to certify that funds are available and the project budget is sufficient to cover anticipated costs. As with any cost estimate, uncertainties should be clearly identified, and a "range of costs" approach should be used that is appropriate to the design stage (conceptual, preliminary, final) to reflect anticipated cost fluctuations.
- The O&M supervisor also has a key responsibility for facility performance. Although the O&M supervisor may appreciate how a reuse project can cut water and sewer costs, the additional staff time required may be a concern. When this position is held by more than one individual (e.g., operations chief and head of maintenance), the complexity of a proposed treatment process and its effect on maintenance resources will also receive additional scrutiny.
- The sustainability officer will want to understand how investment in the proposed project will help the facility meet company-wide sustainability targets. This position may also serve as the project sponsor or a chief advocate if the project's primary purpose is to reduce the facility's overall "environmental footprint," and may be supported by the business line chief executive officer (CEO) responsible for the products manufactured at the facility. This will more likely be the case if the reuse project is being featured as an example of the company's commitment to sustainability. In the final analysis, the sustainability officer—like any project sponsor—will have to communicate the value of the project to other decision makers in terms they value (i.e., the business case).
- The environmental health and safety (EH&S) manager will typically need to approve the use of any hazardous chemicals that may be required to operate or maintain the proposed treatment technology. Their chief concerns will often relate to any permitting needed and the assurance that the project (once finished) will not create a risk of exceeding any discharge permits. In addition, the EH&S manager is often called upon to review construction plans to ensure they do not pose a threat to facility personnel. In the same manner, the government

affairs specialist may be asked to sign off on the company's ability to obtain all the permits required to build and operate the project.

In addition to these staff positions, companies often contract with a number of consultants and service providers who are responsible for various aspects of facility operation. For example, companies commonly outsource the O&M of their cooling towers. This is often managed through performance contracts that pay third-party entities an annual retainer, in exchange for which they are responsible for all chemicals and labor necessary to keep the cooling system in good working order. A reuse proposal that changes the quality of the cooling tower makeup water (and potentially its treatment) would likely require input from the service company, as well.

In their report, *Evaluation of Historical Reuse Applications and Summary of Technical/Regulatory Issues and Related Solutions for Industrial Reuse Projects,* Layton and Bowdan (2015) documented such a charter process, identifying a broad range of key questions stakeholders must address at the outset of each project. They further recommended that the responsible parties should jointly signify their support for the investigation into water recycling and reuse to ensure their cooperation during project development and their willingness to consider the results at the appropriate time. Figure 2.7 presents this charter.

Different companies are structured differently. For example, owners of smaller companies may retain all decision-making authority, allowing them to more easily work together when necessary. This was the case in a water-scarce area of India, where owners of enterprises in two different industries (sponge iron manufacturing and denim washing) formed an association to receive municipal recycled water and avoid the high cost of water delivered by tanker truck.

A successful tool for helping to identify water efficiency opportunities is the WKB process. Key characteristics of a WKB are the following:

- Team structure
- Deep-dive into processes
- Project planning and pre-WKB activities
- WKB live event
- Report out

As with the charter process, teams include site personnel who are experts in the manufacturing facility as well as experts in processes and technologies for reducing water use within manufacturing operations. Teams work together using a variety of continuous improvement tools to identify and develop an array of realistic and cost-effective opportunities. The WKB process has been successfully used at dozens of manufacturing facilities across a range of sectors, including automotive, pharmaceutical, chemical, oil and gas, food and beverage, home goods and appliances, and aerospace to name a few (Moore & Buzby, 2017).

5.4 Goals of the Business Case

A good business case addresses the issues important to each of the decision-makers. It also will successfully unite the interests of "the board room and the boiler room" by demonstrating how corporate goals can be accomplished in a way that is compatible with the efficient, uninterrupted operation and maintenance of the facility. This is critical because almost every team member has "veto authority" when it comes to adopting discretionary process improvements. For this reason, the use of a chartering tool can be particularly helpful because it engages team members from the outset so they can make their concerns known early in the process.

In larger companies, a team of "internal stakeholders" often need to agree to implement a project. Even then, whereas some large companies retain authority on the corporate side, others give significant autonomy to facility managers. From the perspective of funding, some sustainability goals are prefunded such that they are not required to meet a predetermined ROI, or they may have relaxed ROI requirements (relative to other investments). Other times, the CEO may have access to funds that can be allocated for sustainability projects, enabling the facility to implement a reuse project. This is especially true if a reuse project is required to protect the facility's social "license to operate," as in water-stressed areas where implementation of a reuse project is not constrained by its immediate ROI.

Finally, it should be noted that when it comes to implementing an industrial water reuse project, timing is everything. Shortsighted or not, a proposal that would be accepted during the midst of a multiyear drought might be deferred or rejected outright during a rainy year. Similarly, companies are not immune to business cycles, especially manufacturing plants, whose output is determined by market demand. During slow cycles when revenues are down, there may be a markedly lower appetite for process improvements, despite the fact that they may be easier to implement when operational requirements are more accommodating to on-site construction.

5.5 Address Objections in Advance

Over time, it has been observed that certain objections are repeatedly raised in opposition to implementation of industrial water reuse projects. A successful business case will answer these arguments in advance. The following is a

Description
Identify the nature or type of IWR application; e.g., cooling system, boiler system, process; type of plant/industry, location, etc.

Purpose
Identify the business reason for the project (i.e., project drivers/needs analysis). Describe how the project links to major strategies, objectives and drivers.

Key Stakeholders
Identify for whom/by whom this project is being undertaken and other organizations/agencies/entities that will be involved; e.g., regulatory agencies, water district, plant operations, etc.

Decision Maker(s)
Specify for whom the project is being created and list others who eventually may be in the approval loop

Functional Requirements
Identify in general terms what operating conditions the IWR solution must satisfy; e.g., demand/capacity, guaranteed quality range, etc.

Scope/Approach
List and describe in brief the major steps and activities involved in completing the project. For each of the major steps, estimate the amount of effort and time required.

Success Indicators
List the measures – business and operational – by which the project will be considered a success. Example metrics include: amount of fresh water saved, cost factors, median flow rates, operability, etc.

Risk Factors
Identify any and all factors that could undermine the timely and cost-effective delivery of the project; e.g., regulatory and compliance issues, resource availability/bandwidth, plant runtime, maintenance concerns

Resources and Information
Identify the primary project contributors and potential, additional resources or information needed to complete the project on time and within budget.

Organizational Links
Identify other organizations, departments or initiatives that are linked to this project for input / output.

Budget
Is there a designated budget for this project? Can an estimate be developed based on previous projects of similar scope/complexity? Also, funding considerations/ contributor.

Scheduling
Estimate start-to-finish cycle times with consideration given to simultaneous activities and potential critical timing issues. List milestone review and approval dates.

Core Project Team
The Core Team will include organizational leads and support personnel, contractor and vendor representatives, primary contacts for stakeholders, etc.

FIGURE 2.7 Chartering the industrial water reuse project (Layton & Bowdan, 2015).

list of commonly encountered objections and some strategies for responding to them (refer to *A Framework for the Successful Implementation of On-Site Industrial Water Reuse* [Moore & Buzby, 2017] for a more inclusive list with detailed responses):

- "The project is too expensive."
    - Ensure that all other options, including conservation projects, which tend to be less expensive, have been fully vetted.
    - Add to the utility price the cost of "heating and treating" water within the facility to reflect the "true cost of water" (Section 4.1).
    - Investigate grant and low-interest loan opportunities that may be available from federal or state institutions, local utilities, and non-profit organizations.
    - Look for ways to reduce cost without compromising quality. Are there relatively small water uses that are far away and not worth connecting with a new pipeline?
- "Our current water situation is fine: We don't need this project."
    - Utility prices typically fail to include the cost of obtaining new supplies, especially those needed to make up for future reductions in water availability because of climate change. Project a cost escalator based on regional or statewide rates to ensure that they reflect the future value of the project, which will only increase as water stresses increase.
    - Present the financial effect of water cutbacks and the value of having resilient on-site supplies (Section 4.2.3).
- "We aren't a water company."
    - If in-house expertise is lacking, hire third-party contract operators or consultants to provide training and advice on best management practices.
    - Consider partnering with local water utilities during project development to ensure that all options have been vetted and that operational expertise is available.

6.0 CONCLUSION

A successful industrial water reuse project is the result of the combined effort of many people throughout the facility. To accomplish this level of coordination, project sponsors must not only perform traditional engineering tasks, they must also embrace company objectives, work within budget constraints, adapt to operational sensitivities, effectively communicate the value of the project, and ensure that all stakeholders are included in the process. These key tasks are as follows:

- **Clearly define project drivers and objectives and communicate them to all participants.** The drivers—cost savings, water savings, site reliability, sustainability goal, or other—determine the type of project. It is important for all participants to know whether the project stems from a voluntary water conservation goal or the need to keep the facility open in the face of chronic water shortages. If applicable, numerical goals (e.g., 20% reduction of effluent) should also be identified.
- **Identify and confirm participation of key stakeholders and decision-makers.** Most people in a production line will be affected by the implementation of a water recycling program. It is important to determine who these people are and get their input on the project early and often for a project to succeed. Avoid "tunnel vision" by addressing the issues important to each of the stakeholders. If you cannot meet their needs completely, make sure they understand your constraints. They may be able to propose alternatives to make an even better project. Including stakeholders at the outset and confirming their roles will avoid significant confusion later on.
- **Develop a good water balance, understanding the source of variability.** An accurate picture of water and wastewater volume, location, and quality is critical to the development of a suitable reuse project. A full site audit will reveal the greatest water users and the largest wastewater discharge locations, which will be a starting point for actually developing the project. This is a group effort that draws on the collective wisdom of the entire facility.
- **Identify "reduce" opportunities before investigating "reuse" and "recycle."** Before pursuing any recycling project, it is important to investigate opportunities to conserve through process optimization and efficiency measures. This is almost always the most cost-effective approach to cutting water use and reducing wastewater discharge.
- **Focus on the end user and understand their water quality requirements.** Keep the end user in mind at all times to meet water quality and delivery demands. Reuse never takes place in a "black box," and when recycled water fails to meet process needs, even a multimillion-dollar project becomes worthless. Bringing end users in early and keeping them engaged throughout the planning, equipment selection, and design process will also help overcome objections and create a more persuasive business case.

- **Develop a strong business case and communicate the benefits of the project.** Evaluate all quantitative and qualitative costs and benefits to determine the overall effect of the project and create a robust evaluation matrix based on these criteria to decide between project alternatives. Whether evaluating projects based on cost benefit, TBL, or consensus criteria, include an ROI calculation in the presentation, tailor the project to fit within the available budget, and make sure the project aligns with the company's financial goals as well as its reuse objectives.

7.0 REFERENCES

2030 Water Resources Group. (2013). *Managing water use in scarce environments—A catalogue of case studies.* www.waterscarcitysolutions.com

Arceivala, S., & Asolekar, S. (2006). *Wastewater treatment for pollution control and reuse* (3rd ed.). McGraw Hill Education (India) Private Limited.

Barbera, M., & Gurnari, G. (2018). *Wastewater treatment and reuse in the food industry.* Springer International Publishing.

California Building Standards Commission. (2016). 2016 California plumbing code. California Building Standards Commission.

California Department of Water Resources. (2013). *Commercial, industrial, and institutional task force best management practices report to the legislature, v. 2.* https://www.water.ca.gov/wateruseefficiency/sb7/committees/urban/u1/

California State Water Resources Control Board. (2014). *Resolution No. 2014-0038 to adopt an emergency regulation for statewide urban water conservation.* https://www.waterboards.ca.gov/board_decisions/adopted_orders/resolutions/2014/rs2014_0038_regs.pdf

Cashman, S., & Mosley, J. (2016). *Life cycle assessment and cost analysis of water and wastewater treatment options for sustainability: Influence of scale on membrane bioreactor systems* (EPA/600/R-16/243). U.S. Environmental Protection Agency.

Costello, M. (2010). Water, energy conservation in the pharmaceutical industry. *WaterWorld*, 10.1 (January, 2010). https://www.waterworld.com/articles/iww/print/volume-10/issue-1/feature/water--energy-conservation.html

Bertolotti, A., Fancy, T., Pellegrini, G., Mateos y Lago, I. Schulten, A., & Smith. J. (2019). Sustainability: The future of investing, Global insights, February. https://www.blackrock.com/corporate/insights/public-policy/sustainability

East Bay Municipal Utility District. (2019). *Recycled water projects.* https://www.ebmud.com/water/recycled-water/current-recycled-water-users/

Grönwall, J., & Jonsson, A. (2017). Regulating effluents from India's textile sector: New commands and compliance monitoring for zero liquid discharge law. *Environment and Development Journal, 13.1*(13).

Hamouda, M., Anderson, W., & Huck, P. (2009). Decision support systems in water and wastewater treatment process selection and design: A review. *Water Science & Technology, 60*(7).

Harris-Lovett, S., Lienert, J., & Sedlak, D. (2019). A mixed-methods approach to strategic planning for multi-benefit regional water infrastructure. *Journal of Environmental Management, 233*(218–237).

International Association of Plumbing and Mechanical Officials. (2015). *Uniform plumbing code* (Chapter 16). International Association of Plumbing and Mechanical Officials.

International Life Sciences Institute. (2008). Considering water quality for use in the food industry. *ILSI European Report Series.* International Life Sciences Institute Europe.

Kammeyer, C. (2018). Water stewardship in supply chains: The missing piece? *CEO Mandate.* https://ceowatermandate.org/posts/water-stewardship-supply-chains-missing-piece/

Kenway, S., Howe, C., Maheepala, S., Kelly, T., & Davis, C. (2007). *Triple bottom line reporting of sustainable water utility performance.* American Water Works Association Research Foundation.

Layton, R., & Bowden, J. (2015). *Evaluation of historical reuse applications and summary of technical/regulatory issues and related solutions for industrial reuse projects: Managing industrial reuse projects for success* (WRRF Research Report 12-03). WateReuse Research Foundation.

Markardan, J. (2017). *Considering water use in pharma manufacturing.* https://www.pharmtech.com/considering-water-use-pharma-manufacturing

Moore, B., & Buzby, M. (2017). *A framework for the successful implementation of on-site industrial water reuse* (WRRF No. 14-04). Water Research Foundation.

Morrison, J., & Schulte, P. (2010). *Corporate water accounting: An analysis of methods and tools for measuring water use and its impact.* Pacific Institute.

National Academy of Sciences. (2012). *Potential for expanding the nation's water supply through reuse of municipal wastewater: Report of the committee on the assessment of water reuse as an approach for meeting future water supply needs, water reuse.* https://www.nap.edu/read/13303

Nike, Inc. (2018). *FY16/17 sustainable business report.* https://www.globalreporting.org/Pages/FR-Nike-2018.aspx

Obrien, M. (2000). *Making better environmental decisions: An alternative to risk assessment.* MIT Press.

Patel, H., & Vashi, R. T. (2015). *Characterization and treatment of textile wastewater.* Elsevier.

Raucher, R. (2006). *An economic framework for evaluating the benefits and costs of water reuse, final project report and user guidance.* WateReuse Research Foundation.

Rosenblum, E. (2016). *Putting the pieces together: Interagency agreements for integrated water use.* https://watereuse.org/wp-content/uploads/2015/01/Interagency-Agreements-Integrated-Water-Use.pdf

Rosenblum, E., & Anderson, J. (2006). The water reclamation matrix: A framework for sustainable urban water use. In B. Beck & A. Speers (Eds.), *2nd leading edge conference on sustainability in water-limited environments.* International Water Association.

Rosenblum, E., Erickson, K., Freedman, J., Goldman, E., Gluck, S., Majamaa, K., & Yu, J. (2012). Industrial water recycling surges forward. *Water Reuse & Desalination, 3*(3), 12–15.

Sarni, W. (2011). *Corporate water strategies.* Earthscan.

Stanford University Land and Buildings. (2005). *Guidelines for lifecycle cost analysis.* Stanford University.

Stokes, J. & Horvath, A. (2006) Life cycle energy assessment of alternative water supply systems. *International Journal of Life Cycle Assessment 11*(335). https://doi.org/10.1065/lca2005.06.214

Strock, K. (2011). *Water reclamation in Washington state: An analytical review of WAC 173-219.* https://digital.lib.washington.edu/researchworks/bitstream/handle/1773/17024/Strock_Kaylene_Paper_Final.pdf

Toyota. (2018) *Envisioning environmental sustainability.* Online report. https://www.toyotabharat.com/toyota-in-india/environment/sustain-report/

Treweek, G. (1982). Industrial reuse of wastewater: Quantity, quality and cost. In E. Middlebrooks (Ed.), *Water reuse* (pp. 521–548). Ann Arbor Science Publishers.

U.S. Environmental Protection Agency. (2012). *The importance of water to the U.S. economy: Synthesis report.* U.S. Environmental Protection Agency Office of Water.

U.S. Environmental Protection Agency. (2017). *A water efficiency case study—Intel Corporation's Ocotillo campus*. https://19january2017snapshot.epa.gov/www3/region9/waterinfrastructure/docs/Intel-Ocotillo-Arizona.pdf

Victoria University Institute for Sustainability and Innovation. (2007). *Guidance for the use of recycled water by industry.* https://waterportal.com.au/swf/images/swf-files/415-001-guidance-for-the-use-of-recycled-water-by-industry_final-report.pdf

World Business Council for Sustainable Development. (2017). *Business guide to circular water management: spotlight on reduce, reuse and recycle*. World Business Council for Sustainable Development.

3

Fundamentals

Val S. Frenkel; Joseph Cleary; Edward Horai; Mikel E. Goldblatt, PE; Graham Juby, PhD, PE; Tina Arrowood; Jorge Aguinaldo; Erik Desormeaux; Raj Kailasam, PhD; Derek Stevens; & Leila Tootchi

1.0 BIOLOGICAL TREATMENT

1.1 Introduction and Fundamentals

Biological wastewater treatment takes advantage of microbial communities for the intent of converting organic matter within a waste stream. Understanding how these microbes carry out their natural metabolic processes allows us to replicate environments that facilitate their ability to do so for the intent of efficiently producing clean effluent. The conventional activated sludge process, and its many variations, has been the principal choice for the biological treatment of wastewater for more than 100 years because of its effectiveness and ability to be modified. *Activated sludge* can be defined as "sludge floc produced in a raw or settled [wastewater] by the growth of zoogleal bacteria and other organisms in the presence of dissolved oxygen and accumulated in sufficient concentration by returning floc previously formed," whereas the process itself can be defined as "a biological [wastewater] treatment process in which a mixture of a [wastewater] and activated sludge is agitated and aerated. The activated sludge is subsequently separated from the treatment [wastewater] by sedimentation and wasted or returned to the process as needed" (Cleary & Eckenfelder, 2014).

Whether the process is suspended, attached, or a hybrid, each technology has a biological component and, therefore, requires habitable conditions. Some of the factors that dictate the biological activity of a system include the availability of food and nutrients, pH, temperature, and dissolved oxygen concentration. The treatment goals dictate the technology chosen as well as the type of bacteria used to meet those treatment goals. Table 3.1 shows the types of bacteria, their carbon (food) source, and the other necessary compounds that support biomass growth. Outside of carbon (50% of dry weight), other elements needed for cell growth include oxygen (22%), nitrogen (12%), hydrogen (9%), and phosphorus (2%) (Tchobanoglous et al., 2014). Phosphorus and nitrogen can be limiting nutrients, meaning that the addition of one or both is necessary for biomass growth. The remaining 5% of a bacterial cell's dry weight may contain other elements, each consisting of 1% or less of the dry weight. For each environmental factor, there is a habitable and optimum range. Most bacteria prefer a pH between 6.5 and 7.5 for optimal growth.

Regarding temperature, bacteria are separated into three categories: psychrophilic (10 to 30 °C [50 to 86 °F]; optimum range 12 to 18 °C [53.6 to 64.4 °F]); mesophilic (20 to 50 °C [68 to 122 °F]; optimum range 25 to 40 °C [77 to 104 °F]); and thermophilic (35 to 75 °C [95 to 167 °F]; optimum range 55 to 65 °C [131 to 149 °F]) (Tchobanoglous et al., 2014). The bacteria in each system are also susceptible to toxic or inhibitory compounds

TABLE 3.1 Classification of biological reactions and the constituents that support biomass growth (Created using information from Tchobanoglous et al., 2014).

Reaction	Definition	Electron Donor → Electron Acceptor	Byproducts
Aerobic Oxidation	Biological treatment that occurs in the presence of free dissolved oxygen; oxygen is consumed by aerobic microorganisms to drive metabolic processes	Organic compounds → O_2	CO_2, H_2O
Denitrification (Anoxic Reaction)	Biological reduction of nitrate to nitrogen gas and other gaseous end products containing nitrogen	Organic compounds → NO_2^-, NO_3^-	N_2, CO_2, H_2O
Acid Fermentation	Biological process done under anaerobic conditions where longer chained compounds are broken down to shorter chained acids such as VFAs (acidogenesis) and acetic acid (acetogenesis)	Organic compounds → Organic compounds	VFAs
Iron Reduction	Biological reduction of ferric iron to Fe (III) and other end products	Organic compounds → Fe (III)	Fe (II), CO_2, H_2O
Sulfate Reduction	Biological reduction of sulfate to hydrogen sulfide and other end products	Organic compounds → SO_4	H_2S, CO_2, H_2O
Methanogenesis	Biological conversion of acetic acid or hydrogen and carbon dioxide to methane	VFAs → CO_2	Methane
Nitrification	The two-step biological oxidation of nitrogen (mostly in the form of ammonia) to nitrite and then to nitrate	NH_4^+, NO_2^- → O_2	NO_2^-, NO_3^-
Iron Oxidation	The biological oxidation of Fe (III) to ferric iron	Fe (III) → O_2	Ferric iron, Fe (III)
Sulfur Oxidation	The biological oxidation of hydrogen sulfide and other sulfur containing compounds to sulfate	H_2S, S^0, $S_2O_3^{2-}$ → O_2	SO_4^{2-}
Anammox	An anaerobic treatment process in which nitrate is used by specialized planctomycete bacteria as the electron acceptor to convert ammonia and nitrate to nitrogen gas primarily	NH_4^+ → NO_2^-	N_2, NO_3^-

Heterotrophic bacteria (Carbon source = Organic compounds)

Autotrophic bacteria (Carbon source = Carbon dioxide)

VFAs = Volatile fatty acids

found in a waste stream. For this reason, industrial water, in particular, may require pretreatment to protect the biomass further down the treatment train. These pretreatment technologies will vary depending on the type and concentration of toxic or inhibitory compounds.

1.2 Aerobic

The various aerobic treatment processes discussed in this section include technologies that use microorganisms that are suspended in the wastewater, attached to a surface, or are a hybrid of the two. Of the many suspended growth processes available, the ones that will be discussed include the conventional activated sludge process, sequencing batch reactor (SBR), and MBR. The attached growth processes that will be discussed include tricking filters, fluidized beds, and moving bed bioreactors (MBBRs). Incorporating aspects from these two categories allows for innovative hybrid designs that may fit more accordingly into existing or newly constructed treatment facilities. Of these hybrid processes, only integrated fixed film activated sludge (IFAS) will be discussed.

1.2.1 Suspended Growth

Suspended growth processes are biological treatment processes in which microorganisms responsible for the conversion of organic matter or other constituents in the wastewater to gases and cell tissue are maintained in suspension within the liquid (Tchobanoglous et al., 2014). The microbial flocs are grown and maintained in the reactor, being periodically removed or recycled by sedimentation or other methods to keep optimal removal efficiencies.

1.2.1.1 Conventional Activated Sludge

Conventional activated sludge is a process where the sludge floc is produced in a raw or settled wastewater through the growth of zoogleal bacteria and other organisms in the presence of dissolved oxygen. This floc can also be an accumulation of the sludge floc previously formed and recycled back to the activated sludge tank. Mixed liquor (a combination of sludge floc and the wastewater) then leaves the tank and goes through a final sedimentation process in a subsequent clarifier. Part of the sludge floc that is settled in the clarifier is recycled back to the activated sludge tank. Figure 3.1 presents a diagram of a typical industrial aerated conventional activated sludge process with denitrification; however, many activated sludge processes at industrial facilities do not include denitrification before aeration but do have anoxic selectors to minimize potential for filamentous bacteria buildup.

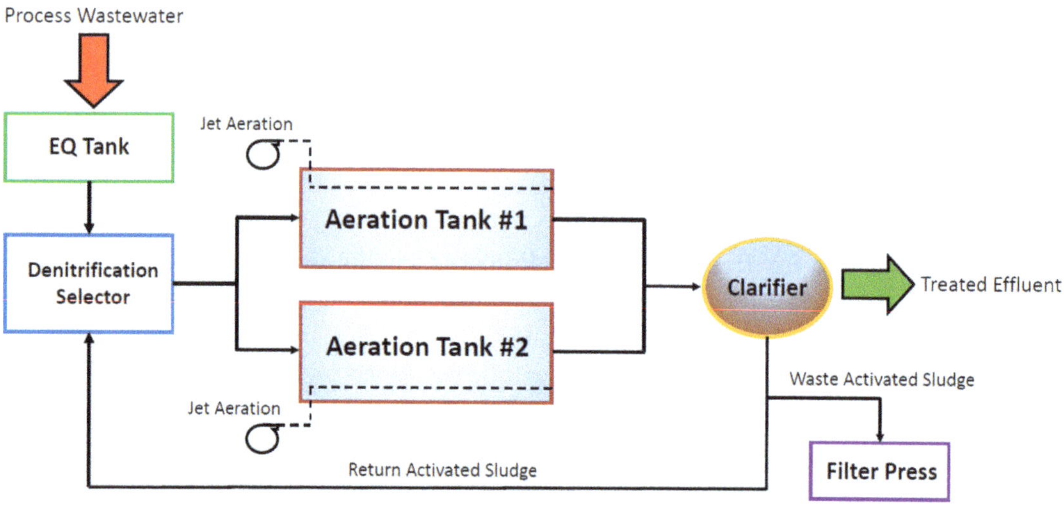

FIGURE 3.1 Aerated conventional activated sludge process with anoxic selector for denitrification. Created using information from Cleary and Eckenfelder, 2014.

The activated sludge process has evolved substantially over the years and is continually improving as the industry progresses. The focus of these improvements includes all aspects of the treatment system as well as the ever-changing world in which these processes take place. Our understanding of these principles has increased because of the experience gained through the design and operation of thousands of facilities and the extensive research and development that goes into their creation. Examples of improvements to these technologies include: increasing biomass concentration and retaining biomass in the system, reducing physical footprint, increasing aeration and oxygen transfer, selecting and deselecting specific microorganisms to achieve better treatment efficiency and performance, protecting more sensitive organisms such as nitrifiers from inhibitory constituents, and improving treated effluent quality. Typical design loading rates are 0.1 to 0.2 kilograms biological oxygen demand (BOD) per day per kilogram mixed liquor volatile suspended solids (MLVSS) (kg BOD/kg MLVSS·d). Bench-scale testing may be required as industrial wastewater has a range of biodegradation and nitrification removal rates and may contain inhibitory and slow degrading compounds. Advantages of the activated sludge process include the following:

- Most proven and oldest of the treatment alternatives continually used today
- Reliable, especially when influent characteristics are consistent
- Odor is reduced because of aeration

- Used in most industries for removal of chemical oxygen demand (COD), BOD, and nitrogen
- Potential to nitrify and denitrify

Disadvantages of the activated sludge process include the following:

- Relatively large footprint compared to other biological options
- Requires secondary clarifier and sludge recycle pumping
- More complex for operator attention and control
- Most susceptible to solids settling problems
- Less biomass inventory (4000 mg/L or less) in the aeration tank resulting in a relatively smaller treatment capacity when compared to the fixed-film and MBR options

1.2.1.2 Sequencing Batch Reactor

In the SBR process, a single tank provides both the biological oxidation and settling processes that are typically associated with multitank conventional activated sludge treatment. The SBR process can provide complete nitrification, substantial denitrification, and biological phosphorus removal with few facility modifications by changing the timing and duration of the aeration and fill-draw cycles. Typical cycle times are 6, 8, and 12 hours for all the processes shown in Figure 3.2. Typical BOD loading rates are 0.1 to 0.2 kg BOD/kg MLVSS·d.

Advantages of the SBR process include the following:

- Process is simple, with as little as one tank required
- Final clarifiers and return activated sludge pumping are not required and easy to automate, leading to less operator attention
- Operational flexibility
- Compact facility
- Good for varying flows
- Lower energy requirements compared to MBR alternative
- Aerated and un-aerated cycles provide ability to incorporate nutrient removal by adjusting cycle times without basin or aeration/mixing modifications
- Provides excellent biological degradation of the wastewater COD
- Additional internal equalization during filling
- Quiescent settling can achieve low effluent total suspended solids (TSS)

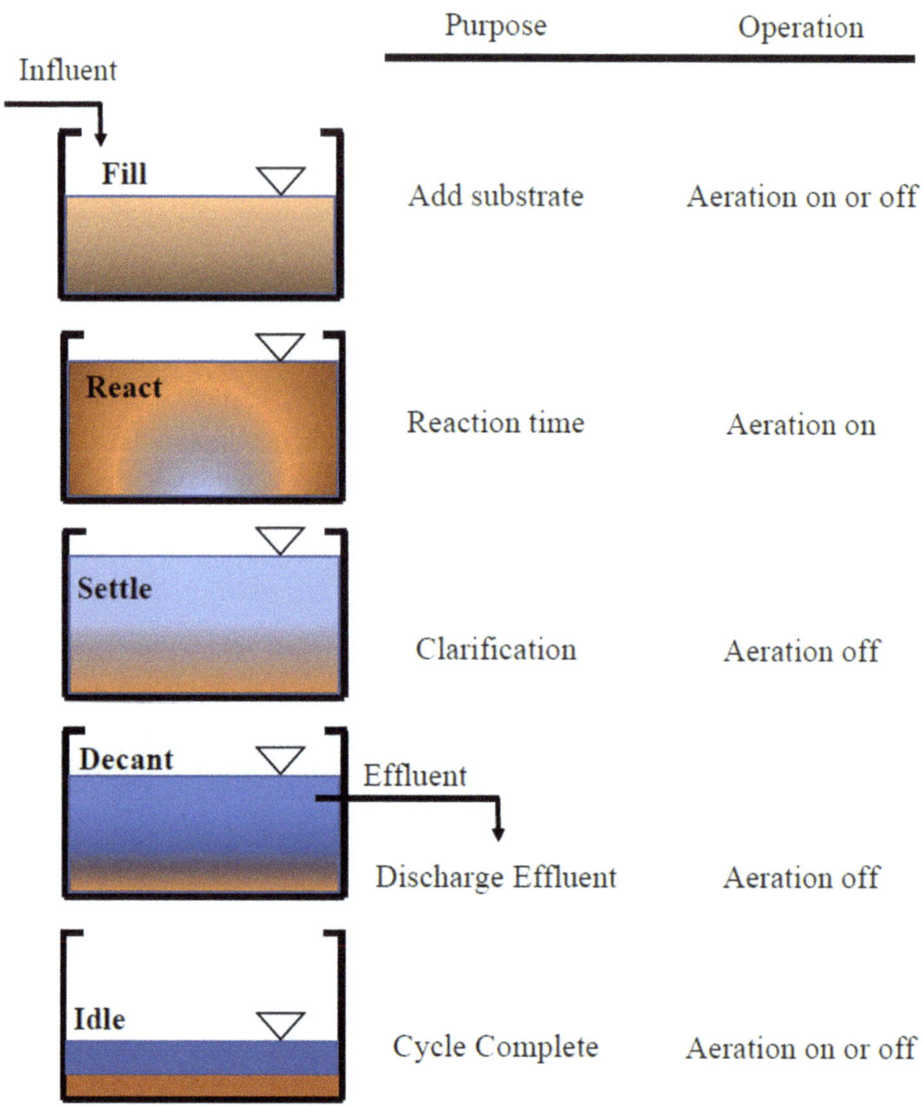

FIGURE 3.2 Aerobic SBR. Created with information from Cleary and Eckenfelder, 2014.

Disadvantages of the SBR process include the following:

- Larger aeration tank volume required
- More complex process controls and instrumentation
- Equalization tank may be required for effluent if there are downstream treatment processes
- Has more process risks because of upsets in the biomass settleability

- Lower effluent TSS can be achieved through other alternative biological processes (e.g., MBR)

1.2.1.3 Membrane Bioreactor

The MBR is regarded as a significant advancement for the activated sludge process because of its proven results and increasing applicability. The MBR system consists of a suspended growth aerobic biological treatment system followed by a membrane filtration system that can either be placed inside or outside of the aeration basin (see Figure 3.3). In short, the MBR system can be thought of as an activated sludge process where the secondary clarifiers are replaced with membranes. These membranes can vary in pore size (i.e., microfiltration, ultrafiltration, or nanofiltration), depending on the treatment objectives. Membrane bioreactor systems can consistently remove organics, TSS, and ammonia as well as nitrogen and phosphorus. Per the U.S. Environmental Protection Agency's Industrial Wastewater Treatment Technology (IWTT) Database (U.S. EPA, 2019), MBR systems are used in 17 distinct industries, ranging from pharmaceutical manufacturing, oil and gas extraction, to meat and poultry products.

As explained previously, the microorganisms present in these treatment systems are the driving force behind their success with regard to the removal of organics. The concentration of these bacteria has a direct correlation with the amount of contaminant removal possible and is a main factor in designing tank sizes and aeration needs. With conventional activated sludge systems, higher percent removal requires larger tanks to accommodate more biomass, which, in turn, requires more aeration. The membrane filtration allows the system to retain a high level of mixed liquor suspended solids (MLSS) (biomass) at typical concentrations of 8000 to 12 000 mg/L. Typical design BOD loading rates are similar to activated sludge, but in a smaller bioreactor with less detention time resulting from the greater active biomass population when compared to the activated sludge. These factors, with the

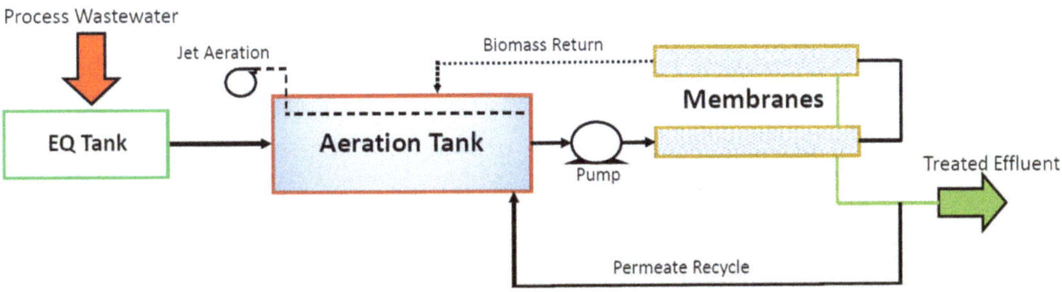

FIGURE 3.3 External membrane filtration systems. Created using information from Cleary and Eckenfelder, 2014.

absence of secondary clarifiers, result in smaller physical footprints than conventional activated sludge systems—an attractive quality for upgrading preexisting treatment facilities. Other advantages of the MBR system include the following:

- Available in packaged systems, allowing for a compact facility with lowest sitefootprint requirements
- Operates at high volumetric loading rates, which result in lower hydraulic retention times (HRTs)
- Operates with longer solids residence times (SRTs) resulting in lower sludge production
- Effluent contains low concentrations of TSS, BOD, low turbidity, silt density index <3, and excellent nutrient removal (nitrogen and phosphorus)
- High-quality effluent may be suitable for reuse in the facility with BOD and TSS less than 5 mg/L, as well as being of sufficient quality to feed directly to reverse osmosis for further purification in a recycling facility.
- Membranes provide physical barrier advantage, less risk of failure in biological system compared to other treatment alternatives

Expandability because of the built-in ability to run at a higher solids concentration and typical inclusion of space for additional cassettes for future module additions as follows:

- Lower operation costs for chemicals, as MBR does not need any chemicals for flocculation and sedimentation but does require chemicals for membrane cleaning.
- Capital and operational costs on decreasing trend, as membrane costs is decreasing.

Some disadvantages of the MBR system include the following:

- Higher energy consumption to overcome trans-membrane resistance, and resulting from increased aeration rate to maintain higher biomass concentration in suspension and to prevent fouling
- More complex process controls and instrumentation
- Higher labor requirements
- Membrane fouling potential
- Cost of membrane replacement after several years of operation
- Cleaning chemicals are necessary

1.2.1.4 Aerobic Granular Sludge

Aerobic granular sludge (AGS) is a novel microbial community that allows simultaneous removal of carbon, nitrogen, phosphorus, and other pollutants in a single sludge system (Nancharaiah & Kiran Kumar Reddy, 2017). Aerobic granular sludge needs to be meticulously cultivated so that larger sludge granules are present in the sludge, allowing for a layered microbial community to take place. These layered granules have a dissolved oxygen gradient, which means a high to low dissolved oxygen concentration from outside layer to granule core. This allows varying types of biological life to exist. The granule structure consists of an aerobic outer layer, an anoxic middle layer, and an anaerobic core, allowing nitrifiers, denitrifiers, and phosphorus accumulating organisms to coexist in the sludge. This complex yet balanced structure creates a lager granule that has excellent settling properties (8 g/L or more sludge volume index; Bates, 2017) and a high MLSS concentration.

Treating wastewater with an AGS system can be done in a single tank without sludge recirculation. It is typically applied for flows greater than 157.73 m^3/h (1 mgd), with tanks deeper than 5.5 m (18 ft), ideally, and requires 6-mm perforated plates for upstream screening (Bates, 2017). An AGS system is an option when retrofitting an existing facility or constructing a new facility, when space is limited, and new regulations call for nutrient removal or in industrial treatment settings. When compared to traditional biological nutrient removal systems, the footprint of an AGS system is 75% less, roughly equivalent to an MBR system. However, when comparing energy consumption, the AGS system requires 50% less energy than an MBR system (Bates, 2017). There is no need for chemical additives, support media, or equipment related to sludge recycling. Like an SBR system, an AGS system follows three main steps: (a) fill/draw, (b) react, and (c) settle. The system is robust against shock loads such as salinity and/or pH fluctuations, chemical spikes, and load variations (Bates, 2017). There are approximately 40 full-scale or pilot facilities worldwide that use an AGS system. The advantages of AGS systems include the following:

- Small physical footprint
- Multiple treatment processes are contained with each sludge granule
- Good sludge settleability
- Reduced energy requirements
- No need for chemicals nor media support and screens
- Can be used in an SBR system with SBR system advantages

A disadvantage is that these systems are not at many industrial facilities yet; mostly municipal wastewater facilities at this time.

1.2.2 Attached Growth (Fixed Film)

Attached growth processes, also known as "fixed film processes," are biological treatment processes in which microorganisms responsible for the conversion of organic matter or other constituents in the wastewater, to oxidized end products and cell tissue, are attached to some inert medium such as rocks, slag, or specially designed ceramic or plastic materials (Tchobanoglous et al., 2014). These inert granular materials can be fixed, expanded, or fluidized and, depending on their configuration, the process can be upflow or downflow. The attached growth treatment processes that will be discussed include the trickling filter, fluidized bed bioreactor (FBBR), and MBBR.

1.2.2.1 Trickling Filter

Trickling filters have been used for biological wastewater treatment for more than 100 years. Trickling filters have a bed of media over which wastewater is continuously distributed. They are classified by hydraulic or organic loading rates. The classifications are as follows: low or standard rate, intermediate rate, high rate, super high rate, and roughing. It is common to use two trickling filters in series to create a two-stage filter configuration, most often with high-strength wastewater or when nitrification is required.

The following factors must be considered in the design of trickling filters: the dosing rate, the type and dosing characteristics of the distribution system, the type and characteristics of the filter medium to be used, the configuration of the underdrain system, provisions for adequate ventilation and natural or forced air, and the design of the required settling tanks. The dosing rate for trickling filters can be approximated by multiplying the organic loading rate in lb BOD_5/1000 cu ft by 0.12. This will give an estimate of dosing rate in inches per pass. The BOD loading rates range from 0.05 to 1.4 kg BOD/m^3·d for a roughing filter before an activated sludge process (Tchobanoglous et al., 2014). Trickling filters are still in operation but are typically not being added to facility upgrades anymore as a result of the availability of other fixed film technologies.

Settling tanks that follow trickling filters are different from clarifiers for activated sludge because there is no sludge recirculation in the former. All the sludge separated in trickling filter settling tanks is removed to sludge processing facilities. The surface loading rate for these settling tanks is based on the facility flow plus the recycle flow minus the underflow. Suggested overflow rates for settling tanks following trickling filters are 16 to 25 m^3/m^2·d (400 to 600 gal/sq ft/d) during average flow and 40 to 50 m^3/m^2·d (1000 to 1200 gal/sq ft/d) during peak flow conditions.

1.2.2.2 Fluidized Bed Bioreactor

The fluidized bed treatment process, or FBBR, is a submerged attached growth process where wastewater is fed upward through a bed of sand or activated carbon; the fluid velocity causes the bed to expand so that a biofilm can grow on the granular material and remove soluble organic matter (Tchobanoglous et al., 2014). The biofilm grown on the granular materials within the reactor allows for a high microbial concentration, leading to higher contaminant removals for a given reactor volume. Because these reactors need to be fluidized (expanded by 30% to 90%), the density of the particles plays a huge role in the successful operation of the system. Loss of the biomass particles can occur if their density changes or the flowrate is altered. However, biofilms have been known to slough off the materials on which they grow after a certain mass has accumulated. Particle collision in the FBBR encourages the biomass sloughing.

Although this "self-regulation" may sound advantageous, variable amounts of biomass can lead to operational difficulties when the system is scaled up. It seems that they may self-regulate the maximum amount of growth on a particular particle but cannot ensure each particle has a minimum amount of biomass. Typical volumetric loading rates for aerated biological filters can be 3.5 to 5.5 kg/m^3·d (0.22 to 0.31 lb/cu ft/d) for BOD removal and 1.8 to 2.5 kg/m^3·d (0.11 to 0.16 lb/cu ft/d) for BOD removal and nitrification (Tchobanoglous et al., 2014).

1.2.2.3 Moving Bed Bioreactor

An MBBR is essentially an activated sludge process where plastic media is added in the aeration basin. Plastic media provides significant surface area expansion for biomass attachment to create a combined suspended and attached growth system. Some MBBR systems do not need a clarifier to meet TSS permit limits for discharge to the water resource recovery facility (WRRF).

The MBBR process can provide the same degree of BOD removal and nitrogen removal as can be obtained using an activated sludge process. Per the U.S. EPA (2019) IWTT Database, MBBRs are used in 20 different industries, most notably petroleum refining, food and beverage, and metal finishing. Typical loading rates are expressed in mass loading as the actual equivalent MLSS is not known because most of the biomass is on or within the media. The biomass per unit volume is higher than activated sludge and SBRs, but less than MBRs. Depending on the loading rates, bacterial communities will vary and the most effective bacteria for a treatment process should be determined by testing. High loading rates (30 g COD/m^2·d) produce compact bacterial biofilms, moderate loading rates (10 to 15 g COD/m^2·d) promote a "fluffy" biofilm, and low loading rates (<5 g COD/m^2·d) promote a more "fluffy" biofilm (Odegaard, 2006). The following are advantages and disadvantages of the MBBR process compared to the activated sludge process:

- Advantages:
    - Small footprint
    - Simpler operation, with no need for sludge recycle, sludge wasting, and SRT control
    - Elimination of concerns with sludge bulking in the secondary clarifiers
- Disadvantages:
    - Higher energy demand because of requirements for elevated dissolved oxygen levels
    - Requirement for use of proprietary media
    - The need for media removal at the time of diffuser maintenance
    - The requirement for improved influent wastewater screening
    - The requirement for a mechanism such as a screen for keeping the media in the bioreactor tank
    - Additional hydraulic headloss caused by flow through screening devices
    - No means of phosphorus removal except for chemical addition

1.2.3 Hybrid

Hybrid treatment processes are those that combine aspects from attached growth and suspended growth biological systems. The biomass attached to inert media, combined with freely suspended biomass, can greatly increase the number of microorganisms in the bioreactor system without expanding the size of a reactor.

The IFAS process is a hybrid treatment technology that combines a fixed film process with conventional activated sludge. This technology aims to bring together the key aspects of suspended and attached growth systems by introducing media to an aeration basin. Figure 3.4 depicts an IFAS process containing a preanoxic basin plus aeration basin with plastic media followed by a clarifier. The media that are added to a conventional activated sludge system can be fixed or free floating. The objective for either type is to increase the amount of surface area available for biological growth.

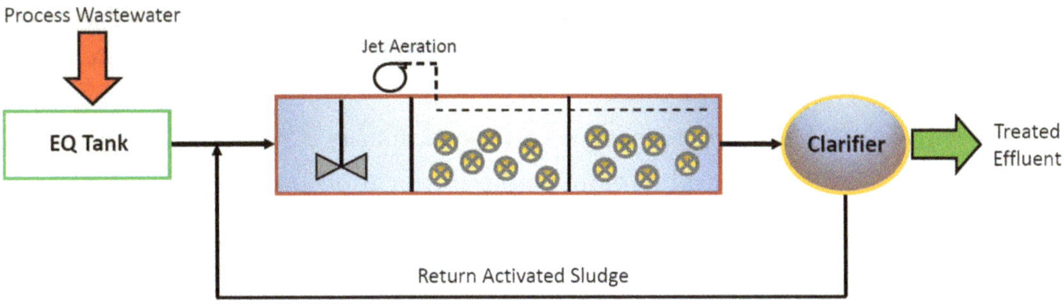

FIGURE 3.4 Integrated fixed film activated sludge system with preanoxic selector for denitrification. Created using information from Cleary and Eckenfelder, 2014.

The fixed media is typically a textile mesh that is lowered into the aeration basin, whereas the free floating media, made up of sponges or plastic shells, is placed directly into the aeration basin. With this increased biomass capacity, comes improved nitrification and, because some of the biomass is attached to media, there can be a decrease in suspended MLSS leading to improved clarification. Integrated fixed film activated sludge systems are commonly used in petroleum refining and in the production of organic chemicals, plastics, and synthetic fibers. The IFAS process is typically used to add benefits of nitrification to an existing activated sludge system without addition of extra aeration zones. Fixed bed biofilm cassettes can also be added directly to an aeration basis to increase capacity, this is referred to as membrane aerated film reactor. The following are examples of drivers for these cases:

- The target effluent requirements have changed from only BOD removal to a low ammonia-nitrogen concentration.
- The system must be converted in a manner that a portion of the existing aerobic tank must be used for a preanoxic zone, resulting in lowering the SRT to below what is needed for nitrification.
- The case where the flowrate and the loading to the activated sludge system is increased, combined with the increased sludge production and aeration tank MLSS concentration limits, result in lower than required SRTs.

The following are advantages and disadvantages of the IFAS process compared to the activated sludge process (Tchobanoglous et al., 2014):

- Advantages:
  - Increased treatment capacity in an identical reactor size

  - Increase in effective mixed liquor concentration to as much as 4000 to 8000 mg/L without an increase in the solids loading rate to the secondary clarifier
  - The ability to increase treatment capacity in phases by adding media
  - A comparably more stable nitrification process because of high attached growth inventory
  - The potential for simultaneous nitrification and denitrification through controlling dissolved oxygen and the loading rate
- Disadvantages:
  - The need for operating at higher dissolved oxygen levels results in a higher energy demand
  - Requirement for use of proprietary media
  - The need for media removal at the time of diffuser maintenance
  - Requirement for improved influent wastewater screening
  - Issue of hydraulic headloss because of flow through screening devices

1.3 Anaerobic

As with aerobic treatment processes, anaerobic treatment processes create environments that facilitate certain bacteria (i.e., acid-forming and methane-forming bacteria) to carry out their natural biological functions. Because there is no oxygen present in these systems, the way in which the bacteria break down the contaminants differs from the typical aerobic system. The anaerobic treatment processes that are discussed in this chapter include the upflow anaerobic sludge blanket (UASB), anaerobic filter, anaerobic fluidized bed, and anaerobic MBR. Figure 3.5 presents a general overview diagram of an anaerobic treatment sequence.

Lower life-cycle costs driven by lower energy input, less sludge production and lower operations and maintenance (O&M) for treating high (organic) strength wastes typically justifies selection of anaerobic over aerobic treatment. The following are advantages and disadvantages of anaerobic treatment systems (Tchobanoglous et al., 2014):

- Advantages:
  - Less energy required
  - Less biological sludge production
  - Less nutrients required
  - Methane production, which is a potential energy source

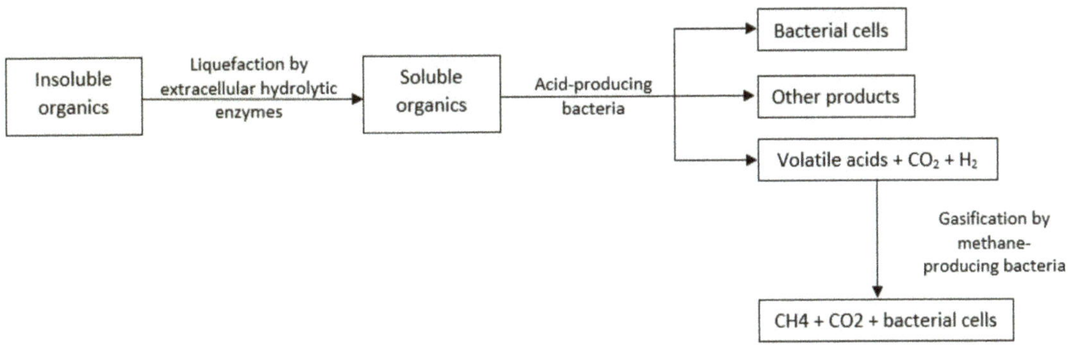

FIGURE 3.5 Anaerobic treatment sequence. Created using information from Cleary and Eckenfelder, 2014.

  - Smaller reactor volume required
  - Anaerobic processes typically include proper venting provisions for methane gas generated, therefore eliminating offgas air pollution potential
  - Able to respond quickly to substrate addition after long periods without feeding
  - Effective pretreatment process
  - Potential for lower carbon footprint
- Disadvantages:
  - Longer startup time to develop necessary biomass inventory
  - May require alkalinity addition
  - May require further treatment with an aerobic treatment process to meet discharge permits limits for nutrient and organics for direct discharge and reuse applications
  - Biological nitrogen and phosphorus removal are not possible
  - Much more sensitive to the negative effect of lower temperatures on reaction rates
  - May be more susceptible to upsets because of toxic substances or wide feeding changes
  - Potential for odor production and corrosiveness of gas

1.3.1 Upflow Anaerobic Sludge Blanket

The UASB is a single-tank process that operates around balancing the settling rate of sludge and the upward flow of the influent waste stream. The wastewater is pumped into the tank from the bottom and, by flowing upward, suspends a sludge blanket that effectively treats the water as it passes through. This sludge blanket consists of microbial granules (1 to 3 mm in diameter) that resist being washed out in the upflow because of their weight (Tilley et al., 2014). As these microbes degrade organic compounds, methane and carbon dioxide are produced, and these gases provide mixing without the need for mechanical assistance. Figure 3.6 presents a diagram of a typical UASB system. As the figure shows, a sloping of the walls below the water surface creates a constriction preventing sludge overflow into the effluent. The fixed cover allows for collection of gases generated during the treatment process.

The following are advantages and disadvantages of the UASB process (Tilley et al., 2014):

- Advantages:
  - High reduction of BOD
  - Can withstand high organic and hydraulic loading rates

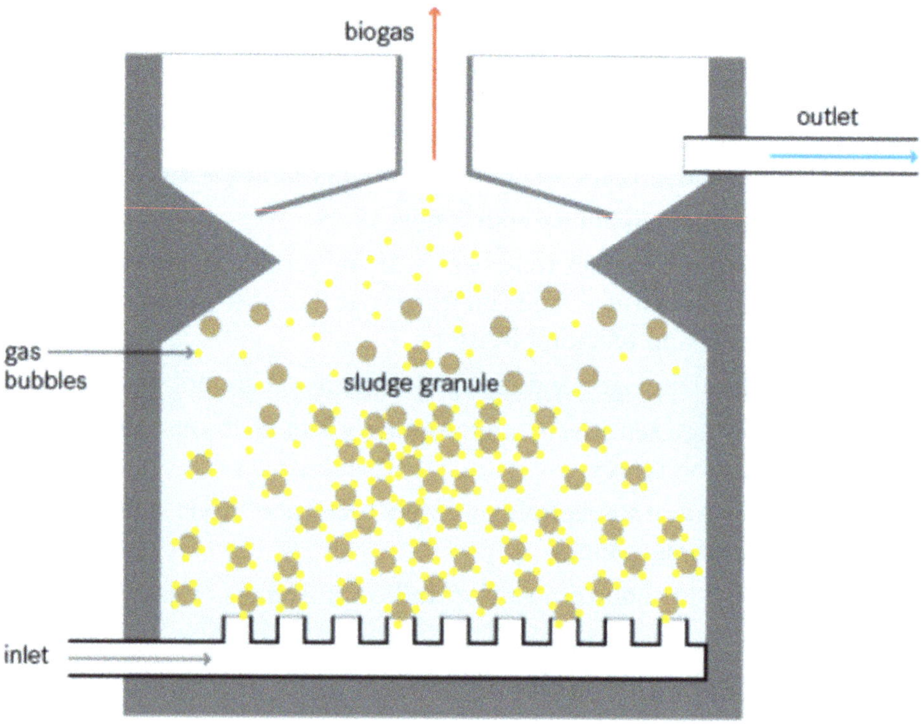

FIGURE 3.6 Upflow anaerobic sludge blanket diagram. Reprinted with permission from Eawag—*Swiss Federal Institute of Aquatic Science and Technology* (Tilley et al., 2014).

- Low sludge production
- Biogas can be used for energy (may require scrubbing)

- Disadvantages:
  - Treatment may be unstable with variable hydraulic and organic loads
  - Requires O&M by skilled personnel; difficult to maintain proper hydraulic conditions (upflow and settling rates must be balanced)
  - Long startup time to work at full capacity
  - A constant source of electricity is required
  - Not all parts and materials may be locally available
  - Effluent and sludge require further treatment and/or appropriate discharge

1.3.2 Anaerobic Filter

Anaerobic filter technology is less common than it was years ago. It is a downflow trickling-type filter technology with fixed crossflow media or other media to which the biomass attaches. It has been used in a variety of industries, including the chemical and beverage industries. For example, it has been in operation since the early 1980s at the Bacardi rum distillery in Puerto Rico. The typical COD loading rate is low relative to other technologies, at less than 3 $kg/m^3 \cdot d$ (0.12 lb/cu ft/d).

Anaerobic filters typically require a startup period of 6 to 9 months to reach full treatment capacity (Tilley et al., 2014), which is attributable to the slow-growing anaerobic bacteria. Flow should be sent through both preliminary and primary treatment processes to remove solids before entering an anaerobic filter. Depending on the scale of the system, units can have an integrated settling compartment or make use of an external primary settling tank (Tilley et al., 2014). Flow through an anaerobic filter is typically upflow to reduce the amount of washout. Typical HRTs can be anywhere from 12 to 36 hours depending on the influent COD for industrial wastewaters up to a COD concentration of 50 000 mg/L or even higher for rum distillery wastewater.

Advantages of anaerobic filter technology include the following:

- Proven technology as evidenced by a record track of facilities that have been in operation for many years
- Low operating cost
- Low sludge production
- High reduction of BOD and solids

Disadvantages of anaerobic filter technology include the following:

- Requires significant and expensive cleaning and potential replacement of media after several years
- As the biomass is attached to the media, and it cannot be easily separated and transferred, the downtime for media replacement could be significant at larger facilities. Long startup period is expected after maintenance as biofilm needs to regrow on the new media.
- Low reduction of pathogens and nutrients

1.3.3 Anaerobic Fluidized Bed

Anaerobic fluidized bed technology consists of a digester tank with sand or activated carbon media, which the biomass grows on. The high surface area of the sand or activated carbon, which is fluidized in an upflow mode, allows for a high biomass per unit volume rate and smaller tank sizes versus other anaerobic treatment alternatives. The anaerobic fluidized bed has been used for many years in a variety of industries, including the food and beverage industries. It has also been used in airport treatment facilities in Albany, New York, and other locations to treat glycol deicing chemicals used during the winter to deice aircraft. Typical COD loading rates are over 20 kg/m^3·d (4.27 lb/cu ft/d). The following are advantages and disadvantages of the anaerobic fluidized bed:

Advantages:

- High COD loading rate and short hydraulic detention time
- Small digester tank volume because of high biomass concentrations on the media
- Can be turned on and off seasonally for food processing and airport deicing applications
    - Relatively quick startup period after being shut down

One disadvantage is that the anaerobic fluidized bed needs downstream TSS removal.

1.3.4 Co-Digestion at Water Resource Recovery Facilities

Co-digestion, which is becoming more popular, is the addition and treatment of high-strength organic liquid industrial wastewater (e.g., winery, dairy, etc.) or fats, oils, and grease to anaerobic digestion at municipal WRRFs. The addition of organic waste is used to increase the biogas generation for not just heating the digester, but to generate electricity in combined heat and power (CHP) reciprocating engines, microturbines, and other alternatives.

A WRRF in upstate New York treats high-strength dairy wastewater (COD, 60 000 mg/L) along with biosolids in two anaerobic digesters. This facility uses the biogas in a CHP system to produce electricity and achieve essentially net-zero, with no need for electricity purchased from the grid.

1.3.5 Anaerobic Membrane Bioreactor

Anaerobic membrane bioreactor (AnMBR) treatment is the combination of conventional anaerobic treatment and physical solids separation using membranes. This technology has become more popular in the last 10 years, with multiple technology providers in the United States. Industries that use AnMBR systems include transportation equipment cleaning, canned and preserved fruits and vegetables processing, and breweries. One of the first instances of use of this technology in the United States is at Ken's Foods near Boston, Massachusetts. It has a completely mixed anaerobic reactor followed by submerged membranes in a separate tank. The membranes in AnMBR systems are cleaned by scour with biogas compared to air in aerobic MBR systems. The membranes are used to maintain anaerobic biomass in the system and allow more biomass per unit volume of anaerobic digester than some of the other anaerobic technologies. Typical COD loading rates are less than fluidized bed technology, in the range of 5 to 10 kg COD/m^3·d (0.31 to 0.62 lb/cu ft/d). The following are advantages and disadvantages of an AnMBR:

- Advantages:
  - Membranes provide low risk of losing biomass to effluent
  - No downstream TSS removal required; produces a low TSS effluent
- Disadvantages:
  - Less full-scale treatment facilities, approximately 10 to 15 versus other anaerobic treatment technologies such as UASB, which have hundreds of full-scale facilities. This is because of AnMBR being a newer technology.
  - Requires membrane cleaning and eventual replacement

2.0 CHEMICAL TREATMENT FOR SOLIDS–LIQUID SEPARATION

2.1 Introduction

This section presents the Stoke's law fluid/solids separability theory, a categorization of solids; basic chemistry; and the fundamentals of precipitation, coagulation, and flocculation; and how application of chemical additives fits with various methods of clarification. Types of coagulants and flocculants are introduced as well as the matching of types of chemical treatment with substrates. Maximizing solids–liquid separation efficacy is one of the steps that allows for maximizing water reuse and minimizing overall consumption of fresh water.

2.1.1 Settling Velocity—Stoke's Law

Stoke's law is a mathematical representation used to describe settling of a discrete spherical particle in a viscous fluid, whereby the drag force on the particle surface is the predominant factor inhibiting gravity separation from the fluid. We can understand factors affecting settleability by looking at the variable terms in Stoke's law, represented in the following equation (3.1):

$$v = \frac{2Gr^2(\rho - \rho_0)}{9\eta} \tag{3.1}$$

Stoke's law shows that the relationship between settling velocity (v, m/h) of a discrete spherical particle through a fluid is proportional to the radius of the sphere squared (r^2, m^2), the density difference between the sphere (ρ, kg/m^3), and the fluid (ρ_o, kg/m^3), and is inversely proportional to the viscosity of the fluid (η, kg/m/s). For the case where viscosity and relative density factors are constant, Stoke's law states that the velocity of settling is proportional to the radius of the sphere squared. If the radius doubles, the settling velocity increases by a factor of 4. Increasing particle size has a pronounced effect on promotion of solids/fluid separation. Holding the size of the sphere constant, the settling velocity increases when the density difference between the particle and the fluid ($\rho - \rho_o$) increases. Note that if the density of the particle is less than the density of the fluid ($\rho < \rho_o$), the velocity term (v) becomes negative, indicating a change of direction. The negative velocity term indicates particles rising through the fluid rather than settling. G is the gravitational constant. Stoke's law does not extend to hindered or compaction settling.

2.1.1.1 Floc Size

The overall goal of coagulation and flocculation is to agglomerate many small particles into fewer larger ones so that separation occurs more quickly.

2.1.1.2 Solid/Fluid Differential Density

One important factor to realize is that most coagulants and flocculants do not have a significant effect on the intrinsic density of the primary particles or the flocs that result. That means that a particle that is naturally more dense than water will sink and is not able to be floated by the action of coagulants or flocculants alone. Flocs that are intrinsically less dense than water cannot be made to sink by the action of coagulants and flocculants alone. Some less-dense-than-water constituents may be incorporated in a settleable floc, as long as the net density of the entire floc particle remains higher than the density of the fluid it is settling in. Higher density agents, such as bentonite clay or colloidal silica, can be added to help slightly lighter than water floc particles sink. To make heavier than water floc particles float, gas bubbles can be introduced that attach to and incorporate into flocs and make the floc/bubble grouping net density less than water, promoting flotation.

2.1.1.3 Viscosity of Fluid as a Function of Temperature

Settling velocity is inversely proportional to viscosity of the fluid. Because the viscosity of water is reduced at higher temperatures, settling velocity increases as the water becomes warmer.

2.1.2 "Solids" in Water

Herein, the term "solids" is used to represent actual solids as well as dissolved ions that could be precipitated, and dissolved ions that would become solids in the absence of water.

2.1.2.1 Suspended Versus Soluble

Figure 3.7 is designed to aid in understanding important classifications of contaminants in water. Reading this graphic from top down and left to right, we first define two broad categories of contaminants: suspended and soluble.

2.1.2.2 Suspended Solids: Settleable Versus Nonsettleable

Next, we break the "suspended" category into settleable (or floatable), and nonsettleable. The settleable (or floatable) contaminants will naturally separate from water by gravity and can be represented by contaminants such as grit or sand (settleable) or free oil (floatable). As particulate matter becomes smaller and smaller, approaching colloidal size, it becomes more prone to suspension as charge repulsion forces overcome gravitational forces; therefore, we call finely divided solids including colloidal material "nonsettleable." Fine oil droplets that are physically or chemically emulsified are typically nonsettleable as well. Figure 3.8 shows a nonsettleable suspension in the jar at right.

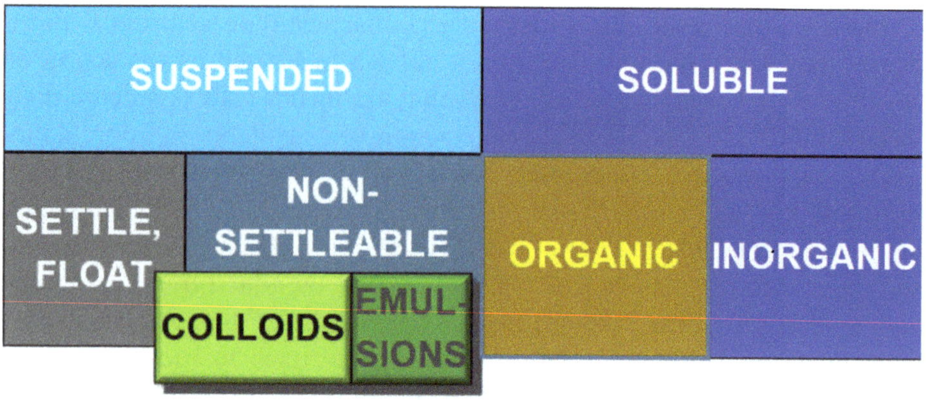

FIGURE 3.7 Classification of wastewater "solids" (Goldblatt & Leitz, 2015).

Nonsettleable suspensions may be destabilized by charge neutralization and interparticulate bridging when chemically treated, allowing separability from the bulk water. Nonsettleable contaminants typically require the use of coagulants or flocculants to neutralize, demulsify, or to build floc size, allowing for settling or flotation of flocs, as illustrated by Stoke's law.

2.1.2.3 Soluble "Solids": Organic Versus Inorganic

Soluble material includes organic and inorganic materials that are truly solubilized. Biological treatment is a standard method used to remove most soluble organic constituents. Chemical precipitation is used to convert inorganic

FIGURE 3.8 Example of nonsettleable solids, turbidity (photo courtesy of Pete Waldrep).

constituents such as calcium and magnesium hardness and some metals into precipitates. The precipitates may be readily separable, or may require chemical flocculation to be physically removed by unit operations such as sedimentation or filtration.

2.1.3 Surface Charge Potential

Figure 3.9 illustrates some of the important interparticle forces at work. Because natural particles in water typically have a net negative charge, particles exhibiting like charges will naturally repel one another. This is called "electrostatic repulsion." van der Waals forces are the relatively weak interparticle forces of attraction. These two opposing forces are used to introduce the concept of coagulation on a molecular level.

2.1.3.1 Fixed Charge Layer

From the depiction of a single particle in water in Figure 3.10, one may observe several layers of ions surrounding the particle. The negatively charged particle will have a layer of positive counterions (cations) at the surface. This is called the "fixed layer" or "Stern layer."

2.1.3.2 Shear Plane, Zeta Potential

Just outside the cationic counterion layer is a more diffuse layer of counter-counter ions, mostly negatively charged. The outer perimeter of this layer

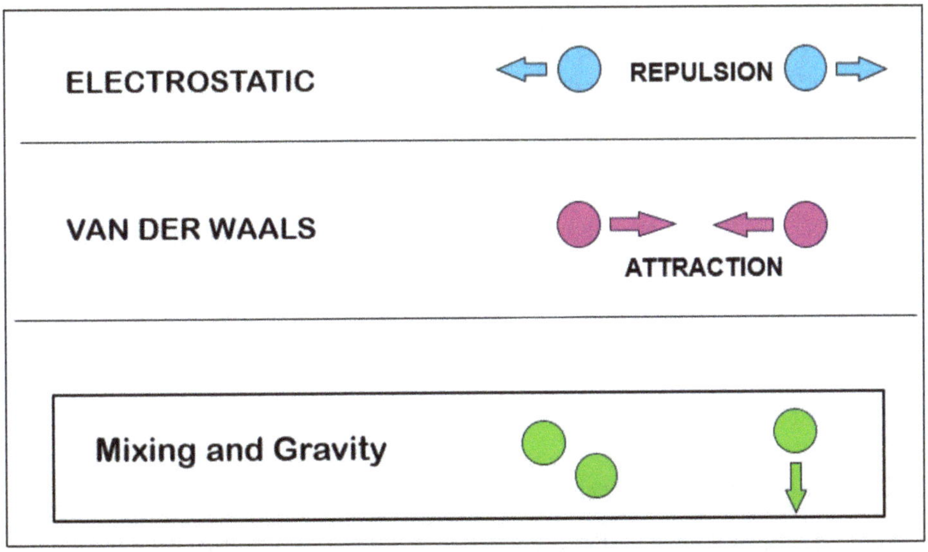

FIGURE 3.9 Key interparticle forces in coagulation, flocculation, and liquid–solids separation (Goldblatt & Leitz, 2015).

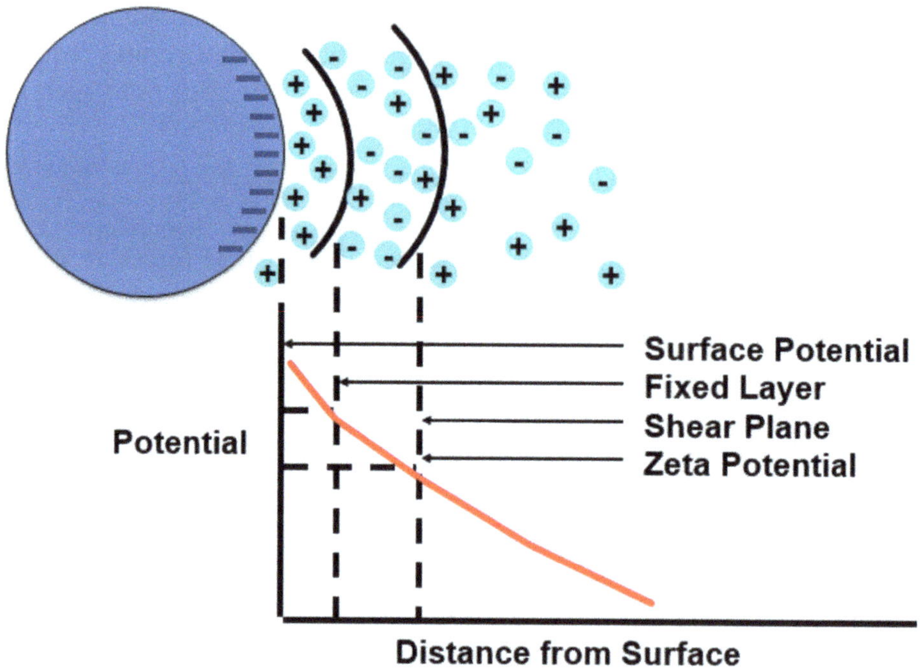

FIGURE 3.10 Ionic charge on and around a single particle in water (Goldblatt & Leitz, 2015).

is called the "slip plane" or "shear plane," and represents the difference between stationary ions on the surface of the particle and movable ions just outside the fixed layer. Zeta potential meters measure charge density in a solution by shearing the movable layer of ions and measuring resultant electrical potential.

Outside of the shear plane is the bulk liquid. These are mathematically derived and modeled layers that describe layer thicknesses. The thicknesses of the layers are inversely related to the electrolytic strength of the liquid solution. Higher electrolytic strength compresses the layers, promoting more interparticle collisions. This is the phenomenon known as "double layer repression."

The use of cationic coagulants promotes the overcoming of interparticle negatively charged repulsive forces, allowing for higher probability of particle collisions and for the van der Waals forces of attraction to predominate. Coagulation by charge neutralization makes initial agglomeration of finely divided particles possible. This is described further in Section 2.3.1.

2.2 Chemical Precipitation

Chemical precipitation is the "formation of a separable solid substance from a solution, either by converting the substance into an insoluble form or by changing the composition of the solvent to diminish the solubility of the substance in" the solution (American Water Works Association, 2011). Lime and lime/soda softening, hydroxide precipitation, and sulfide precipitation are three types of precipitation chemistry.

2.2.1 Lime and Lime/Soda Softening

Lime softening is one of the oldest water treatment methods used for the reduction of total hardness. Co-precipitation of iron, manganese, color bodies, phosphate, and silica can also be made to occur. The addition of hydrated lime, $Ca(OH)_2$, will initiate reactions to precipitate dissolved calcium ions into calcium carbonate, and magnesium ions into magnesium hydroxide, as shown in the following equations (3.2 through 3.4):

$$Ca(HCO_3)_2 + Ca(OH)_2 \rightarrow 2CaCO_3 + 2H_2O \qquad (3.2)$$

$$Mg(HCO_3)_2 + Ca(OH)_2 \rightarrow CaCO_3 + MgCO_3 + 2H_2O \qquad (3.3)$$

$$MgCO_3 + Ca(OH)_2 \rightarrow CaCO_3 + Mg(OH)_2 \qquad (3.4)$$

Carbonate hardness, defined as the cations Ca^{++} or Mg^{++} associated with bicarbonate anions HCO_3^-, is removable to its limit of solubility by lime alone. Figure 3.11 shows a graphic of cations on the left and anions on the right. Using this example, one can see that only slightly more than half of the hardness ions represent carbonate hardness.

The leftover hardness ions are designated as noncarbonate hardness and are associated with the other anions (chloride, nitrate, or sulfate) shown at bottom right of Figure 3.11. The term "permanent hardness" is sometimes used to describe noncarbonate hardness, where hardness ions remain fully soluble unless further chemical intervention is used.

Addition of sodium carbonate (soda ash) allows for further hardness precipitation. The following two equations show precipitation of noncarbonate calcium hardness by soda ash (precipitants are underlined):

$$CaSO_4 + Na_2CO_3 \rightarrow CaCO_3 + Na_2SO_4 \qquad (3.5)$$

$$CaCl_2 + Na_2CO_3 \rightarrow CaCO_3 + 2\,NaCl \qquad (3.6)$$

Magnesium noncarbonate hardness removal requires both lime and soda ash in a two-step reaction as follows:

$$(1)\ MgSO_4 + Ca(OH)_2 \rightarrow Mg(OH)_2 + CaSO_4 \qquad (3.7)$$

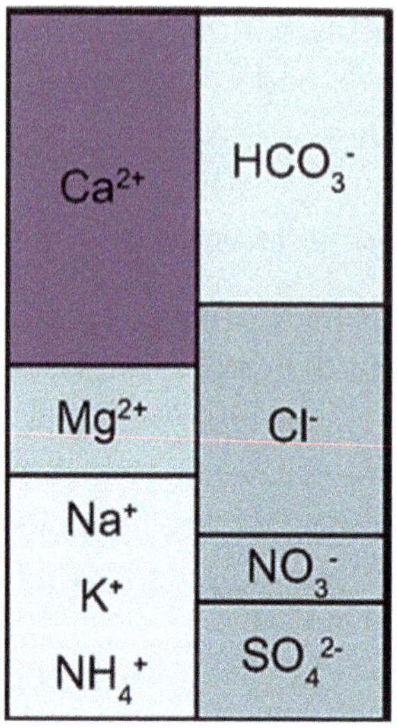

FIGURE 3.11 Cation/anion balance example for lime/soda softening (Goldblatt & Leitz, 2015).

$$\text{(2)}\ CaSO_4 + Na_2CO_3 \rightarrow CaCO_3 + Na_2SO_4$$

Or

$$\text{(1)}\ MgCl_2 + Ca(OH)_2 \rightarrow Mg(OH)_2 + CaCl_2 \qquad (3.8)$$

$$\text{(2)}\ CaCl_2 + Na_2CO_3 \rightarrow CaCO_3 + 2\ NaCl$$

2.2.2 Hydroxide Precipitation

Hydroxide precipitation is commonly used to convert metals to metal hydroxide precipitates, thus allowing physical removal of the precipitated metal hydroxides. Typically, a pH control range of somewhere between 9 and 11 is used, depending on the solubilities of metals to be removed and the extent of removal required. There are many metal solubility versus pH graphs available, such as that shown in Figure 3.12. Note that solubility in natural water or wastewater systems will vary from data generated in a laboratory setting using pure water.

2.2.3 Sulfide Precipitation

Sulfide precipitation may be used to precipitate metals and, in some cases, can achieve a higher level of removal than hydroxide precipitation. However, the transport, storage, and use of sulfide chemicals brings a significant safety consideration; for that reason, it is often not preferred for use in water systems.

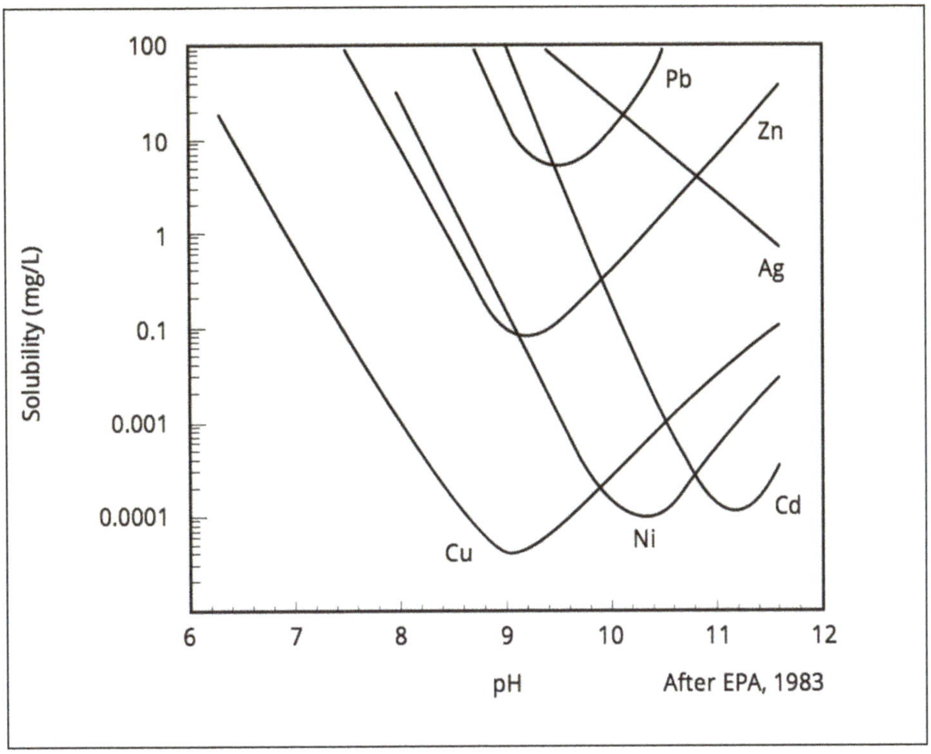

FIGURE 3.12 Solubility of metals in water as a function of pH (Karimi, 2017).

2.3 Coagulation and Flocculation

Coagulation and flocculation are the two functions used for bringing finely divided particulate or emulsified contaminants first into small aggregates that may or may not be visible to the naked eye (coagulation), then into larger groupings that are clearly visible (flocculation). The term "floc" is used in this context to represent these gathered groupings. Joining tiny particles to make larger floc groupings enhances gravity separation as illustrated by Stoke's law.

2.3.1 Definition of "Coagulation"

A simple and useful definition for "coagulation" is the neutralization of microscopic particle surface charge and initial agglomeration of particles into small groupings.

2.3.1.1 Forces of Attraction and Repulsion

Figure 3.13 illustrates the forces between two particles approaching each other. Let the left sphere represent a stationary particle and the right sphere represent a particle approaching the stationary particle, from right to left.

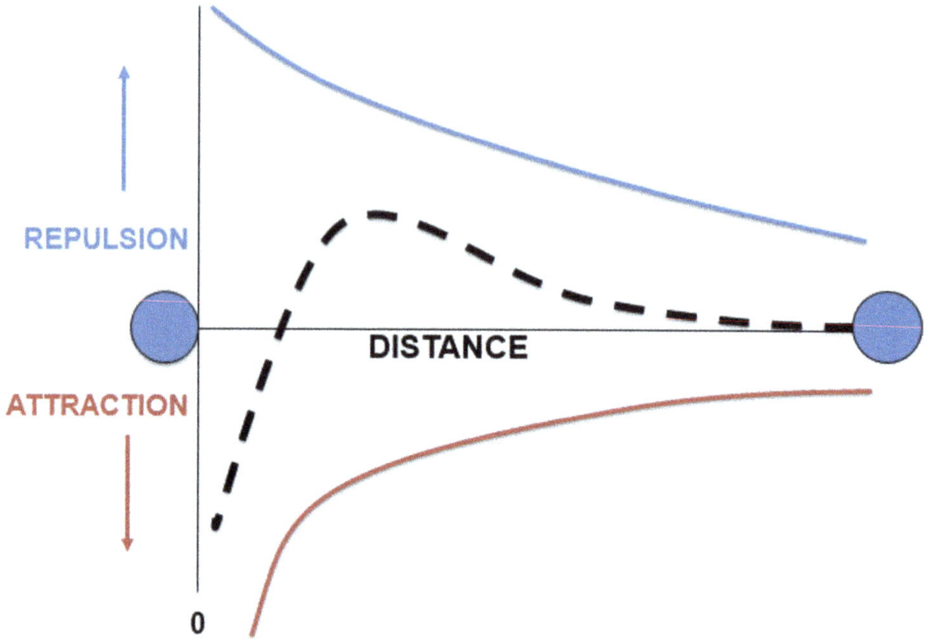

FIGURE 3.13 Interparticle forces: effect of distance between two particles (Solenis, LLC).

The blue curve represents strength of repulsive forces such as electrostatic repulsion. The red curve represents van der Waals forces of attraction between the two particles. The broken-line curve represents the sum of forces, or the sum of the blue and the red curves. One can see that there is a net peak repulsion force at a certain distance as the right particle approaches the left particle. This net repulsive force is responsible for keeping finely divided particles in suspension (i.e., the nonsettleable solids represented in Figure 3.7).

2.3.1.2 Why Coagulate?

Coagulation is the first of a two-step process whereby particles are agglomerated into small groupings to build size to enhance liquid–solid separation according to Stoke's law.

2.3.1.3 How Is Coagulation Accomplished?

Addition of a cationic coagulant neutralizes many of the negative charges on particles, thereby reducing repulsive forces between particles. This effectively lowers the blue repulsion curve in Figure 3.13 so that the hump of net forces (the broken line curve) representing a barrier to interparticle collision is likewise lowered. The result is that the particles will more easily overcome the repulsion barrier and be more likely to pair with each other. This is the basis of coagulation and yields small agglomerations, referred to as "pinfloc."

Coagulants are used in liquid/solids separation applications to neutralize the ionic charges that surround solid particles dispersed in water. Most naturally occurring solid particles in water have a negatively charged surface because of the release of cations such as H^+, Li^+, Na^+, K^+, and Ca^{+2} from the surface of the particle into the surrounding water. When microscopic particles of like charge approach one another, they repel and cannot aggregate to form larger particles. This leads to stable systems of particles in water that will not settle. The higher the anionic charge on the surfaces of the particles, the more stable the colloidal system becomes.

Cationic coagulants adsorb onto the negatively charged particle surfaces, thereby neutralizing the negative charges that are causing repulsion. Optimum coagulation occurs when the particle surface charge is reduced to near zero. A large overdose of cationic coagulant will impart a net positive charge on the particle and restabilization will occur. For this reason, accurate control of the dosage of coagulant is important. Note that there are many more complex theories of coagulation that are beyond the scope of this book.

The first documented use of coagulation for water purification was in ancient Egypt. Figure 3.14 presents a rendering of an ancient Egyptian cave carving showing a device using coagulation and settling to remove fine particulate contaminants.

2.3.1.4 Inorganic Coagulants

The first coagulants used in industry were aluminum sulfate and ferric chloride. What these two compounds have in common is a trivalent metal ion (Al^{+3} and Fe^{+3}) that is strongly attracted to any negatively charged surface. It was found that trivalent cations were much stronger coagulants than divalent cations such as Ca^{+2} and Mg^{+2}. Monovalent cations such as Na^+ and K^+ have weak coagulating abilities. Since the Middle Ages, the most widely used coagulants have been aluminum sulfate, aluminum chloride, ferric chloride, and ferric sulfate. A 20th century development has been the polymerization of aluminum and iron to form inorganic polymers that carry a positive charge. The aluminum compounds go by the names "polyaluminum chloride" and, when they are more fully neutralized, "aluminum chlorohydrate." Table 3.2 presents chemical actives of different forms of trivalent inorganic coagulant chemicals.

FIGURE 3.14 Ancient Egyptian depiction of water purification, pictured on the wall of the tomb of Amenophis II at Thebcs, 1450 B.C. (Ajuha, 2017; Wilkinson, 1879).

There are also many divalent cations used in liquid/solids separation in spite of the fact that they are weaker coagulants (see Table 3.3). One benefit of these is that they do not lower the system pH as much as many trivalent cations.

The final category of inorganic coagulants is the alkaline earth oxides and hydroxides. These are typically used to raise system pH to the point where heavy metals precipitate from solution (e.g., pH 9 to 11), and where hardness removal takes place (normally by alkalinity control; see Table 3.4).

2.3.1.5 Organic Coagulants

Organic coagulants are typically based on nitrogen chemistry because nitrogen, which typically accepts only three bonds, can use its extra pair of electrons to accept a fourth bond in a process known as "quaternization." This induces a positive charge to the nitrogen atom, which is then available to neutralize the negative charges on a particle surface in water. Commercially available products of this type are water soluble polymers containing the quaternized nitrogen moiety. One of the most widely used products is a condensation polymer of epichlorohydrin and dimethylamine (Epi/DMA).

TABLE 3.2 Commonly used inorganic coagulants: water soluble metal salts with +3 valences (Goldblatt & Leitz, 2015).

Common Name	Form	Chemical Formula	Percent as M+3	Percent as M203	Basicity	Other Commercial Names
Aluminum chloride	Dry	$AlCl_3$	20.2	38.2	0	Dry Aluminum chloride
Aluminum chloride	Aqueous	28% $AlCl_3$	5.7	10.7	0	32 Be Aluminum chloride
Aluminum sulfate	Dry	$Al_2(SO_4)_3{*}14H_2O$	9.1	17.2	0	Dry alum
Aluminum sulfate	Aqueous	48.5%$Al_2(SO_4)_3{*}14H_2O$	4.4	8.2	0	Liquid alum; Papermaker's alum
Acidic aluminum Sulfate	Aqueous	X%$Al_2(SO_4)_3$+Y%H_2SO_4	Varies	Varies	0	Acid alum
Sodium aluminate	Aqueous	38%$Na_2Al_2O_4$	12.5	23.6	100	Commercial sodium aluminate
Sodium aluminate	Aqueous	45%$Na_2Al_2O_4$	14.8	28.0	100	High strength sodium aluminate
Low basicity Polyaluminum chloride	Aqueous	30%$Al_2(OH_4)Cl_5{*}4H_2O$	5.1	9.5	17	Polyaluminum chloride (PAC)
Medium basicity Polyaluminum chloride	Aqueous	30%$Al_2(OH_4)_5Cl_3{*}4H_2O$	5.7	10.8	50	Polyaluminum hydroxychloride (PAHC)
High basicity Polyaluminum chloride	Aqueous	50%$Al_2(OH_4)_5Cl_5{*}4H_2Os$	12.2	23.0	83	Aluminum chlorohydrate (ACH)
Ferric chloride	Dry	$FeCl_3{*}6H_2O$	20.7	29.6	0	Dry ferric chloride
Ferric chloride	Aqueous	38% $FeCl_3$	13.1	18.7	0	Liquid ferric chloride
Ferric Sulfate	Dry	$Fe_2(SO_4)_3{*}9H_2O$	19.9	28.5	0	Dry ferric sulfate
Ferric Sulfate	Aqueous	62%$Fe_2(SO_4)_3{*}9H_2O$	12.3	17.6	0	Liquid ferric sulfate

TABLE 3.3 Commonly used inorganic coagulants: water soluble metal salts with +2 valences (Goldblatt & Leitz, 2015).

Chemical Name	Form	Chemical Formula	Percent as M+2	Effect on System pH	Directly Adds Solids?	Other Commercial Names or Comments
Ferrous chloride	Aqueous	20% $FeCl_2$	8.8	DOWN	No	Component of pickle liquor
Ferrous sulfate	Aqueous	15% $FeSO_4$	5.5	DOWN	No	Component of pickle liquor
Calcium chloride	Dry	$CaCl_2*2H_2O$	27.3	DOWN	No	Calcium chloride dihydrate
Calcium chloride	Dry	$CaCl_2*6H_2O$	18.3	DOWN	No	Calcium chloride hexahydrate
Calcium chloride	Aqueous	35% $CaCl_2$	12.6	DOWN	No	Liquid calcium chloride
Magnesium chloride	Dry	$MgCl_2*6H_2O$	12.0	DOWN	No	Magnesium chloride hexahydrate
Magnesium chloride	Aqueous	15%$MgCl_2$	3.8	DOWN	No	Liquid magnesium chloride
Magnesium sulfate	Dry	$MgSO_4*7H_2O$	9.9	DOWN	No	Epsom salts
Magnesium sulfate	Aqueous	25% $MgSO_4$	5.0	DOWN	No	Epsom salts solution
Zinc chloride	Aqueous	50% $ZnCl_2$	24.0	DOWN	No	Rarely used today

TABLE 3.4 Alkaline earth oxides and hydroxides (Goldblatt & Leitz, 2015).

Chemical Name	Form	Chemical Formula	Percent as M+2	Effect on System pH	Directly Adds Solids?	Other Commercial Names
Calcium oxide	Dry	CaO	71.5	UP	Yes	Lime / quicklime/caustic lime
Calcium hydroxide	Dry	$Ca(OH)_2$	54.1	UP	Yes	Hydrated lime/slaked lime
Calcium hydroxide	Aqueous	35% $Ca(OH)_2$	18.9	UP	Yes	Hydrated lime/milk of lime
Magnesium oxide	Dry	MgO	60.3	UP	Yes	Magnesia/magnesium lime
Magnesium hydroxide	Dry	$Mg\ (OH)_2$	41.7	UP	Yes	Hydrated magnesia
Magnesium hydroxide	Aqueous	62% $Mg\ (OH)_2$	25.8	UP	Yes	Liquid magnesia
Magnesium hydroxide	Aqueous	8% $Mg\ (OH)_2$	3.3	UP	Yes	Milk of magnesia (USP)
Calcium + magnesium oxides	Dry	65% CaO + 35% MgO	67.6	UP	Yes	Dolomitic lime
Calcium + magnesium hydroxides	Aqueous	65% $Ca(OH)_2$ + 35% $Mg(OH)_2$	49.8	UP	Yes	Hydrated/slaked dolomitic lime

The two reactants come together in a reactor to form the polymer commonly known as "Epi/DMA polyamine." Figure 3.15 shows the two reactants and the Epi/DMA product.

Another useful monomer is diallyl dimethyl ammonium chloride (DADMAC; also known as DMDAAC). Figure 3.16 illustrates the monomer polymerizing to form poly[DADMAC].

Using addition polymerization, poly[DADMAC] can be made with an average molecular weight between 50 000 and 1 500 000 daltons.

For both poly[DADMAC] and epi/DMA polymers, higher molecular weight often translates into a larger and stronger floc with faster settling and a more compact final sludge volume. The stronger, faster settling floc may, in some cases, result in a more turbid supernatant than a slower settling floc. Selection of the proper molecular weight for a given application requires consideration of clarity requirements, chemistry, and liquid/solids separation process equipment capability. Organic coagulant products are typically available in fully soluble liquid form.

2.3.1.6 Inorganic and Organic Coagulants Compared

In general, inorganic coagulants differ from organic coagulants in the ways shown in Table 3.5. The blue-shaded cells in Table 3.5 indicate an advantageous aspect.

When the use of pH adjustment chemicals and sludge disposal is added into consideration, organic coagulants will, in some cases, have an economic advantage over inorganic coagulants. The corrosion potential from lower pH and lower alkalinity effluent water may also be a detriment caused by the use of inorganic coagulants.

Because there are advantages and disadvantages associated with each type of coagulant, blends of inorganic and organic coagulants are sometimes used to optimize a treatment program. The optimum blend can produce clean effluent water over an existing pH range with the proper floc size and settling rate while minimizing final sludge volume.

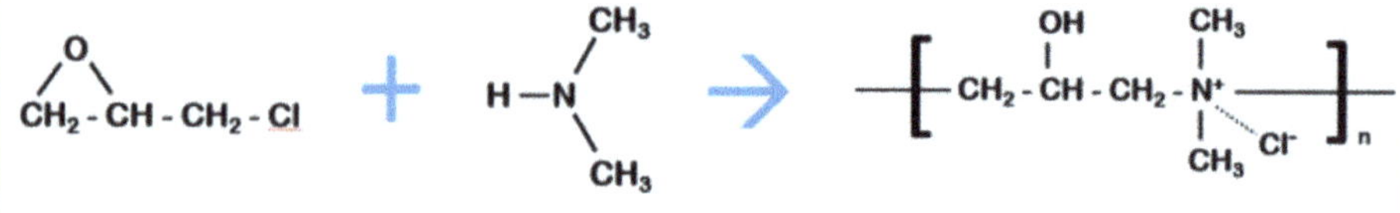

FIGURE 3.15 Epichlorohydrin and dimethylamine form the epi/DMA coagulant building block (Goldblatt & Leitz, 2015).

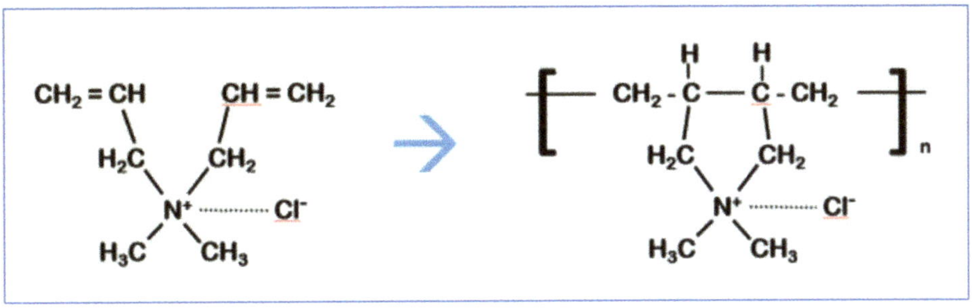

FIGURE 3.16 The DADMAC monomer polymerizes to poly[DADMAC] (Goldblatt & Leitz, 2015).

TABLE 3.5 Comparison of advantages of inorganic and organic coagulant product (Goldblatt & Leitz, 2015).

Inorganic Coagulants	Organic Coagulants
Inorganic coagulants are almost always cheaper than organic coagulants when comparing unit chemical costs only.	Required organic coagulant dosages may be many times lower than inorganic coagulant dosages.
Inorganic coagulants may produce lighter and more voluminous flocs that do not settle as rapidly as organic coagulant-treated flocs, often necessitating the use of a flocculant (aka coagulant aid) to assist settling and capture.	Conversely, organic coagulants can provide more compact flocs that settle more rapidly with less pinfloc and carryover in the effluent. In many cases, the addition of an organic coagulant eliminates the need for a flocculant.
The final sludge volume is almost always greater with inorganic coagulants because they form voluminous metal hydroxides; they may also be more difficult to dewater.	Organic coagulant-treated sludges are typically less voluminous and more dewaterable. Therefore, when sludge disposal costs are added into the economics equation, organic coagulants may be more economical to use than inorganic coagulants.
Inorganic coagulants may produce cleaner supernatant water because the voluminous metal hydroxide flocs settle more slowly while providing a sweep effect that assists in the capture of fine particles.	Faster settling may be necessary in some sedimentation unit operations.
Inorganic coagulants are pH sensitive and function best over a relatively narrow range of system pH values (e.g., 5–7 for Al).	Quaternary ammonium organic coagulants can work over wide pH ranges (2–13).
Inorganic coagulants consume alkalinity and lower system pH.	Organic coagulants have little effect on system pH.

2.3.2 Flocculation

Flocculation is the gathering together of microscopic or barely visible agglomerations, called "pin flocs," into larger, readily visible groupings called "flocs." The larger groupings enhance solids–liquid separation as defined in Stoke's law.

2.3.2.1 Flocculant Types

Flocculants are high to very high molecular weight polyelectrolytes, sometimes referred to simply as "polymers," that can be anionically or cationically charged, or nonionic. Flocculants are also sometimes referred to as "coagulant aids," not to be confused with coagulants. Most commercially available flocculants are based on polyacrylamide and its co-polymers and derivatives. The high reactivity of acrylamide monomer leads to products with average molecular weights between 1 000 000 and 20 000 000 Da. These high molecular weight long chains of molecules have the capability of physically bridging one pin floc, or coagulated grouping, to another.

Flocculant products are available in solid, liquid emulsion, and liquid forms, most of which require specialized equipment for dissolution before application. The upper left box of Figure 3.17 shows the acrylamide monomer, which is the basic building block of most industrial-grade flocculants. Acrylamide carries no net electrical charge (cationicity or anionicity) and can be classified as a nonionic monomer. Acrylamide can polymerize with itself when a free radical initiator is added to a solution of acrylamide monomer. The asterisk (*) in Figure 3.17 indicates a free radical that initiates, then propagates, the reaction. The chain can be as much as 300 000 monomer units long.

To create functionality, there is an anionic monomer known as acrylic acid, the salt of which is known as sodium acrylate. When sodium acrylate is copolymerized with acrylamide, as illustrated in Figure 3.18, an anionic polyacrylamide is formed. Note that the ratio of sodium acrylate to acrylamide charged to the reactor controls the degree of anionicity of the finished polymeric flocculant.

There are also cationic monomers that polymerize quite well with acrylamide. A commonly used one is an ester-based cationic monomer, as shown in Figure 3.19. When this monomer is charged to a reactor with acrylamide and reacted, a cationic co-polymer is formed. As with the anionic co-polymers, the ratio of cationic monomer to acrylamide controls the final cationicity of the polymeric flocculant.

There are many other cationic monomers that may be used in reaction with acrylamide. The polyacrylamide may be referred to as the backbone

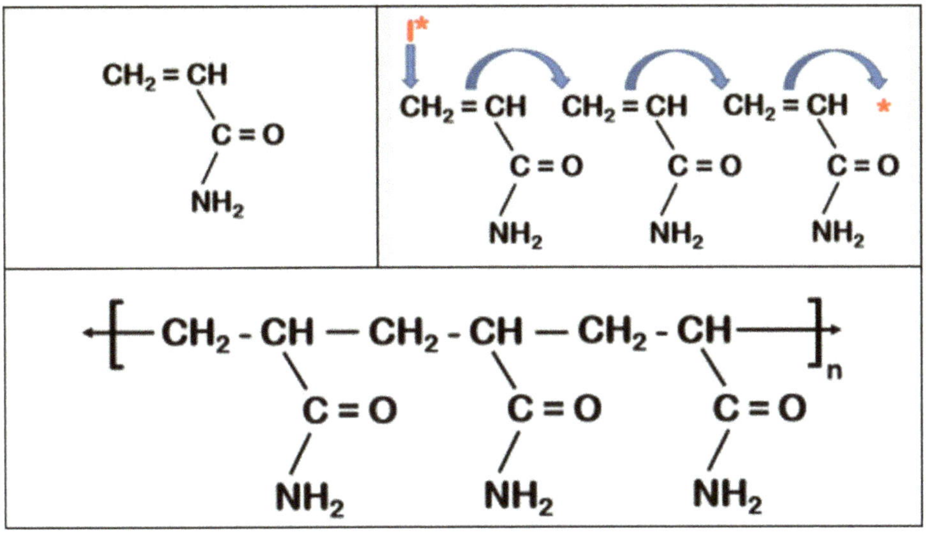

FIGURE 3.17 Upper left: Acrylamide monomer; upper right: addition polymerization; bottom: polyacrylamide (Goldblatt & Leitz, 2015).

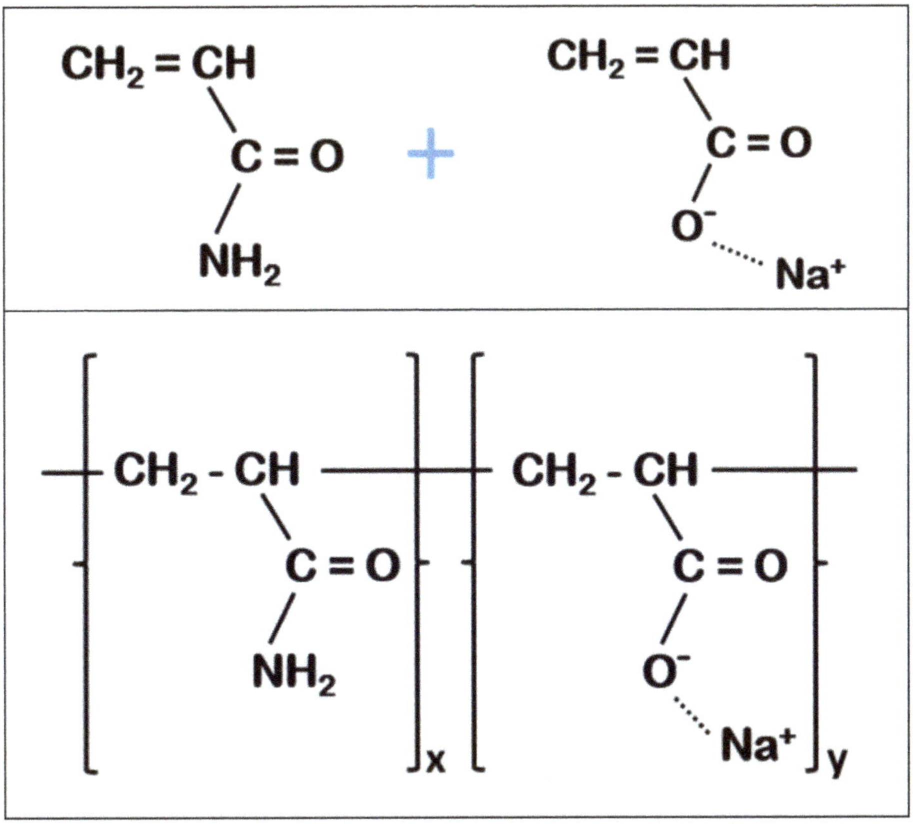

FIGURE 3.18 Upper left: acrylamide monomer; upper right: sodium acrylate; bottom: anionic polyacrylamide polymer (Goldblatt & Leitz, 2015).

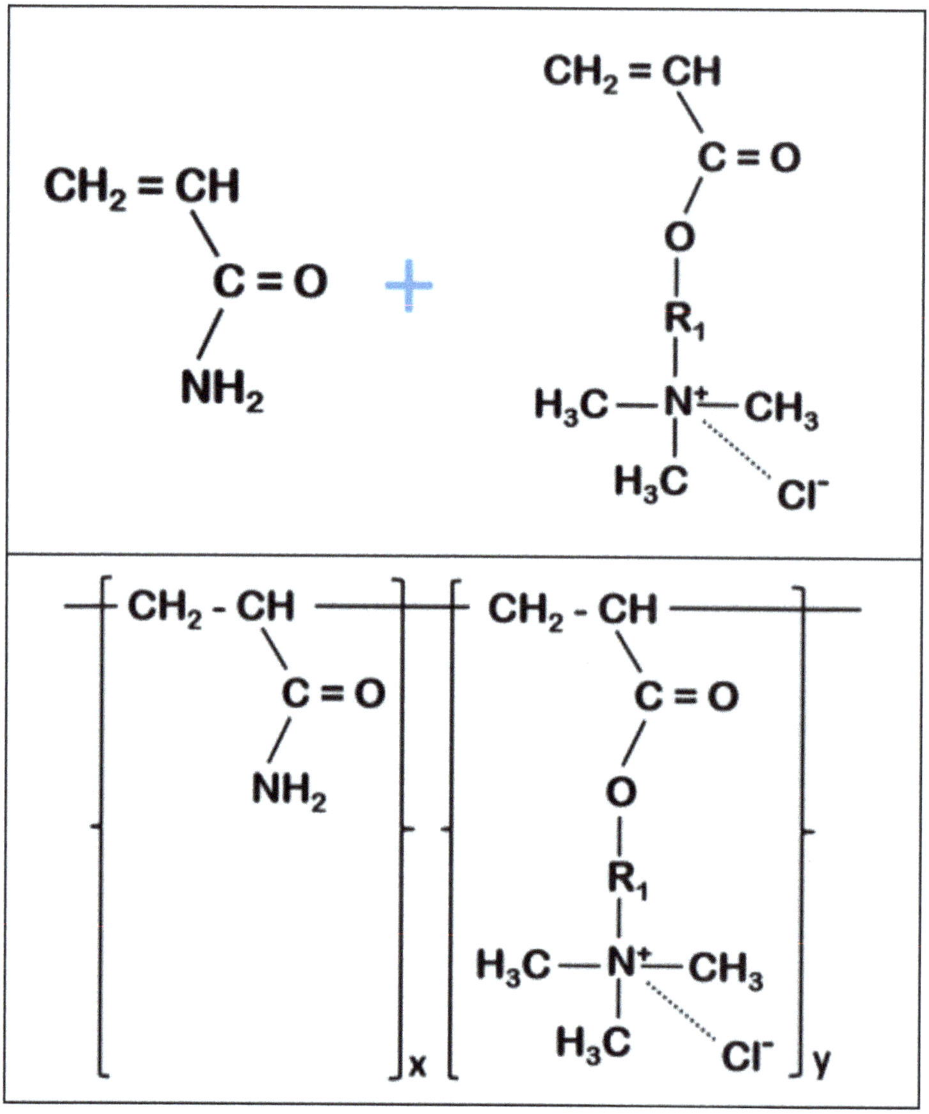

FIGURE 3.19 Upper left: acrylamide monomer; upper right: ester-based cationic monomer; bottom: polyacrylamide ester-based cationic co-polymer (Goldblatt & Leitz, 2015).

of the polymer. When one type of charged monomer is used in conjunction with acrylamide, we call the resultant flocculant a "co-polymer." Two different charged monomer types added to acrylamide makes a "terpolymer." Branching and crosslinking are other characteristics imparted to polymeric flocculant products to improve performance in certain substrates and processing environments.

2.3.2.2 *Flocculants in Clarification*

In practice, influent water clarification and wastewater treatment often include treatment with a coagulant to neutralize negative surface charges to near zero. Then, a flocculant acts to bridge the neutralized agglomerates. A single high molecular weight flocculant molecule is large enough to bridge several pin floc agglomerates simultaneously. The loops and tails that extend from these microflocs can further interact with other microflocs, thereby creating macroscopic floc structures. Relative to coagulation alone, flocculation results in larger structures that settle faster, improve capture of pinfloc, and reduce carryover of solids in the supernatant. Figure 3.20 illustrates, from left to right, results of progressively higher dosing of coagulant or flocculant to turbid water in just-mixed beakers. One may discern some pin floc in the first jar at left, certainly in the second jar, and observe progressively faster settling of larger flocs in the third and fourth jars.

2.3.2.3 *Flocculants in Sludge Dewatering*

Flocculants are also used in sludge dewatering applications to assist in the release of water and concurrent production of higher concentration solids. This results in a lower volume and drier dewatered sludge. Added to sludge, the flocculant bridges smaller agglomerations of particles, forming larger flocs with free water between the flocs. This creates avenues of egress for water drainage from sludge. In the case of biosolids sludges, gravity sedimentation and thickening will produce only a 2% to 5% solids sludge. Dewatering with flocculant chemistry and a dewatering unit operation can

FIGURE 3.20 Jar tester used to determine dosage of coagulant or flocculant products (photo courtesy of Pete Waldrep, Solenis, LLC).

produce a sludge cake of 15% to 20% solids that takes up a small fraction of the original un-dewatered sludge volume.

Industrial sludge generators use a wide variety of dewatering equipment to achieve this goal. Frequently used dewatering machines include belt filter presses, centrifuges, plate and frame presses, and screw presses. Vacuum filters, filter bags, and drying beds are also used. Most sludge dewatering unit operations require the addition of flocculants for maximum dewatering performance.

Each type of sludge and each type of dewatering equipment will require a chemical program that is optimized for that application. For example, biological sludges almost always require a cationically charged, high molecular weight flocculant, but the optimum molecular weight and degree of polymeric flocculant branching and crosslinking will depend on whether the application uses, for example, a belt press (lower shear) or a centrifuge (high shear forces).

Inorganic sludges such as those found in coal preparation plants and mining applications, typically require an anionic or nonionic, high molecular weight flocculant. Paper mills, which have a mixture of biological and inorganic sludges, can use flocculants with lower cationic charges. In all cases, the choice of product depends on the type of sludge, the dewatering equipment in use, and the results desired. Selection of flocculants for dewatering requires skill, knowledge of chemistry and mechanics, testing, and experience.

2.3.2.4 Natural Flocculation

The floc-forming bacteria in biological wastewater treatment systems do just as the name implies. Healthy activated sludge biota naturally flocculate. Under optimal operating conditions, natural flocculation that occurs is sufficient to build floc size so that no chemical additives are required to provide adequate separation. However, in many industrial wastewater treatment activated sludge systems, settling aids are made available onsite as a standby aid for use during upset or nonideal conditions.

2.4 Summary

This section presented a basic introduction to the chemistry part of physical/chemical separation including precipitation, coagulation, and flocculation concepts. Precipitation is used to remove metals or hardness from solution. Coagulation is used to bring finely divided particles into agglomerations, and flocculation is used to increase floc size to enable rapid and economical solids–liquid separation.

There are many types of products used for precipitation, coagulation, and flocculation. Testing appropriate chemistry coagulants and flocculants on substrates, using laboratory or bench testing, is recommended for best chemical application and as a part of process optimization. Optimal application of precipitation, coagulation, and flocculation processes allows for efficient final preparation for industrial water reclamation and reuse to minimize wastewater discharge.

3.0 MECHANICAL TREATMENT FOR SOLID–LIQUID SEPARATION

3.1 Introduction

Separation of nondissolvable solids from the liquid stream is a fundamental unit process in water treatment. The type of solids–liquid separation process that is used largely depends on the type, size, and consistency of the solids that are to be removed. This section deals with the use of static and rotating types of screens, centrifugal force to create separation of solids, sedimentation techniques that use gravity to create separation of solids, dissolved air flotation, and gravity and pressure filtration through static media. Section 4.0 addresses the use of synthetic membranes to separate solids from liquids.

3.2 Screening

Screening is one of the most basic unit processes used to separate solids from liquids. Screening systems are typically located at the head of a treatment train to remove gross solids from the feed water, but they can also be used further down the treatment train for specific solids removal requirements. This section discusses the use of both static and rotating screens. An important part of screening is the handling of the solids material that is retained or removed by the screen, which might include liquid-carrying sluice ways, screw conveyors, or belt conveyors, followed by washing and/or compacting equipment. However, solids handling equipment is beyond the scope of this section and will not be discussed in detail.

3.2.1 Static Screens

As the name implies, static screens are stationary; they do not move. However, they can be, and often are, associated with mechanical equipment that moves to remove solid material that is captured by the screen. Static screens are typically composed of vertical steel bars located in an open channel. The vertical bars are set in a frame to provide a consistent gap between the bars, called the "screen opening," and the feed water flows by gravity through the screen. Solids that are larger than the screen opening are retained on the upstream side of the screen, while those that are smaller, pass through. Typical static screen openings used in the United States are 16 mm (5/8 in.), 12.5 mm (1/2 in.), 9.5 mm (3/8 in.), and 6.3 mm (1/4 in.).

Selection of the screen opening depends on the type of feed water requiring treatment; therefore, knowledge of the range of solids sizes, shape and type, or origin of the solids in the feed water is important. The smaller the screen opening size, the greater the amount of solids that will be captured and the greater the headloss across the screen. Paper, plastics, and rags can be difficult to remove using vertical bar screens because the solids can sometimes align with the screen opening and pass through. The same is true for rodshaped solids.

Industrial effluent can contain items such as labels, hair nets, and bottle tops/caps. These can be effectively removed on a self-cleaning parabolic screen. The feed water is introduced at the top of the screen and flows downward across and through the screen, leaving the solids to be retained on the screen surface. Retained solids can pass into a conveyor or be removed by hand from the lower portion of the screen. Parabolic screens are typically manufactured in stainless steel with screen opening sizes in the range of 0.25 to 2 mm (0.01 to 0.08 in.). Figure 3.21 shows an example of the static bar screen and a static parabolic screen.

3.2.2 Rotating Screens

Rotating screens typically use the motion of the screen to guide captured solids to the solids conveying system. The screening principles are the same, in that solids larger than the opening size will be removed from the feed stream. Rotating screens can be in the form of flat sections, drums, or circular discs and can be configured with bar-type screens, perforated plates, or woven stainless steel wire. These configurations allow for much smaller screen openings to be used. For example, woven stainless steel mesh screens can provide openings down to 100 μ and are often characterized using the mesh tables, as opposed to inches or millimeters. Table 3.6 provides a selection of screen opening sizes in various standards, for comparison.

Rotating screens can be fitted into open channels or can be used as standalone units into which the feed water is pumped. Spray nozzles are typically fitted to clean solids off the drum and direct them to the solids collection system. Figure 3.22 presents an example of a rotating drum screen

FIGURE 3.21 Examples of an automatically raked bar screen (left: reprinted with permission from JWC Environmental) and static parabolic screen (right: reprinted with permission from Kusters Water).

installed in a channel to remove solids from a process or effluent stream, and a rotating disc screen.

3.3 Centrifugal Separation

Centrifugal force is another approach to separating solids from liquids. Two examples are presented in this section.

3.3.1 Hydrocyclones

Hydrocyclones use centrifugal force to separate solids in a feed stream into a more concentrated solids stream that exits the bottom of the unit and a less concentrated stream that exits the top of the unit. The hydrocyclone itself is stationary, and the feed stream is introduced tangentially to the unit by pumping (see Figure 3.23). Two separate vortex flow patterns are set up within the unit—the primary vortex that takes the larger particles to the inside wall of the cyclone and out the bottom spigot, and a secondary vortex that travels upward, carrying most of the water and finer solids out through the vortex finder at the top. It is important that the underflow and overflow discharges are "free" so that air can enter the unit and siphoning is prevented.

Sizing of a hydrocyclone depends on the size distribution of solids in the feed stream and the desired "split" of solids required to report to the overflow, not the feed flow. The smaller the size distribution of the solids in the feed stream, the smaller the diameter of the hydrocyclone that is needed.

TABLE 3.6 Selection of standard screen openings.

American Standard Test Sieve (ASTM)			Tyler Equivalent Mesh Size
Millimeter	Micron	Inch/No.	Mesh
16		5/8 in.	—
12.5		1/2 in.	—
9.5		3/8 in.	—
6.7		0.265 in.	—
6.3		1/4 in.	—
4.75		No. 4	4
4.0		No. 5	5
2.8		No. 7	7
2.0		No. 10	10
1.0	1000	No. 18	18
	500	No. 35	35
	425	No. 40	40
	300	No. 50	50
	250	No. 60	60
	212	No. 70	70
	180	No. 80	80
	150	No. 100	100
	125	No. 120	120
	106	No. 140	140

Selection of the appropriate size for the vortex finder and the spigot controls the split of the flow between each outlet and, therefore, the concentration of the underflow. Multiple units are used in parallel for larger flows.

3.3.2 Centrifugal Filter

Technologies have been developed that use centrifugal force in conjunction with a fine screen to remove higher concentrations of solids. The centrifugal

FIGURE 3.22 Examples of rotating screens for solids removal showing the inside of an inclined drum screen (left) and a rotating disc screen (right).

action created by a tangential feed flow can be combined with a fine screen to remove particles down to the submillimeter range. Solids are discharged periodically. These units need to operate at a specific feed pressure and reportedly have water recovery levels greater than 95%.

3.4 Gravity Separation

Gravity separation, clarification, or sedimentation are almost always preceded by coagulation and flocculation to destabilize and agglomerate solid particles for more efficient settling. (See Section 2.0 for further details.)

3.4.1 Sedimentation

Sedimentation is one of the earliest unit operations used in water treatment (Reynolds & Richards, 1996) to separate solids from the liquid phase on a continuous basis. Solids naturally settle to the bottom of the tank where they are periodically removed, and the liquid continuously flows over weirs at the surface of the sedimentation basin. Sedimentation is often used as a method to remove the bulk of remaining solids from the feed water before filtration (discussed later).

Sedimentation tanks are typically circular or rectangular and can be manufactured in concrete or steel. The shape and materials of construction are often governed by the volume of water to be treated, with larger flows using concrete tanks. The configuration of the settling tanks, either circular or rectangular, is often dictated by the available space at the treatment site and/or operator preference. The goal of a settling tank is that the settling velocity of the bulk of the solids is greater than the upward velocity of the water. This should ensure that most of the solids settle within the clarifier. The controlling parameters in clarifier design are the surface loading rate,

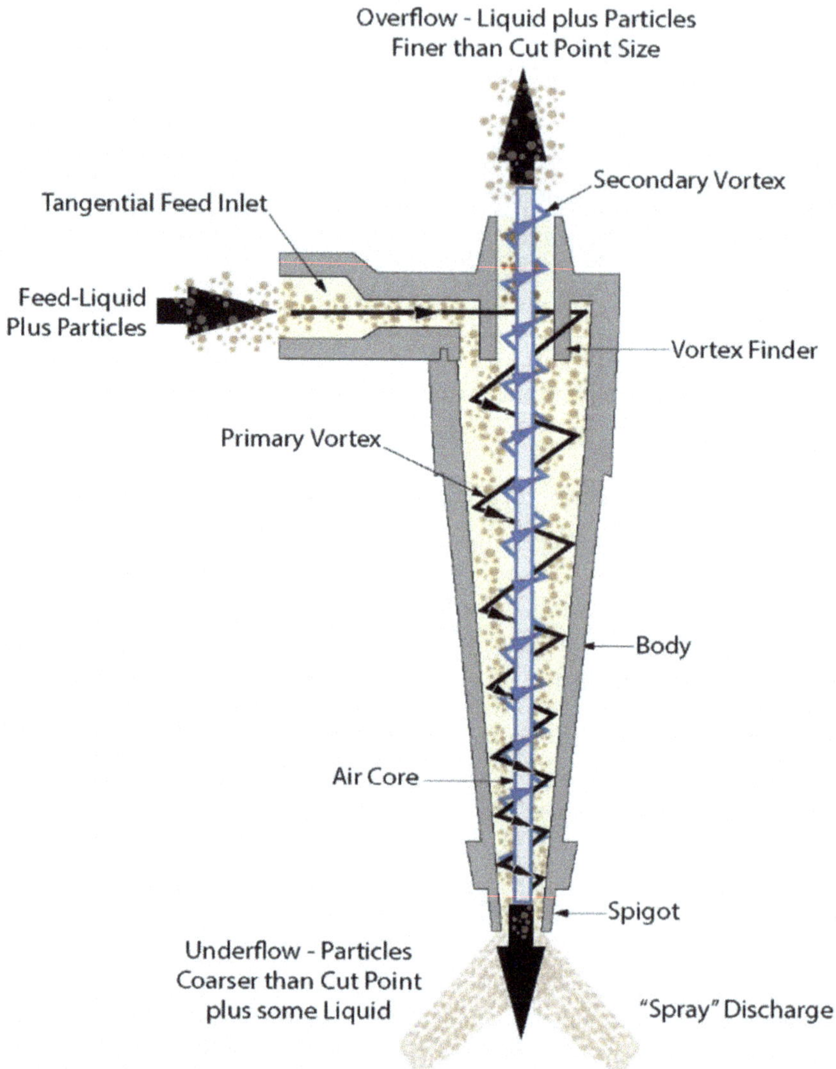

FIGURE 3.23 Overview of hydrocyclone operation

the weir loading rate, the detention time, the clarifier depth, and the solids withdrawal rate. The solids withdrawal rate, on a mass basis, should match the feed solids loading rate, otherwise solids buildup will occur in the clarifier leading to a "failure" condition in which solids begin to flow over the weirs.

Table 3.7 presents some typical operating design criteria for sedimentation clarifiers for different applications. Where operating experience on a specific feed water is not available, particle-size analyses and batch settling tests can be carried out to determine the anticipated range of solids settling velocity that will be required for the specific application.

Circular clarifiers are fitted with a mechanism that continually rotates and directs settled solids toward a central well for removal (Figure 3.24). The drive for the rotating mechanism can be either centrally located or mounted peripherally on the end of the mechanism arm. The feed water is introduced to the central stilling well to avoid generating eddies that travel down and stir up solids that have already settled. Over the years, the design of clarifiers has been modified to include various additions, such as lateral skirts, to improve the fluid dynamics within the unit to limit the agitation caused by continuous operation as well as the separation of the solids and liquid phases over varying feed water conditions and flowrates.

In rectangular clarifiers, the flow pattern is more plug flow, with the feed stream entering at one end of the tank and the clarified effluent leaving at the other. The length of the tank is determined by the settling rate of the particles with the lowest settling velocity. The goal is that such particles will have settled into the sludge layer at the bottom of the tank before the frontal velocity, moving laterally across the tank, reaches the surface overflow

TABLE 3.7 Typical design criteria for clarification tanks.

	Typical Design Values		
Parameter	Water Treatment	Primary Effluent[d]	Secondary Effluent[d]
Surface overflow rate, m/h (gpd/sq ft)	1.2–2.4 (700–1400)[a] 0.8–1.6 (500–1000)[c]	1.0–1.6 (600–1000)[b] 1.4–2.0 (800–1200)[c]	0.7–1.4 (400–800)[c]
Weir loading rate, m^3/h·m (gpd per ft) of weir length	6.2–9.3 (12 000–18 000)[c]		
Clarifier depth, m (ft)	3–4.5 (10–15)[a]	3–3.6 (10–12)	3.6–4.5 (12–15)
Detention time, hr	1.5–4[a] 4–6[c]	<2	

[a]Kawamura (2000).

[b]Can be doubled under peak (4-hour) hydraulic loading conditions.

[c]For coagulation with iron or alum salts (Reynolds & Richards, 1996).

[d]Typical wastewater treatment values.

FIGURE 3.24 Sectional view of a circular clarifier with central drive mechanism (top) clarifier under construction at a WRRF.

weirs. To maintain close to plug flow conditions, the length-to-width ratio of rectangular clarifiers is frequently set at 2 : 1 (Reynolds & Richards, 1996), although 4 : 1 has been suggested by Kawamura (2000) and 8 : 1 has been used in practice. Distribution of the influent flow across the width of the clarifier is important to help create an even flow distribution and promote plug flow like conditions. Baffle systems of various designs can be used for this purpose.

The largest solids entering the clarifier will settle fastest and reach the bottom of the tank first. A system of scrapers moves across the floor of the tank toward the inlet end (see Figure 3.25) to direct settled solids into a hopper for withdrawal. This same mechanism can be used to move scum and other floating material to the opposite end of the tank for removal. The chain-and-flight mechanisms tend to be maintenance intensive. At the outlet end of the clarifier, the clarified water flows over V-notch weirs into a launder (see Figure 3.26) for gravity flow out of the unit. Covers are often placed over the outlet weirs to reduce the buildup of algae, which reduces the cleaning requirements for operations staff.

An enhancement to the performance of sedimentation tanks is the inclusion of lamella plates or tube settlers. Lamella plates create a path through

FIGURE 3.25 View of a rectangular clarifier with baffled inlet wall and chain and flight mechanism to move sludge to the hopper.

which the partially clarified flow passes, reducing the depth of the water column through which a given solids particle must settle before it is removed from solution. Lamella plates are installed at an angle of 45 to 60 deg to create the shortened flow path for fine solids, and they are more suitable for rectangular-shaped tanks because they can be more easily positioned to avoid rotating equipment. The "plates" are manufactured in various

FIGURE 3.26 Rectangular clarifier outlet launder and V-notch weir plates with covers to reduce algal growth.

arrangements and can be formed into honeycomb-style blocks that are made of fiberglass or plastic, and provide a more densely packed Lamella plate system to increase capacity.

Solids that settle on the plates must flow down the plates and then drop to the bottom of the tank for removal, hence the need to install the plates with a high slope angle. These systems are not suitable for high solids loads and are typically installed toward the end of a rectangular clarifier, where the bulk of the solids have already settled out. Periodic flushing of the plates may be needed to maintain the clarified water flow path and remove solids that build up on the plates. Figure 3.27 shows an example of a tube settler system.

3.4.2 Ballasted Clarification

As discussed in the previous sections, the tank design relies on gravity to create the separation between the solids and liquid phases. In ballasted clarification, an external solid (ballast) is added to the feed stream to enhance the settling process, essentially gathering up the fine particles on the ballast and allowing them to settle faster. The majority of the ballast is then recovered from the solids stream that is withdrawn from the unit, separated from the solids, and then recycled to be used again. A small amount of fresh ballast is typically added to the feed stream to make up for losses.

Ballasted clarification systems are effective for reducing the footprint size of the clarification process, which is a real advantage when there is limited treatment facility area, and for increasing the capacity of existing clarification systems. Ballast material includes microsand, recycled solids, and magnetite. In the case of microsand, a hydrocyclone (discussed earlier) is

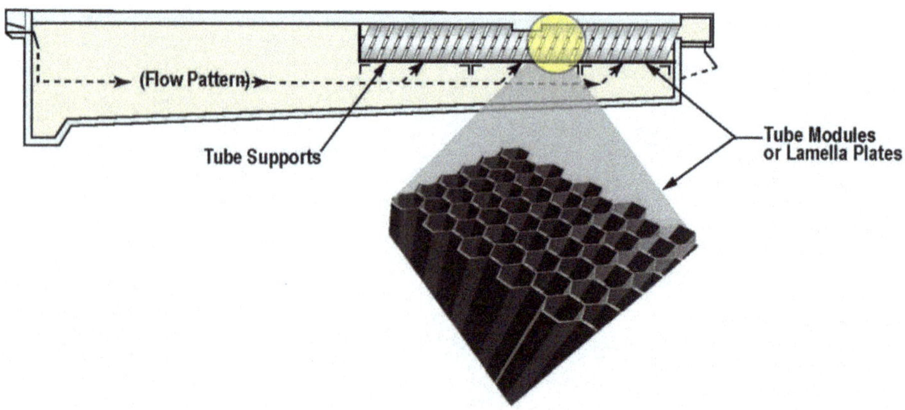

FIGURE 3.27 Schematic of lamella plate (inclined tube settler) in a rectangular clarifier.

used to separate the microsand from the process solids. In the case of magnetite ballast, a drum magnet is used to separate and recover the magnetite. Several manufacturers have developed proprietary systems that include coagulation, ballast addition, settling (including tube settlers/lamella plates), and the ballast recovery system. Examples include the Actiflo® system by Veolia Water Technologies, Inc. (see Figure 3.28); the DensaDeg® system by Suez Water Technologies and Solutions; the CoMag® system by Evoqua Water Technologies; and the RapiSand® system by WesTech Engineering.

3.5 DISSOLVED GAS FLOTATION

Low-density organic solids and oils do not settle well. Allowing gravity to do the job would require excessively large and costly sedimentation tanks and not achieve much in the case of oil. In cases like these, flotation may be a suitable alternative to consider. In flotation, air, or other gases, depending on the circumstances, is introduced to the feed stream under pressure. At high pressure, the air dissolves in the feed stream. The flow enters the process unit though a backpressure valve releasing the pressure, which causes release of the air as millions of micron-sized bubbles that expand and rise to the surface.

For successful dissolved air flotation (DAF) operation, the air bubble size should be in the range of 10 to 100 μ. The high surface area of the expanding bubbles captures suspended solids and oil and floats them to the surface, where they are removed. Examples where DAF is used effectively

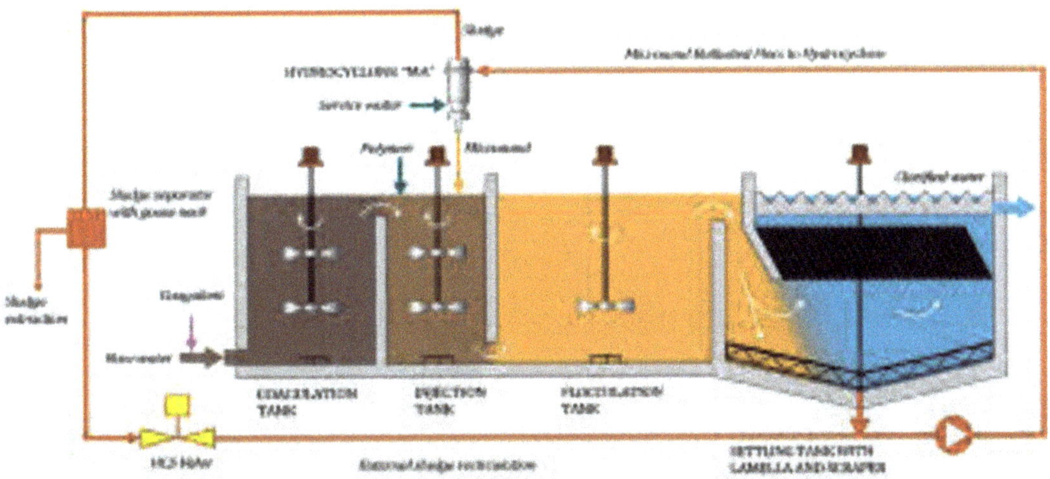

FIGURE 3.28 Schematic of the ballasted clarification process (courtesy of Veolia Water Technologies).

include thickening organic solids before anaerobic digestion and concentrating metal precipitates as part of a metal recovery process. Figure 3.29 presents a schematic of the DAF process. A key design parameter in sizing DAF systems is the air-to-solids weight ratio. Ratios ranging from 0.01 to 0.06 have demonstrated to be effective, and a value of 0.02 is appropriate for many applications (Lash & Kominek, 1980). van Vuuren (1997) reports that, for most clarification applications, as opposed to solids thickening systems, the air requirement seems to be in the range of 5 to 10 mg/L, based on the influent flow. Another key parameter is the portion of the effluent that is recycled (see Figure 3.29). For clarification applications, van Vuuren (1997) notes that the recycle ratio typically varies from 6% to 12% of the raw feed water flowrate.

The addition of polymers or coagulants to the DAF feed can enhance the performance of the unit and provide better solids-liquids separation.

3.6 Filtration

Filtration is a common unit process in water treatment. Barnes and Wilson (1983) note that the first recorded use of a filter unit designed to treat water for domestic use is attributed to John Gibb in Paisley, Scotland, in the 1820s. This success led to filters built and operated in neighboring England. Filtration rates for these filters were in the range of 0.1 to 0.3 $m^3/m^2 \cdot h$ (0.04 to 0.12 gpm/sq ft) (Weber, 1972); they are what we would call "slow sand filters." This technology is still used in parts of Europe and the rest of the world.

The challenge with slow sand filters is the large area required for filtration. Also, when the filters plug, they are taken offline so the surface of the

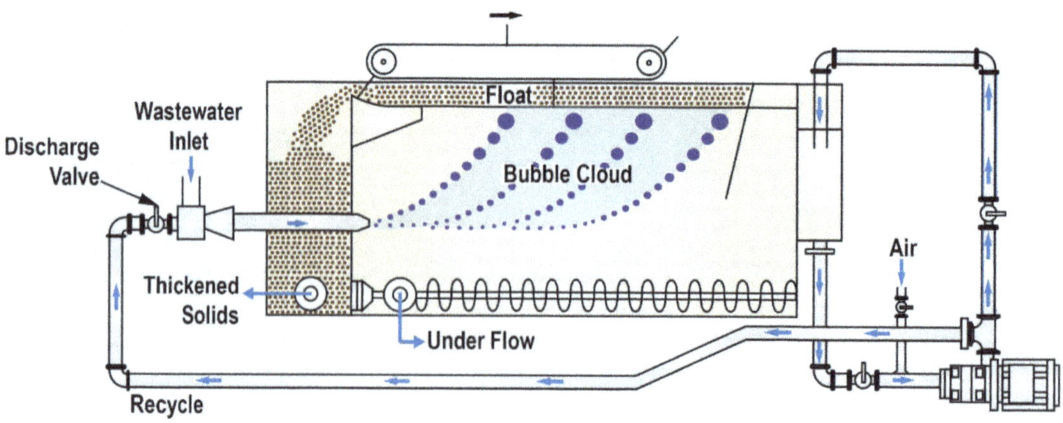

FIGURE 3.29 Schematic of the DAF process.

filter can be scraped to remove the top fouling layer before filtration can resume. The successful use of slow sand filters is dependent on the quality of the feed water, specifically a low suspended solids concentration. Because these filters are typically not used in treatment of industrial effluent, they will not be discussed further.

This section focuses on porous media filtration. Membrane filtration is discussed in Section 4.0. The location of the filtration step in the treatment train is dictated by the quality of the feed water. High loads of suspended solids (e.g., >50 mg/L) in the feed water typically require pretreatment, using one or more of the processes discussed earlier, to avoid blinding the filter media too quickly. Filtration typically serves as one of the final steps in the wastewater treatment train, a process that should result in a low turbidity (typically less than 2 NTU and, in some cases, below 0.5 NTU). In effluent recycle applications, the feedwater source may already have a low turbidity, in which case the filtration step can be at or near the front of the treatment train.

3.6.1 Rapid Gravity Filters

Unlike slow sand filters, rapid gravity filters operate at significantly higher filtration rates. Early filters built in the United States in the late 19th century operated in the range of 2.4 to 9.8 $m^3/m^2{\cdot}h$ (1 to 4 gpm/sq ft) (Weber, 1972). Today, it is not uncommon to find gravity filters operating in a 9.8 to 14.4 m/h (4 to 6 gpm/sq ft) range, and even up to as high as 32.4 $m^3/m^2{\cdot}h$ (13.5 gpm/sq ft), where the feed water quality allows. Coagulation and flocculation (discussed in Section 2.0) are frequently used upstream to increase the size of the particle floc before it enters the filter.

The mechanisms involved with filtration are complex and will not be explored here. Suffice to say that they include surface straining, attachment by adsorption to the media, removal by a cake formed in the top portion of the filter, and interstitial straining deeper in the filter bed. The filtration efficiency is affected by the filter media, the feed water characteristics, and the rate of filtration.

3.6.1.1 Types of Filter Media

Common filter media for gravity filters include sand, anthracite, garnet, and gravel. Filters may include a single media, such as sand, with a gravel underdrain system; or multimedia, which could include two or three of the media listed above. Granular activated carbon can also be used as a filtration media in addition to its role as an adsorbent. Granular activated carbon is also the media of choice for biological active filtration, often referred to as "biological activated carbon" (BAC), because the high surface area of GAC provides an ideal environment for bacterial growth. In this role, GAC acts more as the environment for biological growth, rather than for adsorption. The microbial population feeds off of the biodegradable organics in the feed water, effectively keeping the media "clean." The BAC filter is typically preceded by an ozonation step, which cleaves large organic molecules into smaller parts more assimilable by the biomass on the GAC.

The choice of filter media, its dimensions, and the media depth used is called the "filter design," and it depends on the application and purpose of filtration and quality of the filtrate required. Table 3.8 presents a summary of some filter media types, applications, and typical design criteria (adapted from Kawamura, 2000). Two terms are used in the table—the effective size and the uniformity coefficient. The effective size is the diameter of the sieve opening that will allow 10%, by weight, of the media to pass through (typically expressed at the d_{10}). The uniformity coefficient is the d_{60}/d_{10}.

3.6.1.2 Filter Backwashing

As solids build up in the filter bed, the headloss across the filter increases. There are different ways to control the filter operation, but, at some point, the headloss becomes too great and the filter must be taken offline and backwashed to remove the solids buildup. Filter backwashing is a critically important part of any filtration system because, without good cleaning, the filter media will not be restored to the "clean bed" conditions, and filter backwashing will be more and more frequent.

There are a multitude of filter underdrain designs, the discussion of which is beyond the scope of this book. The key is that a good underdrain system must be able to distribute the backwash water evenly across the entire cross-section of the filter so that the filter bed expands evenly and "channeling" of the backwash water does not occur. If air scour is to be used as well, either separately or in combination with the backwash water, then careful consideration must be given to the airflow rate and means of distribution. Figure 3.30 shows a cutaway of a typical concrete gravity filter, and Figure 3.31 shows a dual-media filter during an air scour.

3.6.2 Pressure Filters

Pressure filters use the same principles of filtration as rapid gravity filters, except that the media is housed in a pressure vessel and the filter itself is not open to the atmosphere. Pressure filter systems tend to be used for lower feed flows, where steel tanks can be used rather than concrete structures. By pressurizing the feed water, pressure filters can operate at higher pressure losses across the filter media, which can allow for longer filter runs. The equipment needed for pressure filters includes feed water pumps, backwash pumps, and air scour blowers.

Figure 3.32 presents an example of pressure filters. Piping is arranged so that the pressurized feed stream enters near the top of the vessel and exits at the bottom. Backwash and air scour lines enter the base of the vessel, and backwash water exits at the top. Various designs of underdrains have been developed to distribute the backwash water and scour air as evenly as possible across the cross-section of the filter. Without good distribution, the media will not clean well, and shorter run times and possible clogging of the filter will occur.

The media used in pressure filters is the same as that used in gravity filters and could be single-, dual- or multi-media, depending on the application as well as GAC.

3.6.3 Precoat Filters

Precoat filters are a form of pressure filter that are used as a polishing step to produce a high-quality, low-turbidity product stream, typically for lower-flow or high-value products. These filters can have candles or filter elements fitted internally (like a pool filter), onto which the filter media is coated before a filtration cycle. Diatomaceous earth or other customized precoat material can be used. The precoat media forms a layer through which the feed water is filtered. Once the filter operating pressure is exceeded, the unit is backwashed, which removes the filtered solids as well as the precoat. New precoat is then applied before the next filtration cycle.

Another example of a precoat filter is a vacuum-drum filter, typically used to recover a valuable solids product from a feed slurry. In this case, the drum is coated with the precoat material, and then the drum is partially submerged in the feed slurry solution. A vacuum is applied from inside the drum, which draws solution through the precoat media as it rotates slowly through the slurry. The product solids build up on the outside of the drum as it rotates, and the liquid is discharged. To avoid the solids layer becoming too thick, the drum rotates past a knife edge, which scrapes off a layer of the product solids that are then recovered. Figure 3.33 presents an example of this.

4.0 MEMBRANE FILTRATION

4.1 Introduction

Membrane filtration in industrial water reclamation and reuse in an important process to minimize liquid discharge. Often, the final treatment to

TABLE 3.8 Filter media types, applications, and typical design criteria (adapted from Kawamura, 2000).

Filter Media	Type of Filter (Loading Rate)	Typical Design Criteria[a]	Comments
Fine sand	Slow sand filter 0.12–0.41 m/h (0.05–0.17 gpm/sq ft)	d_{10}: 0.25–0.35 Uniformity coefficient : 2–3 Depth: 1–1.2 m (3.3–4 ft) Specific gravity: ≥ 2.63	• Simple design and construction • Good effluent quality without pretreatment • Requires large filter bed area • Applicable only for good quality feed water • Requires frequent scraping of surface layer
Medium sand	Rapid sand filter 4.8–7.2 m/h (2–3 gpm/sq ft)	d_{10}: 0.45–0.65 Uniformity coefficient : 1.4–1.7 Depth: 0.6–0.76 m (2–2.5 ft) Specific gravity: ≥ 2.63	• Proven and widely accepted filtration process • Wide applications if pretreatment provided • Short filter runs because of surface filtration • Requires coagulation and auxiliary washing system
Coarse sand	High-rate filter 12–28 m/h (5–12 gpm/sq ft) Direct filtration	d_{10}: 0.8–2.0 Uniformity coefficient : 1.4–1.7 Depth: 0.8–2.1 m (2.6–7 ft) Specific gravity: ≥ 2.63	• Effective high-rate filtration process with long filter runs • Wide application range with polymer pretreatment • Auxiliary wash system limited to air-scour type • Requires deep bed cells with special underdrain

Multimedia • anthracite-sand • anthracite-sand-garnet	High–rate filter 12–24 m/h (5–10 gpm/sq ft) Direct or in-line filtration	*Sand:* d_{10}: 0.45–0.65 Uniformity coefficient: 1.41.5 Depth: 0.3 m (1 ft) Specific gravity: ≥ 2.63 *Anthracite (coal):* d_{10}: 0.9–1.4 Uniformity coefficient: 1.4–1.5 Depth: 0.45 m (1.5 ft) Specific gravity: ≥ 1.5–1.6 *Garnet:* d_{10}: 0.25–0.3 Uniformity coefficient: 1.2–1.5 Depth: 0.08 m (0.25 ft) Specific gravity: ≥ 4.–4.1	• Effective high-rate filtration process with long filter runs • Proven and widely accepted filtration process • Either surface wash or air-scour is required as an auxiliary washing system and polymer is required as a filter aid • Proper selection of each media is important • Requires a high backwash rate for restratification of the filter bed
GAC	Removal of organic contaminants 7.2–14.4 m/h (3–6 gpm/sq ft) Empty bed contact time = 10–20 min	d_{10}: 0.5–1.0 Uniformity coefficient: 1.5–2.5 Depth: 1.8–3.6 m (6–12 ft) Specific gravity: ≥ 1.35–1.37	• Proven and accepted process for specific removal of organic contaminants • Can also operate effectively as a conventional filter • GAC must be replaced or regenerated if used in adsorption application • Can be operated in a biological mode (typically with upstream ozonation process)

[a] d_{10} = effective size

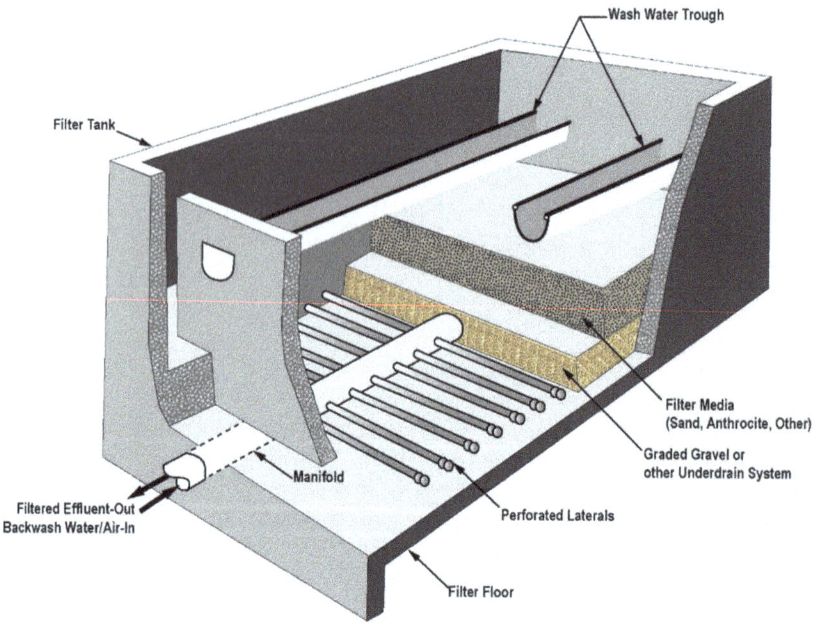

FIGURE 3.30 A cutaway of an example gravity filter arrangement (adapted from U.S. EPA, 1990).

FIGURE 3.31 Dual media (anthracite/sand) filter during air scour.

FIGURE 3.32 Example of a pressure filter arrangement.

minimize liquid waste is reverse osmosis or nanofiltration, which require pretreatment, often using filtration, either media or membrane filtration. The advantage of membrane filtration is it provides consistent filtered water quality even with fluctuating feed or raw water quality. Microfiltration and ultrafiltration are often classified as low-pressure membrane filtration processes, gaining acceptance as processes that provide reliable and

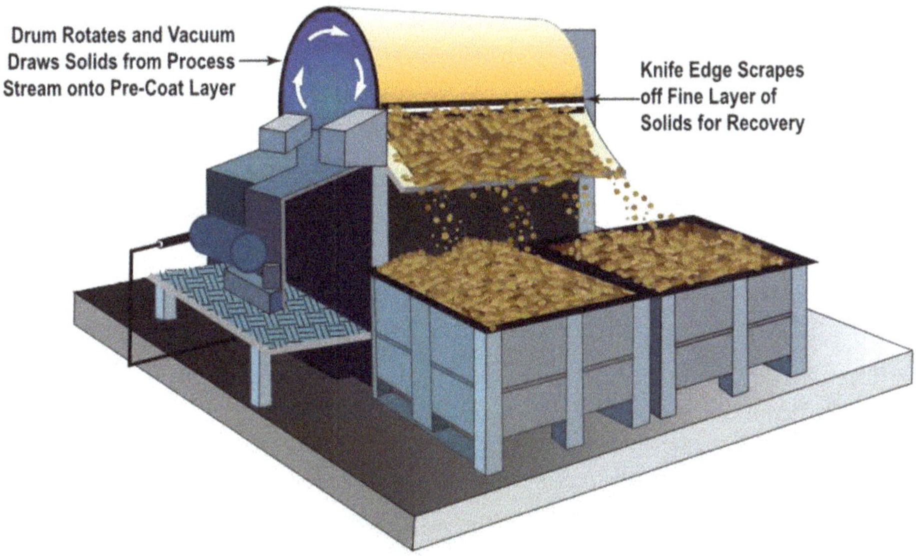

FIGURE 3.33 Schematic example of precoat vacuum drum filter for recovering solids byproduct.

efficient removal of macromolecules, colloids, particles, turbidity, color, and microorganisms. The microfiltration/ultrafiltration membranes are made of organic and inorganic microporous materials that allow passage of water while retaining particles and macromolecules. Dissolved organic and inorganic constituents can be removed with the use of pretreatment processes such as coagulation-flocculation, absorption, and oxidation. The nominal pore size of microfiltration and ultrafiltration are 0.1 to 10 μm and 0.01 to 0.1 μm, respectively. The typical operating pressure of microfiltration is approximately 100 to 400 kPa (15 to 60 psi), whereas, for ultrafiltration, the operating pressure is 200 to 700 kPa (30 to 100 psi).

4.2 Principle of Operation

4.2.1 Water Permeation

Pure water transport across the clean porous membranes will be the function of pore diameter, d_p, (cm, porosity [ε], or total open area of pores per unit membrane surface, applied pressure P_T [Pa], viscosity of the water μ [Pas] and thickness of the membrane skin Δx [cm]). The most frequently used equation for describing flow through pores is based on the Hagen-Poiseuille model for streamline (laminar) flow through channel

$$J = [\varepsilon\, d_p^{\,2} P_T] \,/\, [32 \Delta x \mu] \tag{3.9}$$

where J is the membrane flux given in terms of velocity (cm/sec).

Another pure water volumetric flux (J) using modified form of Darcy's law is

$$J = Q_{total}/A = \Delta P/\mu R_m \tag{3.10}$$

R_m is the hydraulic resistance of the clean membrane to water permeation. The volumetric flowrate of pure water Q_{total} (L/h) across a given surface area of clean membrane A (m^2) is expected to increase linearly with increasing transmembrane pressure (TMP) and decrease with increasing absolute viscosity and ΔP (Pa) in the transmembrane pressure. Transmembrane pressure is defined as the average feed pressure minus filtrate pressure. The typical units of water flux, J is liters of water per square meter of membrane per hour (LMH) or gallons of water per square foot of membrane per day (gpd/sq ft). The hydraulic resistance of the membrane (R_m) can be related to physical properties of the membrane, as follows:

$$R_m = [8\tau \Delta z]/[\pi r^4 P_{\text{pore}}] \tag{3.11}$$

where P_{pore} = pore density, the number of pores per unit membrane area (Cheryan, 1998),

r = pore radius,

τ = tortuosity factor, and

Δz = pore length or skin thickness

The primary effect of temperature on the water flux is the influence on viscosity, which increases with decreasing temperature and can be estimated by the following equation:

$$\mu_T = 1.784 - 0.0575T + 0.0011T^2 - 10^{-5}T^3 \quad (3.12)$$

where μ_T = viscosity of water (cP) at T and

T = water temperature (°C).

4.2.2 Transmembrane Pressure

Transmembrance pressure is another parameter in characterizing membrane filtration. The feed pressure drives the water through the membranes and influences membrane flux. The higher the feed pressure, the more water goes through the membranes as well as the more solids are deposited on the membrane surface and in the pores, shortening the filtration cycle. The TMP of microfiltration/ultrafiltration membranes vary depending on membrane materials of construction, configuration, and operating conditions.

The TMP, ΔP, can be calculated as a function of inlet pressure (P_i), outlet pressure (P_o), and permeate pressure (P_p), as follows:

$$\Delta P = [(P_i + P_o)/2] - P_p \quad (3.13)$$

Transmembrane pressure used in wastewater treatment applications is typically expressed as bars, pounds per square inch, or kilopascals. Polymeric pressurized microfiltration/ultrafiltration membranes used in wastewater treatment typically operate at a TMP between 20 and 300 kPa (3 and 43.5 psi). Submerged microfiltration/ultrafiltration membranes operate under vacuum, the maximum theoretical pressure in 100 kPa, to minimize possible collapsing of membrane fibers by excessive vacuum. The practical maximum TMP of submerged systems is typically 70 to 80 kPa (10 to 11 psi).

4.2.3 Membrane Permeability

Performance of microfiltration/ultrafiltration membranes is also measured using membrane permeability, which is the ratio of the membrane flux and TMP. Membrane permeability, sometimes called "specific flux," is often expressed as LMH/kPa or gpd/sq ft/psi. Most of the polymeric membranes used in wastewater treatment operate at membrane permeability of 0.75 to 5.0 LMH/kPa (3 to 20 gpd/sq ft/psi).

4.2.4 Particle Separation

Microfiltration/ultrafiltration have actual pores on the membrane surface and the separation of impurities is by size exclusion. The microfiltration/ultrafiltration design cannot rely on feed water constituents' chemical analysis alone, but also physical characteristics of the feed stream. Additional information used in designing microfiltration/ultrafiltration is typically from a library of data collected by membrane systems integrators and manufacturers from similar feed streams, previous projects, and actual facility operating logs or data from bench-scale or pilot test results and membrane manufacturers' design software.

Membrane filter surfaces have pores, which exclude solids, ranging from 0.002 to 01 μm for ultrafiltration and 0.03 to 10 μm for microfiltration. In terms of molecular weight cutoff as defined by membrane manufacturers (Cheryan, 1998), it is approximately 10 000 to 100 000 daltons for ultrafiltration and greater than 100 000 daltons for microfiltration. Membrane filtration can reduce turbidity in water streams consistently to less than 0.1 NTU. The positive barrier provided by microfiltration and ultrafiltration can achieve this level of turbidity, independently of the feed water quality. Particle removal for microfiltration/ultrafiltration membranes ranges from 2 to 5 log removals. For microorganisms like Giardia and Cryptosporidium, greater than 4 log has been reported, whereas for bacteria, the removal varied from 0.7 to 9.8 log removal. Based on membrane pore size, ultrafiltration membranes having a pore size of 0.01 μm can reject typical water-borne viruses from 0.02 to 0.10 μm (Ministry of Health, NZ, 2007), which are smaller in size than microfiltration membranes having a pore size of 0.10 μm.

4.2.5 Membrane Compaction

It has been observed that the resistances of polymeric membranes increase with increasing TMP (Chellam & Jacangelo, 1998). In these studies, the membrane resistance has been modeled as being straight line or power law functions of TMP. Membrane compaction will decrease the productivity of a microfiltration/ultrafiltration system operated at constant flux as the pressure is varied to maintain the flux on fouling. Membrane compaction will cause increase in TMP.

4.2.6 Concentration Polarization

"Concentration polarization" is the accumulation of rejected particles near the membrane surface causing the concentration near the surface to be greater than the bulk stream. The effect of concentration polarization can be observed during crossflow filtration of colloidal suspensions. When TMP is applied, particles are transported to the membrane and accumulate near the membrane surface. The accumulation of particles creates a concentration gradient and, as a result, particles diffuse back into the liquid suspension. At steady state, the convection of particles toward the membrane surface is balanced by the diffusion away from the membrane surface.

The limiting flux, J_{lim}, can be calculated (Fane, 1986) by the following equation:

$$J_{lim} = K\, ln(C_m/C_b) \tag{3.14}$$

where K = the mass transfer coefficient,
C_b = the concentration at membrane surface, and
C_m = the maximum concentration at membrane surface.

4.2.7 Flux Decline Mechanisms

The low-pressure membrane filtration process is a solids–liquids separation that goes through a cycle of allowing continuous passage of water through the membrane until the solid particles deposited on the membrane's surface hinder the flow of water. The deposited particles are removed by backwashing with filtered water (or with water and air) and are flushed out, effectively restoring the passage of water through the membrane. This operation is typical of the dead-end filtration configuration.

Another approach is to allow the continuous flow of water stream through the membrane surface, allowing some water to pass though the membrane while the bulk of the flow and particles are recirculated. As the water stream is recirculated, the concentration of particles increases until it reaches a critical concentration at which the rate of water passage through the membrane stabilizes. Particles are continuously removed by bleeding off part of the recirculating water stream or back-pulsing with air. This process is called "crossflow filtration."

Depending on the properties of the particles or other factors, the passage of water through the membranes sometimes cannot be restored to its original clean conditions. This is called "flux decline." The causes of flux decline can be attributed to the following: nature and characteristics of the particles; membrane materials of construction; and operating conditions such as pressure, temperature, water viscosity, age of the membrane, and presence of other substances that can change the chemical and physical properties of the membrane.

4.2.8 Fouling

Fouling of a low-pressure membrane is characterized by gradual reduction of water flow through the membrane, when operating at constant feed pressure or experiencing an increase in TMP when operating at constant flux as a result of adsorption or deposition of particles on the membrane surface, blinding the membrane pores or hindering the flow of water. These fouling mechanisms are categorized as "pore absorption," "pore blocking," and "cake formation."

Pore absorption occurs when the diameter of the particle is smaller than that of the pore. The particles are deposited in the pore walls along their entire length, which effectively decreases the pore diameter, causing an increase in hydraulic resistance. Pore blocking occurs when the diameter of the particles is approximately equal to the membrane pore diameter and blocks the passage of water to the pores, causing a reduction in the number of effective pores and increasing hydraulic resistance.

Cake formation occurs when the particles are too large to enter the pores. The increase in the mass of particles deposited on the surface of the membrane or cake formation provides additional resistance to filtration that increases with operating time, causing an increase in TMP or permeate flux decline. Chemically and hydraulically irreversible fouling may also occur.

Microfiltration/ultrafiltration performance can be modeled by the resistance-in-series approach and, using Darcy's law as a theoretical basis, the flux decline can be determined by the following equation:

$$J = (1/A)(dV/dt) = \Delta P/[\mu(R_m + R_t)] \tag{3.15}$$

$$R_t = R_{cp} + R_a + R_b + R_c$$

where J = permeate flux,
A = membrane surface area,
V = volume of the permeate,
t = operation time,
ΔP = TMP drop,
μ = water viscosity,
R_m = membrane resistance,
R_t = total fouling resistance,
R_{cp} = concentration polarization resistance,
R_a = pore absorption resistance,
R_B = pore blocking resistance, and
R_c = cake layer resistance.

With the wide spectrum of foulant in wastewater streams that contribute to microfiltration/ultrafiltration fouling, the resistance-in-series concept offers a simple means to group the fouling parameters while allowing each parameter to be determined individually. For further information on resistance-in-series concept applied to microfiltration/ultrafiltration, please refer to *Manual of*

Water Supply Practices—Microfiltration and Ultrafiltration Membranes for Drinking Water (AWWA, 2005).

In dead-end filtration, when the particles removed are larger than the pores, they are removed on the surface as cake. During unstirred filtration at constant pressure under conditions of perfect retention, the following expression can be derived relating to the cumulative filtrate volume, V, to the filtration time, t (Cheryan, 1998):

$$t/V = (\mu R_m / A\Delta P) + (\mu \alpha C_o V) / (2A^2 \Delta P) \tag{3.16}$$

where A = membrane area,
μ = absolute viscosity,
ΔP = TMP,
C_o = feed water particle concentration, and
α = cake-specific resistance on mass basis.

Rigid particles can form compressible cakes. As filtration pressure is increased, particles may rearrange to reduce the permeability and porosity. The cake resistances must be determined over a wide range of pressures expected to be encountered during filtration. The pressure dependence of specific resistance is often modeled as power law, as follows:

$$\alpha = \alpha_o \Delta P^n \tag{3.17}$$

where α_o = specific resistance at unstressed state (zero pressure) and
n = compressibility index.

Cake compression and membrane compaction may reduce membrane productivity as pressure is increased to maintain a constant flux.

4.2.9 Crossflow Filtration Fouling Reduction

Feed water with high suspended solids, pressurized membrane filtration with backwashing mode is not possible, and crossflow filtration via tubular membranes is used. In crossflow filtration of streams with suspended particles, the tangential shear is expected to scour materials deposited on the membrane, thereby reducing fouling. When calculating particle deposition and possible subsequent fouling, the back-transport caused by Brownian and shear-induced diffusion needs to be included. Brownian diffusion has an inverse relationship on particle size and shear-induced diffusion is directly proportional to the square of the particle radius and the shear rate. It was also demonstrated that specific resistance of cakes formed under crossflow filtration decreases with increasing shear. However, crossflow filtration mode may require higher energy compared to dead-end filtration mode.

4.2.10 Colloidal Fouling

Colloidal interactions play an important role in particle fouling, which is often the cause of productivity decline in microfiltration/ultrafiltration membrane processes. When colloids approach solid surfaces in aqueous media, this results in several interactions, including the van der Waals interaction, electrical double-layer interaction, hydration force, hydrophobic interaction, and stearic interaction, which collectively reduce the passage of water through the membrane. This fouling can typically be minimized by periodic backwashing.

4.2.11 Organic Fouling

In wastewater specifically from biological processes or chemical plants, effluent organic matter is the main cause of organic fouling. Lysis of bacterial cells can release many macromolecules that include carbohydrates, proteins, nucleic acids, and cell fragments. Characteristics of the organic matter alone are not the only determining factors for fouling. Membrane properties, characteristics of feed water, and the system's hydrodynamic conditions influence fouling. Charge repulsion may mitigate fouling if both fouling materials and membranes carry the same charge. The pH value may also influence fouling because many of those charges are pH-dependent. High ionic strengths increase fouling because high ionic strength reduces the charge repulsion through double-layer compression. Presence of counterions can promote fouling through surface charge neutralization and bridging.

Properties of the membranes also affect fouling. Important membrane properties include surface roughness, pore size rating, hydrophobicity, surface charge, and pure water permeability of the membrane. Membranes with smaller pore size, higher pure water permeability, and lower (negative) surface charge tend to have higher fouling potential.

4.2.12 Biofouling

"Biofouling" refers to biological activities on the membrane surface that result in unacceptable permeability loss. The microbes retained by the membranes would remain active, grow, and multiply to a point of significantly reducing membrane permeability. The biofouling process includes attachment of the microorganism on the membrane surface and formation of biofilm through cell growth and multiplication. The most important nutrient for microbial growth and multiplication is carbon. Carbon source can be dissolved organics and assimilable organic carbon in the feed stream. Assimilable organic carbon concentration in the feed stream is often considered as a measurement of biofilm formation potential.

4.2.13 Coagulant Aid Polymer Fouling

Membrane filtration is sometimes preceded by conventional pretreatment processes that use coagulant aid polymers, which can cause rapid and potential irreversible fouling of the microfiltration/ultrafiltration membranes. The risk of fouling is significant with cationic polymers because most of the microfiltration/ultrafiltration polymeric membrane materials are negatively charged. Any pretreatment processes in front of polymeric microfiltration/ultrafiltration membranes should be designed to minimize or eliminate polymer carryover.

4.3 Membrane Filtration Units

4.3.1 Materials of Construction

Membrane filtration materials can be polymeric or inorganic. Each has its advantages and disadvantages, depending on application and environment. Newer commercial membranes have been made of both polymeric and inorganic materials. The following sections describe the characteristics of each material.

4.3.1.1 Polymeric Membranes

Most low-pressure polymeric microfiltration/ultrafiltration membranes that are commercially available are manufactured from various varieties of polymers including polypropylene, polyvinylidene difluoride (PVDF), cellulose derivatives, polyether sulfone, polyether sulfone/polyvinylepyriolodione (PES/PVP) blends, polysulfone, polyacrynitrile, polyvinyl chloride (PVC), and a variety of other polymeric materials, as shown in Table 3.9. These materials vary with respect to mechanical strength, fouling resistance, hydrophobicity, and chemical resistance. An important consideration when selecting microfiltration/ultrafiltration membrane for treating industrial wastewater is tolerance of oxidants to allow for high-concentration chlorine for cleaning or continuous chloroamines dosing to control organic and biological fouling. Of the aforementioned materials, PES/PVP blends and PVDF offer the highest oxidant tolerance, while polypropylene is the least oxidant tolerant. Typically, polypropylene membranes are not subjected to free chlorine.

4.3.1.2 Inorganic Membranes

Ceramic membranes are metal (aluminum or titanium) and nonmetal (oxides, nitrides, or carbides) that can be used across a wide pH range: from highly acidic to highly basic environments. Although ceramic membranes offer several advantages including high durability (chemical, mechanical, and operation high temperature operation), high water productivity (permeability/flux), long service life, and microbial fouling resistance, the capital cost is still significantly higher. Aside from being higher cost, an additional key disadvantage of ceramic membranes is their high sensitivity to temperature gradient, which could lead to membrane cracking.

Metallic membranes are made from sintering metal powders such as tungsten, palladium, or stainless steel and depositing the sintered material in porous substrate. Metallic membranes have similar advantages as the ceramic membranes. The main disadvantage of metallic membranes is the inactivation or destruction of the membrane surface by introduction of chemicals or physical process to alter the characteristics of the membrane surface.

4.3.2 Membrane Geometries

The microfiltration/ultrafiltration membrane comes in various geometries: hollow fiber, capillary, tubular, flat sheet, spiral wound, or plate and frame. Examples of the membrane geometries are shown in Figure 3.34. Most commercial polymeric hollow fiber or capillary microfiltration/ultrafiltration membranes range in size from 0.5 to 2.0 mm outside diameter, 0.3 to 1.0 mm inside diameter, with wall thickness of 0.1 to 0.6 mm and fiber length of 1 to 2 m. Typically, 3000 to 20 000 hollow fibers or capillaries are bundled together longitudinally, potted in a resin (typically cured epoxy or urethane) on both ends, and encased in a pressure vessel (module). Hollow fiber and capillary microfiltration/ultrafiltration membranes come in either single bore or multibore. Single bore are constructed like a tube, while multibore consists of a number of tubes inside a single porous casing. The advantages of hollow fiber include: high packing density (high surface-to-volume ratio), filtration at lower feed pressure, and lower pressure drop across the membrane modules.

Microfiltration/ultrafiltration membranes with larger outside diameter (typically from 6 to 50 mm) are considered tubular. Individual membrane tubes are placed inside fiberglass reinforced plastic, PVC, chlorinated polyvinyl chloride (CPVC), or stainless steel casing, and the two ends of the tube are fixed with potting material, or end plates; the connections are flanged, threaded, clamp type, or quick-connect couplings. The flow pattern in tubular membranes is inside-out; the feed water is introduced to the tube lumens under pressure and filtrate flows through the walls of the membrane tubes into the outside shell of the membrane casing or module. The advantages of tubular membranes include large channel diameters that allow treatment of water streams with higher solids concentrations and operation at high crossflow velocity, which help control fouling and deposition of scales.

TABLE 3.9 Membrane materials.

Hydrophilic Polymers			
Polyacrylonitrile	PAN	Cellulose Acetate	
Polyvinyl Chloride	PVC	Cellulose Triacetate	CTA
Polyvinyl Chloride	PVC	Cellulose Acetate Butyrate	CAB
Polyamide	PAN	Cellulose Acetate Propionate	CAP
Polyacrylic Acid	PAA	Cellulose Nitrate	CN
Polyvinyl Acetate	PVAC	Cellulose Acetate Propionate	CAP
Polyvinyl butyral	PVB	Cellulose Propionate	CP
Polyethylene Oxide	PEOX	Ethtyl Cellulose	
		Carboxymethyl Cellulose	CMC
Hydrophobic Polymers			
Polysulfone	PSF	Polytetrafluoroethylene	PTFE
Polyethersulfone	PES	Polyethylene	PE
Polyvinylidene Fluoride	PVDF	Silicone	Si
Polycarbonate		Polypropylene Oxide	PPO
Polypropylene		Polypropylene Sulfide	PPS
Polymethyl methylacrylate	PMMA	Polystyrene	
Ceramic Membrane			
Aluminum Oxide/-Alumina	Al_2O_3	Titanium Dioxide	TiO_2
Silicon Carbide	SiC	Zirconium Dioxide/Zirconia	ZrO_2
Sintered Metals			
Stainless Steel		Nickel	
Hastelloy		Titanium	
Inconel		Alloy 20	

Source: J.T. Aguinaldo

Flat sheet microfiltration/ultrafiltration membranes consist of membrane leaves sandwiched between a filtrate carrier, where the water flows across both membrane surfaces through the membrane into the filtrate carrier. Three sides of the membrane leaves are sealed, forming an envelope, and the filtered water collected in the filtrate channel is directed to a collection manifold. The flat sheet microfiltration/ultrafiltration membrane envelop or membrane leaf terminates in a common filtrate collector. To increase the packing density, flat sheet membranes are formed into leaves and rolled spirally, then enclosed in a vessel. In submerged configuration, the membrane sheets or leaf are submerged in a tank. The solids are retained in the bulk stream in the tank or in the circulating liquid stream for crossflow filtration.

In plate and frame configuration, each membrane and filtrate carrier or collector is installed in the parallel frames arranged to allow water to flow across the membrane surface and filtrate through the membrane sheets into the filtrate carrier, while the reject is collected on the opposite side of the feed header. Each membrane is fed from a common header and reject is collected in the reject manifold, while the filtrate is collected from the filtrate collector into a common filtrate manifold. The ceramic membranes can have hollow fiber, capillary, tubular, flat sheet, and spiral wound configurations.

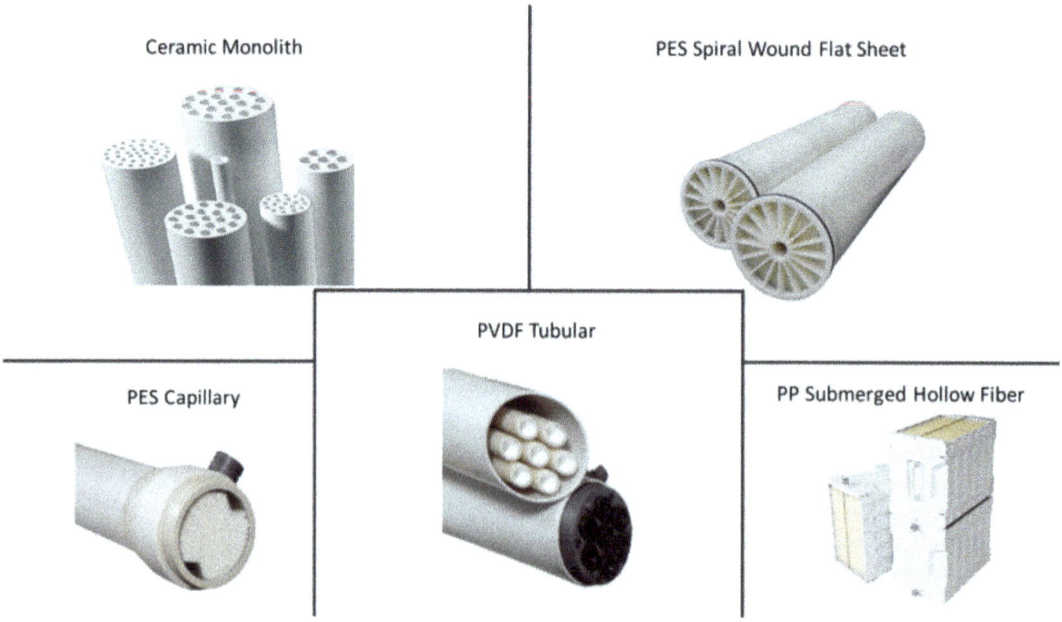

FIGURE 3.34 Typical membrane geometries from membrane manufacturers. *Source:* J.T. Aguinaldo

4.3.3 Membrane Process Configuration

Microfiltration/ultrafiltration systems come in two basic process configurations: pressurized and submerged.

Pressurized Microfiltration/Ultrafiltration Membrane Configuration

- In pressurized microfiltration/ultrafiltration membrane configuration, the membrane fibers, tubes, or sheets are housed inside pressure vessels.
- The feed stream is fed under pressure through a feed pump and the filtrate flows through the membrane into the filtrate collector under pressure or at atmospheric pressure.
- The backwash water and cleaning solution are also fed under pressure and the spent stream is collected under pressure or atmospheric pressure before discharging to the waste drain.
- In pressurized crossflow configuration, the feed stream is fed under pressure and the filtrate is collected under atmospheric pressure, while the recirculating stream under pressure goes to the recirculating tank.
- It is also possible to operate without recirculating to the tank. The backwash or backflush and cleaning streams are also fed under pressure, and the spent stream is collected under pressure or at atmospheric pressure before being discharged to the waste drain.

Submerged or Immersed Microfiltration/Ultrafiltration Membrane Configuration

- In submerged or immersed microfiltration/ultrafiltration membrane configurations, the membranes are mounted in a containment tank or vessel and are open to the atmosphere.
- The filtrate is collected through a vacuum pump/positive displacement pump.
- The backwash and cleaning solution is fed under pressure and the spent streams discharge to the membrane tank with the bulk stream, or the tank is emptied to the drain after each cleaning.

4.3.4 Membrane Flow Configuration

Depending on the direction of filtrate flow, the microfiltration/ultrafiltration membranes are further categorized as inside-out or outside-in, as shown in Figure 3.35. In inside-out, the water stream is fed into the filter lumens and is collected on the outside of the fibers. One misconception in operating inside-out microfiltration/ultrafiltration is it is prone to clogging; however,

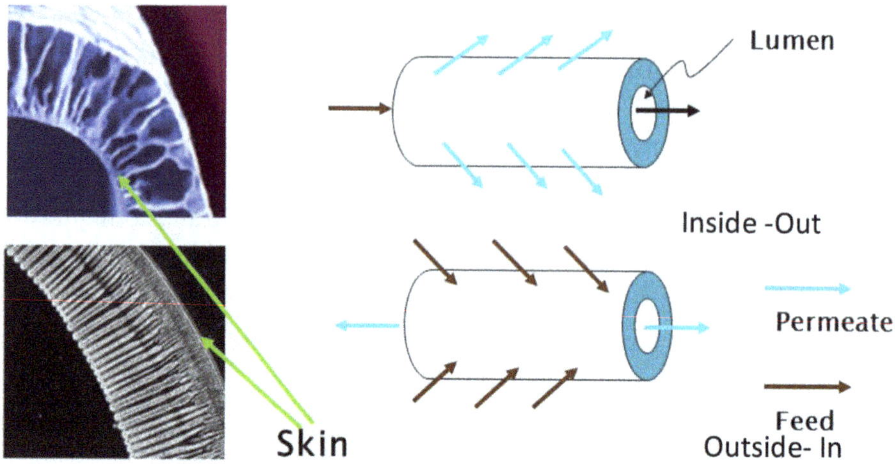

FIGURE 3.35 Membrane flow configuration (Sandia National Laboratories/U.S. State Department, 2011).

with proper pretreatment and operation, clogging can be prevented. The deposition of solids in inside-out hollow fiber or capillary membranes can be controlled, and backwashing or backflushing is predictable.

For outside-in, the feed water is distributed around the filter fiber, the water goes through the membrane, and the filtrate is collected through the fiber lumen. Submerged microfiltration/ultrafiltration is considered outside-in. Pressurized outside-in membranes are difficult to clean with water backwash only and often require air backwash to restore productivity. Foulants tend to accumulate outside the fiber ends, reducing the membrane effective area over time. The accumulated foulant can be removed by aggressive chemical cleaning and intense agitation, typically with air. Outside-in membranes operate at higher feed pressure for the same flux as inside-out membranes or lower fluxes at the same design TMP. Pressurized membranes are operated at 20 to 250 kPa (3 to 35 psi).

4.3.5 Membrane Operating Configuration

There are three microfiltration/ultrafiltration operating configurations: dead-end, crossflow filtration, and submerged filtration.

4.3.5.1 Dead-End Filtration

In dead-end (sometimes called "direct flow"), 100% of the feed water is driven through the membrane and solids accumulate on the membrane surface as cake during filtration mode. The accumulated solids are removed at timed intervals by stopping the filtration and backflushing or backwashing with filtrate, raw water, or air to flush out the accumulated solids and discharged to waste. The dead-end filtration is illustrated in Figure 3.36.

4.3.5.2 Crossflow Filtration

For pressurized microfiltration/ultrafiltration membranes operating at crossflow configuration, a certain velocity across the microfiltration/ultrafiltration membrane surface is maintained, reducing the formation of cake. The feed stream to the microfiltration/ultrafiltration is circulated and the solids accumulate in the recirculating stream. A portion of the recirculating stream is bled off to maintain a solids concentration in the system.

4.3.5.3 Submerged Filtration

The filtrate is drawn through the membrane by a pump operating below atmospheric pressure (i.e., vacuum). All submerged systems are outside-in. Submerged systems are typically equipped with air systems for agitation and bleed systems to control solids buildup in the containment tank and at the membrane surface. The air may be continuous or intermittent.

Submerged systems operate under slightly negative or vacuum pressure, typically 20 to 80 kPa (3 to 12 psi). Submerged membranes are typically used when treating feed water with variable quality or with turbidity fluctuations or high suspended solids. Submerged membranes are commonly used in MBRs with high MLSS up to 20 000 mg/L, but the typical values range from <10 000 to 16 000 mg/L. The maximum TMP (theoretical) for

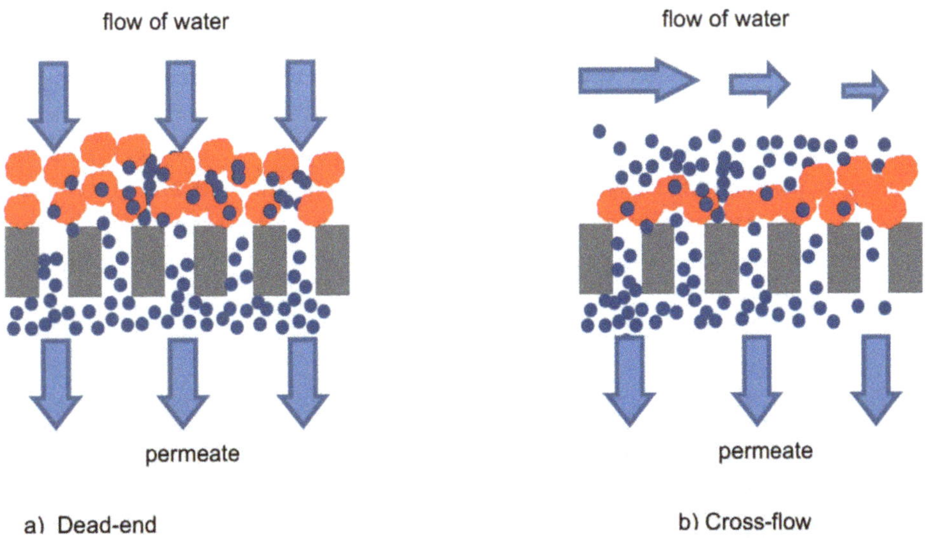

FIGURE 3.36 Filtration mode (Adapted from U.S. EPA, 2005).

submerged membranes is approximately 100 kPa of vacuum or lower TMP of 70 to 80 kPa (10 to 12 psi).

Submerged membranes offer more surface area per unit facility footprint compared to pressurized membranes. Both the dead-end and crossflow microfiltration/ultrafiltration membrane configuration are cleaned at timed intervals or when the pressure drop across the membrane reaches a preset value.

4.4 Membrane Systems Design

4.4.1 Facility-Specific Conditions

The following sections summarize specific conditions that affect membrane systems design.

4.4.1.1 Feed Water Characteristics and Quality Variability

Feed water characteristics and quality variability are the primary considerations when designing the microfiltration/ultrafiltration facility. The facility flows, capacities, and constituent concentrations should be well defined, including peak flow, average daily flow, future flow or demand, maximum and minimum concentrations of constituents in feed water and required filtrate quality, and physical and chemical characteristics of the feed stream such as temperature, turbidity, pH, suspended and dissolved solids concentration, size, and settling characteristics of particles. Ideally, feed water sample analyses will be taken over an appropriate period of time that reflect changes in season or facility operating schedules.

In terms of feed water quality, thorough review of the constituents and their characteristics in the feed water should be made to ensure they are compatible with the materials of construction of microfiltration/ultrafiltration membranes. For industrial wastewater, temperature is sometimes an issue. The selected membrane materials should be compatible with the operating temperature. There are cases where the treated water needs to maintain its original temperature, examples are boiler feed and hot water process applications. Understanding how the feed waste stream is generated is important because it may provide some explanations on some anomalies or missing feed water data.

4.4.1.2 Treated Water Quality and Quantity

The microfiltration/ultrafiltration membrane filtration process typically produces consistent quality filtered water in terms of turbidity, silt density index, and TSS. Overall treatment requirements for a microfiltration/ultrafiltration facility may include water recovery, wastewater discharge quality and quantity, and the presence of specific constituents, both in filtered water and wastewater generated by the microfiltration/ultrafiltration. To meet the overall treatment requirements dictated for industrial water reuse, further posttreatment may be necessary.

4.4.1.3 Facility Availability and Reliability

Facility availability and reliability requirements should be taken into consideration when designing the microfiltration/ultrafiltration membrane filtration facility to make sure offline conditions and planned and unplanned facility shutdowns are taken into account in planning for system redundancy. This applies to membrane backwashing, cleaning, integrity testing, routine facility maintenance, and emergencies.

4.4.1.4 Facility Automation and Control

The microfiltration/ultrafiltration facility is effectively operated when it is fully automated, with data logging and report generation programmed into the supervisory control and data acquisition (SCADA) system. Frequency of backwashing and chemical cleaning, along with operating cycle times, can be correspondingly optimized.

4.4.2 Specific Membrane System Design

The important design parameters for microfiltration/ultrafiltration systems are pretreatment, fouling control, feed stream conditioning, sustainable operating membrane flux, backwashing and cleaning efficiencies, wastewater minimization, and disposal.

4.4.2.1 Pretreatment to Microfiltration/Ultrafiltration

Pretreatment and feed conditioning for membrane treatment processes are categorized into the following general treatment strategies and processes:

- Remove the materials and substances that may interfere with the membrane treatment processes, causing physical and chemical damage or alteration of the membrane materials that will significantly affect the performance of membranes.
- Provide conditions that will enhance or maximize the performance of the membrane treatment in terms of flux and rejection. Among these are pH control to maintain solubilities of salts and retard crystallization, and oxidation or reduction of certain chemical species.
- Minimize accumulation of foulants on the surface of the membranes that can block the membrane pores and hinder the flow of water.

Screens and strainers are installed in the feed stream to the microfiltration/ultrafiltration system to protect the membranes from debris and particulates that can clog the membrane pores or damage the membrane surface. The typical screen size opening ranges from 100 to 300 μm. Certain constituents in wastewater will permanently alter the chemical composition and physical integrity of the membrane materials. Most commercially available membranes are made of organic polymeric materials. Polymeric materials are susceptible to chemical attacks such as solvents, hydrocarbons, and oxidizing agents.

The other materials in the membrane modules such as spacers, casing, channels, and glue need to be checked for compatibility with the constituents present in the feed stream or chemicals added as part of the treatment process. In addition to the membrane module materials, it is also critical to evaluate the compatibility of the associated materials and equipment, such as pipe, valves and fittings, pumps, tanks, instruments, and so on. When in doubt, check with the membrane or equipment manufacturers.

Significant concentrations of solvents, such as ketones, aldehydes, alcohols, gasoline, hydrocarbons, and other organic compounds that will have a direct reaction with the membrane materials, should not be present in the feed water at all times. Treatment of solvents and other organic chemicals in feed water should be dealt with separately using conventional processes such, stripping, oxidation, as activated carbon adsorption, or other specialty and proprietary processes.

Oil and grease, in general, are not compatible with most membranes used in water treatment and should be avoided or removed in the feed water; however, most membranes can tolerate certain amounts of oils and grease. It is important to contact the membrane manufacturer because there are no general guidelines for allowable oil or grease concentrations in microfiltration/ultrafiltration feed. There are, however, specialized membranes that can separate oil from water. These specialty oil removal microfiltration/ultrafiltration membranes can also be used to pretreat feed water for nanofiltration and reverse osmosis. It is better to consult the membrane manufacturer's guidelines.

Cationic polymers that are often used in pretreatment processes have detrimental effects on the membranes and must be avoided. Some esters have detrimental effects on the membrane. Aromatic hydrocarbons are detrimental to polysulfone membranes, while most microfiltration/ultrafiltration membranes have good resistance to aliphatic hydrocarbons.

Adsorption of organic substances on the membrane surface causes permeability loss, or flux loss at constant pressure. Organic compounds present as emulsion may form an organic film on the membrane surface, and they must be removed or eliminated. Adsorption of organics onto the membrane tends to occur most preferentially with high molecular mass compounds when these are hydrophobic or positively charge.

Although ceramic, zirconium oxide, carbon, and sintered metal microfiltration/ultrafiltration are resistant to most organic and inorganic chemicals and can be operated at elevated temperatures, they also have their limitations. For example, ceramic membranes are damaged by hydrofluoric acid, while some sintered metals are attacked by some inorganic acids. It is important to investigate the compatibility of the ceramic or sintered metal materials with the stream being treated.

4.4.2.2 Foulant, Particulate, and Suspended Solids Control

Foulant is any material that will physically blind the membrane pores or block the flow of water through the membrane. Major foulants are particulates, colloids, metal oxides, biological materials, or natural organic matter. Suspended solids in feed will negatively affect most membranes. The suspended solids will plug or blind the pores on the surface of the membranes and water channels, resulting in an increase in pressure drop and hindering the flow of water, thereby reducing productivity. The increase in pressure drop requires increased feed pressure to maintain membrane productivity. However, excessive pressure drop will physically damage the membrane elements. Other sources of fouling come from the chemicals added during pretreatment, which includes coagulant, flocculant aid, some polymers (especially cationic), and dissolved matters present in the feed water, which may precipitate or solidify as a result of physical and chemical pretreatment processes.

Colloids are small particles commonly defined by membrane manufacturers (Cheryan, 1998) as ranging from 0.01 to 1.0 μm in size. The colloids are electrically charged and are in suspension because they repel one another. The sources of colloids in water reuse include: feed water passing through or over clay geological structures can pick up aluminum silicate colloids; corrosion products from carbon steel pumps, piping, and filters may occur as iron colloids; feed water pretreated with alum and improperly clarified may contain aluminum hydroxide; the presence of colloidal sulfur from hydrogen sulfide; organic colloids naturally occurring in surface waters or that are resultant of certain manufacturing operations; and bacterial debris from a biological treatment process preceding the membrane treatment. Colloidal fouling in the membranes is caused by coagulation of the colloids and deposition of the coagulated material on the membrane.

Aluminum fouling is caused by the precipitation of aluminum hydroxide. The aluminum ion is amphoteric and has minimum solubility at a pH of approximately 6.5 to 6.7. If alum coagulation is part of the pretreatment, it should be operated at pH of 6.5 to 6.7 to obtain the minimum solubility of aluminum.

The suspended solids or particulates in the feed water can be settleable solids or nonsettleable solids, which may include colloids. Settleable solids can be removed by plain settling if the feed water is allowed to stand for some time, whereas nonsettleable solids require the addition of chemicals, coagulant, or flocculant for settling. Within the category of nonsettleable solids, the size of colloids ranges from 0.01 to 1.0 μm and they are charged, so they can pass through microfiltration or ultrafiltration membranes with larger nominal pore sizes. To remove the colloids, it is important to coagulate and flocculate them to form larger particles (as discussed earlier in this Chapter, Section 2) and to also neutralize their charges. The upper limit for suspended solids concentration recommended by one of the major and largest microfiltration and ultrafiltration manufacturer operating at dead-end filtration mode is 1000 mg/L, which was adopted by various manufacturers. Relative to industrial water reuse, the ultimate aim of suspended solids and particulate removal is often to have a pretreated feed to nanofiltration and reverse osmosis membranes with a silt density index of less than 3.0, turbidity <0.5 NTU, and TSS below a detectable limit. The recommended pretreatment processes for microfiltration/ultrafiltration for suspended solids removal will depend on the physical and chemical properties of the materials to be removed. Coagulation-flocculation is the most flexible pretreatment process, when followed by clarification and/or filtration, because it can work with variable suspended solids loads. Coagulation can be performed upfront of microfiltration/ultrafiltration systems to enhance membrane performance by removing foulants or other contaminants from the feed water before treatment by membranes. Coagulation can be performed via conventional methods (coagulation-flocculation/sedimentation processes) or in-line. Direct filtration as applied in media filtration can also be applied in microfiltration/ultrafiltration. Direct filtration or in-line coagulation offers the benefit of reduced footprint compared to conventional coagulation processes. However, the use of high-rate tube plate settlers has also been used upstream of microfiltration/ultrafiltration in advanced water treatment applications.

Common metallic salt coagulants include alum, ferric chloride, polyaluminum chloride, and aluminum chlorohydrate, which are compatible with membrane material. Coagulation can improve membrane performance by increasing membrane productivity, enhancing dissolved organic carbon and/or nutrient (i.e., phosphorus) removal, and reducing membrane cleaning requirements. The specific mechanisms responsible for these enhancements include the following:

- Reduction of particles that enter the membrane pores. The size or number of particles will depend on specific application.

- Conditioning of the cake layer in the membrane, to prevent blinding of the membrane pores. The cake layer should allow unhindered passage of water to the membrane surface.
- Enhancement or facilitation of back transport (Brownian diffusion). The back-transport of smaller particles is the result of Brownian movement, while larger particles (macrocolloids) are controlled by an orthokenitic mechanisms such as shear-induced diffusion.

The appropriate type and dose of coagulant is dependent on several factors including feed water quality, temperature, and pH, and these factors are best optimized by performing jar testing. In general, optimum concentrations during treatment may be lower than typical jar test results based on removal rates of $SUVA_{254}$ or turbidity and, at best, to adjust during piloting. $SUVA_{254}$ is specific ultraviolet absorbance calculated by dividing UV absorbance at 254 nm (cm^{-1}) by dissolved organic carbon (DOC) concentration, (mg/L). Roughly 1 ppm DOC is equivalent to -0.5 metal ion (3+), such as Fe^{+3}.

Ferric salts are popular because they work well in water with pH range of 5 to 8.5. The most common ferric salt used is ferric chloride ($FeCl_3$). The dosage for $FeCl_3$ will depend on the quality of feed water, particularly on alkalinity and suspended solids concentration (as confirmed by jar testing), which ranges from 0.2 to 20 mg/L as Fe^{3+}. At low turbidity and low alkalinity water, sweep floc will be required, which means a relatively higher concentration of Fe^{3+} will be required, although there is a risk of metal hydroxide fouling. In microfiltration/ultrafiltration system, it is not necessary to form large flocculated particles but formation of microflocs is sufficient to be separated by the membranes. Hydrolysis reactions for ferric salts are fast, requiring mixing times of less than 1 second, preferably 0.5 seconds. Ferric chloride is also effective at removing small amounts of oil and grease in feed water. Alum and poly aluminum chloride are also used as coagulant in water reuse. In general, the contact time for in-line coagulation is approximately 30 to 60 seconds.

Coagulation is always followed by flocculation. Flocculation chambers with vertical-shaft turbines are preferred because of the minimum area of metals exposed to the water. The retention time in the flocculation chamber ranges from 5 to 30 minutes, with velocity gradient values of 10 to 80/s. Alternatively, proprietary coagulation/flocculation systems are available from various manufacturers. Most of them are in circular configuration. Conventional coagulation and flocculation are typically followed by clarification and media filtration but in microfiltration/ultrafiltration pretreatment streams with flocs with TSS concentration <100 mg/L can go directly to microfiltration/ultrafiltration system. Exception to this are integrated chemical precipitation-membrane filtration systems, which can handle high suspended solids.

Coagulation-flocculation (followed by filtration) must be applied when oils (hydrocarbon and silicone based) and greases are present in the feed water stream. For more details on coagulation-flocculation, refer to the *Manual of Water Supply Practices—Operational and Control of Coagulation and Filtration Processes* (AWWA, 2010).

When the concentration of suspended solids after coagulation-flocculation is relatively high (>100 mg/L), it is recommended to be followed by clarification. Clarification or sedimentation can be achieved in both circular and rectangular (square) tanks. Rectangular (or square) tanks with Lamella or tilted parallel plates are recommended to minimize the clarifier footprint. Using ferric salt as a coagulant, the resulting flocs or sludge settle well in Lamella or tilted parallel plate clarifiers and tube settlers. The typical surface loading rate (overflow rate) for clarifier (Lamella or tilted parallel plate) is 1.25 to 2.5 m/h (0.5 to 1.0 gpm/sq ft), with a retention time of 1.5 to 4 hours.

For horizontal flow rectangular clarifier tanks with plate and tube settlers, the surface loading rate is 2.5 to 3.7 m/h (1.0 to 3.0 gpm/sq ft), with a retention time of 6 to 10 minutes for tube settlers and 15 to 25 minutes in plate settlers. For additional information on clarifier design, refer to *Water Treatment Plant Design* (AWWA & ASCE, 1997).

The DAF process uses a clarifier to separate suspended solids or oils that do not readily settle because they are lighter than water. Algae, aquatic organisms, plants, and solids with strong affinity to air can float in DAF. One advantage of DAF is its ability to remove oil and grease as well. Starting with a TSS of 100 mg/L, DAF can reduce TSS down to less than 10 mg/L. Dissolved air flotation can be a good pretreatment to microfiltration/ultrafiltration systems. For more information on the selection and design of DAF, refer to *Dissolved Air Flotation for Water Clarification* (Edzwald & Haarhoff, 2012).

4.4.2.3 Feed Water Stream Conditioning

The pretreatment and conditioning feed water must involve a total system approach for continuous and reliable operation. In addition to the pretreatment processes described in previous sections, conditioning of the feed water such as pH adjustment, addition of antiscalant and dispersant chemicals, disinfectant or biocide, and specialty chemicals to improve the performance of the membrane are necessary.

4.4.2.4 Operating Sustainable Membrane Flux

The design sizing of microfiltration/ultrafiltration membranes is based on sutainable membrane flux, which is a measure of filtered water through specified membrane area per unit time, typically expressed as liters per square meter per hour (Lmh) or gallons per square feet per day (gfd). Every membrane manufacturer provides their recommended design and operating membrane flux, depending on the type of microfiltration/ultrafiltration membrane, configuration, material of construction, and mode of operation. Other factors that affect the membrane flux are the type, characteristics, and quality of the feed water stream. Typically, TSS or turbidity are the primary criteria used in preselecting membrane flux, although other factors such as temperature, total organic carbon (TOC), alkalinity, source of water, and presence of some contaminants are taken into consideration. Membrane manufacturers typically have a database of various applications with recommended design and operating sustainable membrane flux, which are included in their proprietary design software.

Maintaining a sustainable membrane flux is the basis for successful operation of microfiltration/ultrafiltration facilities. Solids and particulates accumulate on the membrane surface and in the membrane pores and are required to be flushed out or backwashed at given time intervals to maintain acceptable flux. Operational membrane flux has to be selected to experience minimum fouling rate during the filtration cycle, to ensure the pressure loss as result of solids accumulation is within the specified range of operating pressures, and to thereby extend the cycle of filtration and backwashing.

When operating at elevated concentration of solids and high fouling potential of feed water, lower sustainable flux is selected.

4.4.2.5 Membrane Backwashing

Solids and particulate accumulate in the membrane surface or pores during the filtration cycle, resulting in an increase in TMP. The solids and particulates are removed periodically by backwashing the membrane with filtered water. Backwashing is typically initiated by a timer or differential pressure switch. The differential pressure switch is set at a specific preset TMP. Operating beyond the specified TMP, the membrane flux will be decreased, the filtered water quality will deteriorate, potentially leading to irreversible and permanent fouling. Most microfiltration/ultrafiltration membrane systems' backwash is triggered by preset time, which is adjusted based on solids in the feed and fouling potential of the feed water.

Filtered water is typically used in backwashing the microfiltration/ultrafiltration, although some manufacturers recommend water and air in sequence. The backwashing rates are designed to maximize the removal of solids accumulated on the membrane surface as well as from within the membrane pores during the filtration cycle. Backwash is intended to remove solids, but, over time, membrane surfaces will also accumulate organic matters and biofilms that are not addressed by backwashing. The accumulation of organic matter and biofilm on the membrane surface if not removed will increase TMP, productivity, and can lead to permanent fouling. This type of membrane fouling can be controlled by chemically enhanced backwashing (CEB). Membrane manufacturers recommend the frequency of CEB. Chemically enhanced backwashing is conducted like regular backwashing, but cleaning chemicals are injected to the backwash water. Sometimes in CEB, the membranes are soaked for several minutes in chlorine and/or other cleaning chemicals, then backwashed.

4.4.2.6 Membrane Cleaning

Regular membrane backwashing and CEB do not completely remove the foulant on membrane surfaces and in the pores. After backwashing and CEB, the membrane TMP will slightly increase above the TMP at the start of the filtering cycle. Over time, the TMP after backwashing and CEB will have degraded sufficiently that it is necessary for the membranes to be taken offline and cleaned with chemicals to restore the TMP to an acceptable level. This type of cleaning, referred to as "membrane clean-in-place" (CIP), is aimed at recovering membrane permeability.

Membrane CIP is typically needed every few months and is performed using a combination of low-pH solution followed by high-pH solution. For low-pH CIPs, citric acid, sulfuric acid, or hydrochloric acid are used, whereas for high pH CIPs, a solution of sodium hydroxide and sodium hypochlorite are used. During CIP, the membranes are soaked in cleaning chemicals, during which the chemicals are recirculated through the membranes for a period recommended by microfiltration/ultrafiltration manufacturer. The membranes are then flushed and rinsed before returning to the normal filtration cycle.

Inorganic foulants, such as those resulting from iron and manganese compounds and colloids accumulated in membrane surface and pores, are cleaned with citric, sulfuric, or hydrochloric acids. Foulants of organic origin can be cleaned with sodium hydroxide, whereas foulants of biological origin are cleaned using disinfectants such as sodium hypochlorite, peracetic acid, and surfactants. Some membrane manufacturers recommend that cleaning chemicals be heated; however, most chemicals do not require heating. Other chemicals and enzymes are added to specifically remove certain foulants.

4.4.2.7 Membrane Life and Warranty

The microfiltration/ultrafiltration membranes age with use and passage of time, which results in loss of productivity because of chemical degradation or permanent irreversible fouling. The useful life of the membrane can be 5 to 10 years or more, but relies heavily on the membrane materials, pretreatment process and feed water conditioning, backwash and cleaning techniques, and manufacturer's support.

Most new microfiltration/ultrafiltration membranes come with a two-stage warranty during their useful life. The initial stage is full replacement without any cost to the buyer, which is 1 to 2 years. The next stage of warranty is prorated, where the cost of membrane replacement is prorated up to 100% at the end of membrane useful life. The prorated membrane warranty is typically continued for 5 to 10 years after the delivery of the membranes.

4.4.3 Membrane Treatment Facility Design

4.4.3.1 Pilot Testing

A significant amount of design and operating data are available on different microfiltration/ultrafiltration membrane systems, but each facility treating specific industrial wastewater is different because of the wide variations of feed water quality that is reflective of upstream manufacturing operations. An important consideration during the design of a microfiltration/ultrafiltration system is the difference in each manufacturer's product offering. Unlike reverse osmosis and nanofiltration membranes, where the products are more "standardized" and performance between different manufacturers tend to be rather similar, microfiltration/ultrafiltration membrane selection can be a challenging exercise.

To minimize risks, it can be valuable to conduct pilot testing. Pilot testing typically mimics the full-scale facility's design conditions, operation, and performance. It also provides an opportunity to confirm and verify design and operating parameters made during the preconstruction or feasibility stage of the project. For more details on pilot testing, refer to Chapter 7 of this book, as well as to *Manual of Water Supply Practices—Microfiltration and Ultrafiltration Membranes* (AWWA, 2005).

4.4.3.2 Building Layout

In an industrial WRRF intended to minimize liquid waste discharge, the microfiltration/ultrafiltration system is frequently a subprocess to an overall facility process. The location, orientation, and footprint of the microfiltration/ultrafiltration system should be taken into consideration, as well as for the pre- and post-treatment processes. The footprint of the microfiltration/ultrafiltration system and sometimes the height of the building is determined by the membrane system and membrane flux. As a general rule, the microfiltration/ultrafiltration membrane facility will require an area of approximately 2.45 $m^3/(m^2{\cdot}h)$ (1 gpm/sq ft), but there is wide variability in this number—for example, some facilities have capacities per floor area in the range of 4.9 to 7.35 $m^3/(m^2{\cdot}h)$ (0.3 to 2 to 3 gpm/sq ft).

Some facilities such as immersed or vacuum-type microfiltration/ultrafiltration systems have a smaller footprint, but higher ceiling clearance because of the need to access the tank from above. Horizontal ultrafiltration systems will require a smaller footprint than vertical systems, but higher ceiling clearance; although, for some newer vertical ultrafiltration membranes with more compact arrangement, the difference is not as significant as the older generation of membranes.

If microfiltration/ultrafiltration is used as pretreatment to reverse osmosis/nanofiltration, it is worth considering direct coupling the two processes, although the entire microfiltration/ultrafiltration system needs to be designed for a higher pressure rating. Direct coupling means the filtrate line from the microfiltration/ultrafiltration feed directly to the reverse osmosis/nanofiltration high-pressure pump eliminates the need for a separate microfiltration/ultrafiltration filtrate tank, low-pressure reverse osmosis/nanofiltration feed pump, and reverse osmosis/nanofiltration cartridge filters.

4.4.3.3 Materials and Construction

In industrial environments, proper selection of materials and construction methods should be carefully taken into consideration. The materials should be compatible with the liquid stream being treated as well as the effects of the environment on the microfiltration/ultrafiltration systems. The common materials used in pipelines and manifolds in microfiltration/ultrafiltration systems are plastic materials such as PVC, CPVC, high density polyethylene, and fiber-reinforced plastic as well as stainless steel. Some of the pipelines are subjected to cyclic variations in pressure, which causes fatigue over time, or installed pipes are subjected to excessive stress because of their piping. Similarly, selection of valves, fittings, tanks, and pumps should always take into consideration materials' compatibility.

4.4.3.4 Instrumentation and Control

The successful operation of microfiltration/ultrafiltration systems relies heavily on automation, requiring simultaneous monitoring and control of a large amount of input/output points, which makes manual operation impossible. Most microfiltration/ultrafiltration systems rely on a programmable logic controller or distributed control system with SCADA interfaces. The control system must also include redundant control or fail-safe standby systems. The design of the control system, which is often provided by a third party, must be coordinated closely with the membrane systems provider.

4.4.3.5 Clean-in-Place

Membrane cleaning is one of the most important maintenance procedures in the operation of microfiltration/ultrafiltration systems. A typical CIP system includes a circulating pump, mixing/cleaning solution tank with or without a heater, piping, isolation valves, and an optional cartridge filter. The membrane manufacturer recommends the flowrate of the cleaning solution based on the membrane area. Typical cleaning solutions are either acidic, basic, or proprietary formulations depending on the foulant on/in membranes.

4.4.3.6 Wastewater Disposal

Dirty backwash water and spent cleaning solution should be disposed of properly to avoid pollution of the receiving water. The wastewater from microfiltration/ultrafiltration systems must be treated or neutralized before disposal according to local or national laws.

4.4.3.7 Equipment Selection

Equipment associated with the microfiltration/ultrafiltration systems are selected based on the general guidelines established by good engineering practices or industry standards. The membrane manufacturers' recommendations on the selection of the equipment should also be taken into consideration. Pumps are selected based on their best efficiencies. It is common practice to have variable frequency drives in pumps to allow for efficient operations and the ability to operate at various duty points, as required by the process.

4.5 Membrane Filtration Facility Operation

4.5.1 Controlling Microfiltration/Ultrafiltration Facilities

The basic mode of operation for microfiltration/ultrafiltration systems includes filtration, backwash, air scour (optional), and CEB (sometimes called "maintenance wash"), which are automatically controlled based on setpoints. Clean-in-place is manually initiated. While in operation, the microfiltration/ultrafiltration system is in filtration mode most of the time. As the filtration progresses, the particulates accumulate in the membrane units, which are to be removed and flushed out from the system through backwashing. Backwashing is initiated based upon time, volume of water filtered, or change in TMP over time.

Microfiltration/ultrafiltration systems operate as constant-rate filters, where the control system adjusts the operating pressure or flow to maintain a given flux or flow setpoint, taking into account membrane fouling, temperature, and volumetric demand.

4.5.2 Data Collection, Recordkeeping, and Process Monitoring

Data collection and recordkeeping are good practices to ensure successful performance of the microfiltration/ultrafiltration systems, allowing effective process monitoring and good sources of troubleshooting information should problems occur in the future. Data collection should start immediately when the facility is put into operation. The initial membrane data will be used as the referenced true clean membrane operating criteria. Important data collected on a regular basis include pressures and flow of the feed, flow of filtrate, temperature, permeability, and feed and filtrate turbidity.

Transmembrane pressure is typically calculated from the pressure and tank level data. The permeability is normalized at standard operating temperature (typically 20 °C, or 68 °F). In most microfiltration/ultrafiltration systems, data collection and process monitoring are automated through SCADA systems. The facility operator reviews and evaluates the data and makes crucial decisions if the facility performance deviates from what is expected.

4.5.3 Maintaining Productivity and Filtrate Quality

The most important task in the operation of a microfiltration/ultrafiltration system is maintaining productivity. Although microfiltration/ultrafiltration systems provide consistent filtrate quality, independent of feed water quality, anticipating the variations in feed quality or operating conditions will provide the operator with warning so they can react effectively. The frequency, timing, and duration of backwashing and CEB are important parameters the operator has to set to achieve efficient facility operation.

Filtrate quality can be monitored using turbidity meters, particle counters, or integrity verification. In potable water application, integrity testing and verification are important operating procedures because of regulatory reporting; however, in industrial wastewater recycle systems the applicability and type(s) of testing will be dependent upon the needs of downstream recycle processes (e.g., reverse osmosis) and/or any facility processes that reuse the microfiltration/ultrafiltration permeate.

4.6 Membrane Filtration Applications

4.6.1 Pretreatment to Reverse Osmosis and Nanofiltration

The most common application of microfiltration/ultrafiltration in industrial water reuse is pretreatment of reverse osmosis and nanofiltration feed to reduce fouling potential. Some of the full-scale and pilot case studies covered in Chapters 6 and 7, respectively pertain to this type of recycle system configuration. The primary advantages of microfiltration/ultrafiltration compared to conventional clarification with media filtration are as follows:

- In most cases, there is no need or reduced consumption of chemicals (coagulant, flocculants, disinfectants, pH adjustment)

- Size-exclusion filtration as opposed to media depth filtration
- Constant quality of the treated water in terms of particle and microbial removal at varying feed TSS concentration
- Process and facility compactness
- Simple automation
- Low to no risk of breakthroughs
- Superior TSS removal; superior effluent quality.

4.6.2 Standalone Treatment Processes

Standalone microfiltration or ultrafiltration systems are used where impurities in wastewater are particulates, suspended solids, and colloids. These systems are typically used in recycling ultrapure water in wafer-slicing processes in semiconductor facilities or recovering paint in automotive plants (i.e., ELPO/E-Coat process). One of the earliest applications of ultrafiltration in wastewater treatment was to remove and/or recover oil and grease (McDonald, 1973; Porter, 1973). The membrane manufacturers who pioneered in these applications Koch, Membrex, Carbosep, Amicon, and Romicon (both Amican and Romicon were later acquired by Koch Membranes).

4.6.3 Integrated Chemical Precipitative Process

This process combines chemical precipitation such as lime or lime-soda softening with tubular crossflow or flat sheet microfiltration or ultrafiltration processes to reduce hardness, phosphorus, fluoride, radionuclides, arsenic, heavy metals, sulfate, and silica. In this process, formation of crystalline precipitate scours the membrane surface, reducing fouling. The solids concentration can be as high as 50 000 mg/L or 5%.

One application of this process is the treatment of reverse osmosis or nanofiltration brine reject to allow subsequent processing of the treated brine through reverse osmosis or nanofiltration to increase the overall facility recovery. It is also possible to treat the brine from reverse osmosis or nanofiltration with integrated precipitative softening and ultrafiltration and recover the filtrate without subsequent reverse osmosis or nanofiltration treatment to achieve high recovery. An example of this is the City of Palm Coast nanofiltration brine recovery using integrated chemical precipitative to achieve 98.5% overall facility recovery (Locke et al., 2015). The dirty backwash from the ultrafiltration is sent back to the inlet of the facility. The only water loss is the water in the dewatered sludge. The advantage of this process is significant reduction of facility footprint compared to conventional clarification. This process is described in a case study in the *Manual of Water Supply Practices—Inland Desalination and Concentrate*

Management (AWWA, 2018). Non-municipal applications of integrated chemical precipitative-ultrafiltration membrane (Aguinaldo, 2009) are treatment of cooling tower and boiler blowdown (Lander & Chan, 2012), flue gas desulfurization wastewater, mining wastewater, oil refinery wastewater, and produced water.

4.6.4 Polyelectrolyte-Enhanced Ultrafiltration and Micellar-Enhanced Ultrafiltration

Other current applications of ultrafiltration include polyelectrolyte-enhanced ultrafiltration (PEUF) and micellar-enhanced ultrafiltration (MEUF). Polyelectrolyte-enhanced ultrafiltration is applied when a water stream contains low molecular weight inorganic solute or metal; using a polyelectrolyte, the metal ions can be bound to the charge locations of the polymer. Metal ion-bound polyelectrolytes, being larger in size, are rejected by the ultrafiltration membrane. In MEUF, a surfactant is selected to generate surfactant micelles that have a hydrophobic core and positive or negative hydrophilic outer surface. Trace amounts of organic solutes can be solubilized within the hydrophobic core of surfactant micelles and oppositely charged ionic contaminants are attached on the outer surface of the micelles by electrostatic attraction.

Both PEUF and MEUF are applicable in heavy metals treatment removal. Microfiltration/ultrafiltration, coupled with coagulation and flocculation, includes membrane lime-soda softening (sometimes called "precipitative softening") and as an alternative to conventional clarification.

4.6.5 Polishing Microfiltration/Ultrafiltration

Microfiltration/ultrafiltration can be used as a final treatment process or as a polishing stage. Production of ultrapure water from wastewater may require a polishing filter to remove particulates, TOC, and organic matter. Ultrafiltration membranes can offer physical barriers to some microorganisms and viruses when conventional disinfection processes cannot be used.

4.7 Cost of Membrane Filtration Facilities

Figures 3.37, 3.38, and 3.39 provide generalized Class 5 cost curves for membrane equipment, membrane filtration facilities, and O&M. The membrane equipment is often intended to be incorporated to a treatment facility. The membrane equipment (Figure 3.37) includes the screens, pumps, valves, variable frequency drives, strainers, air compressor (for pneumatic air supply), backwashing facilities, and chemical dosing facilities. The membrane filtration facilities' costs (Figure 3.38) include construction and installation. The O&M costs (Figure 3.38) include energy, chemical cleaning, and membrane replacement. The cost curves were based on the costs described in

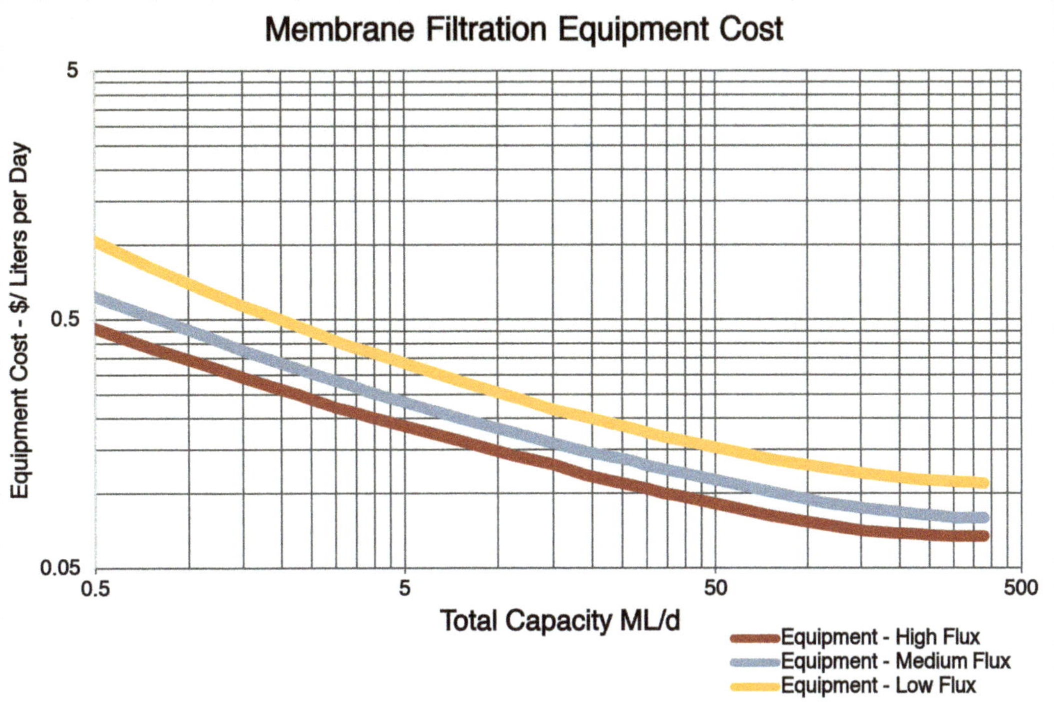

FIGURE 3.37 Membrane filtration equipment cost. *Source:* J.T. Aguinaldo

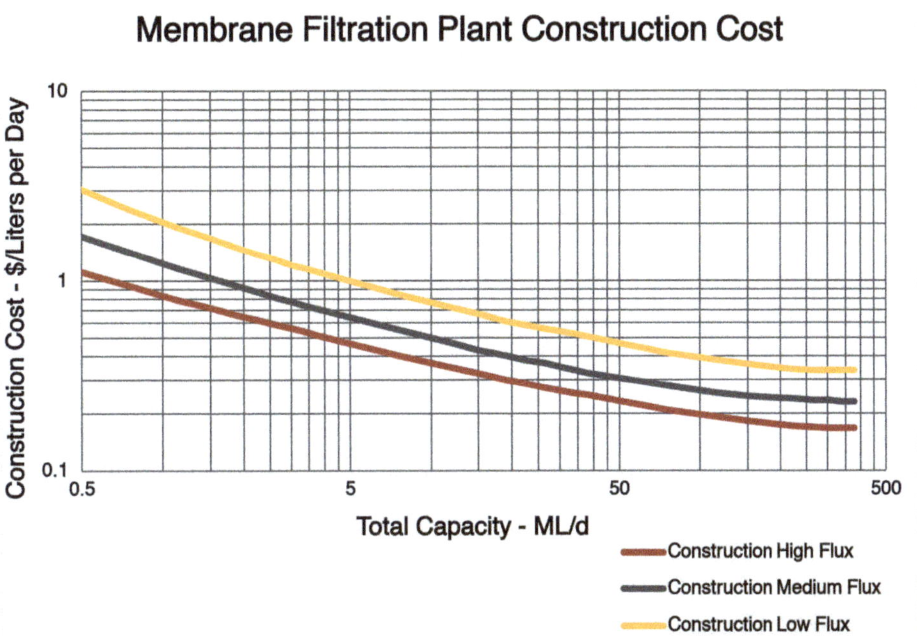

FIGURE 3.38 Membrane filtration facility cost. *Source:* J.T. Aguinaldo

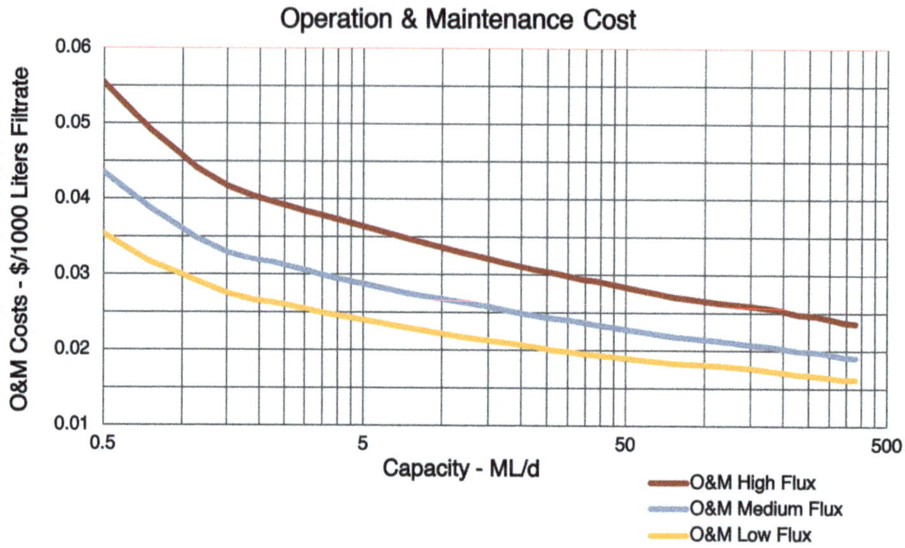

FIGURE 3.39 Membrane filtration facility operations and maintenance cost. *Source:* J.T. Aguinaldo

Manual of Water Supply Practices—Microfiltration and Ultrafiltration Membrane for Drinking Water (AWWA, 2005), updated and adjusted to 2018 cost using the Engineering News Record (ENR) indices and verification from several membrane systems integrators and membrane facility construction companies. It is important to note that the cost data provided in the above-mentioned AWWA reference were developed for drinking water facilities, however the cost for comparable capacity in industrial plants operating in the same design parameters will be similar. Note that these costs indications are for the membrane equipment and membrane facilities, not total project costs. Of course, project costs in municipal and industrial setting will be different. In the context of this book, an industrial recycling facility will be receiving water that tends to be higher in total dissolved solids (TDS) and/or COD and/or TSS than feed to a drinking water facility, sometimes by a sizable margin. As a result, membranes used for industrial recycle will tend to have lower flux rates than would be expected for a drinking water application. For this reason, it may be most appropriate to associate typical industrial recycle membrane systems with the "low flux" curves found in Figures 3.37, 3.38, and 3.39.

Published bid prices from recent membrane filtration projects were also compared. The membrane equipment, membrane filtration facilities, and O&M costs are based on various operating fluxes: low, <60 LMH (35.4 gpd/sq ft); medium, 60 to 80 LMH (35.4 to 47.3 gpd/sq ft); and high, 100 to 150 LMH (59 to 88.5 gpd/sq ft).

5.0 ADSORPTION AND ION EXCHANGE

5.1 Introduction

Adsorption and ion exchange share many common features in terms of application for separation, as such they are frequently grouped together as "sorption." Both these processes involve transfer of one or more solutes from a fluid phase (gas or liquid) to a solid phase. The most general definition for these two technologies is the enrichment of chemical species from a fluid phase onto the surface of a solid. For either of these technologies to work efficiently, some important criteria to keep in mind when designing and developing contaminant removal systems include

- High capacity for contaminant of concern
- High selectivity for contaminant of concern
- Favorable mass transfer kinetics
- Broad operational pH range
- Economically viable
- Feasible operation

5.2 Adsorption

Adsorption, as opposed to absorption, is essentially a surface phenomenon where ions, atoms, or molecules from a liquid or a gas phase adhere to the surface of a solid; in contrast, "absorption" can be defined as the complete transfer of contaminants from one phase into another phase; dissolution of gases in liquids is an example of absorption. Adsorption techniques are widely used in water treatment to remove contaminants from aqueous phase. For instance, adsorbents are used to remove organic contaminants in the field of water purification. The contaminant adsorbed is called the "adsorbate" or "solute" and the adsorbing phase is the "adsorbent."

During adsorption, the solid adsorbent will eventually become saturated or nearly saturated with the adsorbate. Because adsorption is a surface process, surface area of the adsorbent is a vital parameter when selecting an appropriate adsorbent. Changing the properties of the system can cause the adsorbates to be transferred back into the fluid phase; this reverse process is known as "desorption."

5.2.1 Types of Adsorption

Forces of attraction exist between the adsorbate and adsorbent. These attractions can be weak or electrostatic in nature (van der Waal forces) or caused by chemical interactions between adsorbent and adsorbate, which are stronger attractive forces. Hence, adsorption can be broadly classified into two types: physisorption and chemisorption.

5.2.1.1 Physisorption

Physisorption (or "physical adsorption") is the type of adsorption governed by weak electrostatic or intermolecular forces (van der Waals forces) and does not involve a significant change in the electronic characteristics of the species involved. An equilibrium is typically established between the adsorbent and the fluid phase under appropriate conditions of pressure and temperature. Even though physisorption is a weaker attractive phenomenon (low enthalpy of adsorption), it plays an important role in nature and in the field of water treatment. The adsorption of gases (e.g., hydrogen, nitrogen, etc.) on the surface of adsorbents, like activated carbon at ambient conditions, is an example of physisorption (Figure 3.40).

The following are some characteristics of physisorption:

- Activation energy is not needed or minimal.
- It is an exothermic process, in general, and is reversible.
- The change in electronic states of the adsorbent and adsorbate is minimal.

5.2.1.2 Chemisorption

Chemisorption (or chemical adsorption) is a type of adsorption that is characterized by stronger attractive forces (higher enthalpy of adsorption); the forces involved are valence forces of the same kind as those operating in the formation of chemical compounds (Figure 3.40). The distinction between chemisorption and physisorption is similar to that of distinguishing between chemical and physical interaction in general.

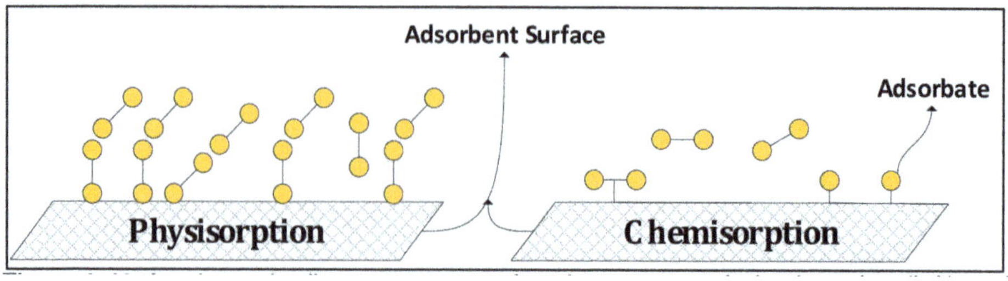

FIGURE 3.40 A schematic diagram representing the concept of physisorption (left) and chemisorption (right).

The following are some of the characteristics of chemisorption:

- It is typically an irreversible phenomenon and is an exothermic process.
- It often requires activation energy.
- It typically forms a monolayer of adsorbate on the active sites of the adsorbent.
- The changes in the electronic states of both the adsorbate and adsorbent can be detected using appropriate analytical means.
- Because of its nature, chemisorption can be specific with respect to the adsorbate.

5.2.2 Types of Adsorbents

Adsorbents are typically categorized by their physical characteristics such as base material, porosity, and particle size. Most adsorbents are human-made (e.g., activated carbon), but are also readily found in nature (e.g., zeolites). Because the adsorption forces are dependent on the surface of the adsorbent material and the nature of the adsorbate, different substances are adsorbed with varying affinities. The type of adsorbent is sometimes also classified by their mode of operation—equilibrium controlled and kinetically controlled (De Hann & Bosch, 2013).

More than one type of adsorbent can be suitable for a given application using either equilibrium or a kinetic removal scheme. Table 3.10 presents an overview of some common types of adsorbents and their applications (it is important to note that this is not a comprehensive list of adsorbent types and their applications).

The following are some important characteristics to keep in mind when choosing the appropriate adsorbents:

- Particle size
- Porosity and pore size distribution
- Compatibility with contaminants of concern or removal conditions (pH, temperature, etc.)
- Surface morphology

Pore size and distribution are some of the most important factors for adsorbent performance and are most likely distributed throughout the sorbent. Pore sizes are typically classified into three ranges: macroporous (>50 nm), mesoporous (2 to 50 nm), and microporous (<2 nm). Many adsorbent materials, such as activated carbons, silica gels, and alumina, are typically amorphous, comprising a complex network of interconnected pores.

TABLE 3.10 Example of commercially used adsorbents and their possible applications.

Adsorbent Type	Applications
Activated Carbon	Gases from air; removal or odors; dissolved organics from water
Alumina	Drying of gases and organic solvents; fluorine removal from chemical processes
Organoclay	Organics removal from water; suspended matter removal; treatment of oil and grease
Polymeric Adsorbent Resins	Removal of colors, organics as part of water purification; separation of aromatics from aliphatics; protein purification
Functionalized Silica Gel	Removal of moisture from packaging; gas drying; removal of water from organic solvents; selective removal of contaminants from a multi contaminant stream
Zeolites	Removal of water; drying of gases; removal of oxygen, inert gases, CO_2; hydrogen purification; general pollution and moisture control; various organics separation

5.2.3 Characterization Techniques

Adsorbent characterization is a crucial aspect when designing contaminant removal technologies. Many techniques for adsorbent characterization exist, but no standard set of analytical methods is used for all types of adsorbents. "Particle characterization" is a broad term that comprises identification of adsorbent particles by size, particle shape, surface properties such as charge or zeta potential, pore size, pore distribution, apparent density, abrasion number, pH_{PZC}, and mechanical integrity. Although there are a wide range of commercially available characterization techniques, each has its advantages and disadvantages depending on the type of adsorbent being characterized. The following subsections contain commonly used techniques for particle characterization (note that this is not a comprehensive list of characterization techniques).

5.2.3.1 Dynamic Light Scattering

Dynamic light scattering (DLS) is a commonly used technique and non-invasive, allowing complete sample recovery. It is typically used to measure particles with a broad range of size (nanometers to millimeters), and the sample quantity requirement is typically low. The analysis can be either carried out in a suspension or as is with good repeatability. Dynamic light scattering is also used for measuring particle size distribution in water/wastewater treatment applications.

5.2.3.2 Laser Diffraction Analysis

Laser diffraction analysis is a widely used technique for analyzing particle size of materials that are in the range of microns up to several millimeters in size. The measurements are fast and provide good repeatability with high throughput. This technique is similar to DLS and some of its applications are similar to DLS in the water treatment industry.

5.2.3.3 Particle Imaging

Particle imaging measures particle size using digital imaging, including particle shape and size distribution and morphology and, in some instances, particle color. This technique is typically good for particles with higher average size (millimeters) and can also be used to measure suspended solids in wastewater.

5.2.3.4 Sieve Analysis or Gradation Test

Sieve analysis is a crude technique for characterization of particle size distribution that is typically used for granular material by allowing the material to pass through a series of sieves with decreasing particle or mesh size. This technique is not suitable for materials that are prone to shearing or abrasion.

5.2.3.5 Porosimetry

Porosimetry is the most widely used analytical technique for the quantitative determination of various parameters of a given material's pore structure (viz., pore volume, surface area, average pore diameter, absolute/bulk densities). In general, this technique involves the intrusion of a non-wetting fluid, typically in vacuum into the material under investigation (Figure 3.41).

The two most common porosimetry techniques used are mercury porosimetry and gas adsorption-desorption (typically nitrogen or argon); both these techniques are used in conjunction to characterize void structure of porous materials (Conner et al., 1986). The pore size distribution from the gas adsorption method is measured at the boiling temperature of the gas (e.g., 77K for N_2, 87K for Ar), and pore sizes up to submicron ranges can be quantitatively determined. Because mercury porosimetry uses pressure intrusion for pore size determination, it is possible to detect the pore sizes over a broad range (nanometers to millimeters) (Adamson & Gast, 1997). Because higher pressures (up to 20 000 psi) may be required for pore sizes

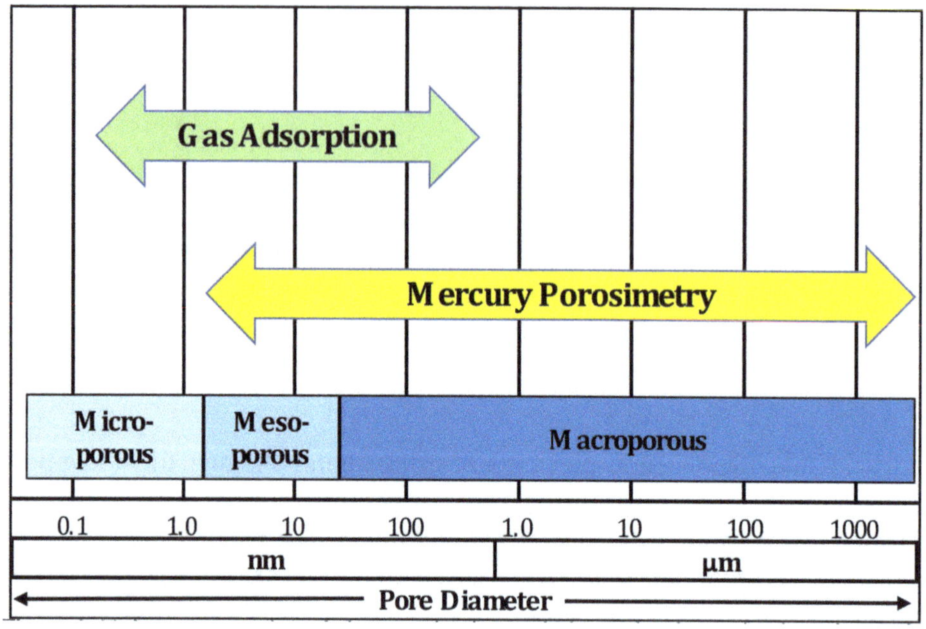

FIGURE 3.41 A representation of characterization techniques for determination of porosity of adsorbents (note: gas permeation is mostly used for membrane characterization; hence, it is not depicted here).

below 10 nm, it is imperative to ensure that the material under test can withstand such high pressures.

Another common way to measure pore size is via gas permeation; typically, this technique is used to measure pore sizes and distribution of membranes or polymers at high pressures and temperatures. This technique is optimal for measuring pore diameters in the range of 1 nm up to 100 nm. Gas permeation can also be used for fractionating a mixture of gases using polymeric membranes having selective permeability to a gas.

Various other characterization techniques exist to investigate and understand special adsorbents such as zeolites and functionalized silica gel because of their ordered pore structures and geometry (Jacobs et al., 2001). The following is a compilation of characterization techniques used for measurement of specific adsorbent parameters (this is not a comprehensive list):

- **X-ray diffraction:** This is an effective nondestructive technique for characterization of crystalline materials; it can provide information on material structure, chemical composition, average grain size, and other structural parameters such as crystal defects.

- **X-ray photoelectron spectroscopy (XPS):** As a surface sensitive technique, XPS, is used to measure elemental composition at extremely low levels. Because of its sensitivity, this technique can be used to probe any surface interactions, electronic state of active sites, and possible defects.
- **X-ray fluorescence spectroscopy (XFS):** Another nondestructive technique that uses x-rays or gamma rays to excite material surface and obtain characteristic secondary x-ray emission from a material surface is XFS. These secondary x-rays are used to characterize materials for chemical or elemental analysis; they are also used for a wide range of materials, from glasses and ceramics to metals and paintings.
- **Electron microscopy:** "Electron microscopy" is a broad term for imaging material surfaces; many types of electron microscopy are commercially available, and their uses vary by the type of sample and the information required. Some common types of electron microscopy techniques are transmission electron microscopy, scanning electron microscopy, reflection electron microscope, and scanning transmission electron microscopy. This technique provides information on general material structure, pore characteristics (size, distribution, etc.), and, in some instances, chemical composition of the material.
- **Nuclear magnetic resonance:** Not commonly used to characterize adsorbents, this technique can provide information on the surface functionality of the material and the chemical composition, in certain cases.

In addition to the analytical techniques mentioned in subsection 5.2.3, there are a few more that are used to characterize structure, functional groups, chemical composition, and pore characteristics. These are thermogravimetric analysis, differential scanning calorimetry, scanning tunneling microscopy, atomic force microscopy, vibrational spectroscopy, and adsorption isotherms.

5.2.4 Characteristics and Performance

The performance of adsorbent materials can be evaluated either using simulations (calculated) or by experimental means, although each methodology has its strengths and challenges. As mentioned earlier in this chapter, an ideal adsorbent is desired to have high selectivity, high capacity, broad operational range, and low energy of adsorption. A common technique used for determining adsorbent performance are adsorption isotherms, defined as a relation between the amount of adsorbate bound on the surface of the adsorbent and constant pressure and temperature (Adamson & Gast, 1997; Conner et al., 1986).

Some general factors that affect adsorption include temperature; because adsorption is an exothermic process, best results are obtained at lower temperatures. Higher pressure increases adsorption until saturation is reached; once saturation is reached, no additional adsorption is achieved, irrespective of the additional pressure applied. Different adsorption isotherms are studied throughout the scientific community, with two of them studied extensively: Langmuir adsorption isotherm and Freundlich adsorption isotherm. Both of these models have their strengths and challenges, depending on the situation. A third adsorption isotherm that is typically used to explain physisorption of gases onto a solid surface is the Brunauer–Emmett–Teller theory; however, this theory will not be discussed in this chapter.

5.2.4.1 Langmuir Adsorption Isotherm

The Langmuir adsorption isotherm model determines the relationship between the concentrations of surface adsorbed species to the number of active sites on that surface at equilibrium. This model takes into account three broad assumptions:

- All adsorbent sites are equivalent.
- Each active site can only bind with one adsorbate molecule/ion.
- There can be no interactions between adjacent adsorbate ions/molecules.

If an adsorption process follows this model, then it can be concluded that adsorption takes place as a monolayer on a homogeneous surface without other surface interactions between the adsorbed species. The simplified version of the Langmuir model can be rearranged as a linear relation between the concentration of adsorbed species and concentration of species in equilibrium, as follows (Kailasam et al., 2009):

$$C_e/Q_e = 1/Q_m K_{ads} + C_e/Q_m \tag{3.18}$$

where

C_e = the concentration of adsorbate ions/molecules in solution at equilibrium,
Q_e = the concentration of ions/molecules adsorbed onto the surface of adsorbent,
Q_m = the measure of the maximum capacity of the adsorbent, and
K_{ads} = the intensity of adsorption.

The concentration of ions/molecules adsorbed onto the surface of adsorbent (Q_m) can be calculated from the slope of the straight-line plot of C_e/Q_e versus C_e. The constant, K_{ads}, can be derived from the slope and the intercept of the same straight line ($1/Q_m K_{ads}$). If the plot is linear, then Q_m provides an estimate of the active sites on the adsorbent surface and K_{ads} provides an estimate of the driving force or the equilibrium constant for the process.

5.2.4.2 Freundlich Adsorption Isotherm

Herbert Freundlich is credited with providing the first isotherm model that describes the relationship between the concentration of solute on the adsorbent surface to the concentration of solute remaining in the continuous phase. Freundlich adsorption isotherm is commonly used to investigate adsorption of organics from water. As with the Langmuir isotherm model, the Freundlich isotherm model can be simplified into a linear relationship, as follows:

$$\log (x/m) = \log K + (\log C)/n \tag{3.19}$$

where

x = the mass of adsorbate ions/molecules in solution at equilibrium;
m = the mass of adsorbent;
C = the concentration of the adsorbate in solution at equilibrium (in the case of gaseous adsorbate, pressure [P] is used instead of concentration); and
K and n = constants for the adsorbate/adsorbent system at a given temperature.

A graph obtained by plotting "log(x/m)" with respect to "log C" (or) "log P" provides information on the isotherm model. A linear line indicates that the system under study obeys the Freundlich isotherm model; the slope and the intercept provide information on the model's dependence on adsorbate concentration or adsorbate pressure.

5.3 Ion Exchange

In the context of this book, ion exchange describes a water purification process using a specialized polymeric material, known as ion exchange resin. These resins are designed with functional groups that are quite effective at attracting target ions from water. The electric charge of these target ions give them an affinity for the functional groups on the resin. In this process, target ions from water (having a high affinity for the resin) are "exchanged" with (weakly held) ions at the resin functional groups, releasing the weakly held ion into solution in the process (Helfferich, 1962). For example, in water treatment, hardness-causing ions (like calcium) are replaced by non-hardness-causing ions (like sodium or potassium) to impede or stop scale formation (Equation 3.20). To have an advantage over other technologies, ion exchange reactions need to be reversible because this allows potential reuse of the ion exchanger media.

$$Ca^{2+}_{(aq)} + 2\ NaZ_{IX} \qquad CaZ_{IX} + 2\ Na^{+}_{(aq)} \tag{3.20}$$

where, Z_{IX} represents a site on the ion exchange material surface.

Ion exchange resembles sorption; however, the main difference between ion exchange and sorption is the stoichiometric nature of ion exchange (Wheaton & Lefevre, 2000). Sorption removes species from solution but does not replace that species by donating to the solution. In contrast, ion exchange must satisfy balance of charge by the stoichiometric transfer of a species back into solution (Figure 3.42).

5.3.1 Ion Exchange Resin Types

Polystyrene crosslinked with divinyl benzene (DVB) is the most widely used ion exchange matrix, and the degree of DVB crosslinking depends on the desired stability of the resin/exchanger. Although other matrices have also

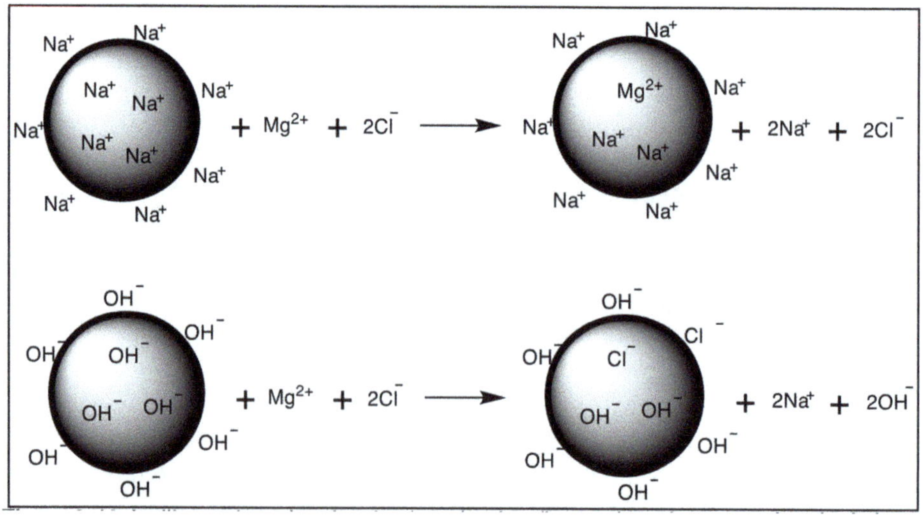

FIGURE 3.42 An illustration of cation (top) and anion (bottom) exchange resins depicting surface ion exchange while maintaining charge balance.

been used (e.g., alumina, coal, membranes, acrylamides, methacrylates, phenol-formaldehyde, silica, zeolite, and liquid ion exchangers), depending on the compatibility with the waste stream, they are much less common (Helfferich, 1962). Nevertheless, the surface active functional groups are the active moiety involved in ion exchange and, depending on the specific ion being exchanged, the type of functional groups can be varied in two categories: cation exchangers (acidic) and anion exchangers (basic) (Figure 3.42). These two can be further classified into the following four types: strongly acidic, weakly acidic, strongly basic, and weakly basic moieties (Helfferich, 1962; Wheaton & Lefevre, 2000). A brief description of each type is presented in the following subsections.

5.3.1.1 Strong Acid Cation Exchanger

Ion exchangers that use sulfonic acid functionality are typically classified as a strong acid cation (SAC) exchanger. When used for raw water treatment, SAC exchangers remove nearly all raw water cations and replace them with hydrogen or sodium ions. Strong acid cation exchangers have a wide operational pH range. Strong acid cation resins made of high-quality polystyrene-DVB typically have a long life expectancy as long as they are not regularly exposed to chlorine, huge thermal swings, or foulants that can hinder efficient mass transfer kinetics. The affinity of sulphonic acid resins for cations varies with the ionic size and charge of the cation. The general affinity toward cation increases with the increasing cation charge and increasing atomic number. For SAC-type exchangers, an example for the affinity series is as follows (Hubicki & Kolodynska, 2012):

$$Th^{4+} > Al^{3+} > Ca^{2+} > Na^{+} \tag{a}$$

$$Tl^{+} > Ag^{+} > Cs^{+} > Rb^{+} > K^{+} > NH_4{}^{+} > Na^{+} > Li^{+} \tag{b}$$

$$Ba^{2+} > Sr^{2+} > Ca^{2+} > Mg^{2+} \tag{c}$$

$$Fe^{3+} > Al^{3+} \tag{d}$$

5.3.1.2 Weak Acid Cation Exchanger

Ion exchangers that use carboxylic acid functionality are typically classified as weak acid cation (WAC) exchangers. Weak acid cation exchangers are typically in operation to remove cations that are associated with alkalinity. Unlike SAC exchangers, WAC exchangers do not remove all the cations from water systems. Their high affinity toward hydrogen ion is exploited for high-efficiency regeneration of the exchanger. The general affinity series of this type of cation exchanger is as follows (Hubicki & Kolodynska, 2012):

$$H^+ > Mg^{2+} > Ca^{2+} > Sr^{2+} > Ba^{2+} > Li^+ > Na^+ > K^+ > Rb^+ > Cs^+$$

Both the cation exchangers (SAC and WAC) previously discussed are commercially available with sodium or hydrogen as the counter ion (exchangeable ion). The sodium ion form is typically used to remove hardness and to soften the water and the hydrogen form is typically used to remove alkalinity. Within industrial facilities, it is most common to see the softening application, which uses a concentrated NaCl stream to regenerate the resin. In water treatment, WAC exchangers are frequently used in conjunction with SAC exchangers to maximize efficiency.

5.3.1.3 Strong Base Anion Exchanger

Ion exchangers with quaternary amine functionalities are known as strong base anion (SBA) exchangers. There are two types of SBA exchangers—Type I and Type II. Type I exchangers are ones with the active site completely methylated. In Type II exchangers, one of the methyl groups is replaced with an ethanol group (Figure 3.43). Type I exchangers are typically used to remove any weakly ionized acids. Type II exchangers work similarly to Type I ion exchangers and also provide greater regeneration efficiency than Type I exchangers.

When in its hydroxide form (as counter ion), SBA exchangers remove all common anions (sulfate, chloride, carbonate, silicate, etc.) from aqueous streams. Regeneration using a strong alkali (typically sodium hydroxide or ammonium hydroxide) is most common because it returns the hydroxide ion back to the exchanger. Similar to the cation exchangers, the affinity for the exchanger is affected by the target anion. The general affinity series for this type of anion exchanger is as follows (Hubicki & Kolodynska, 2012):

$$PO_4^{3-} > AsO_4^{3-} > ClO_4^- > SCN^- > I^- > S_2O_3^{2-} > CrO_4^{2-} > SO_4^{2-} > SO_3^{2-} > HPO_4^{2-} >> NO_3^- > Cl^- > HCO_3^- > OH^-$$

FIGURE 3.43 Strong base anion exchangers (Type I [left] and Type II [right]).

5.3.1.4 Weak Base Anion Exchanger

Ion exchangers that do not have a quaternary ammonium functionality are known as a weak base anion (WBA) exchanger. The weak base functionality can be primary, secondary, or tertiary amine groups. Weak base anion exchangers are predominantly used to remove sulfates, chlorides, and nitrates. Both the anion exchangers (SBA and WBA) discussed here are commercially available with hydroxide or chloride as the counter ion (exchangeable ion) and are typically used in conjunction with cationic exchangers to act as scavengers. The hydroxide form is typically used to remove all mineral acids produced by cationic exchangers, and the chloride form is typically good for alkalinity reduction/removal (Flynn, 2009).

5.3.2 Characterization of Resins

Some of the analytical techniques discussed earlier in this chapter (5.2.3) can also be used to characterize ion exchange resins, although not all of them are required because ion exchange resins are typically not as porous as the adsorbents previously discussed. Some of the parameters that are investigated using analytical techniques for ion exchange resins are physical structure of the resin, chemical composition of the resin matrix, and resin bead size and distribution.

The general physical nature of resins is twofold: macroporous and gels. Macroporous resins are more rigid for higher strength and to withstand higher pressures. Gels are more common because they are homogeneous polymer with varying degrees of crosslinking (8% up to 20%) within its polymeric structure (Helfferich, 1962; Wheaton & Lefevre, 2000). As to the chemical structure of the resin matrix, two of the most common polymer backbones used are acrylates and styrene, with DVB as a common crosslinking agent. In most industrial water treatment applications, ion exchange resin particle size ranges between 15 to 60 mesh (1.25 to 0.25 mm), with varying uniformity coefficient depending on whether the chosen resin is a standard resin or uniform particle size resin. Some niche applications use specialized resin sizes but are not common for generic applications because of the specialized treatment and care required (Flynn, 2009; Kailasam et al., 2009; Zhao & SenGupta, 2000).

5.3.3 Characteristics and Performance

Performance of ion exchange resins is typically determined using experimental techniques; the most common parameter used is resin capacity, which is expressed in mass per unit volume (grams per liter [g/L] or equivalents per liter [eq/L]). Detailed analysis of the water before and after treatment can provide information on the capacity of a specific contaminant of interest or as a total ionic load of the resin (Flynn, 2009; Wheaton & Lefevre, 2000). For example, to determine capacity of cation removal, all the cations in the feed water need to be converted and expressed as their calcium carbonate equivalents (g/L) before calculating total cation load or capacity. The removal capacity of a given resin is typically based on the type of resin and the design of the system, and capacity also varies based on a number of factors (co-contaminants or fouling, resin degradation, empty bed contact time, regeneration scheme, etc.).

5.4 Design of Equipment

Design of equipment for adsorption or ion exchange requires some basic knowledge of removal type—equilibrium based or kinetic based (De Hann & Bosch, 2013). Most removal design for water- or wastewater-related treatment is based on equilibrium selective materials. A lot of factors need to be considered when designing equipment for adsorption or ion exchange because this is not a "one-size-fits-all" kind of solution that can be used for all applications. One of the most important factors in designing ion exchange/adsorbent systems is to know the capability and the capacity of the material and the metallurgy in use and its limitations (such as pH, temperature, pressure, chemical compatibility, etc.). A treatability study is recommended to investigate and evaluate potential pitfalls or challenges associated with scaling up to a commercial application.

5.4.1 Configurations

There are many configurations of contaminant removal systems used commercially that involve sorbents in general (similar setup for adsorption or ion exchange). Some of the common configuration types are fixed bed and fluidized bed design; although both have their advantages and disadvantages, the most common type used in the field of water treatment are fixed bed design systems. Depending on the specific requirements or applications of a treatment train, ion exchange systems may include a single resin or multiple resins with many variations (Flynn, 2009).

Treatment systems comprising multiple removal columns of beds can be arranged either in series or parallel, depending on the need. One of the most common bed configurations is the "lead-lag" sequence, where there are two or more beds in series, with the first bed known as the "lead bed" and the consecutive beds known as "lag beds." During operation, once the lead bed gets saturated or exhausted, it is taken off service for regeneration or replacement, while the lag bed becomes the lead bed and the newly regenerated or replenished bed becomes the lag bed. Another common configuration for routine contaminant extraction is the carousel system, where continuous contaminant removal and recovery are achieved using multiple beds on a rotating carousel. This type of system is typically used for special applications where the influent or feed water chemistry is relatively consistent (Flynn, 2009; De Hann & Bosch, 2013).

5.4.2 Regeneration

Regeneration is typically only used onsite at an industrial facility for ion exchange systems, but has also been used for adsorption/desorption cycles in some selected applications; this section will discuss regeneration in ion exchange systems. As it is clear from the previous section, ion exchange is a semibatch process and, once the resins are saturated with contaminants of concern, the bed needs to be taken offline for regeneration of the resins. The regeneration process typically reverses the ion exchange process that occurs in service to recharge the resin back to the desired form. The regeneration cycle or step may be one of two following types: cocurrent (or coflow) or counter-current (or reverse flow) (De Hann & Bosch, 2013; Flynn, 2009).

In cocurrent regeneration, the regenerant solution takes the same path as the feed solution, typically from the top to the bottom of the ion exchange bed (downflow). This is typically used when quick regeneration is needed in conjunction with lower flowrates; however, it is not recommended because this may not fully regenerate the ion exchange bed. Conversely, in counter-current regeneration, the regenerant solution flows in the opposite path as the feed solution, with the regenerant solution coming in contact with the less-contaminated section of the resin bed first, making the regeneration process more efficient (Flynn, 2009). This mode typically requires less regenerant and only works efficiently if the resin bed stays in place and does not get fluidized in the regeneration process. Regardless of the regeneration mode used, the regeneration step is typically packed in between backwash and rinse cycles to ensure that any ions trapped within the void volume are removed and do not interfere with the subsequent step in the process (Flynn, 2009; Kailasam et al., 2009).

5.5 Applications and Challenges

Adsorption and ion exchange are widely used technologies having a presence in just about every industrial sector, including pharmaceutical, hydrometallurgy, petrochemical, and food and beverage industries, to name a few. At an industrial facility, ion exchange and sorbents are most frequently used for the treatment/purification of water or wastewater. The applications can either be general removal or for removal of specific contaminants.

5.5.1 General

Some general applications of adsorbents and ion exchange systems include removal of dissolved organic species from contaminated streams (adsorbents) and demineralization to obtain ultrapure water for boiler operations (resins); a general biochemistry application of ion exchange is separation of proteins. Some of the challenges with general applications of ion exchange are the reduced capacity of the resin because all the contaminants are co-removed without specificity to any one of the constituents and fouling of the resin bed. Some of the challenges with general applications of sorbents can be that each type of organic will have a different affinity for the sorbent, and thus each will be removed with different efficacy, and weakly adsorbed organics can be desorbed for the sorbent by more strongly adsorbed organics.

5.5.2 Contaminant Specific

Ion exchange and adsorption have been used for many applications that are specific to a particular industry or problem. For resins, most of the contaminant-specific applications are in the field of separation and recovery of target metals using surface functionalized materials; hence, they are widely used in hydrometallurgy and rare-earth metal extractions (Zhao & SenGupta, 2000). Chelating groups are attached to the surface of ion exchange resins for attaining specificity toward a targeted contaminant; for example, phosphonate-modified chelating material is used to remove anions of arsenic in the presence of other common interfering species such as sulfate, chloride, and so on (Kailasam et al., 2009; Richardson et al., 2016). Some of the challenges of contaminant-specific applications for resins are their narrow operational range with respect to pH or concentration of contaminants in the influent matrix.

For adsorbents, examples of contaminant-specific applications for water purification pertain at an industrial facility include dechlorination of water before reverse osmosis treatment, or removal of a specific contaminant that may be present in a given source water. Adsorbents will sometimes be used within industrial WRRFs as well, and a common application is for removing specific chemical(s) from the wastewater that are associated with upstream manufacturing operations, with the aim to prevent release of the compound(s) to the environment.

5.6 Future Research Direction

Adsorption and ion exchange materials are primarily used in the field of water and wastewater treatment. Increases in the purification requirements for municipal and industrial water use necessitate advancements in the efficiency of this technology. To this end, research is ongoing to optimize ion exchange technologies by developing contaminant-specific materials or media (Kailasam et al., 2009; Moyer et al., 2012; Zhao, 2000). Some of the future directions for contaminant removal using sorbent-based technologies are modification of pore structure and surface chemistry to enhance adsorption of specific key contaminants such as perfluoroalkyl substances, chelating ion exchangers, molecular imprinted polymers, ion recognition, and molecular organic frameworks (Liu et al., 2015). These advanced techniques typically use receptors that specifically target contaminants of interest and immobilize them onto the sorbent. The specificity and removal capacity of most of these approaches depend on a variety of factors such as pH, temperature, other interfering species in the matrix, and so on. Most of these advanced technologies are at the research stage, but some large-scale niche applications exist in the mining sector (Alexandratos, 2009; Kailasam et al., 2009).

6.0 REVERSE OSMOSIS AND NANOFILTRATION FUNDAMENTALS AND USE IN MINIMAL LIQUID DISCHARGE PROCESSES

6.1 Introduction to Reverse Osmosis and Nanofiltration Membrane Treatment Systems

Reverse osmosis and nanofiltration membranes are semipermeable polymeric materials. They are among the top separation technologies used for the removal of salts and/or dissolved organic contaminants from water (Gude, 2016). Advances in semipermeable membrane performance over the last 30-plus years have resulted in improved energy efficiency and preference of membrane technology over traditional thermal treatment (Bennett, 2013; Ghaffour et al., 2013; Glater, 1998; International Atomic Energy Agency, 2000; Lonsdale, 1982). To understand the role of semipermeable membranes in producing purified water, it is important to be familiar with the following terms:

- Permeate—the purified product water produced by a membrane system;
- Recovery—the percentage of incoming feed water that emerges from a membrane system as permeate.
- Rejection—the percentage of incoming solute or contaminant concentration removed from feed water by the membrane.
- Passage—The opposite of rejection, passage is the percentage of incoming solutes or contaminant concentration allowed to pass through the membrane into the permeate.

- "Flow"—the rate of feed water introduced to the membrane element or membrane system. Concentrate flow, also known as "reject," is the flowrate of nonpermeated feed water that exits the membrane element or membrane system. This concentrate stream contains most of the dissolved constituents originally carried into the element or into the system from the feed source.
- "Flux"—the rate of permeate flow through a unit area of membrane, typically quoted by the industry in units of liters per square meter per hour or gallons per square foot per day.

In reverse osmosis, a high removal TDS is important, while in nanofiltration, specific solutes of interest are designed to be selectively removed over others. For example, many nanofiltration membranes exhibit low rejection for monovalent ions and high rejection for divalent ions and organic matter. An overview of the solute–water separation process using semipermeable reverse osmosis and nanofiltration membranes is provided in this chapter. The materials science of membranes as well as the practical application of the membrane assembled in a module and used in a system are also reviewed.

6.2 Semipermeable Thin-Film Composite Membranes

Reverse osmosis and nanofiltration membranes are most commonly thin-film composite (TFC) materials comprising a nonwoven polymeric fabric, a porous polysulfone support layer, and a polyamide barrier film (Cadotte & Petersen, 1981). The TFC membranes have nearly replaced asymmetric cellulose acetate reverse osmosis membranes and offer higher rejection and improved tolerance to a wide pH range for more effective membrane cleaning (Johnson & Busch, 2010). Each layer of the membrane serves a function. The nonwoven web provides support for the membrane architecture when used under applied pressure. It also provides a platform for casting the polysulfone solution and enabling diffusion-induced phase inversion of the polysulfone to achieve a highly controlled microporous structure. The porous polysulfone layer further serves as a support platform for the thin semipermeable polyamide film.

The polyamide barrier film in reverse osmosis membranes is formed by interfacial polymerization of, most commonly, metaphenylenediamine in an aqueous phase and trimesoylchloride in an oil phase (Baker, 2004; Cadotte, 1981). The aqueous phase is applied to the preformed porous polysulfone support and the oil phase is placed above the support, forming an oil-water interface at the polysulfone support surface. Reaction of the amine and acid chloride takes place at the interface and, as the polyamide forms, the reaction becomes self-limiting because of reduced transport of monomer through the growing polyamide film. Like reverse osmosis membranes, nanofiltration membranes are also interfacially polymerized polyamide thin-film composites, but use aliphatic diamine monomers such as piperazine in place of the aromatic metaphenylenediamine (Paul & Jons, 2016). Interfacial polymerization of the polyamide results in a 20- to 200-nm-thick film that is permeable to water, but rejects larger dissolved constituents such as salts. The characteristics of this reaction are carefully controlled to tailor to desired performance, setting one membrane product apart from another.

The process of osmosis is governed by thermodynamics, which provide a natural driving force to equalize the chemical potential between two phases. Across a semipermeable membrane, osmosis occurs by the migration of water from the side with lower osmotic pressure, or lower TDS, to the side with higher osmotic pressure, or higher TDS. For a single component solution, the osmotic pressure (Π) is the product of moles of ions per mole of solute (i), solute molar concentration (M), ideal gas constant (R), temperature (T), and osmotic coefficient (ϕ), as follows:

$$\Pi = iMRT\varphi \tag{3.21}$$

where φ represents the concentration dependent activity of the salt in solution, which allows this equation to be useful even at high concentrations (Robinson & Stokes, 2002). For multisolute solutions at varying temperatures, more complex approaches such as the Pitzer model can be used to determine activity and accurately calculate osmotic pressure (Pitzer, 1991; Wolfe et al., 2011).

A hydrostatic pressure difference applied across a membrane can be used to counteract the thermodynamic driving force of osmosis. Reverse osmosis occurs when this hydrostatic pressure is greater than the osmotic pressure difference, causing water to instead flow from a region of high TDS (the feed) to a region of low TDS (the permeate). The transport of water and solutes in reverse osmosis is generally well described by the solution-diffusion model (Rautenbach & Albrecht, 1998; Wijmans & Baker, 1995). According to this model, the volumetric flux of water through a membrane, J_w, is the product of the intrinsic water permeability of the membrane (*A-value*) and the net driving pressure (NDP) across the membrane interface, from feed (P_f) to permeate (P_p), as follows:

$$J_w = A \cdot NDP = A(\Delta P - \Delta\Pi) = A[(P_f - P_p) - (\Pi_f - \Pi_p)] \tag{3.22}$$

Water permeability is an intrinsic property of the membrane material, a result of the interfacial polymerization process and barrier film creation. Numerous commercial membranes are sold on the market that cover a wide range of permeabilities, which are typically quoted in units of LMH, kPa, or gpd/sq ft/psi. The NDP is the difference between the net applied and osmotic pressures on either side of the film, between the feed and permeate. In many cases of reverse osmosis, the hydrostatic and osmotic pressures of the permeate are negligible, and NDP becomes the difference between the applied and osmotic pressures in the feed stream.

Solutes are rejected by semipermeable membranes based on their size and ability to sorb into, diffuse through, and desorb out of the membrane. Features are designed into the membrane barrier film, such as charge and crosslinking, to tailor the selectivity. In general, reverse osmosis and nanofiltration membranes reject large ionic solutes more than small neutral ones. Solute passage in the solution-diffusion model is described by the following equation:

$$j_s = B(C_{s,F} - C_{s,P}) \tag{3.23}$$

The mass flux of dissolved solute, j_s, is the product of the salt permeability (*B-value*) and concentration difference of solute (C_s) across the membrane. The *B-value* of a membrane is a function of both size and ionic state of the solute and is typically expressed in common units of flux such as liters per meter squared per hour or gallons per square foot per day (gpd/sq ft). Solute passage, *SP*, can be derived from Equations 3.22 and 3.23 as a simplified expression depending only on the *B-value* for the solute in question and the operating water flux, J_w, as follows:

$$SP = \frac{C_{s,P}}{C_{s,F}} = \frac{j_s}{j_w C_{s,F}} = \frac{B}{B + J_w} \tag{3.24}$$

Solute passage in reverse osmosis is reduced when using membranes with a low *B-value*, and/or when operating under high rates of water flux (i.e., high feed pressure). The *A*-value and *B*-value both increase with temperature, with the *B*-value typically increasing at a faster rate. Consequently, whether NDP or flux is maintained in a system, in practice one finds the solute passage of a membrane will increase with temperature.

In actual operation, considerations must be made to account for losses that occur in the form of pressure drop within the filtration module as well as concentration polarization of solutes against the membrane surface. Concentration polarization is a boundary layer effect that causes the local osmotic pressure on the feed side of the membrane to be greater than that of the bulk fluid. Higher operating fluxes and/or lower crossflow velocities in the feed channel both contribute to higher concentration polarization. Comprehensive summaries of these effects can be found elsewhere (Johnson & Busch, 2010; Shrivastava & Stevens, 2018).

The feed pressure required and quality of permeate water produced depend on the type of element chosen to treat the water. The desire is to have a membrane with a low salt permeability and high water permeability. Having the best of both these characteristics is typically not possible because the polymeric membrane structural features used to reduce salt permeability also reduce water permeability. Likewise, the polymeric features used to increase water permeability also tend to increase salt permeability (Geise et al., 2011; Stevens et al., 2017). Because of these inherent tradeoffs, membrane manufacturers offer a portfolio of products with varied flow/rejection combinations that can be chosen to treat feed water to the desired permeate quality at the lowest feasible feed pressure.

6.3 Spiral Wound Module

Water and solute permeability are key characteristics used to design a membrane for a specific application. However, the membrane alone is not useful without being packaged in a module that enables efficient flow of water over the membrane surface and channels to collect purified permeate and discharged concentrate streams.

Spiral wound modules have been adopted by the industry for this purpose. The TFC reverse osmosis/nanofiltration membrane sheets are assembled in spiral wound modules by winding membrane leaves, sometimes referred to as "envelopes," around a central permeate collection tube (Johnson & Busch, 2010). Membrane leaves are formed by sandwiching a permeate spacer between two membrane sheets, where the nonwoven support of the TFC composite faces the permeate spacer. The perimeter of this sandwich is sealed on three sides using glue while leaving one side open. The open side of the envelope is glued to a perforated permeate tube. Several of these envelopes are attached to the permeate tube to achieve the desired amount of active membrane area for the module. A feed spacer is placed between each leaf of membrane and, finally, the leaf and feed spacer layers are wound around the permeate tube to form a cylinder. Final processing steps to attach endcaps and protect the assembly in a fiberglass shell are performed to provide the finished product.

Spiral wound modules are operated in crossflow configuration, where the feed water enters through one scroll face of the module via the channel defined by the feed spacer and exits the other face. The spacer itself is a square-like patterned polyolefin net with defined thickness, typically ranging 0.71 to 0.86 mm. While defining the height of the feed channel, the spacer also adds turbulence, which serves to reduce concentration polarization at the membrane surface. As water passes over the active surface under pressure, a portion permeates through the membrane, and the rejected solutes become more concentrated as the remaining water continues down the feed channel until it exits out the opposite scroll end. From a simple mass balance of the feed, permeate, and reject streams, the following expression for solute concentration in the reject ($C_{s,R}$) can be derived. For a given recovery, r,

$$C_{s,R} = C_{s,F}\left(\frac{1}{1-r}\right) - \frac{1}{1-r}SP \tag{3.25}$$

where r is defined as the ratio of permeate flow, Q_{perm}, to feed flow, Q_{feed}, as follows:

$$\text{Recovery} = r = \frac{Q_{perm}}{Q_{feed}} \tag{3.26}$$

In cases of reverse osmosis, the last term is quite small, and reject concentration increases exponentially with recovery. This buildup of concentration, along with pressure drop losses, causes NDP and, correspondingly, the water flux, to drop down the length of a module. The water that permeates the membrane eventually leaves the module through the central permeate tube. In systems, these modules are connected in series, where the concentrate from one module becomes the feed for the next and the permeate tubes are all interconnected into one stream.

Spiral TFC modules used by industry are typically 1.02 m (40 in.) in length, but come in different common diameters, including 6.4 cm (2.5 in.), 10.2 cm (4 in.), 20.3 cm (8 in.), and 40.6 cm (16 in.). The most common diameter seen in wastewater treatment systems is 20.3 cm; however, in minimal liquid discharge (MLD) systems, the volume of the stream can become reduced to the point at which smaller active areas, and hence smaller diameters, are needed. Because modules are constructed primarily with polymeric materials, there are operating pressure and temperature limits for their use. Typically, the pressure limits are less than 83 bar (1200 psi) for seawater-type membranes and less than 41 bar (600 psi) for brackish-water-type membranes. Temperature is typically limited to less than 45 °C (113 °F) but can vary depending on the choice of plastics used in the module construction. There are some specially designed modules that enable operation beyond these limits.

6.4 Reverse Osmosis System Design

Membrane modules are operated within pressure vessels. Each pressure vessel has a feed, reject, and permeate port and can host from one to eight modules per vessel. Wastewater reuse systems typically use no more than six to seven elements per pressure vessel to limit the amount of pressure drop and provide better flux balance. The vessels are configured in a system design that carefully considers the objectives of the WRRF, including feed water source, volume, composition and temperature, the desired amount and purity of the treated water, and the desired volume and concentration of the reject stream. These factors can be influenced by the effectiveness of the pretreatment, the type of reuse application, and the degree of local water scarcity and/or strictness of discharge regulations.

A reverse osmosis or nanofiltration system is divided into "stages" and "passes." The feed/reject stream connects one stage to the next, while the permeate stream connects one pass to the next. Sequential stages are added to increase recovery and are constructed by consolidating the rejected water from a previous stage upstream into a smaller number of vessels. A booster pump may be placed between stages to provide sufficient NDP downstream to account for pressure drop and the increasing osmotic strength of the feed/reject. Alternatively, backpressure may be applied to the permeate in the initial stage(s) to lower NDP and avoid overfluxing while still maintaining acceptable NDP in the final stage(s). Multiple passes are used to purify the permeate water from the first pass to achieve higher quality.

The number of elements used in a system is determined by the desired average operating flux, which is a design parameter that depends on the type of water that the reverse osmosis system is treating and the quality of the pretreatment ahead of the reverse osmosis. High operating fluxes reduce the number of required modules, saving on capital expenses and replacement element costs, but require larger feed pressures (higher NDP) and have higher concentration polarization, which makes the membrane more susceptible to fouling (Chong et al., 2008; Goosen et al., 2005; She et al., 2016). For example, low TDS waters with minimal potential for fouling (e.g., second pass reverse osmosis) can operate at high fluxes of approximately 35 to 42 LMH.

Conversely, wastewaters with high fouling propensity and moderate TDS levels typically require lower fluxes ranging from 12 to 25 LMH) (FilmTec, 2019). The fouling potential of a given wastewater is difficult to predict without conducting pilot trials or having experience operating systems treating similar wastewater types. Choosing too high of an operating flux may result in frequent system cleanings and significant downtime, which can reduce the throughput of the system.

The operating flux can also be limited by the TDS of the water. The water of the final stage of a MLD design will have high osmotic pressure. Smaller NDPs, and hence lower fluxes, may be required in these final stages to remain below the maximum operating pressure of the reverse osmosis element. For these reasons, the final stages of MLD systems are typically designed to operate at low fluxes of approximately 8.5 to 11 LMH.

Once an average operating flux has been determined, the modules must be arranged into pressure vessels and stages to meet desired recovery of the raw water and permeate quality while adhering to remaining design guidelines (such as maximum operating flux, minimum reject flow, and maximum feed flow) to ensure pressure drop, fouling, and scaling are manageable. Membrane manufacturers provide software programs that simulate all of these metrics in customizable water treatment trains. Within the first pass, it is not uncommon for each stage to be designed for 50% water recovery, which typically maintains an acceptable flux balance down the length of the vessel and reject flow. A two-stage system would be designed in a 2:1 array where the first stage will have twice as many pressure vessels as the second stage, and the overall system would achieve 75% water recovery (Figure 3.44).

The consolidation of pressure vessels between stages increases the feed flow to each second-stage vessel, reducing polarization. Figure 3.45 illustrates the evolution of feed and permeate water in a two-stage design operating at 75% recovery. In MLD designs aimed at higher recovery and not limited by solute solubility limits, stages are added to the system until the composition of the reject stream reaches an osmotic pressure that is near or equal to the maximum operating pressure limit of the system.

The desired recovery dictates the number of stages used. Take as an example a wastewater system producing 200 m^3/h (800.5 gpm) of purified water. Design operation at 15 LMH (8.82 gpd/sq ft) would require approximately 360 modules, each with 37 m^2 (400 sq ft) of active area. Assuming 7 elements per pressure vessel, the system would require 51 pressure vessels. Organized in a two-stage, 2:1 (34:17) array of 357 elements with a feed flow of 8 m^3/h (35.2 gpm) per first-stage vessel, the system would achieve 73% recovery. The same number of vessels could instead be organized in a three-stage, 30:15:6 array. Operated under the same permeate flow (200 m^3/h), average flux (15 LMH), and feed flow per first-stage vessel, this three-stage system would achieve 83% recovery.

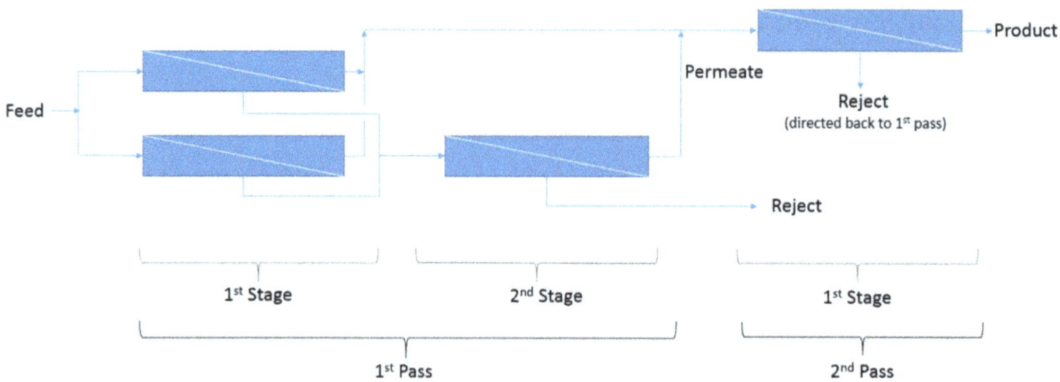

FIGURE 3.44 Schematic of a reverse osmosis system (courtesy of Derek Stevens and Tina Arrowood).

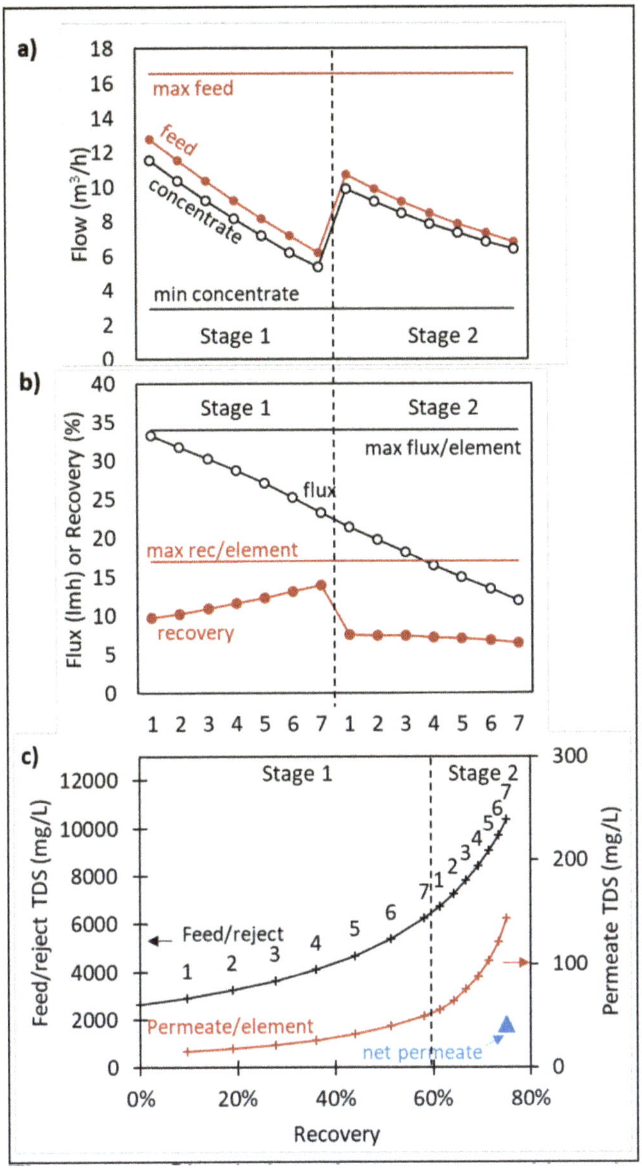

FIGURE 3.45 Simulation of a 2:1 array, with 7 elements per pressure vessel, of brackish water reverse osmosis elements treating 2600 mg/L surface water to 75% recovery: (a) feed and concentrate flow entering and leaving each element, (b) flux and recovery of each element, and (c) concentration of TDS in the feed/reject and permeate with recovery. (Figure created by Derek Stevens and Tina Arrowood; however, design guidelines provided by the membrane manufacturer for operation in this water are shown for reference [FilmTec, 2019]).

6.5 Brine Management and Minimal Liquid Discharge

When designing a reverse osmosis membrane treatment system for wastewater reuse, one must consider the following basic parameters: feed water volume, composition, hardness, low solubility solutes such as silica, pH, temperature, required permeate quality, required permeate volume, and the fate of the reject stream. Most commonly, membrane-based wastewater reuse systems are designed to recover 70% of the water. The remaining 30% discharged has a higher concentration of dissolved contaminants (−3.3x) than the feed water to the reverse osmosis system, and there is growing concern over the effect this discharge can have on the environment.

Increasingly limited outlets for the concentrate stream have led to higher surcharges for discharging it. This cost and the reduced availability of fresh feed water sources has driven an increase in demand to reduce discharge volumes and increase recovery in wastewater treatment systems. A typical concentrate stream from a WRRF, however, has high TDSs (TDS = 4000 to 10 000 mg/L) and organic contaminants (COD = 40 to >500 mg/L). Because of the challenges of treating this water composition, thermal evaporation and condensation or large footprint evaporation ponds in arid climates were once considered the only option. Even with the most advanced energy-efficient evaporators, however, operation is expensive and can be cost prohibitive. Additionally, the upfront capital cost to install evaporators is significant. Using pressure-driven, membrane-based MLD lowers both the overall cost of zero liquid discharge (ZLD) by 60% and the environmental effect of thermal evaporation by concentrating and purifying up to 95% of wastewater (Bond and Veerapanenei, 2008).

6.6 Special Reverse Osmosis/Nanofiltration Design Considerations to Achieve Minimal Liquid Discharge

Water recovery using reverse osmosis is bounded by Equation 3.24. A system must apply pressure high enough to overcome the osmotic barrier, with a sufficient combination of water permeability (*A*-value) and remaining NDP to achieve desired rates of flux. In a system designed for high water recovery, the water concentration significantly changes along the feed channel from the entrance to the exit of the system. Additionally, the resistance to pass water through the membrane (1/*A*-value) may be dynamic because of fouling. As such, high-recovery MLD systems must be carefully designed with the proper configuration and selection of reverse osmosis/nanofiltration element types.

To achieve high recovery, reverse osmosis elements with high *A*-value membranes are desired in the tail end of the concentration process because they offer the least resistance to pass water. Common commercially available seawater reverse osmosis (SWRO) membranes range in *A*-value from 1.2 to 2.5 LMH·bar. Table 3.11 summarizes the effect membrane choice can have on flux when treating a highly concentrated brine stream. All else equal, a lower *A*-value membrane will produce less water and achieve lower water recovery.

A balance between water recovery and permeate quality must be considered when optimizing the design of a system. Concurrent with the TDS of a feed increasing down the length of a stage (or stages), the TDS of the permeate will rise as well. As discussed, membranes with higher water permeability also have higher salt passage and further add to the TDS of the permeate when they are used in MLD systems. Designs hosting more than one element type are commonly used with rejection membranes at the front end and high *A*-value membranes at the back end, with the transition taking place either between or within stages. The latter, known as an "internally staged design," first gained popularity in municipal seawater desalination (Garcia Molina et al., 2009). Both approaches balance flux, reduce energy consumption, optimize recovery, and can be designed to produce a blended permeate stream that meets quality requirements for use.

Standard SWRO elements have a typical maximum allowable operating pressure of 82 bar (1200 psi). Recently, specialty reverse osmosis elements designed for higher pressure operation, up to 120 bar (1740 psi), have been made commercially available. Figure 3.46 illustrates the expanded treatment range such ultra-high-pressure reverse osmosis elements are able to achieve. Using these elements, more concentrated brines can be realized, and water recovery can be increased. The 120-bar operation increases the theoretical osmotic limit to approximately 140 000 mg/L sodium chloride. These limits are further increased in solutions containing higher molecular weight salts, such as sulfates, because osmotic pressure depends on molar concentration, not mass.

TABLE 3.11 Effect of membrane *A*-value on water flux when treating a concentrated feed containing 8% sodium chloride at room temperature.

	Case 1	Case 2
A (LMH/bar)	0.10	0.07
P_{feed} (bar)	68.9	68.9
P_{feed} (bar)	67.9	67.9
NDP_{feed} (bar)	1.0	1.0
Flux (LMH/bar)	2.46	1.73

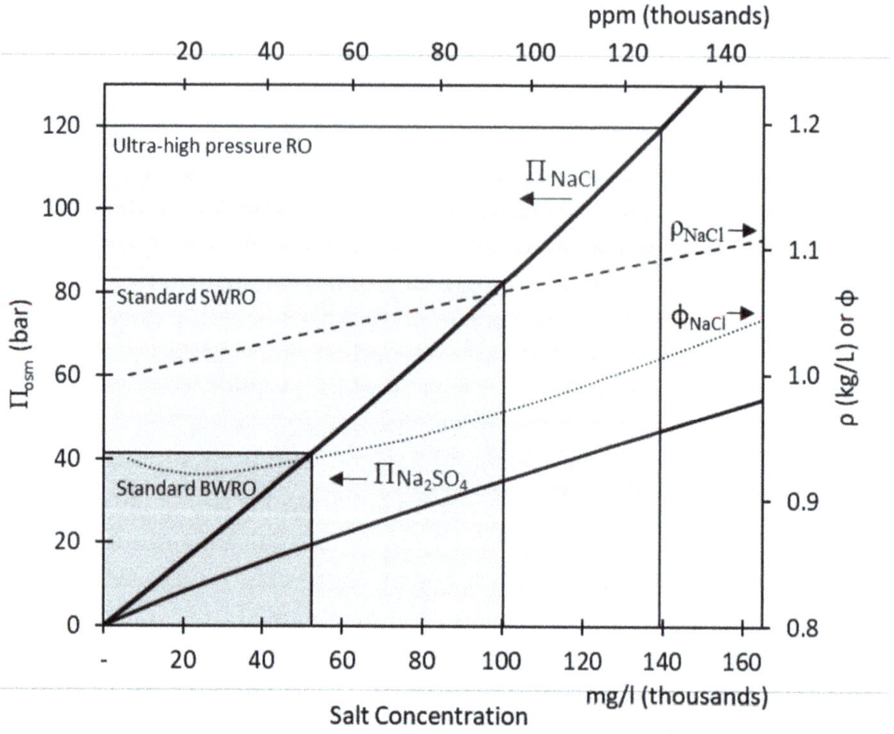

FIGURE 3.46 Osmotic pressure of pure solutions of aqueous sodium chloride (NaCl) and sodium sulfate (Na_2SO_4) at 25 °C as a function of concentration and overlay of the operating windows for standard and ultra-high-pressure reverse osmosis element designs. *Π*osm was calculated from Equation 3.23 using density and osmotic coefficient information available in the literature. (Figure created by Derek Stevens and Tina Arrowood; based on data from Perry & Green, 1973; Robinson & Stokes, 2002).

In the face of these limitations, an alternative to maintaining positive net driving pressure in concentrated feeds is to increase the osmotic pressure of the permeate (see Equation 3.24). $\Pi_{permeate}$ is not typically considered an accessible design parameter because reverse osmosis membranes reject the vast majority of salts, driving $\Pi_{permeate}$ toward zero, and reuse applications typically dictate stringent permeate quality. However, systems facing ZLD requirements must consider the cost of disposal of the waste solid salt mixture. If the TDS of the wastewater is high in sodium chloride, an ion-selective, nanofiltration membrane that passes monovalent ions, but rejects divalent ions and organic contaminants, can be used to further concentrate a reverse osmosis reject stream. This can be accomplished at pressures less than the reverse osmosis system because $\Pi_{permeate}$, enriched by sodium chloride, can be significant in size and increases the NDP. Nanofiltration membranes for MLD are designed to produce a low hardness permeate stream with a high concentration of monovalent ions in the permeate, offering brine suitable for reuse in many applications (e.g., textile dyeing, chloro-alkali).

Figure 3.47 illustrates a combined reverse osmosis/nanofiltration membrane purification scheme. Different membrane types should be considered in each of the defined stages to maximize recovery and produce product water of adequate quality for reuse.

6.7 Managing Membrane Fouling

Membrane fouling can arise from various components in the feed water, including colloids, particles, biological matter, organic matter, and inorganic scale. Raw wastewaters can have high COD, near −2000 mg/L, and typical TDS ranges of 4000 to 10 000 mg/L. Pretreatments with ultrafiltration, lime softening, and/or ion exchange resins are effective technologies to manage colloidal fouling, particulate fouling, and inorganic scaling. Hardness and

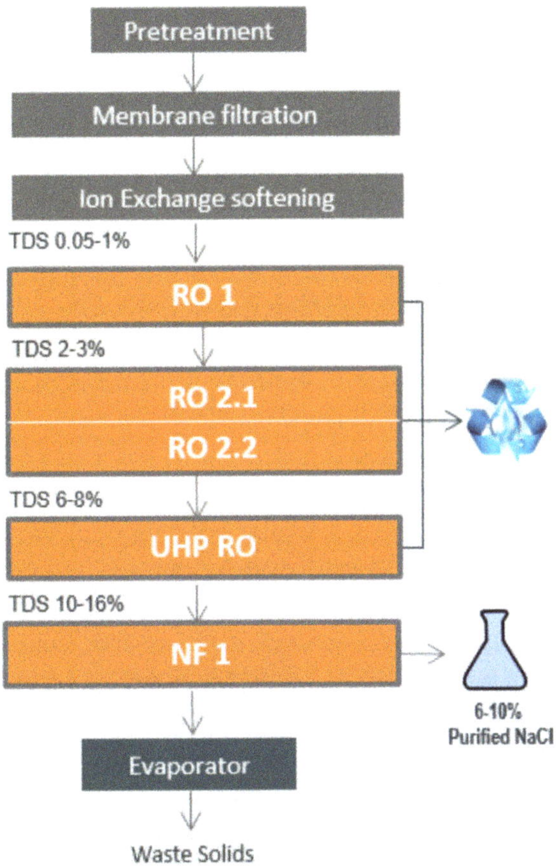

FIGURE 3.47 Simple schematic of an MLD membrane design offering using reverse osmosis and nanofiltration membranes (courtesy of Derek Stevens and Tina Arrowood).

silica can be particularly problematic in a reverse osmosis MLD system if they are not adequately addressed during pretreatment because their concentrations increase exponentially with reverse osmosis recovery and can quickly reach solubility limits for precipitation (Antony et al., 2011; Bond & Veerapaneni, 2008; Milne et al., 2014).

Biological or chemical pretreatment followed by clarification are used to reduce the color and COD level, but even after such treatment, the COD in the feed to an MLD system can commonly be between 60 and 200 mg/L. This creates a significant biological and organic fouling potential for reverse osmosis elements, which worsens as the contaminants in the water are concentrated up to 10 times during the pure water recovery process (see Equations 3.25 and 3.26).

Fouling in a reverse osmosis element results in either a decrease in membrane permeability, an increase in feed-to-concentrate pressure drop, or a combination of both. The decline in membrane permeability is caused by deposition of foulant on the membrane surface, which creates a resistive layer for water to travel through the membrane. This foulant can be biofilm, natural organic matter, synthetic organic matter, inorganic scale, or a combination of any of these.

As a general guideline for multistage reverse osmosis, biological fouling primarily occurs in the first stage of a system, whereas scaling occurs in later stages. Loss of membrane permeability causes either the system recovery to decline or the need for higher feed pressures to achieve the same recovery. A decline in performance of >20% should be avoided to eliminate the potential for irreversible fouling (FilmTec, 2009).

An increase in the feed-to-concentrate differential pressure is often a sign of biological fouling, especially if the increase is isolated to the first stage of a treatment system. Bacteria in the water settle on the surfaces of the reverse osmosis membrane and/or feed spacer and use the organic contaminants in the water as nutrients to feed their colony. As a survival mechanism, bacteria excrete extracellular polymeric substances, also known as "biofilm," which are sticky hydrogels composed of carbohydrates and proteins (Matin et al., 2011). The biofilms serve to protect the bacteria from the harsh turbulent environment of the reverse osmosis element, but also collect the organic contaminants in the water to serve as food for the bacteria colonies to consume and thrive. These biofilms accumulate in the feed channel of the reverse osmosis elements and block water flow, creating a high feed-to-concentrate pressure difference.

A differential across a pressure vessel of greater than 3.5 bar must be avoided because it can result in mechanical damage of the reverse osmosis elements. Furthermore, the larger the biofilm, the more difficult it is to clean. To avoid irreversible fouling, it is recommended to clean the system when the differential pressure increases by 20% (FilmTec, 2019).

There are several techniques used to manage biological and organic fouling within reverse osmosis systems such as the following: (a) conduct a CIP or offsite cleaning to regain performance, (b) dose nonoxidizing biocides into the feed water, and (c) use fouling-resistant membranes. When either the membrane permeability or feed-to-concentrate differential pressure reaches a cleaning trigger, the reverse osmosis system is taken out of service and a CIP with strong base followed by strong acid is conducted. This routine is commonly used as the primary means to manage fouling in reverse osmosis systems when fewer than six cleanings are needed per year.

However, the fouling rate of systems treating challenging wastewater can result in CIPs occurring more frequently than once a month. This not only creates a significant hassle for the operators, but it also reduces system capacity and the reverse osmosis element lifetime and increases cleaning chemical costs. To reduce the downtime of frequent CIPs, some installations will allow the performance to deteriorate beyond the manufacturer recommended cleaning trigger. This strategy can provide a short-term remedy, but a buildup of irreversible foulant can result in even higher frequency of cleanings or shorter element lifetimes in the long term.

Chlorine treatment is a cost-effective biocide used to prevent biofouling in systems upstream of reverse osmosis, but, because of the incompatibility of chlorine and the polyamide barrier layer of reverse osmosis membranes (Glater et al., 1994), it is neutralized before the feed water reaches the reverse osmosis system. To protect reverse osmosis systems, fast-killing nonoxidizing biocides such as 2,2-dibromo-3-nitrilopropionamide, DBNPA, are used (Bertheas et al., 2009). However, these materials are more expensive than chlorine and are typically only shock-dosed for a short period of time each day. Such treatments can help control, but not altogether eliminate, biofouling. Today, biofouling is the number one, primary nuisance limiting the theoretical performance of most reverse osmosis wastewater reuse systems.

Fouling-resistant reverse osmosis modules enable easier management of reverse osmosis systems. Fouling resistance is designed in reverse osmosis elements by pairing a fouling-resistant membrane chemistry with a fouling-resistant feed spacer design. Membrane chemistry is made fouling resistant by controlling the charge, roughness, and hydrophilicity of the membrane surface. The feed spacer design can control the mixing of contaminants at the membrane surface, reduce feed channel plugging, and lower the initial pressure drop of a system to provide a wider operating window before reaching the maximum allowed pressure drop.

In addition to mitigating fouling, it is also critically important for these reverse osmosis modules to be cleanable. Performance must be recoverable when cleaning triggers are reached, and the membrane must be resilient enough to withstand repeated extreme pH cleaning cycles over its desired lifetime. Fouling-resistant brackish water reverse osmosis modules have been commercially available for more than a decade. Until recently, however, only standard seawater modules have been available for treating high TDS waters. Seawater modules are designed for the high osmotic pressures of a typical seawater desalination system, which is within the range of MLD design; however, these modules are not optimized to resist fouling and withstand the frequent cleanings faced in wastewater MLD designs.

7.0 FORWARD OSMOSIS

7.1 Background and Purpose

7.1.1 Introduction

Forward osmosis is a membrane separation process that gently removes water from a feed solution and with minimal impact to that feed solution. It uses a natural process driven by chemical energy. Water flows across a semipermeable membrane from a liquid with lower osmotic pressure (feed) into a working fluid with higher osmotic pressure (draw).

A forward osmosis membrane is housed in a forward osmosis module, which effectively acts as an osmotic pressure exchanger. Forward osmosis concentrates the feed and dilutes the draw. In some forward osmosis systems, the draw is reconcentrated and reused. The diluted draw remains relatively clean and is typically easier than the feed stream to concentrate/regenerate via other membrane or thermal processes.

Forward osmosis differs from reverse osmosis and most membrane separation processes because it uses osmotic pressure instead of hydrostatic pressure to power the separation of fluids. This means that feed pressure can effectively be decoupled from the filtration rate (i.e., membrane water flux), which, in turn, gives forward osmosis distinct advantages over other types of membrane filtration and separation technologies. One advantage is that the maximum osmotic pressure during separation is not limited by hydrostatic feed pressure as in the case of reverse osmosis. Other advantages may include higher rejection of dissolved contaminants, lower rates of fouling, reduced degradation of the feed solution, and the ability to use new methods for membrane integrity monitoring.

Forward osmosis is often mislabeled as a membrane technology to replace reverse osmosis membranes for desalination; however, a more accurate description of forward osmosis is as a unique tool to provide solutions not feasible or desirable with current separation processes. In some cases, this means that forward osmosis may compete directly with other processes (e.g., reverse osmosis or evaporators) or, in other cases, that forward osmosis may complement these processes by expanding potential applications and making them more viable or cost-effective.

In other words, forward osmosis is not simply a competing process, but a tool that, at its best, can remove water from a product or waste stream where other membrane processes typically fail. Forward osmosis can also result in significant savings in the form of capital expenditure, operating cost, energy use, or, more often and more importantly, the cost of transporting and storing this additional water in a waste or product stream.

Based on this introduction, a common question is when to consider forward osmosis over competing processes. This section includes specific examples from forward osmosis membrane manufacturers or users that recommend forward osmosis-based solutions over other options for applications in water reuse, waste minimization, product concentration, and, in some specific cases, seawater desalination.

7.1.2 How Forward Osmosis Works

Forward osmosis is an osmotic process that uses a semipermeable membrane to separate water from a feed solution. Unlike pressure-driven membrane separation processes, such as reverse osmosis, nanofiltration, and so on, forward osmosis requires two solutions of a different nature on each side of the membrane to generate the driving force for filtration. The driving force for this separation is an osmotic pressure gradient between a solution of high osmotic pressure, referred to as the "draw solution," or simply the "draw," and a solution of lower osmotic pressure, referred to as the "feed." The osmotic pressure gradient is used to induce a net flow of water through the membrane into the draw, thus concentrating the feed.

The draw solution can consist of a single or multiple solutes that may be simple salts (e.g., sodium chloride, magnesium chloride, magnesium sulfate, etc.); seawater; an onsite brine (e.g., geothermal brines); common food-safe additives (e.g., sugars, polyols, juice concentrates, etc.); or a substance specifically tailored to forward osmosis applications, often referred to as a "designer draw" (e.g., ammonium bicarbonate, magnetic nanoparticles, switchable polarity solvents, etc.). The feed solution can be any dilute product stream (e.g., fruit or vegetable juice, milk, other beverage, waste stream, seawater, or high-salinity brine).

7.1.3 Relationship Between Forward Osmosis and Related Osmotic Processes

Figure 3.48 shows the relationship between forward osmosis, pressure retarded osmosis (PRO), and conventional reverse osmosis—three closely related separation processes (Lee et al., 1981).

As explained earlier, a semipermeable forward osmosis membrane separates two solutions of different concentrations, such that water flow from the feed side into the draw solution occurs because of the osmotic pressure

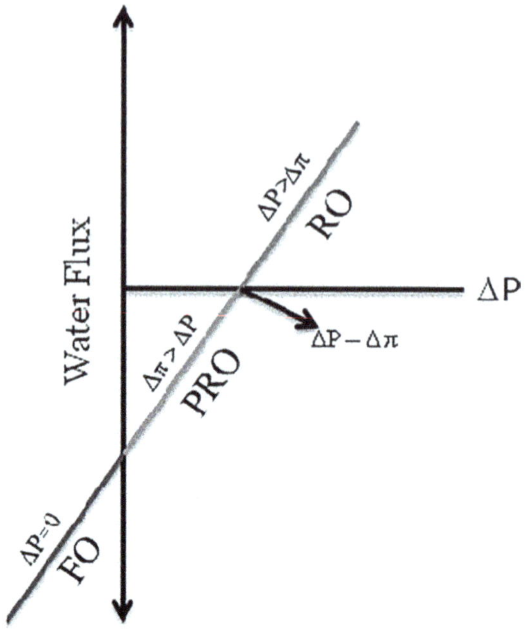

FIGURE 3.48 Illustration showing the relationship between forward osmosis, PRO, and reverse osmosis for a membrane. Modified image adopted from Lee et al. (1981).

difference between the two solutions. However, when a gradually increasing pressure is applied to the salt solution (in the direction from draw to feed), the rate of the water flow (i.e., flux) begins to decrease and will eventually become zero once the applied pressure equals the osmotic pressure difference across the membrane.

The process in which water flows by osmosis across a membrane into a pressurized salt solution is known as "PRO." Once the applied hydrostatic pressure into the salt solution exceeds the osmotic pressure difference across the membrane, the direction of water flow reverses and is referred to as "reverse osmosis." In reverse osmosis, hydrostatic pressure exceeding the osmotic pressure is applied to the feed to produce a clean water permeate.

7.1.4 Applications and Benefits

A forward osmosis process, when designed and used appropriately, can provide new solutions for food and beverage concentration or "difficult to treat" water reuse and waste minimization projects. For example, forward osmosis can provide the following benefits when concentrating a natural beverage product, such as fruit juice:

- A higher sugar concentration than a reverse osmosis process, which is impeded by a hydrostatic pressure limitation. This is because forward osmosis is driven by osmotic pressure instead of hydrostatic pressure, as in the case of reverse osmosis.

- A higher quality product than competing evaporators that use heat that degrades flavors, colors, and nutrients; forward osmosis does not use heat, thereby preserving the quality of the product.

As for "difficult to treat" wastewater, forward osmosis offers the following benefits:

- Potential for reduced fouling, cleaning frequency, system downtime, and process complexity for wastes that have high oil and grease contents (not suitable for other types of polymeric membranes); fibrous materials (e.g., algae, pulp, and paper wastes, etc.); sugars; COD; or other organics that cause rapid fouling of other types of polymeric membranes (e.g., microfiltration, ultrafiltration, nanofiltration, reverse osmosis, membrane distillation, etc.).
- Forward osmosis can be used as a dilution step (i.e., osmotic dilution) to extract clean water from a contaminated waste stream (e.g., tertiary treated wastewater) into seawater to reduce salinity by dilution, which may reduce the energy use and costs of desalination.
- Forward osmosis can incorporate new methods of real-time integrity monitoring by using a marker in the draw solution and operating the draw solution at higher pressures than input and output streams.

Forward osmosis can also provide the following benefits when treating high-salinity wastes (e.g., near ZLD) that exceed the limitations of reverse osmosis membranes:

- Forward osmosis can concentrate aqueous solutions to higher salinity than reverse osmosis which is impeded by a hydrostatic pressure limitation. This is because forward osmosis is driven by osmotic pressure instead of hydrostatic pressure, as in the case of reverse osmosis.
- Forward osmosis can also use a membrane (modified nanofiltration or reverse osmosis, membrane distillation, etc.) or a thermal draw recovery process that can operate more efficiently on a clean draw stream than a high-fouling, high-salinity feed in a suitable way to reduce capital, operating, and/or energy costs.
- Forward osmosis can occasionally use an onsite draw stream or brine, when available, without draw recovery, thus significantly reducing capital and energy costs.

Forward osmosis has been shown to reduce fouling rates and prevent irreversible fouling when compared to pressure-driven membrane processes.

Because osmosis is a spontaneous, chemically driven process, forward osmosis requires zero or low hydraulic pressure. While it is often theorized that low hydraulic pressure reduces membrane fouling, recent studies suggest that a different mechanism(s) may be the reason for reduced fouling observations (Tow et al., 2017). Whatever the cause of reduced fouling, forward osmosis has shown advantages in treating challenging feeds where competing membranes are limited by either foulants or hydrostatic pressures.

Forward osmosis processes have gained attention in numerous applications, including water reuse and desalination, food and beverage (Jiao et al., 2004), mining, oil and gas (Bryan et al., 2015; Kerri et al., 2013), power (Bruce & Elimelech, 2012), and pharmaceutical (Kong et al., 2015; Xue et al., 2012) industries. Most forward osmosis applications fall into two broad categories: product concentration and water reuse. Product concentration is more commonly used in the food and beverage, fragrance, dye industrial chemical, and mining industries. Water reuse encompasses both high-salinity and near-ZLD applications as well as low-salinity industrial and potable reuse purposes.

In some cases, a "forward osmosis only" process can be used to concentrate the feed if there is an existing draw solution available for dilution. In such cases, the diluted draw may be disposed of or be useful at a diluted concentration. Such cases typically require that a site have multiple process streams and/or an infrastructure suitable for this arrangement, which may be rare. However, most forward osmosis applications require that the draw be recovered for reuse. This additional step concentrates the draw to power the forward osmosis process and generates a clean water permeate. This typically involves a combination of two different separation processes such as forward osmosis+reverse osmosis, forward osmosis+membrane distillation, or forward osmosis+thermal (evaporator) combinations, with the second process acting as the draw recovery step. In some cases, forward osmosis can concentrate a waste (e.g., high COD waste), turning it into a product (e.g., feedstock or fertilizer) while simultaneously producing clean water for reuse.

7.1.5 Commercial Adoption

Although the theory of forward osmosis has been circulated since the 1960s, more recent advances in forward osmosis membranes and modules have been necessary for its commercial applications. Over the last few years, commercial forward osmosis-based systems have begun operating around the world, including in several waste minimization systems in China (e.g., FTSH2O, Oasys) and food and beverage processing in North America (Porifera). Additionally, pilot-scale systems are being evaluated for industrial and municipal water reuse (e.g., Porifera, Aquaporin, and Trevi Systems) and seawater desalination (e.g., Hyorim Industries, Trevi Systems). Details can be available from listed technology providers. The following sections summarize the advancements in forward osmosis membranes, housings, and systems that have led to greater commercial adoption.

7.2 Membranes and Desired Properties

7.2.1 Introduction

The ideal forward osmosis membrane would have similar rejection characteristics to a reverse osmosis membrane, but different physical characteristics than reverse osmosis in an effort to improve flux efficiency. Rejection is important because it is desirable to (a) retain as much draw solute as possible, (b) maintain the purity of the draw, and (c) maintain the osmotic pressure difference between the feed and draw. The following sections will discuss the importance of rejection on feed and draw chemistry and concentration polarization in more detail.

The physical characteristics of a forward osmosis membrane are important in terms of flux efficiency. While reverse osmosis membranes need to be thick and sufficiently robust enough to handle high hydrostatic and transmembrane pressures, forward osmosis membranes should be thin and open to maximize porosity and minimize concentration polarization in an effort to maximize performance (typically measured as specific flux or flux per unit of average osmotic pressure differential between the feed and draw).

7.2.2 Overview of Commercial Forward Osmosis Membranes

Commercial forward osmosis membranes are available in both asymmetric and TFC forms. Asymmetric membranes are typically made in a single-step phase inversion process from cellulose triacetate (CTA) polymer solution. Cellulose triacetate is an often-selected material for forward osmosis membranes because of its hydrophilic nature, a higher chlorine resistance compared to TFC membranes, and potential antifouling properties because of neutral surface charges. On the other hand, these membranes have relatively low water permeability and higher reverse salt flux values compared to TFC membranes, which affects commercial system footprint and cost. Additionally, CTA membranes degrade when exposed to certain draw solutions (Baker, 2004) and undergo hydrolysis in harsh environments, including alkaline conditions.

Alternatively, commercial TFC-FO membranes exhibit higher salt rejection and water flux values with better chemical stability and mechanical strength. The TFC-FO membranes are also reportedly more stable than CTA membranes over a broad pH range of 2 to 12 as has been learned from several tests by PORIFERA. The TFC membranes are relatively hydrophobic and may have a higher contact angle of surface charge. They will also have different fouling properties depending on the composition of the feed and draw solutions. Additionally, TFCs are sensitive to free chlorine. There are several companies that manufacture forward osmosis membranes (Bryan et al., 2014),), including Porifera, Fluid Technology Solutions (formerly known as HTI), and Aquaporin, most of which sell commercial membranes, modules, and/or systems.

As previously mentioned, the properties of the draw solution (high concentration) and feed solution (low concentration) are important in terms of characterizing membrane performance. Deionized water is commonly used as the feed to compare performance with different draw solutions. Draw solution compounds, according to their physio-chemical properties, can be classified into inorganic salts, organic salts, volatile compounds, nonvolatile compounds, polymers, and oligomers. The compounds in these classifications may overlap with each other.

The other property often referred to in terms of membrane performance is the forward osmosis membrane orientation. The forward osmosis process can be operated with the membrane in two different membrane orientations: forward osmosis mode and PRO mode. In forward osmosis mode, the active membrane (or skin layer) will be in direct contact with the feed solution and the draw solution flows against the porous support layer. When the draw solution is in direct contact with the membrane active layer and the feed flows against the membrane's porous support, the process is referred to as "PRO mode." In general, permeate flux values during PRO mode are higher than in forward osmosis mode; however, the membrane is typically much more prone to fouling in PRO mode. The flux difference between forward osmosis mode and PRO mode operations is primarily the result of external concentration polarization (ECP) and internal concentration polarization (ICP). The following section presents a detailed description of concentration polarization.

7.2.3 Key Concentration Polarization Considerations

To fully understand the properties of the ideal forward osmosis membrane, it is also important to understand the concepts of concentration polarization. Concentration polarization is a phenomenon that explains the reasons for reduced membrane performance associated with increased concentrations of solutes in the feed and draw. For example, it often explains the difference between the theoretical and observed membrane flux when calculating clean water flux based on osmotic pressure difference (i.e., driving force) across the forward osmosis membrane (assuming membrane fouling has not occurred). Concentration polarization is highly dependent on the thickness and porosity of the membrane and mixing of the feed and draw solutions.

The general equation to quantify the water flux (J_W) through a semipermeable membrane is

$$J_W = A\sigma(\Delta\pi - \Delta P) \tag{3.27}$$

where A = the membrane water permeability constant,
σ = the reflection coefficient,
Π = the osmotic pressure difference across the membrane, and
P = the applied pressure.

For forward osmosis, ΔP is typically assumed to be zero, such that Equation 3.27 simplifies to

$$J_W = A\sigma\Delta\pi \quad (3.28)$$

Assuming that $\sigma = 1$ would mean that the membrane completely rejects the solutes, which is typically assumed in forward osmosis studies because of the high rejection property of the membrane (Baker, 2004). With this assumption, Equation 3.28 can be expressed as shown below

$$J_W = A\Delta\pi = A(\pi_{D,b} - \pi_{F,b}) \quad (3.29)$$

From Equation 3.29, it is clear that there is typically a linear relationship between the osmotic pressure difference ($\Delta\pi$) of the bulk draw ($\pi_{D,b}$) and the bulk feed ($\pi_{F,b}$). However, actual water flux values are lower than theoretical water flux values calculated from osmotic pressure difference (Loeb et al., 1997; Mehta & Loeb, 1978a). This is primarily because the effective osmotic pressure difference across the membrane ($\pi_{D,m} - \pi_{F,m}$) is lower than the bulk fluid osmotic pressure difference ($\pi_{D,b} - \pi_{F,b}$) referenced in Equation 3.29. This phenomenon is called "concentration polarization," and it may take several forms depending on the type of membrane and the operating conditions used during separation.

7.2.4 External Concentration Polarization

Consider a simple example of water and salt solutions used in a forward osmosis process, where both water (from feed to draw) and salt (from draw to feed) pass through the membrane in opposite directions. Movement of water from the feed to the draw effectively dilutes the draw side salt solution. Assuming a dense membrane, the osmotic power of the bulk draw ($\pi_{D,b}$) is higher than the draw concentration adjacent to the membrane surface ($\pi_{D,m}$) (i.e., $\pi_{D,b} > \pi_{D,m}$).

Similarly, the diffusion of salts from the draw to the feed causes the accumulation of salts on the feed side, hence $\pi_{F,m} > \pi_{F,b}$ where $\pi_{F,b}$ and $\pi_{F,m}$ are feed osmotic power at bulk and adjacent to membrane surface, respectively. The changes in osmotic power occurring on the outer membrane surfaces are termed ECP. Because of this, ECP can be alleviated by mixing, which is typically created by increasing flowrates to create turbulence.

The process of the draw being diluted on the membrane surface by feed water transport through a membrane is called dilutive external concentration polarization (DECP). Similarly, solutes diffuse through the membrane from the draw to the feed surface of the membrane, which is referred to as concentrative external concentration polarization (CECP). The effects of DECP and CECP are limited to the active membrane surface where solute rejection is occurring.

In forward osmosis, ECP occurs at the surface of the membrane active skin layer; hence, the models mentioned below are applicable to dense membranes and double skin membranes only. Understanding CECP and DECP is important to accurately estimate the effective osmotic driving force to precisely determine forward osmosis membrane flux performance. However, it is not easy to estimate the effective feed concentration at the membrane surface. In depth models for ECP in forward osmosis using boundary-layer film theory to calculate concentration polarization in pressure-driven membrane processes can be found in forward osmosis-related literature (McCutcheon & Elimelech, 2006; McCutcheon & Elimelech, 2007). Both concentrative ECP and dilutive ECP are considered in these models.

However, for most of the osmotic processes, asymmetric membranes are commonly used today; thus, the usefulness of these flux models are limited. Flux models for asymmetric membranes are, therefore, more important and it is also where significant ICP effects are observed.

7.2.5 Internal Concentration Polarization

Most commercial forward osmosis membranes are less than 100-mm thick. Thinner TFC membranes are more permeable and consist of an ultra-thin active skin layer on a relatively thick porous support known as a "TFC." Considering the two different membrane orientations for forward osmosis operation (forward osmosis mode and PRO mode), the concentration polarization concept becomes much more complex to understand for forward osmosis operations using TFC membranes. In this case, the existence of two distinct boundary layers creates ICP on both sides of the membrane and reduces the effective flux. The second boundary layer occurs within the porous support layer of the membrane and is typically referred to as "ICP."

Contrary to ECP, ICP is inversely proportional to membrane thickness. Flow turbulence has little effect in minimizing ICP because it is hard to remove or mix the solute molecules trapped inside the pores. Internal concentration polarization will contribute to significant water flux decline in forward osmosis (McCutcheon & Elimelech, 2006; Mehta & Loeb, 1978b). Forward osmosis studies have reported that ICP could reduce water flux by more than 80%; however, the details are important because using an efficient forward osmosis housing and system design is key to limiting the effects of ICP.

Similar to ECP, ICP can also be subdivided into two types depending on membrane orientation. In PRO mode, where the draw solution is in contact with the active skin layer of a TFC membrane, water enters the porous support layer and permeates across the active layer into the draw solution, which can be explained as DECP. At the same time, the salt in the feed freely enters the open structure of the membrane porous support and can become trapped. Salt cannot easily penetrate the pores in the active layer that is basically designed to reject salts; thus, the concentrations in the pores will increase. This process is typically referred to as concentrative internal concentration polarization (CICP).

Lee et al. (1981) and Loeb et al. (1997) describe the following expression for CICP and how it relates to water flux and other membrane constants:

$$K = \left(\frac{1}{J_W}\right) ln \frac{(B + A\pi_{D,m} - J_W)}{B + A\pi_{F,b}} \tag{3.30}$$

where B = the salt permeability coefficient of the active skin layer of the membrane, and

K = the diffusion resistivity of solute within pores, defined by

$$K = \left(\frac{t\tau}{D\varepsilon}\right) = \frac{S}{D} \tag{3.31}$$

Here, D is the diffusion coefficient of the solute, whereas t,τ, and ε, respectively, represent the thickness, tortuosity, and porosity of the porous support layer of the membrane, and is the structural parameter of membrane. The importance of the S value on membrane performance will be discussed in following sections.

The assumptions are that (a) the membrane is operated at a high flux and (b) the solute transport from the feed to the draw side is negligible, indicating a high percentage of solute (salt) rejection by the membrane, which means B is negligible compared to the other terms in Equation 3.30. With these assumptions and upon rearranging the parameters in Equation 3.30, the membrane flux (J_W) is derived as:

$$J_W = A[(\pi_{D,m} - \pi_{F,b} exp(J_W K)] \tag{3.32}$$

Therefore, the product of the water permeability coefficient and the effective osmotic driving force gives the membrane water flux. The exponential term in Equation 3.34 refers to the CICP, defined as follows:

$$\left(\frac{\pi_{F,i}}{\pi_{F,b}}\right) = exp(J_W K) \tag{3.33}$$

where $\pi_{F,i}$ is the osmotic pressure of the feed underneath the active layer within the porous support. The effect is concentrative when $\pi_{F,i} > \pi_{F,b}$.

By substituting $\pi_{D,m}$ values from Equation 3.28 into Equation 3.32, we can obtain a model for the effect of ICP and ECP on permeate flux, as follows:

$$J_W = A\left[\pi_{D,b}\, exp\left(-\frac{J_W}{k}\right) - \pi_{F,b}\, exp(J_W K)\right] \tag{3.34}$$

One can solve Equation 3.34 by incorporating the inputs obtained from experiments or calculations to predict water flux through an asymmetric membrane in PRO mode. Expression 3.34 is coupled with CICP and DECP contributions. In PRO-mode operation, dilutive internal concentration polarization (DICP) has little effect; hence, it has not been included in Equation 3.34.

7.2.5 Dilutive Internal Concentration Polarization

In the case of forward osmosis-mode operation, where the feed solution is in contact with the active skin layer of the membrane and the draw solution is in contact with the porous support side of the membrane, DICP typically dominates water flux. In this case, water permeates through the membrane, dilutes the salt in the pores in boundary layers, drastically reducing the net driving force between the feed and the draw. In forward osmosis mode, DICP plays a major role and CECP can be ignored ($\pi_{F,b} = \pi_{F,a}$). Loeb et al. (1997) considered the effect of DICP on water flux in the forward osmosis-mode operation, as follows:

$$K = \left(\frac{1}{J_W}\right) ln \frac{B + A\pi_{D,b}}{B + J_W + A\pi_{F,m}} \tag{3.35}$$

The permeate water flux expressed in Equation 3.36 was obtained by rearranging Equation 3.35 and assuming that salt permeability is negligible (i.e., $B = 0, \sigma = 1$), as follows:

$$J_W = A[(\pi_{D,b} exp(-J_W K) - \pi_{F,m}] \tag{3.36}$$

in which $\pi_{D,b}$ is given by

$$\frac{\pi_{D,m}}{\pi_{D,b}} = exp(-J_W K) \tag{3.37}$$

Here, $\pi_{D,m}$ is the concentration of draw on the inside of the active layer within the porous support. The negative exponent indicates that $\pi_{D,b} > \pi_{D,m}$.

By substituting Equation 3.27 into Equation 3.36, we get the following:

$$J_W = A[\pi_{D,b}\, exp(-J_W K) - \pi_{F,b}\, exp(\frac{J_W}{k})] \quad (3.38)$$

Equation 3.38 models water flux for asymmetric membranes during forward osmosis-mode operation. The equation includes both DICP and CECP terms. The expressions clearly indicate that both ECP and ICP contribute negatively to the overall osmotic driving force. The severity of negative effect increases with an overall increase in driving force (i.e., the higher the membrane flux, the higher the concentration polarization effect). Estimation of various parameters from the expressions subsequently helps to minimize the concentration polarization contributions by changing operating conditions, membrane characteristics, and boundary conditions.

7.2.6 Forward Osmosis Membranes

7.2.6.1 Ideal Forward Osmosis Membrane Properties

An efficient forward osmosis membrane should possess (a) high chemical resistance to potential feed and draw solutions for stability and (b) high water permeability and solute rejection for performance. To maximize performance, the ideal membrane structure is built on a thin, microporous substrate to minimize ICP with hydrophilic material, with a neutral charge to enhance water flux and improve antifouling properties. Researchers have considered a wide variety of materials and structures for forward osmosis membranes in their pursuit of the optimal forward osmosis membrane. Early commercial deployments of forward osmosis used flat-sheet CTA forward osmosis membranes. As with the case of the first CTA reverse osmosis membranes, these CTA membranes suffered from poor water permeability and were susceptible to degradation resulting from their low chemical stability. Polyamide TFC-forward osmosis membranes of 50 to 100 μm thickness, specifically designed for forward osmosis, have been introduced commercially as an alternative to the existing asymmetric CTA membranes because they exhibit higher water flux and salt rejection.

7.2.6.2 Thin-Film Composite-Forward Osmosis Membrane Structure and Modification

Early TFC-FO membranes, typically reverse osmosis membranes used as forward osmosis membranes, were too thick for efficient forward osmosis performance such that ICP would significantly affect performance. Newer TFC-FO membranes are both thin for performance reasons and strong for stability reasons.

Thin-film composite membrane preparation involves two steps: first, making a porous support layer by phase inversion and, second, incorporating an ultra-thin dense active layer on the porous support by interfacial polymerization. The properties of the support layer dictate ICP, and the interfacial polymerization layer dictates water permeability and salt rejection. The ideal forward osmosis membrane would be manufactured in such a way to minimize ICP and to maximize water permeability and salt rejection. The best parameter used to describe these ideal characteristics is a low membrane structural parameter (S).

Recent efforts to evaluate forward osmosis membrane performance have been primarily focused on reducing the structural parameter, S. The solute's ability to diffuse into or out of the membrane support layer and membrane structural parameter are derived from Equation 3.31, as follows:

$$S = \frac{t\tau}{\epsilon} \tag{3.39}$$

Where t,τ, and ε are the thickness, tortuosity, and porosity of the membrane, respectively. The structural parameter is an important intrinsic parameter of a membrane and it is the primary variable that affects ICP in the membrane support layer. The S value of a membrane can be obtained according to Equation 3.39 and by correlating forward osmosis test results (Phillip et al., 2010; Tiraferri et al., 2013). Equation 3.39 indicates that a smaller S value can be obtained by reducing tortuosity, increasing porosity, and/or reducing the thickness of the support layer structure. The phase inversion step, or procedures involved during the phase inversion process, play an important role in controlling S values of the final membrane.

In addition to reducing the S value, alternative methods may reduce the effects of ICP, including modifying the polymeric substrate by incorporating nanomaterials using electrospun nanofiber substrates, casting porous supports onto vertically aligned nanotube substrates, and preparing double-skin forward osmosis membranes. In the forward osmosis process, reverse solute diffusion through the membrane from the draw to the feed solution is inevitable. This reverse solute diffusion not only reduces water flux through the membrane, but may also enhance membrane fouling when the chemistry is not compatible (Lee at al., 2010). It can also alter the characteristics of the concentrated product, especially when used in food and beverage applications.

The specific reverse solute flux, which is defined as the ratio of the reverse solute flux to the forward water flux, is considered a primary measure of forward osmosis membrane performance (Hancock & Cath, 2009). Many membranes achieve high flux by reducing the membrane selectivity, resulting in a high reverse solute flux. However, in most cases, the high reverse solute flux results in either increased operational costs or reduced product quality. The ideal forward osmosis membrane should have both a high water flux and a low reverse solute flux.

7.2.6.3 Hydrophilic Thin-Film Composite-Forward Osmosis Membranes for Enhanced Antifouling Properties

Many studies have focused on posttreatment methods for improving membrane properties. One way to reduce membrane fouling is to increase the hydrophilicity, or reduce the hydrophobic nature, of the membrane. Hydrophilization is a membrane posttreatment process that makes the membrane more hydrophilic, increasing the membrane permeability and chlorine resistance. There are multiple methods (e.g., synthesis, solvent treatment, coatings, etc.) that can be used; however, some methods perform better than others.

Synthesizing modified membranes during the interfacial polymerization step has had little success because the additives used to enhance the hydrophilicity of the interfacial polymerization layer typically affect the quality of interfacial polymerization adversely, which reduces membrane performance by increasing the reverse solute flux. Other posttreatment procedures, such as treating the membranes with various water-soluble solvents (e.g., acids such as hydrofluoric acid and hydrogen chloride and alcohols such as ethanol and isopropanol), have been shown to increase membrane flux, but only temporarily. Hydrophilization has also been achieved by coating the membrane with more hydrophilic compounds through chemical reactions. In such cases, changes in membrane performance typically last longer than solvent treatment.

Recent developments of high-flux TFC-FO membranes have led to the increased use of forward osmosis in various applications. However, the effect of operating at higher permeation fluxes can potentially increase the fouling and/or scaling rates on the membrane surface (Gimun & Seungkwan, 2017; Xie et al., 2014). This creates the need for adapting antifouling strategies, including modifying the membrane surface to minimize fouling. Membrane surface modification procedures reported in the literature for reverse osmosis and nanofiltration membranes can similarly be used on forward osmosis membranes to enhance antifouling properties. However, there is no single modification or method that can address all fouling problems associated with a wide variety of different foulants.

It is commonly reported that increasing hydrophilicity makes the membrane surface charge neutral, reduces membrane surface roughness, and optimizes flow and flux rates to consequently reduce membrane fouling (Li et al., 2017; Revanur et al., 2007). In some reports, it was demonstrated that double-skin-layer forward osmosis membranes have a lower fouling propensity than conventional single-skin membranes (Duong et al., 2014). Membrane surface modification using polyethylene glycol (Revanur et al., 2007) or incorporating nanoparticles (Nguyen et al., 2014) also showed a substantial reduction in fouling.

A quick and effective coating method using an aqueous dopamine solution was proposed by Lee et al. (2007) and Kim et al. (2013). A nonselective coating of thin polydopamine on any surface, including polymer membranes, is also gaining attention. Dopamine coatings, unlike other antifouling coating methods, will not drastically reduce membrane flux because they enhance the wetting property of polysulfone-polyamide hydrophobic TFC membranes. A major disadvantage of dopamine coating is that it cannot withstand harsh chemicals and may delaminate under high crossflow conditions (McCloskey et al., 2010).

7.3 Membrane Housings and Configurations

7.3.1 Introduction

Forward osmosis membranes are often mass produced and then inserted into a housing called a "forward osmosis element" or "forward osmosis module," which is designed to allow fluids to flow across the feed side and draw sides of the forward osmosis membrane. For the purpose of simplicity, a forward osmosis element will be a single unit and a forward osmosis module will be a combination of multiple forward osmosis elements connected within a single housing to operate in series.

7.3.2 Overview of Commercial Forward Osmosis Elements

There are several different types of forward osmosis elements on the market, and recent innovations have reduced system cost and improved forward osmosis system efficiency compared to conventional element types. The ideal membrane elements and modules maximize the membrane surface-to-volume ratio to enhance crossflow velocity across the membrane surface for both feed and draw channels. This high crossflow minimizes ECP on both sides of the membrane, resulting in higher specific flux. To achieve high flowrates, the element should exhibit low headloss, allowing multiple elements to be connected within a module in series without a high pressure drop across the module. To achieve an efficient forward osmosis process, it is important for the module to be able to operate in both cocurrent and counter-current configurations, which will be discussed in the next section.

Early forward osmosis element designs were based on conventional reverse osmosis designs, where spiral wound membrane elements are most commonly used because of their popularity in seawater desalination and the wide availability of low-cost pressure vessels (Bamaga et al., 2011; Gu et al., 2011). To convert the reverse osmosis spiral wound element with three ports (feed, draw, permeate) into a forward osmosis element with four ports (feed in, feed out, draw in, draw out), an extra glue line is needed to create the required flow path. As a result, forward osmosis spiral elements contain a lower packing density than reverse osmosis spiral elements. The added glue lines also can create low-flow zones within the module that reduce mixing and contribute to higher headloss than in similar reverse osmosis systems. While spiral-wound elements are relatively easy to operate in cocurrent mode, it is difficult to operate them in counter-current mode, especially at high recovery because of high headloss and difficulties in designing and operating multistage units.

The other type of element commonly found in reverse osmosis is the hollow fiber membrane element (Liu et al., 2013; Setiawan et al., 2011). Hollow fiber elements have the potential to offer the highest packing density. However, there is a trade-off between packing density and other key design considerations including: 1) fouling potential, 2) viscosity, and 3) feed to draw flow ratios. This is because small diameter fibers allow a high packing density, but also intrinsically have a small flow path for solids, foulants, and fluids to pass through. Small flow paths increase headloss, which increases energy use and restricts mixing at the membrane surface. Reduced mixing enhances fouling, scaling and concentration polarization. Hollow fibers have also traditionally been limited to using asymmetric membranes with lower flux and rejection than TFCs. Thin-film composite hollow fiber membranes are an active area of research showing some success (Panu & Tai-Shung, 2012); however, they have only recently been commercially available from a supplier named Aquaporin.

Some hollow fiber elements have been designed with relatively large fibers, which nullifies some of the advantages of high packing density compared to flat-sheet membrane elements. One benefit is that small hollow fibers can often withstand relatively high hydraulic pressures without deformation, which could be useful for certain high-pressure applications, such as PRO or pressure-assisted forward osmosis.

Unique forward osmosis elements and modules have recently been developed specifically for forward osmosis applications (Gu et al., 2011). These modules are similar to plate and frame designs, but have been optimized for forward osmosis and have a higher packing density and smaller footprint than traditional plate and frame elements and modules. The advantage of these modified plate and frame designs is that they provide large flow paths without dead zones on both the feed and draw sides of the membrane.

Additionally, these modules can be designed to operate at high crossflow velocities with significantly lower rates of headloss (pressure drop) than other types of forward osmosis elements and modules. This improves mixing, which is the key to reducing concentration polarization and maximizing membrane performance, as previously described. Another benefit of these types of elements and modules is that they can be designed to have a uniform surface velocity across an entire array of membrane elements and modules operating in series and/or in parallel, which minimizes concentration polarization and is virtually impossible to achieve with other module types because of flow path, housing, and system design considerations. These modules are also relatively easy to operate in both cocurrent and counter-current modes.

7.3.3 Cocurrent Versus Counter-Current Operation

Cocurrent operation occurs when both the feed and draw solutions enter and exit on the same sides of a module. A counter-current operation occurs when the feed and draw enter and exit on opposite sides of a module. A cocurrent configuration is suitable for low osmotic pressure feeds and low water recovery applications, while counter-current has advantages for high osmotic pressure feeds and high water recovery applications. The counter-current operation results in constant flux across the membrane module as the feed gets concentrated and the draw is diluted. This is ideal for forward osmosis processing because it reduces the required average draw concentration to achieve a target feed concentration.

A lower average draw concentration and consistent osmotic driving force reduces fouling and reverse solute flux (RSF)and improves solute rejection compared to cocurrent operation. This is because the flux is high in the first few elements and low in the final few elements in cocurrent operation. The higher flux exacerbates fouling, the lower flux can allow higher solute passage, and the higher average draw concentration increases RSF.

7.3.4 Ideal Forward Osmosis Draw Properties

The ideal forward osmosis system would include the ideal forward osmosis membrane, the ideal forward osmosis element and module configuration, and the ideal draw solution. The previous sections described different types of forward osmosis membranes and forward osmosis elements and modules, and the following paragraphs present ideal draw solution parameters and considerations.

Selection of a suitable draw solution is a key factor in selecting forward osmosis operational conditions. In general, a good draw solute should be nontoxic, inexpensive, readily available, relatively easy to regenerate, and highly soluble in various conditions to avoid precipitation. Among many varieties of available draw solutes, sodium chloride is used most often because of its high diffusivity (high specific flux) and because it is relatively easy to re-concentrate using conventional desalination processes such as reverse osmosis. Apart from inorganic salts and sugars, recent publications indicate that polymers, poly-electrolytes, nanoparticles, dendrimers, volatile solutes, switchable polarity solvents/polymers, and organic ionic salts are also being used to make draw solutions (Lutchmiah et al., 2014).

Zhao and Zou (2011) reported that forward osmosis water flux through membranes was different with various draw solutes, even under the same net osmotic driving force. Using sucrose as the draw solution caused a severe dilutive ICP effect, while a sodium chloride draw provided higher flux values because of lesser ICP effects. These differences, in reference to the dilutive ICP and effects of different draw solutions, may be reasonably explained on the basis of diffusivity and viscosity of the draw solution (Hancock & Cath, 2009). A solution with higher viscosity would be more difficult to diffuse into and out of the membrane porous layer, leading to more severe ICP. By considering variations in solute molecular size, solution viscosity and diffusion of solutes, Equation 3.31 can be modified by introducing the concept of constrictivity, as follows:

$$K = \frac{t\tau}{\delta \varepsilon_{eff} D} \tag{3.40}$$

Here, a new parameter, δ, is the constrictivity factor and ε_{eff} is the effective transport through porosity. The value of ε_{eff} is always less than the overall porosity of the membrane. In other words, the ratio of the solute diameter to the pore diameter (i.e., ϕ) influences δ. The empirical relationship between ϕ and δ can be expressed as (Beck & Schultz, 1970)

$$\delta = (1 - \phi)^4 \tag{3.41}$$

and

$$\phi = \frac{molecular\ diameter}{pore\ diameter} < 1 \tag{3.42}$$

From Equations 3.40 and 3.41, it is inferred that larger solute size will result in larger ϕ, but smaller δ. It also reduces the effective porosity of the membrane because of size exclusion and restricted diffusion. As a result, the solute resistivity, K, will become higher based on Equation 3.40.

These relationships, along with constrictivity (as described earlier), indicate that a draw solution with lower viscosity and higher diffusivity allows the draw solute to diffuse into and out of the membrane support layer more easily, thus reducing ICP. Therefore, when selecting a draw solution, it is necessary to consider solute molecule/ion size, diffusivity, and viscosity. Otherwise, permeate flux will be limited significantly by ICP while operating in forward osmosis mode. Additional factors that also need to be considered when selecting and comparing draw solutions for commercial applications include, but are not limited to, the following:

- Membrane versus thermal draw recovery methods and related efficiencies
- Chemistry (toxicity, corrosiveness, scaling potential, ion exchange between feed and draw, etc.)

- Point of entry (e.g., cocurrent versus counter-current operation) and pressure drop
- Surface velocity of the feed and draw especially for feeds that have high osmotic pressure or are highly viscous
- Integrity monitoring (e.g., dye addition, draw overpressure, or other monitoring method)
- Draw conditioning (e.g., blowdown, oxidant, antifoulant, and/or antiscalant addition)
- Availability and ease of use for adding draw solute that is lost from the system via RSF, draw blowdown, maintenance activities, and draw regeneration

8.0 REFERENCES

Adamson, A. W., & Gast, A. P. (1997). *Physical chemistry of surfaces* (6th ed.). John Wiley & Sons, Inc.

Aguinaldo, J. T. (2009). Integrated chemical precipitation and ultrafiltration treatment of wastewater for zero liquid discharge. *Proceeding of the AIChE Central Florida Section 2009 Clearwater Conference*. June 8–9. AIChE CF Section, Florida.

Ajuha, S. (2017). *Chemistry and water the science behind sustaining the world's most crucial resource*. Elsevier. https://doi.org/10.1016/C2015-0-04748-7

Alexandratos, S. D. (2009). Ion-exchange resins: A retrospective from Industrial and Engineering Chemistry Research. *Industrial and Engineering Chemistry Research*, *48*, 388–398. https://www.separationprocesses.com/Adsorption/AD_Chp01a.htm

American Water Works Association & American Society of Civil Engineers. (1997). *Water treatment plant design* (3rd ed.). McGraw-Hill.

American Water Works Association. (2005). *Manual of water supply practices, M53—Microfiltration and ultrafiltration membrane for drinking water.* American Water Works Association, Membrane Process Committee.

American Water Works Association. (2010). *Manual of water supply practices, M37—Operational control of coagulation and filtration processes.* American Water Works Association Coagulation and Filtration Committee.

American Water Works Association. (2011). *Operational control of coagulation and filtration processes* (AWWA Manual M37, 3rd ed.). American Water Works Association.

American Water Works Association. (2018). *Manual of water supply practices. M69–Inland desalination and concentrate management.* American Water Works Association, Membrane Process Committee.

Antony, A., Low, J. H., Gray, S., Childress, A. E., Le-Clech, P., & Leslie, G. (2011). Scale formation and control in high pressure membrane water treatment systems: A review. *Journal of Membrane Science, 383*, 1–16.

ASTM International. (2019). *ASTM D2035-19, Standard practice for coagulation-flocculation jar test of water.* https://www.astm.org

Baker, R. W. (2004). *Membrane technology and applications* (2nd ed.). John Wiley & Sons.

Bamaga, O. A., Yokochi, A., Zabara, B., & Babaqi, A. S. (2011). Hybrid FO/RO desalination system: Preliminary assessment of osmotic energy recovery and designs of new FO membrane module configurations. *Desalination, 268*, 163–169.

Barnes, D., & Wilson, F. (1983). *Chemistry and unit operations in water treatment.* Applied Science Publishers Ltd.

Bates, B. (2017). *Aerobic granular sludge technology.* Aqua-Aerobic Systems, Inc. Presented at Ohio Biosolids Nereda Semiar_2017. http://www.ohiowea.org/docs/Ohio_Biosolids_Nereda_Seminar_2017.pdf

Beck, R. E., & Schultz, J. S. (1970). Hindered diffusion in microporous membranes with known pore geometry. *Science, 170*, 1302–1305.

Bennett, A. (2013). 50th anniversary: Desalination—50 years of progress. *Filtration Separation, 50*, 32–39.

Bertheas, U., Majamaa, K., Arzu, A., & Pahnke, R. (2009). Use of DBNPA to control biofouling in RO Systems. *Desalination and Water Treatment, 3*, 175–178.

Bond, R. G., & Veerapaneni, S. (2006). Zeroing in on ZLD. *Journal of the American Water Works Association, 100*, 76–89.

Bruce, E. L., & Elimelech, M. (2012). Membrane-based processes for sustainable power generation using water. *Nature, 488*, 313–319.

Bryan, D. C., Almaraz, N., & Tzahi, Y. C. (2015). Forward osmosis desalination of oil and gas wastewater: Impacts of membrane selection and operating conditions on process performance. *Journal of Membrane Science, 488*, 40–55.

Bryan, D. C., Pei, X., Edward, G. B., Jack, H., Keith, L., Nathan, T. H., & Tzahi, Y. C. (2014). The sweet spot of forward osmosis: Treatment of produced water, drilling wastewater, and other complex and difficult liquid streams. *Desalination, 333*, 23–35.

Cadotte, J .E. (1981). *Interfacially synthesized reverse osmosis membrane and its use in removing solute from solute-containing water* (U.S. Patent No. 4277344). U.S. Patent and Trademark Office.

Cadotte, J. E., & Petersen, R. J. (1981). Thin-film composite reverse-osmosis membranes: Origin, development, and recent advances. In A. F. Turbak (Ed.), *Synthetic membranes: ACS symposium series* (Vol. 153). American Chemical Society.

Chellam, S., & Jacangelo, J. G. (1998). Existence of critical recovery and impacts of operational mode on potable water microfiltration. *Journal of Environmental Engineering, 124*, 1211–1219.

Cheryan, T. (1998). *Ultrafiltration and microfiltration* (2nd ed.). CRC Press.

Chong, T. H., Wong, F. S., & Fane, A. G. (2008). The effect of imposed flux on biofouling in reverse osmosis: Role of concentration polarisation and biofilm enhanced osmotic pressure phenomena. *Journal of Membrane Science, 325*, 840–850.

Cleary, J. G., & Eckenfelder, W. W. (2014). *Activated sludge technologies for treating industrial wastewaters: Design and troubleshooting*. DEStech Publications Inc.

Conner, W. C., Cevallos-Candau, J. F., Weist, E. L., Pajares, J., Mendioroz, S., & Cortes, A. (1986). *Langmuir, 2*, 151–154.

De Hann, A. B., & Bosch, H. (2013). *Industrial separation processes*. De Gruyter GMBH.

Duong, P. H. H., Chung, T. S., Wei, S., & Irish, L. (2014). Highly permeable double-skinned forward osmosis membranes for anti-fouling in the eulsified oil-water separation process. *Environmental Science and Technology, 48*, 4537–4545.

Edzwald, J. K., & Haarhoff, J. (2012). *Dissolved Air Flotation for Water Clarification* (AWWA Manual M37, 3rd ed.). American Water Works Association & McGraw Hill.

Fane, A. G. (1986). Ultrafiltration: Factors influencing flux and rejection. In R. J. Wakeman (Ed.), *Progress in filtration and separations, 4*, 101–179.

FilmTec. (2019). *FILMTEC™ reverse osmosis technical manual* (Document No. 609-00071). (available online).

Flynn, D. J. (Ed.). (2009). *The Nalco water handbook* (3rd ed.). McGraw-Hill Companies, Inc.

Garcia Molina, V., Busch, M., & Sehn, P. (2009). Cost savings by novel seawater reverse osmosis elements and design concepts. *Desalination and Water Treatment, 7*, 160–177.

Geise, G. M., Park, H. B., Sagle, A. C., Freeman, B. D., & McGrath, J. E. (2011). Water permeability and water/salt selectivity tradeoff in polymers for desalination. *Journal of Membrane Science, 369*, 130–138.

Ghaffour, N., Missimer, T. M., & Amy, G. L. (2013). Technical review and evaluation of the economics of water desalination: Current and future challenges for better water supply sustainability. *Desalination*, 309, 197–207.

Gimun, G., & Seungkwan, H. (2017). New approach for scaling control in forward osmosis by using in antiscalant-blended draw solution. *Journal of Membrane Science, 530*, 95–103.

Glater, J. (1998). The early history of reverse osmosis membrane development. *Desalination, 117*, 297–309.

Glater, J., Hong, S.-K., & Elimelech, M. (1994). The search for a chlorine-resistant reverse osmosis membrane, *Desalination, 95*, 325–345.

Goldblatt, M. E., & Leitz, C. R. (2015). Chemistry and application of coagulants and flocculants in industrial wastewater treatment systems. *Proceedings of the 87th Annual Water Environment Federation Technical Exhibition and Conference* (pp. 2007–2028). Water Environment Federation.

Goosen, M. F. A., Sablani, S. S., Al-Hinai, H., Al-Obeidani, S., Al-Belushi, R., & Jackson, D. (2005). Fouling of reverse osmosis and ultrafiltration membranes: A critical review, *Separation Science and Technology, 39*, 2261–2297.

Gu, B., Kim, D. Y., Kim, J. H., & Yang, D. R. (2011). Mathematical model of flat sheet membrane modules for FO process: Plate- and frame module and spiral-wound module. *Journal of Membrane Science, 379*, 403–415.

Gude, V. G. (2016). Desalination and sustainability—An appraisal and current perspective. *Water Research, 89*, 87–106.

Hancock, N. T., & Cath, T. Y. (2009). Solute coupled diffusion in osmotically driven membrane processes. *Environmental Science and Technology, 43*, 6769–6775.

Helfferich, F. (1962). *Ion exchange.* McGraw-Hill Book Company.

Hubicki, Z., & Kolodynska, D. (2012). *Ion exchange technologies*. InTech Publishing.

International Atomic Energy Agency. (2000). *Examining the economics of seawater desalination using the DEEP code* (Report No. IAEA-TEC-DOC-1186). International Atomic Energy Agency.

Jacobs, P. A., Flanigen, J. C., Jansen, J. C., & van Bekkum, H. (2001). *Introduction to zeolite science and practice* (2nd ed.). Elsevier Science B.V.

Jiao, B., Cassano, A., & Drioli, E. (2004). Recent advances on membrane processes for the concentration of fruit juices: A review. *Journal of Food Engineering, 63*, 303–324.

Johnson, J., & Busch, M. (2010). Engineering aspects of reverse osmosis module design. *Desalination and Water Treatment, 15*, 236–248.

Kailasam, V., Rosenberg, E., & Nielsen, D. (2009). *Industrial & Engineering Chemistry Research, 48*(8), 3991–4001.

Karimi, H. (2017). Effect of pH and initial pb(II) Concentration on the lead removal efficiency from industrial wastewater using Ca(OH)2. *International Journal of Water and Wastewater Treatment, 3*(2): doi http://dx.doi.org/10.16966/2381-5299.139

Kawamura, S. (2000). *Integrated design and operation of water treatment facilities* (2nd ed.). John Wiley & Sons.

Kerri, L. H., Nathan, T. H., Nathan, R. H., Eric, W. A., Edward, G. B., Pei, X., & Tzahi, Y. C. (2013). Forward osmosis treatment of drilling mud and fracturing wastewater from oil and gas operations. *Desalination, 312*, 60–66.

Kim, H. W., McCloskey, B. D., Choi, T. H., Lee, C., Kim, M. J., Freeman, B. D., & Park, H. B. (2013). Oxygen concentration control of dopamine-induced high uniformity surface coating chemistry. *ACS Applied Material Interfaces, 5*, 233–238.

Kong, F.-X., Yang, H.-W., Wu, Y.-Q., W., Wang, X.-M., & Xie., Y. F. (2015). Rejection of pharmaceuticals during forward osmosis and prediction by using the solution–diffusion model. *Journal of Membrane Science, 476*, 410–420.

Lander, J., & Chan, M. T. (2012). Challenges of recycling cooling tower blowdown in a power plant. *Proceedings of the Ultrapure Water Asia 2012 Conference*. Singapore, July 4. Tall Oaks Publishing.

Lash, L. D., & Kominek, E. G. (1980). Primary-waste-treatment methods. In V. Cavaseno (Ed.), *Industrial wastewater and solid waste engineering*. Chemical Engineering McGraw-Hill Publishing Company.

Lee, K. L., Baker, R. W., & Lonsdale, H. K. (1981). Membranes for power-generation by pressure retarded osmosis. *Journal of Membrane Science, 8*, 141–171.

Lee, S., Boo, C., Elimelech, M., & Hong, S. (2010). Comparison of fouling behavior in forward osmosis (FO) and reverse osmosis (RO). *Journal of Membrane Science, 365*, 34–39.

Lee, H., Dellatore, S. M., Miller, W. M., & Messersmith, P. B. (2007). Mussel-inspired surface chemistry for multifunctional coatings. *Science, 318*, 426–430.

Li, X., Hu, X., & Cai, T. (2017). Construction of hierarchical fouling resistance surfaces onto poly(vinylidene fluoride) membranes for combating membrane biofouling. *Langmuir, 33*, 4477–4489.

Liu, C., Fang, W., Chou, S., Shi, L., Fane, A. G., & Wang, R. (2013). Fabrication of layer-by-layer assembled FO hollow fiber membranes and their performances using low concentration draw solutions. *Desalination, 308*, 147–153.

Liu, X., Demir, N. K., Wu, Z., & Li, K. (2015). *Journal of the American Chemical Society, 137*, 6999–7002.

Locke, P. J., Greiner, F., Popko, R., & Cooke, N. (2015) ZLD Process utilizes ultrafiltration to achieve 98.5% overall recovery. *Proceedings of the American Water Works Association/American Membrane Technology Association Membrane Technology Conference & Exposition* [CD-ROM]; Orlando, Florida, March 2–6; American Water Works Association.

Loeb, S., Titelman, L., Korngold, E., & Freiman, J. (1997). Effect of porous support fabric on osmosis through a Loeb–Sourirajan type asymmetric membrane. *Journal of Membrane Science, 129*, 243–249.

Lonsdale, H. (1982). The growth of membrane technology. *Journal of Membrane Science, 10*, 81–181.

Lutchmiah, K., Verliefde, A. R. D., Roest, K., & Rietveld, L. C. (2014). Forward osmosis for application in wastewater treatment: A review. *Water Research, 58*, 179–197.

Matin, A., Khan, Z., Zaidi, S. M. J., & Boyce, M. C. (2011). Biofouling in reverse osmosis membranes for seawater desalination: Phenomena and prevention. *Desalination, 281*, 1–16.

McCloskey, B. D., Park, H. B., Ju, H., Rowe, B. W., Miller, D. J., Chun, B. J., Kin, K., & Freeman, B. D. (2010). Influence of polydopamine deposition conditions on pure water flux and foulant adhesion resistance of reverse osmosis, ultrafiltration, and microfiltration membranes. *Polymer, 51*, 3472–3485.

McCutcheon, J. R., & Elimelech, M. (2006). Influence of concentrative and dilutive internal concentration polarization on flux behavior in forward osmosis. *Journal of Membrane Science, 284*, 237–247.

McCutcheon, J. R., & Elimelech, M. (2007). Modelling water flux in forward osmosis: Implications for improved membrane design. *American Institute of Chemical Engineers, 53*, 1736–1744.

Mcdonald, D. F. (1973). Ultrafiltration joins the ranks of industrial separation processes. *Process Engineering*, 79, 76–77.

Mehta, G. D., & Loeb, S. (1978a). Internal polarization in the porous substructure of a semi-permeable membrane under pressure-retarded osmosis. *Journal of Membrane Science, 4*, 261–265.

Mehta, G. D., & Loeb, S. (1978b). Performance of permasep B-9 and B-10 membranes in various osmotic regions and at high osmotic pressures. *Journal of Membrane Science, 4*, 335–349.

Milne, N. A., O'Reilly, T., Sanciolo, P., Ostarcevic, E., Beighton, M., Taylor, K., Mullett, M., Tarquin, A. J., & Gray, S. R. (2014). Chemistry of silica scale mitigation for RO desalination with articular reference to remote operations. *Water Research*, *65*, 107–133.

Ministry of Health. (2007). *Pathogens and pathways, and small drinking water-water supplies: Resources for the drinking-water assistance programme*. Wilmington, New Zealand.

Moyer, B. A., Custelcean, R., Hay, B. P., Sessler, J. L., Bowman-James, K., Day, V. W., & Kang, S. (2012). *Inorganic Chemistry*, *52*, 3473–3490.

Nancharaiah, Y. V., & Kiran Kumar Reddy, G. (2017). Aerobic granular sludge technology: Mechanisms of granulation and biotechnological applications. *Bioresource Technology*, *247*, 1128–1143. https://doi.org/10.1016/j.biortech.2017.09.131

Nguyen, A., Zou, L., & Priest, C. (2014). Evaluating the anti-fouling effects of silver nanoparticles regenerated by TiO2 on forward osmosis membrane. *Journal of Membrane Science, 454*, 264–271.

Odegaard, H. (2006). Innovations in wastewater treatment: the moving bed biofilm process. *Water Science and Technology, 53*(9), 17–33.

Panu, S., & Tai-Shung, C. (2012). High performance thin-film composite forward osmosis hollow fiber membranes with macrovoid-free and highly porous structure for sustainable water production. *Environmental Science and Technology, 46*, 7358–7365.

Paul, M., & Jons, S. D. (2016). Chemistry and fabrication of polymeric nanofiltration membranes: A review. *Polymer*, *103*, 417–456.

Perry, R. H., & Green, D. W. (1973). *Perry's chemical engineers' handbook* (6th ed.). McGraw Hill.

Phillip, W. A., Yong, J. S., & Elimelech, M. (2010). Reverse draw solute permeation in forward osmosis: Modeling and experiments. *Environmental Science and Technology, 44*, 5170–5176.

Pitzer, K. S. (1991). *Activity coefficients in electrolyte solutions* (2nd ed.). CRC Press.

Porter, M. C. (1973) Membrane ultrafiltration for pollution abatement and by-product recovery. *AIChE Symposium Series 69,* 129, 100–122.

Rautenbach, R., & Albrecht, R. (1989). *Membrane processes*. John Wiley & Sons.

Revanur, R., McCloskey, B., Breitenkamp, K., Freeman, B. D., & Emrick, T. (2007). Reactive amphiphilic graft copolymer coatings applied to poly(vinylidene fluoride) ultrafiltration membranes. *Macromolecules, 40*, 3624–3630.

Reynolds T. D., & Richards P. A. (1996). Unit operations and processes in environmental engineering (2nd ed.). PWS Publishing Company.

Richardson, J., Djukanovic, V., & Henderson, W. (2016). Compositions and methods for selective anion removal (U.S. Patent No. WO 2016054644 A1). U.S. Patent and Trademark Office.

Robinson, R. A., & Stokes, R. H. (2002). *Electrolyte solutions* (2nd ed.). Dover Publications, Inc.

Sandia National Laboratories/U.S. State Department. (2011). *The role of chemistry and material science in advanced water treatment. SAND 2011-5958P*. Sandia National Laboratories.

Setiawan, L., Wang, R., Li, K., & Fane, A. G. (2011). Fabrication of novel poly(amide-imide) layer. *Journal of Membrane Science, 369*, 196–205.

She, Q., Wang, R., Fane, A. G., & Tang, C. Y. (2016). Membrane fouling in osmotically driven membrane processes: A review. *Journal of Membrane Science*, *499*, 201–233.

Shrivastava, A., & Stevens, D. M. (2018). Energy efficiency of reverse osmosis. In V. G. Gude (Ed.), *Sustainable desalination handbook: Plant selection, design and implementation*. Butterworth-Heinemann/Elsevier.

Stevens, D. M., Shu, J. Y., Reichert, M., & Roy, A. (2017). Next-generation nanoporous materials: Progress and prospects for reverse osmosis and nanofiltration. *Industrial & Engineering Chemistry Research, 56*, 10526–10551.

Tchobanoglous, G., Burton, F. L., Stensel, H. D., & Metcalf & Eddy. (2014). *Wastewater engineering: Treatment and resource recovery* (5th ed.). McGraw-Hill Education.

Tilley, E., Ulrich, L., Lüthi, C., Reymond, Ph., Schertenleib, R., & Zurbrügg, C. (2014). *Compendium of sanitation systems and technologies* (2nd ed.). Swiss Federal Institute of Aquatic Science and Technology (Eawag).

Tiraferri, A., Yip, N. Y., Straub, A. P., Romero-Vargas Castrillon, S., & Elimelech, M. (2013). A method for the simultaneous determination of transport and structural parameters of forward osmosis membranes. *Journal of Membrane Science, 444*, 523–538.

Tow, E. W., Lienhard, V., & John, H. (2017). Unpacking compaction: Effect of hydraulic pressure on alginate fouling. *Journal of Membrane Science, 544*, 221–233.

U.S. Environmental Protection Agency. (1990). *Technologies for upgrading existing or designing new drinking water treatment facilities* (EPA/625/4-89/023). U.S. Environmental Protection Agency.

U.S. Environmental Protection Agency. (2019). *Industrial wastewater treatment technology database*. https://watersgeo.epa.gov/iwtt/guided-search

U.S. Environmental Protection Agency. (2005). *Membrane filtration guidance manual* (EPA 815-R-06_009). U.S. Environmental Protection Agency.

van Vuuren, L. R. J. (1997). Dissolved air flotation. In F. A. van Duuren, (Ed.), *Water purification works design*. Water Research Commission. ISBN 1 86845 345 6.

Weber, W. J. (1972). *Physicochemical processes for water quality control*. John Wiley & Sons.

Wheaton, R. M., & Lefevre, L. J. (2000). *DOWEX ion exchange resins—Fundamentals of ion exchange*.

Wijmans, J. G., & Baker, R. W. (1995). The solution diffusion model: A review. *Journal of Membrane Science, 107*, 1–21.

Wilkinson, J. G. (1879). *The manners and customs of the ancient Egyptians*. New ed., rev. and cor. New York: Scribner and Welford.

Wolfe, T. D., Bourcier, W. L., & Metcalfe, P. (2011). Osmotic pressure calculation using Pitzer equation modeling. International Desalination Association World Congress. Perth, Australia, September 4–9.

Xie, M., Nghiem, L. D., Price, W. E., & Elimelech, M. (2014). Impact of organic and colloidal fouling on trace organic contaminant rejection by forward osmosis: Role of initial permeate flux. *Desalination, 336*, 146–152.

Xue, J., Junhong, S., Can, W., Jing, W., & Chuyang, Y. T. (2012). Rejection of pharmaceuticals by forward osmosis membranes. *Journal of Hazardous Materials,* 227–228, 55–61.

Zhao, D., & SenGupta, A. K. (2000). *Industrial & Engineering Chemistry Research, 39*, 455–462.

Zhao, S., & Zou, L. (2011). Relating solution physicochemical properties to internal concentration polarization in forward osmosis. *Journal of Membrane Science, 379*, 459–467.

9.0 SUGGESTED READINGS

American Water Works Association. (2007). *Manual of water supply practices, M6—Membrane applications for water reuse*. American Water Works Association Membrane Process Committee.

American Water Works Association. (2007). *Manual of water supply practices. M46–Reverse osmosis and nanofiltration*. American Water Works Association, Membrane Process Committee.

Betz Laboratories, Inc. (1991). *Betz handbook of industrial water conditioning* (9th ed.). Betz Laboratories, Inc.

De, S., & Modal, S. (2012). *Micellar enhanced ultrafiltration*. CRC Press.

Duffus, J. H. (2002). *Pure applied chemistry*, *74*, 793–807.

ENR Cost Indices. (2020). https://www.enr.com/economics/historical_indices/construction_cost_index_history_2020.html

Edzwald, J. K. (2012). *Dissolved air flotation for water clarification*. McGraw Hill.

Encyclopedia Britannica. *Chemical precipitation*. https://www.britannica.com/science/chemical-precipitation

Jankovic, A. (2015). Developments in iron ore comminution and classification technologies. In L. Lu (Ed.), *Iron ore*. Elsevier Ltd. ISBN 978-1-78242-156-6.

Mulder, M. (1996). *Basic principles of membrane technology*. Kluwer Academic.

PR News Wire. (2018). Ion exchange resins market—global industry analysis, size, share, growth, trends, and forecast 2018–2026. https://www.prnewswire.com/news-releases/ion-exchange-resins-market—global-industry-analysis-size-share-growth-trends-and-forecast-2018---2026-300647640.html

Table of Stability Constants. *Metal chelates*. Dojindo Molecular Technologies, Inc. https://www.dojindo.com/Images/Product%20Photo/Chelate_Table_of_Stability_Constants.pdf

Water Environment Research Foundation. (1993). *Guidance manual for polymer selection in wastewater treatment plants* (Project 91-SIP-5). Water Environment Research Foundation.

Water Environment Federation. (2006). *Membrane system for wastewater treatment*. Water Environment Federation.

XFlow (2004) *Technical Bulletin – Selection criteria XIGA™ vs AquaFlex™* XFlow BV., Enchede, Netherlands.

4

Advanced Oxidation Technologies

Ali Safarzadeh-Amiri

1.0 INTRODUCTION

Fresh water is essential for food security, human health, and environmental vitality (http://www.who.int/mediacentre/factsheets/fs391/en/). The demand for fresh water increases with population growth, increasing industrial and agricultural activities, and energy generation, and it is exacerbated by climate change. Water conservation, reuse, and recycling are absolutely necessary to preserve environmental quality and food security, grow industrial output, and meet future energy needs; currently these are being practiced in the United States (California and Texas), Singapore, India, Mexico, and Australia, among others.

There does not exist a technology that can remediate an industrial wastewater and produce recyclable/reusable water. Treatment of industrial wastewater for recycling/reuse is complex and may involves pretreatment (e.g., dissolved air flotation, coagulation/flocculation/sedimentation, biological [aerobic and/or anaerobic] treatment, and posttreatment [sedimentation, nanofiltration or ultrafiltration, and reverse osmosis filtration]). Advanced oxidation processes (AOPs) could serve crucial rues in the above treatment train. For example, AOP may be used at (a) pretreatment stage to remove biocides, improve biodegradability, and so on; and (b) polishing stage to oxidize biocides that do degrade biologically (Laera et al., 2012), reduce reverse osmosis filtration fouling (Souza et al., 2016), and so forth.

Advanced oxidation processes are defined (Glaze, et al., 1987) as "those that rely on the generation of hydroxyl radicals (•OH)," one of the most powerful ($E_0 = 2.75$ V) and highly reactive oxidizing reagents (Legrini et al., 1993) (Table 4.1). The •OH has a short lifetime and reacts indiscriminately and fast ($k = 10^7 - 10^{10}\ M^{-1}\ s^{-1}$; Buxton et al., 1988) with most organic compounds in water either by abstracting a hydrogen atom (Equation 4.1) or adding to a double bond (Equation 4.2), as follows:

$$\bullet OH + RH \rightarrow H_2O + R\bullet \qquad k = 9.7 \times 10^8\ M^{-1}\ s^{-1} \quad (4.1)$$

$$\bullet OH + {>}C{=}C \rightarrow C^{\bullet}{-}C(OH){<} \qquad k = 2.9 \times 10^9\ M^{-1}\ s^{-1} \quad (4.2)$$

The •OH can be generated by a number of processes, including (a) photolysis of hydrogen peroxide (H_2O_2) with ultraviolet (UV) light, (b) O_3 decomposition catalyzed by hydroxide ion, (c) UV irradiation of ozone (O_3), (d) reaction of O_3 with H_2O_2 and hydroxide ion (OH^-), (e) UV irradiation of a mixture of H_2O_2 and O_3, (f) irradiation of titanium dioxide (TiO_2) with

TABLE 4.1 Standard reduction potential of a few oxidants compared with the standard potential of hydrogen electrode (E_0 = 0 V) (Legrini et al., 1993; Wardman, 1989).

Oxidants	Couple	Standard Reduction Potential/ V	pH
Fluorine (F_2)	$F^{\cdot}/F$	3.03	
Hydroxyl Radical (•OH)	$^{\cdot}OH, H^+/ H_2O$	2.75	0
Ozone (O_3)	$O_3/O_3^{\cdot -}$	2.07	
Hydrogen Peroxide (H_2O_2)	$O_2\bullet, 2H^+/H_2O_2$	1.76	0
Hydroperoxyl radical ($HO_2\bullet$)	$HO_2\bullet, H^+/H_2O_2$	1.495	0
Chlorine Dioxide (ClO_2)		1.50	
Chlorine (Cl_2)		1.36	

UV-visible light, (g) irradiation of water (H_2O) with vacuum UV (VUV), and (h) the Fenton and photo-Fenton processes. Of these AOPs, UV/H_2O_2, O_3 and UV/O_3, O_3/H_2O_2 and the Fenton and photo-Fenton AOP processes are by far the most common and applicable for industrial water recycle and covered in some details, and the rest are described briefly. The space limitation and the number of papers published in this area prevent an of the literature.

1.1 The Ultraviolet/Hydrogen Peroxide Advanced Oxidation Process

Hydrogen peroxide (H_2O_2) is a colorless liquid that is similar to, but 40% denser/heavier than, water. It is a relatively powerful oxidizing reagent (E_0 = 1.77 V) and an antiseptic. It is used for bleaching in the pulp and paper industry and controlling sulfide emission in sewer lines. However, H_2O_2 does not react with, and degrade, organic compounds in water in the absence of a catalyst because these reactions have high activation energy. Hydrogen peroxide absorbs in the UV region and generates •OH when irradiated with UV light.

1.2 The Mechanism and Kinetics of the Ultraviolet/Hydrogen Peroxide Process

Hydrogen peroxide weakly absorbs UV light (Figure 4.1) and homolyzes into •OH when irradiated with UV light. The photodecomposition of H_2O_2 was studied extensively in the 1950s by Hunt and Taube (1952), Weeks and Matheson (1956), and Baxendale and Wilson (1957), and this reaction is agreed to proceed as follows:

$$H_2O_2 + h\nu \rightarrow 2\ \bullet OH \qquad \text{Rate} = F_{H_2O_2} I_0 \varphi_{\bullet OH}/V \qquad (4.3)$$

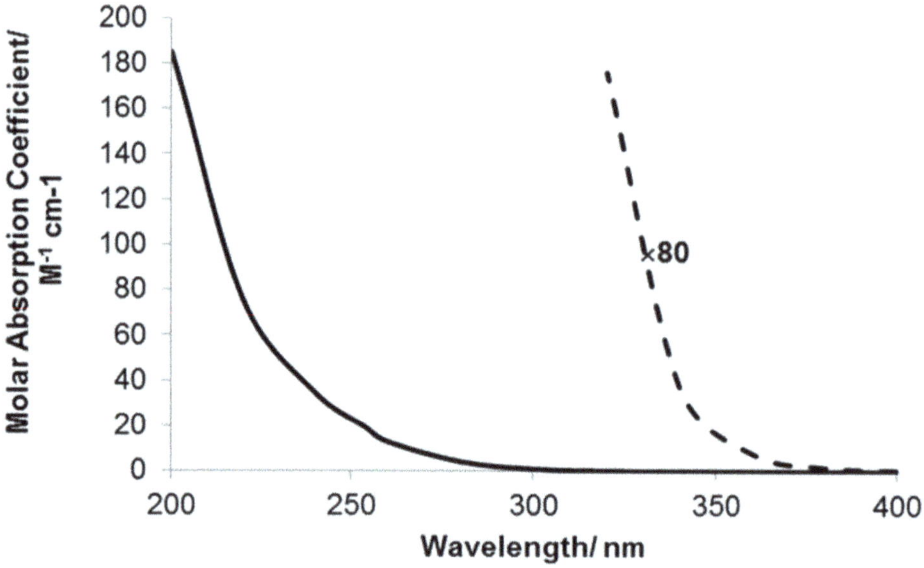

FIGURE 4.1 Plots showing molar absorption coefficient of H_2O_2 as a function of wavelength in Milli-Q water.

$$\bullet OH + H_2O_2 \rightarrow H_2O + HO_2\bullet \qquad k = 2.7 \times 10^7\ M^{-1}\ s^{-1} \tag{4.4}$$

$$HO_2\bullet + \bullet OH \rightarrow H_2O + O_2 \qquad k = 6 \times 10^9\ M^{-1}\ s^{-1} \tag{4.5}$$

$$HO_2\bullet + HO_2\bullet \rightarrow H_2O_2 + O_2 \qquad k = 8.3 \times 10^5\ M^{-1}\ s^{-1} \tag{4.6}$$

where hv—h is Plank's constant (6.626×10^{-34} J/s) and v is the light frequency (s^{-1})—represents photon energy (J) and $F_{H_2O_2}$ is the fraction of light absorbed by H_2O_2,

I_0 = the total incident UV photon flux,
$\varphi_{\bullet OH}$ = the quantum yield of •OH formation, and
V = the total volume of irradiated water.

The net result of the aforementioned reactions, in the absence of •OH scavenger, is conversion of H_2O_2 into H_2O and O_2, as follows:

$$2\ H_2O_2 + hv \rightarrow 2\ H_2O + O_2 \tag{4.7}$$

The overall quantum yield of generation of •OH is unity (1.0) (Baxendale & Wilson, 1957; Hunt & Taube, 1952; Weeks & Matheson, 1956). The •OH generated from photolysis of H_2O_2 (Equation 4.1) reacts fast with both the target organic pollutant (P) and nontarget organic and inorganic compounds (•OH scavengers, S) and degrades them, as follows:

$$\bullet OH + P \rightarrow \text{product(s)} \tag{4.8}$$

$$\bullet OH + S \rightarrow \text{product(s)} \tag{4.9}$$

Applying steady-state approximation to •OH concentration (the H_2O_2 concentration used is typically high enough to generate •OH at a steady-state level), the rate of removal P can be expressed by the following equation:

$$\frac{-d[P]}{dt} = \frac{k_{P,OH}[P] I_0 F_S F_{H_2O_2} \varphi_{OH}/V}{k_{P,OH}[P] + \Sigma_i k_{S_i,OH}[S]_i + k_{H_2O_2,OH}[H_2O_2]} \tag{4.10}$$

where $k_{P,OH}$ = the rate constant of the reaction between •OH and P,
$k_{S_i,OH}$ = the rate constant of the reaction between •OH and S,
$k_{H_2O_2,OH}$ = the rate constant of the reaction between •OH and H_2O_2, and
F_S = the fraction of light absorbed by the solution.

The above Equation 4.10 simplifies to following equations (4.11) and (4.12) under low P and high P values, respectively.

$$-d[P]/dt = k'_1[P] \tag{4.11}$$

$$k'_1 = \frac{k_{P,OH} I_0 F_S F_{H_2O_2} \varphi_{OH}/V}{\Sigma_i k_{S_i,OH}[S]_i + k_{H_2O_2,OH}[H_2O_2]} \tag{4.12}$$

$$\frac{-d[P]}{dt} = k'_0 = I_0 F_S F_{H_2O_2} \varphi_{OH}/V$$

where k'_1 (min^{-1}) is the pseudo-first-order rate constant.

1.3 Figures of Merit

The UV/H_2O_2 degradation of organic pollutants in aqueous solution could consume considerable amount of energy. It is for this reason that a standard of measurement has been defined that can be used to assess electrical efficiency and cost associated with UV/H_2O_2 AOP and compare efficiency of the UV systems provided by different vendors. The measurement standard used to evaluate the electrical efficiency of UV systems is figures of merit (Bolton et al., 2001). It is defined, for the first-order reaction, as the electrical energy in kilowatt hours (kWh) required to reduce the concentration of a pollutant by 1 order of magnitude (90%) in one cubic meter or 1 m^3 (1000 gal)—termed E_{EO}—of contaminated water. The E_{EO} (kWh m^{-3} $order^{-1}$) for the batch and flow-through processes can be calculated using Equations 4.13 and 4.14, as follows:

$$EEO = \frac{P(\mathrm{kW}) \times t \times (\mathrm{min}) \times 1000(\mathrm{L m^{-3}})}{V(\mathrm{L}) \times 60(\mathrm{min\ h^{-1}}) \times \log(c_i/c_f)} \tag{4.13a}$$

$$EEO = \frac{38.4 \times P}{V \times k'_1} \tag{4.13b}$$

$$EEO = \frac{1000 \times P}{60 \times F \times log(c_i/c_f)} \tag{4.14}$$

where P = the power input (kW) from the wall to operate the UV lamp(s),
t = the irradiation time (min),
V = the volume (L) of the water in the reactor,
c_i and c_f = the initial and final concentrations of the pollutant being oxidized,
k'_1 = pseudo-first-order rate constant, and
F = the flowrate (m^3/h).

The factor 38.4 in Equation 4.13b is the conversion factor 1000 × ln(10)/60. Implicit in the aforementioned definition is that UV/H_2O_2 oxidation of P follows pseudo-first-order kinetics.

The figure of merit for the zero-order reaction (i.e., when pollutant concentration is high) is defined as the electric energy in kilowatt-hours required to degrade a unit mass (e.g., 1 kg) of a given contaminant in polluted water, termed EEM. The EEM (kWh/kg) for the batch and flow-through processes can be calculated using Equations 4.15 and 4.16, as follows:

$$E_{EM} = \frac{1000\, Pt}{VM\,(c_i - c_f)} \tag{4.15}$$

$$E_{EM} = \frac{P}{FM\,(c_i - c_f)} \tag{4.16}$$

The term M in these equations refers to the molar mass of the pollutant.

1.4 The Advantages and Disadvantages of the Ultraviolet/Hydrogen Peroxide Advanced Oxidation Process

The UV/H_2O_2 process has both advantages and disadvantages. Some of the advantages, which have previously been alluded to, are as follows:

- The process generates •OH, one of the most powerful oxidants that reacts quickly with most organic and inorganic pollutants and degrades and destroys them.
- The rate of oxidation and removal of pollutants is typically fast, requiring short hydraulic retention time and relatively small space.

- The pollutants are not transferred from one phase to another for later treatment, which is the case in air stripping, granular activated carbon (GAC), and reverse osmosis treatment processes.
- The vast majority of organic pollutants can be destroyed by this process.
- The treatment can be carried out at circumneutral pH.

The disadvantages of the UV/H_2O_2 process are as follows:

- The process is not efficient for the treatment of industrial wastewater contaminated with moderate to high concentrations of target an non-target pollutants and with moderate to high UV absorbance, which is a common feature of most industrial effluents.
- The treatment costs, both capital and operating, could be high for water with moderate alkalinity and background chemical oxygen demand (COD), which are rather commonplace traits for industrial wastewater.

Other disadvantages worth noting are possible mercury spills in cases of lamp breakage and interruption of the treatment process.

1.5 Applications

The UV/H_2O_2 AOP can be used to remediate industrial wastewater, pretreat them to improve their biodegradability, or to polish those that have been treated biologically. Examples of these applications are provided in Table 4.2. All tests were performed employing bench-scale units. UV/H_2O_2 AOP was used to polish the first two examples in Table 4.2, remediate the next example, and pretreat the last two to improve their biodegradability.

The pharmaceutical wastewater—entry 1, Table 4.2—is contaminated with an antibacterial drug, nalidixic acid, its intermediate products, and reagent —mostly acetic acid (920 mg/L). It is characterized with 2000 mg/L COD, 620 mg/L total organic carbon (TOC), 50 mg/L nalidixic acid, and pH 4 (Laera et al., 2012). The water can be treated biologically to remove ≥90% COD but not nalidixic acid, a bactericide that can be degraded efficiently with an AOP. The two processes can be incorporated into a recirculation loop in which AOP is placed in the effluent stream of bioreactor. The EEO for UV/H_2O_2 oxidation of 50 mg/L of nalidixic acid—using 17 mg/L H_2O_2—was found to be approximately 7.8 kWh/m^3. The UV/H_2O_2 AOP was used to reduce silt density, reduce reverse osmosis fouling, and generate high-quality reuse water from a petroleum wastewater that was collected after biotreatment and sand filtration processes (Souza et al., 2016). It was characterized by high absorbance ($\varepsilon_{254\ nm}$= 0.5), high conductivity (1130 to

1950 μS/cm), 12 to 19 mg/L TOC, 7–57 mg/L TSS (total suspended solids), and 5 to 28 mg/L VSS (volatile suspended solids). The UV/H_2O_2 AOP was able to mineralize. The EEO for the oxidation and removal of TOC was found to mineralize 64% of TOC, reduce its silt density, and improve reverse osmosis flux with 136 to 215 and an EEO of approximately 31 kWh/m^3. The treatment cost is high compared to other AOP as will be discussed in the next section.

The UV/H_2O_2 AOP was found remarkably efficient for treating effluent from an acetate and polyester fiber dyeing process. The EEO—using 300 mg/L of H_2O_2—for the oxidation and removal of COD was found to be approximately 8 kWh/m^3. The treatment cost is approximately \$1.28 m^{-3}, which may not be the cheapest because the wastewater is highly biodegradable (BOD_5-to-COD is 0.4). In contrast, treatment of olive mill effluent with this process is inefficient and requires a large UV-dose and a high concentration of H_2O_2 and is not economically viable. The inefficiency can be attributed to high COD and the complex nature of organics present in this wastewater (Bedoui et al., 2008).

The last example provided in Table 4.2 is a petroleum wastewater that cannot be treated biologically without dilution (Philippopoulos & Poulopoulos, 2003). The UV/H_2O_2 AOP, as a pretreatment option to improve its bio-treatability, was explored. The oxidation and removal of COD is slow, requires high concentration of H_2O_2 and UV-dose and results from the nature of organic pollutants in this effluent.

2.0 THE OZONE AND ULTRAVIOLET/OZONE ADVANCED OXIDATION PROCESS

2.1 Ozonation

Ozone is a pale blueish gas and a powerful but selective oxidant with a pungent smell, limited solubility, and stability in water. The solubility of O_3 in water is determined by the Henry's constant or solubility ratio (R_L).

$$R_L = C_l/C_g$$

Where C_l and C_g are ozone concentration in liquid and gas phases, respectively. For example, R_L value was found to be approximately 0.3—between pH 2 and 6—for the C_l value of 3% volume percent (Erogova et al., 2015).

Ozone is an electrophile (i.e., it is electron deficient and accepts electrons) and reacts fast with nucleophiles (i.e., chemicals that can donate electrons). For example, O_3 reacts fast with amines (i.e., antibiotic drugs and azo-dyes) and phenol and its derivatives. The rate of reaction increases with the increasing electron donating ability of organic compounds (see Table 4.3).

TABLE 4.2 Examples of UV/H_2O_2 oxidation and removal of specific and nonspecific organic contaminants from industrial wastewater.

Source	Pollutant	C_i/mg/L	C_f/mg/L	pH	EEO/kWh $Order^{-1}$ m	Operating Cost[a]/$ $Order^{-1}$ m^{-3}	References
Pharmaceuticals	nalidixic acid	50	2	4	7.8	0.80	Laera et al., 2011
Petroleum	TOC	39–81	12–29	7	31	2.26	Souza et al., 2016
Textile	COD/Color	930	93	3[b]	8.0	1.26	Azbar et al., 2004
Olive mill	COD	2160	175	5	220	29.91	Bedoui et al., 2008
Petroleum	Remediation	9000	45	7.8	34.7[c]	12.34	Philippopoulos & Poulopoulos, 2003

[a]Includes electricity ($0.1 for one kWh) and H_2O_2 ($3.0 for a gallon of 50% H_2O_2) costs.
[b]This is the optimum pH for UV/H_2O_2 treatment. The solution pH is 9.2.
[c]This an estimated value using data presented in a graph.

TABLE 4.3 Rate constants for the reaction of selected groups of organic compounds with O_3 (Hoigne and Bader, 1983a, 1983b) and •OH (Buxton et al., 1983).

Compound	$k_{O_3}/M^{-1}\ s^{-1}$	$k_{OH}/M^{-1}\ s^{-1}$
Amines	$<280 - 10^8$	$(1.7 - 34) \times 10^9$
BTEX	$2 - 140$	$(5.1 - 7.9) \times 10^9$
Chlorobenzene	0.75	5.4×10^9
Phenol	1.3×10^3	$(1.2 - 7.6) \times 10^9$
Chlorophenol	600	$(1.2 - 7.6) \times 10^9$
Primary Alcohols	$0.024 - \leq 0.8$	$(9.7 - 42) \times 10^8$
Aldehydes	0.1	$(7.3 - 10) \times 10^8$
Ketones	0.02 – 0.07	1.1×10^8
Chloromethanes	$<0.005 - \leq 0.1$	$<10^5 - 5.8 \times 10^7$

Some researchers also consider ozonation to be an AOP and this is because decomposition of ozone catalyzed by hydroxide ion generates •OH (Merenyi et al., 2010a; Staehalin & Hoigne, 1982). The rate of generation of •OH increases with increasing solution pH and ozonation, as an AOP, it is most effective at pH $\geq$ 11.0. Even then, the rate of generation of •OH is slow.

2.2 The Ultraviolet/Ozone Advanced Oxidation Process

Ozone strongly absorbs UV light ($\varepsilon_{254\ nm} = 3200\ M^{-1}\ cm^{-1}$; Vogna et al., 2004). The primary products of irradiation of O_3—dissolved in water—with UV light are singlet oxygen, $O(^1D)$, and molecular oxygen in the ground ($^3\Sigma_g^-$) or excited state ($1\Delta_g$) (Reisz et al., 2003; von Sonntag, 2008).

$$O_3 + h\nu \leftrightarrow O(^1D) + O_2 \rightarrow \phi = 0.64 \pm 0.05 \quad (4.18)$$

$$O(1D) + H_2O \rightarrow (H_2O_2)\text{-hot} \rightarrow (HO\bullet\ \bullet OH) \rightarrow H_2O_2 \quad (4.19)$$

$$(HO\bullet\ \bullet OH) \rightarrow 2\ \bullet OH\ \phi = 0.1 \quad (4.20)$$

$O(^1D)$ is highly energetics and reacts fast with all substrate, including water and generates "hot" H_2O_2, which decomposes homolytically into a pair of geminates •OH radicals, (HO• •OH), in a solvent cage, that mostly recombines to form H_2O_2 and only a small fraction escapes from the solvent cage (Reisz et al., 2003; von Sonntag, 2008). Therefore, H_2O_2 is the major product of UV-irradiation of O_3 and •OH a minor product.

2.3 The Advantages and Disadvantages of Ozonation and Ultraviolet/Ozone Advanced Oxidation Process

Ozonation and UV/O_3 AOP have some of the advantages of the UV/H_2O_2 AOP. The advantages include

- Ozone is generated easily from air and oxygen.
- O_3 reacts with, and degrades, a variety of contaminants—both organic and inorganic—and both the ozonation (at an elevated pH) and UV/O_3 process generates •OH.
- They do not—for the most part—transfer pollutants from one medium to another, and can be used to treat the vast majority of organic pollutants in water.
- Typically, the toxicity of toxic compounds oxidized by ozone does not increase.
- The UV/O_3 process may exhibit "synergistic" effect; that is, the combination of O_3 and UV irradiation would be more efficient than the sum of the effects of O_3 and UV irradiation when applied individually for reasons that are discussed above. However, this effect is seen in specific cases.

The disadvantages of ozonation and UV/O_3 AOP are as follows:

- The electrical energy required to generate O_3 is high because of low ozone generator yield (6% to 12% from O_2 and 4% to 6% from air) and has to be generated onsite.
- The yield of •OH—from both the hydroxide catalyzed decomposition process and UV irradiation—is low.
- The capital cost is high, higher than UV/H_2O_2 AOP.
- The solubility of O_3 in water is low (-110 mg/L).
- The process is not suitable for the treatment of industrial wastewater.

The majority of •OH is generated indirectly—by the H_2O_2 catalyzed decomposition of O_3—in this process. As a result, Glaze et al. (1987) concluded that it would be much easier and economical to use H_2O_2 rather than UV light to generate •OH.

2.4 Application

The ozonation and UV/O_3 AOP, like UV/H_2O_2 AOP, have also been used for remediation and pretreatment of industrial wastewaters as well as polishing wastewaters that have been treated biologically. Examples of these applications are presented in Table 4.2. Pilot-scale units were used to treat phenol-contaminated wastewater and the raw refinery effluent (last example, Table 4.4) and the rest were treated using bench-scale unit.

The data presented in Table 4.4 show that O_3 and UV/O_3 AOP to be efficient and economical—compared to UV/H_2O_2 AOP (see Table 4.2)—for the oxidation of nalidixic acid and mineralization of TOC in biotreated pharmaceutical (example 1) and refinery effluents (example 2), respectively. Similarly, conclusion can also be drawn for the polishing of biotreated wastewater—Table 4.4, entry 5—with O_3.

Entries 3 and 4 in Table 4.4 show that O_3 oxidation and removal of COD and color from one textile effluent is approximately one order of magnitude higher than the other. This may be attributed to the nature of organic contaminants and differences in other water quality parameters; for example, dissolved solids, alkalinity, and others. Ozone reacts fast with phenol (see Table 4.3) and example 5 shows that it can be used to oxidize phenol in industrial wastewater. It should be mentioned, however, that phenol concentration in this case was reduced from approximately 37.5 to 4.5 mg/L and remained at that level with further ozone addition. This may be attributed to accumulation of intermediate byproducts, which compete effectively with phenol for O_3.

Finally, O_3 and UV/O_3 AOP have also been used to pretreat a pharmaceutical and a refinery effluent—the last two entries in Table 4.4—and improve their biodegradability. The pharmaceutical wastewater is characterized with 1500 mg/L of berberine hydrochloride, a biocide and a BOD_5/COD of approximately 0.015 (Qin at al., 2015) and the refinery wastewater is characterized with 18 000 to 22 000 mg/L COD, 200 to 2000 mg/L of sulfide, and low BOD_5/COD (Boczkaj et al., 2017). Ozone and UV/O_3 AOP were able to oxidize 49.53% and 58.61% of berberine hydrochloride in pharmaceutical effluent. The improvement of solution biodegradability with these pretreatment processes was not reported; however, the BOD_5/COD increased to approximately 0.29 after treating berberine hydrochloride contaminated wastewater with UV/O_3/H_2O_2 AOP and it is not unreasonable to assume O_3 and UV/O_3 AOP also improves biodegradability. The biodegradability of petroleum refinery wastewater decreased by approximately 9% after O_3 pretreatment.

3.0 THE OZONE/HYDROGEN PEROXIDE ADVANCED OXIDATION PROCESS

Hydrogen peroxide is a weak acid and dissociates into hydroperoxide (HO_2^-) and hydrogen ions (H^+). The former reacts fast [$k = (9.6 \pm 2) \times 10^6$] M^{-1}

TABLE 4.4 Characteristics of a few industrial wastewaters[a] that were treated with O_3 and/or UV/O_3 AOP[b] and associated treatment costs.

Source	Pollutant	pH	C_i/mg/L	C_f/mg /L	O_3_Dose/ mg /L	Operating cost/$ Order^{-1} m^{-3}	Reference
Pharmaceuticals	Nalidixic acid	4	50	1	34.5	0.12	Laera et al., 2011
Petroleum	TOC	7	39–81	2–4	60	(1.32)	Souza at al., 2016
Textile	COD & Color	9.2	920	74 (56)	3647 (2455)	12.85 (9.24)	Azbar et al., 2004
Mixed[c]	COD & Color	7.9	515/108	170/4	434/179	1.52/0.63	Yasar et al., 2006
Olefin Plant	Phenol	6.1	37.5	4.5	275	0.96	Mohammadzadeh et al., 2004
Pharmaceuticals	Berberine	7	1500	757 (621)	—	6.07 (7.47)	Qin et al., 2015
Refinery[d]	COD	10.5	20 000	13 800	6800	23.8	Boczkaj et al., 2017

[a]Both refinery wastewaters in this table are from plants that refine bitumen.

[b]Assumed 35 kWh/kg of O_3. Values in parentheses are for UV/O3 AOP.

[c]The wastewater is a mixture of effluents from 100 textile units, five dairy, two sugar, three flour mills; COD values and O_3-doses and costs are for the raw and biotreated wastewater.

[d]The COD and BOD concentrations varied from 18 000 to 20 000 mg/L and 4000 to 6000 mg/L; respectively, and there is approximately 200 to 1000 mg/L of sulfide in this water.

s^{-1} with O_3 (Staehalin and Hoigne, 1982) and generate •OH according to the following:

$$H_2O_2 \leftrightarrow HO_2^- + H^+ \qquad pK_a = 11.6 \tag{4.21}$$

$$HO_2^- + O_3 \rightarrow HO_2\bullet + O_3^{\bullet-} \tag{4.22}$$

$$HO_2\bullet \leftrightarrow O_2^{\bullet-} + H^+ \qquad pK_a = 4.8 \tag{4.23}$$

$$O_2^{\bullet-} + O_3 \rightarrow O_2 + O_3^{\bullet-} \tag{4.24}$$

$$O_3^{\bullet-} + H_2O \leftrightarrow \bullet OH + OH^- + O_2 \tag{4.25}$$

According to the above reactions, 1 mol of H_2O_2 reacts with 2 mol of O_3 and generates 2 mol of •OH.

$$H_2O_2 + 2\,O_3 \rightarrow 2\,\bullet OH + 3\,O_2 \tag{4.26}$$

The above mechanism has commonly been accepted until fairly recently (Merenyi et al., 2010b) when it was found that the •OH yield is only 67% of that predicted by the stoichiometry of reaction 4.26. According to the above mechanism, the first step in the generation of •OH is an electron transfer (Equation 4.21) from hydroperoxide ion (HO_2^-) to O_3. In the revised mechanism, O_3 ads to HO_2^- and generates HO_5^-, which either decomposes homolytically and generates $HO_2\bullet$ and $O_3^{\bullet-}$ in equilibrium with HO_5^- or converts to molecular oxygen and a hydroxide ion (Merenyi et al., 2010b). The overall reaction for the H_2O_2 catalyzed decomposition of O_3 in this revised mechanism is as follows:

$$H_2O_2 + 3/2\,O_3 \rightarrow \bullet OH + ½\,H_2O + 5/2\,O_2 \tag{4.27}$$

According to this reaction, 1 mol of H_2O_2 reacts with 1.5 mol of O_3 and generates 1 mol of •OH.

Generation of •OH is highly efficient by the above process, making O_3/H_2O_2 AOP—also known as Peroxone—an important AOP for remediation of contaminated waters.

3.1 The Advantages and Disadvantages of Ozone/Hydrogen Peroxide Advanced Oxidation Process

The O_3/H_2O_2 AOP has all the advantages of UV/H_2O_2 AOP plus the following advantages:

- Generation of •OH in this process is not affected by the UV absorbance of solution, and

- The process can be used to pretreat some but not all industrial wastewaters to increase their biodegradability.
- It is more efficient than UV/H_2O_2 AOP for treating/pretreating highly contaminated waters (Safarzadeh-Amiri, 2001).

The disadvantages of O_3/H_2O_2 AOP are as follows:

- It is not useful for treating industrial wastewater contaminated with high concentrations of target and nontarget contaminants.
- O_3 has to be generated onsite.
- Applied O_3 may not be totally consumed.
- The overall unit footprint is large, and larger than UV/H_2O_2 footprint.

3.2 Applications

The efficacy and efficiency of Peroxone for the oxidation and the oxidation of organic have mostly been demonstrated treating synthetic wastewater using a bench-scale unit; a few pilot-scale tests treating emerging micropollutant in drinking and/or municipal wastewater have also been conducted. However, few tests have been carried out employing industrial wastewater. Table 4.5 provides summary of O_3/H_2O_2 treatment of four industrial wastewaters along with their associated treatment costs.

Ozone and O_3/H_2O_2 AOP were used to remove color from a swine production wastewater after biotreatment and nanofiltration or ultrafiltration (Table 4.4, entry 1). Ozonation was more efficient and less expensive—both in capital and operations and maintenance (O&M) costs—removing color after nanofiltration, and O_3/H_2O_2 AOP was more effective and economical for color removal after ultrafiltration (Yoon et al., 2014).

As discussed in Section 4.4, biodegradability of the pharmaceutical (contaminated with BH) and the refinery (refining bitumen) wastewaters (Table 4.4, entries 2 and 3) is low and require pretreatment before bioremediation. The biodegradability of both effluent increases with O_3/H_2O_2 AOP but the treatment costs are high to very high, an indication that, perhaps, this AOP is not a good option for pretreatment of these wastewaters.

Finally, the last entry in Table 4.4 is a wastewater from a chemical production plant; the treatment goal was to reduce solution COD, TOC, and TSS and to meet environmental regulations and make the plant effluent reusable. The O_3/H_2O_2 AOP is able to completely eliminate COD and decrease TOC and TSS by approximately 66% each and meet reuse requirements with H_2O_2 and O_3 dose of approximately 1215 mg/L and 68 000 mg/L, respectively. The high treatment cost (approximately \$8.11 m^{-3}) again indicates that this is not the best option for the pretreatment of this wastewater.

TABLE 4.5 The treatment costs—calculated using published data and equations given in previous section —of treating various industrial wastewater with O_3/H_2O_2 AOP.

Source	Pollutant	pH	C_i/mg/L	C_f/mg/L	O_3-Dose/g order^{-1} m^{-3}	Costs/$ order^{-1} m^{-3}	Reference
Agricultural wastewater	Color[a]	8.6–9	280	28	0.069	0.64	Yoon et al., 2014
Pharmaceuticals	Berberine	7	1500	427	—	6.36[b]	Qin et al., 2015
Petroleum wastewater	COD	10.5	20 000	11 400	9800	39.89	Boczkaj et al., 2017
Chemical	COD/TOC	6.8	1100/273	0.01/95	64.8	8.11	Amaral-Silva et al., 2012

[a]Absorbance at 456 nm (Pt-Co)
[b]Given in the paper

4.0 THE ULTRAVIOLET/HYDROGEN PEROXIDE/OZONE ADVANCED OXIDATION PROCESS

The UV/H_2O_2/O_3 AOP is essentially O_3/H_2O_2 AOP plus UV light. It is believed that that the efficiencies of O_3/H_2O_2 and UV/O_3 AOPs increase with addition of UV light and H_2O_2, respectively, which has been demonstrated treating various organic pollutants spiked in distilled and deionized water. For example, Beltran-Heredia et al. (2001) have shown that the rate of oxidation of p-hydroxybenzoic acid in a synthetic wastewater with UV/H_2O_2/O_3 process is approximately 2.9 and 2.4 times faster than with H_2O_2/O_3 and UV/O_3 AOPs, respectively. Synthetic wastewaters are devoid of background COD, alkalinity, dissolved and suspended solids, and absorbs weakly in the UV region and the results obtained treating synthetic waster would hardly be indicative of UV/H_2O_2/O_3 AOP performance for an industrial wastewater treatment.

Industrial wastewaters are characterized with high concentrations of target and no-target contaminants, dissolved and suspended solids, moderate to high alkalinity, and moderate to low transmission in the UV region (see Safarzadeh-Amiri et al., 1996). Under these conditions, the efficiency of O_3/H_2O_2 AOP may not improve noticeably with the addition of UV light. In addition, O_3 is a much stronger absorber of UV light ($\varepsilon_{254\ nm}$ = 3200 M^{-1} cm^{-1}) than H_2O_2 ($\varepsilon_{254\ nm}$ = 18 M^{-1} cm^{-1}) and absorbs majority of UV light (≥86% assuming a steady-state O_3 concentration of approximately 60 mg/L and H_2O_2 concentration of 340 mg/L) that penetrate into the solution and—as discussed earlier—converts to H_2O_2. Therefore, it is not unreasonable to assume that addition of UV light to O_3/H_2O_2 AOP provides little benefit and only increases capital and O&M costs. For example, the oxidation and removal of COD from a linear alkyl benzene production effluent by UV/H_2O_2/O_3 AOP (58%) was found to be marginally more efficient than O_3/H_2O_2 AOP (53%) (Zangeneh et al., 2014). The UV/H_2O_2/O_3 AOP is not being marketed at this point and may not be used for industrial wastewater treatment and reuse.

5.0 THE ULTRAVIOLET/TITANIUM DIOXIDE ADVANCED OXIDATION PROCESS

Titanium dioxide (TiO_2) is characterized by a filled valence band (VB) and an empty conduction band (CB); the energy difference between the VB and CB, E_g (3.2 eV; the light energy of one Einstein at approximately 388 nm), is known as the "band gap." Absorption of light with an energy hν (λ ≤388 nm) that is equal to, or greater than, E_g promotes an electron (e^-_{cb}) from VB to CB and leaves a hole (h^+_{vb}) in the VB (Schneider et al., 2014).

$$TiO_2 + h\nu \rightarrow e^-_{cb} + h^+_{vb} \quad (4.28)$$

The lifetime of the excited electron (e^-_{cb}) is short, and the excitation energy is mostly converted to heat via electron–hole recombination. The electrons and holes that do not recombine move to the surface of the semiconductor in separate pathways, where they either react with electron acceptors (e.g., O_2) and donors (e.g., OH^-) adsorbed on the catalysts' surface, respectively, or migrate into trapping sites before surface reactions or recombination (Sun and Bolton, 1996).

$$h^+_{vb} + OH- \rightarrow \bullet OH \quad (4.29)$$

$$e^-_{cb} + O_2 \rightarrow O_2^{\bullet -} \quad (4.30)$$

The quantum yield of generation •OH in distilled and deionized water is approximately 4% and increases to approximately 22% in presence of H_2O_2 (Sun and Bolton, 1996). The yield of •OH decreases significantly in water contaminated by nontarget organic contaminants and inorganic ions, for example, $Fe^{2/3+}$ that can scavenge e^-_{cb} and h^+_{vb} (Tong et al., 2012).

The UV/TiO_2 AOP—also known as the photocatalytic water treatment (PWT)—treatment of wastewater has attracted a lot of attention over the past three decades. The major driving forces behind this interest are: (a) the process could be "chemical free," and (b) TiO_2, in principle, could be activated by solar light. However, the fraction of solar irradiation between 300 to 400 nm that reaches the surface of the earth is—as discussed later—approximately 4.5%, and the fraction that can be absorbed by TiO_2 is approximately 3.8% UV light from sun light (Bolton et al., 1995), which is insignificant for the activation of TiO_2.

The vast majority of studies on PWT were performed using synthetic—i.e., distilled, deionized water spiked with low concentration of the desired pollutant(s)—wastewater. Those studies provide little, if any, useful information about the efficacy and efficiency of PWT for remediation of industrial wastewaters that are characterized with moderate to high concentrations of target and nontarget contaminants, alkalinity, UV transparency, and turbidity, which have profound influence on PWT's efficiency. Both organic and inorganic contaminants, for example, iron and copper, can inactivate TiO_2 altogether.

Benotti et al. (2009) compared the degradation efficiency of 32 pharmaceuticals, endocrine disruptors (ECD), and fire retardants spiked at ng/L concentration into Colorado River water from Lake Mead by UV, UV/H_2O_2, and UV/TiO_2 processes using a Purifics PhotoCat™ reactor that was operated with or without TiO_2. The lake water TOC, alkalinity, pH, and absorbance at 254 nm were 2.6 mg/L, 137 mg/L, 8.0, and 0.036 (cm^{-1}), respectively. The experimental results show that EEO (based on the amount of energy input of UV lamps) for the destruction pharmaceuticals, EDC, and fire retardants by UV/TiO_2 is similar to UV photolysis but much higher than UV/H_2O_2 with 10 or 20 mg/L of H_2O_2 and UV/TiO_2 efficiency decrease by a factor of approximately four when energy input for pumping and catalyst recovery are included. The low efficiency of UV/TiO_2 compared to UV/H_2O_2 can be attributed to catalyst loading and alkalinity and natural organic matter in lake water.

The above example shows that the efficiency of UV/TiO_2 AOP is much lower than UV/H_2O_2 AOP and, for now at least, it is not an economically favorable option for remediation of industrial wastewaters.

6.0 THE VACUUM ULTRAVIOLET ADVANCED OXIDATION PROCESS

Water absorbs strongly in the vacuum-UV (VUV) region ($\lambda < 190$ nm) and the average absorption coefficient (base e) of water is approximately 2.90 $\pm$ 1.25 cm^{-1} and 1409.6 $\pm$ 125.4 cm^{-1} at 185 nm and 172 nm, respectively (Weeks et al., 1963). As a result, the VUV light at these wavelengths can penetrate to a depth of approximately 0.7 and 0.0014 cm, respectively, before 99.9% of the incident light is absorbed by water. The actual distance travelled by 172-nm light produced by a Xe excimer lamp is much longer than 0.0014 cm, but it is less than 0.1 cm because it has a broad bandwidth. The sources of VUV light are liquified petroleum gas lamps (emitting at 185 and 253.7 nm) and Xe excimer lamps ($\lambda_{max} = 172 \pm 14$ nm).

Irradiation of water with VUV light generates •OH and hydrogen atom (H•) and a small quantity of hydrated electrons (e^-_{aq}) in a small zone (Getoff & Schenck, 1968; Heit et al., 1998).

$$H_2O + h\nu_{vuv} \leftrightarrow \bullet OH + H\bullet \qquad \Phi_{\bullet OH} = 0.33/0.42 \qquad (4.31)$$

$$H_2O + h\nu_{vuv} \rightarrow \bullet OH + H^+ + e^-_{aq} \qquad \Phi\, e^-_{eq} = 0.045 \qquad (4.32)$$

The quantum yield of generation of •OH (Equation 4.40) from 185 nm irradiation is 0.33 (Gettoff & Schenck, 1968) and from 172 nm irradiation is 0.43 (Heit et al., 1998) at room temperature.

The degradation of organic pollutants with VUV AOP is energy intensive and uneconomical. In addition, •OH generated with VUV irradiation of water, as previously described, reacts fast with organic compounds and generates high concentration of C-centered radicals, which is short lived and recombine to produce oligomers and polymers (Oppenlander, et al., 2005).

Vacuum-UV irradiation of perfluorinated compounds (PFCs), such as, perfluorooctanoic acid (PFOA) and perfluorooctane sulfonate (PFOS) can lead to their complete defluorination. For example, VUV irradiation of PFOA—$C_8F_{15}COOH$—generates $C_7F_{15}\bullet$ and •COOH; the former reacts with water and generates perfluoroheptanoic acid.

$$C_nF_{2n+1}COOH + h\nu_{vuv} \rightarrow C_nF_{n+1}\bullet + \bullet COOH \quad (4.33)$$

$$C_nF_{2n+1}\bullet + H_2O \rightarrow C_{n-1}F_{2n}COOH + 2\,F^- + 2\,H^+ + H\bullet \quad (4.34)$$

The above processes can ultimately lead to total defluorination of PFOA. The process is at its infancy now and the energy require for the degradation of PFOA, and not total defluorination, high in ultra-pure water (UPW). Table 4.6 presents the EEO for the degradation of PFOA in UPW, tap water, (TW), and in the effluent from two different wastewater treatment facilities (WWE1 and WWE2), along with some of their water quality data. The wastewaters were diluted 59% for the test (Giri et al., 1981). The fraction of fluorine released from PFOA after 3 hours of irradiation was approximately 20.54%, 10.60%, 6.68%, and 6.44% in UPW, TW, WWE1, and WWE2, respectively.

The VUV defluorination of PFCs is inefficient and is not economically a viable option for remediation of PFCs contaminated water or industrial wastewaters at this point.

7.0 THE FENTON ADVANCED OXIDATION PROCESS

The oxidation of maleic acid by a mixture of ferrous ions—Fe(II) or Fe^{2+} —and H_2O_2, also known as the "Fenton Reagent", was reported by Fenton in 1894. Approximately 4 decades later, the reactive species responsible for the oxidation of maleic acid in the aforementioned mixture was shown to be •OH (Haber & Weis, 1934), which was generated from the decomposition of H_2O_2 catalyzed by Fe(II) and Fe(III). The mechanism for the formation of •OH was revised (Barb et al., 1951) later and is commonly known as the "classical" or "free-radical" Fenton reaction because •OH is one of the main products of iron-catalyzed decomposition of H_2O_2. The reaction between Fe(II) and H_2O_2 could also generate ferryl ion ($Fe = O^{2+}$ or Fe^{IV}) depending on the solution pH (Bataineh et al., 2012) or presence of chelating agents (Pignatello et al., 2006; Walling, 1998).

TABLE 4.6 Water quality data for UPW, TW, WWE1, WWE2, and EEO for the degradation of PFOA in these waters (EEO is calculated using data provided by Giri et al., 1981).

Parameter (Unit)	UPW	TW	WWE1	WWE2
TDS (mg/L)	6.0	7.62	7.82	7.84
TOC (mg/L)	0.038	1.42	3.99	4.99
Alkalinity (mg/L)	2	37	56.67	46.67
EEO (kWh $Order^{-1}$ m^{-3})	40.3	98.5	139.0	131.6

7.1 The Mechanism of Generation of the Hydroxyl-Radical

The iron-catalyzed decomposition of H_2O_2, according to "classical" mechanism, proceeds as follows (Pignatello et al., 2006):

Reaction	Rate Constant	Eq. No.
$Fe(II) + H_2O_2 \rightarrow Fe(III) + OH^- + \bullet OH$	$k = 40 - 80\ M^{-1}\ s^{-1}$	(4.35)
$Fe^{3+} + H_2O_2 \rightarrow Fe^{2+} + HO^\bullet_2 + H^+$	$2.7 \times 10^{-3}\ s^{-1}$	(4.36)
$HO\bullet + H_2O_2 \rightarrow HO^\bullet_2 + H_2O$	$2.7 \times 10^{7}\ s^{-1}$	(4.38)
$HO\bullet + Fe^{2+} \rightarrow Fe^{3+} + OH^-$	$5.0 \times 10^{8}\ M^{-1}\ s^{-1}$	(4.39)
$Fe^{3+} + HO^\bullet_2 \rightarrow Fe^{2+} + HO_2^+$	$<2.0 \times 10^{3}\ M^{-1}\ s^{-1}$	(4.40)
$Fe^{2+} + HO^\bullet_2 + H^+ \rightarrow Fe^{3+} + H_2O_2$	$1.2 \times 10^{6}\ M^{-1}\ s^{-1}$	(4.41)
$HO^\bullet_2 + HO^\bullet_2 \rightarrow H_2O_2 + O_2$	$8.3 \times 10^{5}\ M^{-1}s^{-1}$	(4.42)

The hydroperoxyl radical, $HO^\bullet_2$, generated in Equation 4.36 is a weak acid (pK_a= 4.8; Sawyer & Valentine, 1981) and dissociates into H^+ and $O_2^{-\bullet}$ (superoxide ion); the latter could also participate in Equations 4.40 to 4.42. According to the aforementioned mechanism, iron cycles between +2 and +3 oxidation states and decomposes hydrogen peroxide to H_2O and O_2 in the absence of •OH scavengers (Barb et al., 1951; Haber & Weis, 1934). It should be noted that both the ferrous and ferric ions form hex-aqua complexes, $[Fe(H_2O)_6]^{2+/3+}$ with water molecules and the coordination water molecules are eliminated for reasons of simplicity.

The rate of reaction of Fe(II) with H_2O_2 (Equation 4.35) is independent of pH below pH 3 and increases at higher pH. This is because Fe(II) hydrolyzes and forms $Fe(OH)^+$ and $Fe(OH)_2$ species (Bataineh et al., 2012; Wells

& Salam, 1968a), which react faster with H_2O_2. However, decomposition of H_2O_2 catalyzed by Fe^{3+} (Equation 4.36) stops at pH above 3 because Fe^{3+} is insoluble at pH > 3. The rate of Equation 4.36 is affected by the presence of organic and inorganic ligands and increases by a factor of 2 to more than an order of magnitude (Wells & Salam, 1968b) in presence of inorganic ligands and by two to three orders of magnitudes in the presence of organic ligands (Croft et al., 1992). The products of these reactions could be high-valent iron species such as Fe^{IV} rather than •OH (Bataineh et al., 2012).

The decomposition of H_2O_2 catalyzed by Fe^{3+} and the formation of ferrous ions is more complex than shown in Equation 4.36. Hydrogen peroxide forms complexes with different species of Fe(III) (Equations 4.43 and 4.46) first and then decomposes slowly to form Fe(II) and $HO_2^{\bullet}$, as follows:

$$Fe^{3+} + H_2O_2 \rightarrow [Fe(HO_2^-)]^{2+} + H^+ \quad (4.43)$$

$$[Fe(HO)]^{2+} + H_2O_2 \rightarrow [Fe(OH)(HO_2^-)]^+ + H^+ \quad (4.44)$$

$$[Fe(HO_2^-)]^{+2} \rightarrow Fe^{2+} + HO^{\bullet}_2 \quad (4.45)$$

$$[Fe(OH)(\ HO_2^-)]^+ \rightarrow Fe^{2+} + OH^- + HO^-_2 \quad (4.46)$$

The rates of reductive decomposition of $[Fe(HO_2^-)]^{+2}$ and $[Fe(OH)(HO_2^-)]^+$ (Equations 4.45 and 4.46) are slow and the rate-determining steps for generation of OH-radicals in the Fenton process after the initial fast Fenton (Equation 4.35). Ligands that coordinate and/or form insoluble complex with Fe(III) (e.g., chloride, phosphate, and sulfate) retard the catalytic effect of iron (Pignatello, 1992).

The potential use of the Fenton reagent for treating wastewater was discovered in the 1960s (Pignatello et al., 2006; Safarzadeh-Amiri et al., 1996) and its ability to oxidize a variety of organic pollutants in water and soil and destroy COD in industrial wastewaters has been demonstrated subsequently (Eisenhouer, 1964; Kulik et al., 2008; Martinez et al., 2003; Pignatello, 1992; Tekin et al., 2006). The treatment process that uses the Fenton reagent is commonly known as the "Fenton Process" and works best at pH 2.8 (Murphy et al., 1989; Pignatello, 1992) because iron is insoluble at higher pH.

The interest in the Fenton process stems from the fact that iron is non-toxic, hydrogen peroxide is harmless to the environment at low to moderate concentrations, and both are commercial commodities, cheap, and are handled easily (Pignatello et al., 2006; Safarzadeh-Amiri et al., 1996). It is for these reasons that the Fenton process is most often used for the treatment of industrial wastewater. The process is particularly effective for treating highly polluted industrial wastewater contaminated, at least partly, with aromatic compounds (Eisenhouer, 1964; Kulik et al., 2008; Martinez et al., 2003; Tekin et al., 2006) for the following reasons:

- The oxidation of aromatic compounds generates intermediate byproducts (e.g., hydroquinone from phenol that catalyzes the redox recycling of iron between +2 and +3 oxidation states) (Chen & Pignatello, 1997; Pignatello, 1992).
- The concentration of iron required is relatively low—compared to that required for treating industrial wastewater without aromatic contaminants—and generates a relatively low volume of sludge.

The iron redox recycling rate will be slow in the absence of aromatic compounds, requiring high concentration of iron (see example), and the rate of oxidation of organic contaminants becomes slow after the oxidation of Fe(II) to Fe(III) (Bishop et al., 1968; Murphy et al., 1989). However, heating the solution can increase the rate of oxidation of contaminants considerably (Perez et al., 2002).

The Fenton process can be initiated with both the ferric and ferrous iron (see example). The process that uses Fe(III) is commonly referred to as the "Fenton-like process" in the literature (Murphy et al., 1989). In any practical application of the Fenton process, as Pignatello et al. (2006) argued, the concentration of Fe(II) is much lower than H_2O_2 and reacts fast with H_2O_2 and oxidizes to Fe(III) in a few seconds to a minute or so; afterward, the process relies on the Fe(III)-catalyzed decomposition of hydrogen peroxide. Furthermore, the two processes cannot be differentiated mechanistically either.

In practice, however, there are important practical differences between the two. First, the Fenton process is instantaneous and has no, or a short, induction period and this is important when treating industrial wastewater because the absence of the induction period translates into a shorter hydraulic residence time (HRT), smaller reactor size, less working space, and lower capital and operating costs. Secondly, the decomposition of H_2O_2 catalyzed by Fe(II), but not with Fe(III), can be initiated at circumneutral pH or even at an elevated pH (Kulik et al., 2008; Safarzadeh-Amiri, 2001) (see examples), and this is also important when treating industrial wastewater because many industries do not want to store strong acid in their premises.

The fast reaction between Fe(II) and H_2O_2 generates a burst of •OH that (a) oxidizes and transforms organic compounds to organic acids and rapidly depresses solution pH to ≤ 3 and (b) reacts with, for example, phenol and generates hydroquinones and other intermediate byproducts (when treating industrial wastewater contaminated, at least partly, with aromatic compounds) that accelerate the redox recycling of iron (Baxendale et al., 1951; Chen & Pignatello, 1997; Menaseti et al., 1976), shortens HRT, requires lower concentration of Fe(II), and reduces the overall treatment costs.

The Fenton process can oxidize and degrade organic pollutants efficiently, but cannot completely mineralize them. This is because the oxidation of organic pollutants generates highly oxidized (or oxygenated) and recalcitrant intermediates that do not react reasonably fast with •OH and cannot compete effectively with other contaminants for •OH (Pignatello, 1992). For example, the oxidation of benzene and phenol by the Fenton process generates oxalic acid, which reacts slowly ($k = 4.7 \times 10^7\ M^{-1}\ s^{-1}$; Buxton et al., 1988) with •OH that is difficult to oxidize and convert to CO_2.

7.2 The Advantages and Disadvantages of the Fenton Process

In addition to discussing the Fenton process, the aforementioned discussion also pointed out its advantages and disadvantages. Briefly, the advantages of this process are ease of operation, low capital and operating costs, and the ability to treat industrial wastewater with a high concentration of pollutants and low transmission in the UV region. The major disadvantages of this process are pH adjustment, generation of sludge, and onsite storage of H_2O_2.

7.3 Ferryl Ion

Although the above discussion concentrated only on hydroxyl radical as the primary oxidant produced by the Fenton reaction, the formation of other oxidizing species, such as $Fe(OOH)^+$ and $Fe = O^+$ (ferryl ion), cannot be discounted. This is because the water molecule coordinated with Fe(II) can be replaced fast ($k = 2 \times 10^6\ s^{-1}$) with hydrogen peroxide (Equation 4.47) and generate $[Fe(OH)(H_2O_2)(H_2O)_4]^+$ complex, which reacts with water and generates either a hydroxyl radical or transforms to a ferryl ion (Bossmann et al., 1998), as follows:

$$[Fe(OH)(H_2O)_5]^+ + H_2O_2 \xleftrightarrow{k \cong 2 \times 10^6 s^1} [Fe(OH)(H_2O_2)(H_2O)_4]^+ \quad (4.47)$$

$$[Fe(OH)(H_2O_2)(H_2O)_4]^+ + H_2O \rightarrow [Fe(OH)(H_2O)_5]^{2+} + HO\bullet + OH^- \quad (4.48)$$

$$[Fe(OH)(H_2O_2)(H_2O)_4]^+ \quad [(H_2O)_4\ Fe^{IV}O]^{3+} + H_2O \quad (4.49)$$

Ensing et al. (2002) proposed somewhat different pathways for the formation of ferryl ion and Hage et al. (1995) suggested that the reaction of chelated-Fe(II) ($[FeL_x]^{2+}$, L = polyamino-carboxylic acids, polycarboxylic acid, etc.) with hydrogen peroxide can generate $[Fe(L)_x(OOH)(H_3O^+)]^+$ as a reactive intermediate instead of •OH (Hage et al., 1995).

$$[FeL_x]^{2+} + HOOH + H_2O \rightarrow [Fe(L)_x(OOH)(H_3O^+)]^+ \quad (4.50)$$

It was surmised that $[Fe(L)_x(OOH)(H_3O^+)]^+$ can transfer a hydroxyl group to aromatic compounds (e.g., benzene) that is not possible with HO• (Hage et al., 1995), although this is not quite true (Walling, 1998); the hydroperoxyl intermediate formed in Equation 4.50 can transform slowly to ferryl ion or react with a water molecule and generate hydroxyl radical. The formation of ferryl ion has been proven by stopped-flow experiments (Wink et al., 1994).

The reduction potential of ferryl ion (1.4 *V* at pH 7; Bossmann et al., 1998) is much smaller than •OH (2.32 *V* at pH 7; Wood, 1988), which makes ferryl ion more selective and less reactive than •OH. For example, the rate constant for the abstraction of hydrogen atom from methanol and formaldehyde by ferryl ions is orders of magnitudes smaller than that by hydroxyl radical (Jacobson et al., 1998). Whether ferryl ion does or does not play a crucial role in the degradation of organic pollutants has not been settled in the literature. It is possible that some contaminants or their oxidation byproducts could form complexes with iron and generate ferryl ion. However, it is doubtful, for the aforementioned reasons, that ferryl ion would be the dominant oxidizing species.

The decomposition of H_2O_2 catalyzed by Iron(III)-tetra amido macrocyclic ligand (TAML) or Fe(III)-TAML complexes/catalysts (Collins & Ryabov, 2017) also generates ferryl ion (Fe^V) (Chanda et al., 2008), as follows:

$$[Fe^{3+} - TAML(H_2O)_2]^- + H_2O_2 \rightarrow [O{=}Fe^V\text{-}TAML]^{2-} + 2\ H_2O \quad (4.51)$$

The Fe^V produced in the previous reaction can react with, and degrade, organic contaminants; it can also react with H_2O_2 and generates Fe^{IV}-O-Fe^{IV}, which, in turn, can react with and oxidize organic contaminants.

Most Fe^{3+}-TAML catalysts work best at mildly basic pH 10 to 12, and their catalytic activity decreases sharply at higher or lower pH and becomes negligibly small for most catalysts around neutral pH (Chahbane et al., 2007). This is because $Fe^{(III)}$-TAML complexes lose metal ions in a mildly acidic environment in presence of some buffering reagents (e.g., $H_2PO_4^-$, HSO_4, CH_3CO_2H, and $HO_2CCH_2CO_2H$ at pH 4 to 9 and in the presence of picolinic acid, imidazole, and azide). These compounds are thought to coordinate with iron, deliver protons in an intramolecular fashion to one of the amido-N, and cleave a Fe-N bond, followed by the fast cleavage of the remaining Fe-N bonds through protolyses/hydrolyses.

The $Fe^{(III)}$-TAML complexes have been used to: (a) degrade, for example, chlorophenols (Gupta et al., 2002), pesticides (Chanda et al., 2006), and many other contaminants in aqueous solution; (b) decolorize effluents from textile plants (Chahbane et al., 2007) and pulp and paper mills (Wingate et al., 2004); and (c) deactivate bacteria (Banerjee et al., 2006). The concentration of catalysts required is remarkably low and on the order of a few micromolar (μM). It should be mentioned that the ligand (i.e., TAML) in these complexes is unstable and degrades in the presence of hydrogen peroxide; however, their stability is much higher than other Fe(III) chelating agents, such as nitrilotriacetic acid (NTA) and ethylenediaminetetraacetic acid (EDTA), and the catalysts recycle numerous times before being disintegrated and destroyed.

The rate of degradation of organic contaminants with the Fe(III)-TAML/H_2O_2 process depends on the chemical structure of contaminants and the solution pH and the nature of contaminant (Chahbane et al., 2007). The process is unable to oxidize compounds generated from the oxidation and ring cleavage of aromatic compounds unless the reaction condition is changed and catalyst is recharged repeatedly. For example, the oxidation of trichloro- and pentachloro-phenols (TCP and PCP, respectively) by Fe(III)-TAML catalyzed decomposition of H_2O_2 generates chloro-maleic and -malonic acids, chlorobenzoic acid, and so on, which were not degraded further at pH 11. However, complete or near-complete degradation of chlorinated byproducts of TCP and PCP can be achieved at a neutral pH and with successive addition of catalyst over a period of approximately 12 hours (Gupta et al., 2002).

The costs of Fe(III)-TAML catalysts have not been reported yet. The catalysts are expected to be expensive, and those that are active near neutral pH are expected to cost more because they are more difficult to synthesize (Ghosh et al., 2008). It is believed that the costs would be competitive for most applications (Collins & Walter, 2006); however, this has to be demonstrated in practice.

7.4 The Fenton-Like Process at Neutral pH

The Fenton process, as previously discussed, works best at pH 2.8 because of the negligible solubility of Fe(III) in pure water at pH > 3. Formation of chelated iron(III)—Fe(III)*L*—with suitable chelating agents (*L*) such as aminopolycarboxylic acids (PACs) (e.g., nitrilotriacetic acid, dihydroxybenzoic acid, humic substance). (Sun & Pignatello, 1992, 1993a) increases Fe(III) solubility in water at neutral pH. These complexes decompose H_2O_2 catalytically (Walling & Goosen, 1973), as follows:

$$\mathrm{Fe(III)}L + H_2O_2 \leftrightarrow (HO_2^-)\mathrm{Fe(III)}L + H_3O^+ \qquad (4.52)$$

$$(HO_2^-)Fe(III)L \leftrightarrow Fe(II)L + HO_2^{\cdot} \quad (4.53)$$

$$Fe(II)L + H_2O_2 \rightarrow Fe(III)L + \bullet OH + OH^- \quad (4.54)$$

Chelating agents not only solubilize Fe(II) and Fe(III), they also decrease the reduction potential of iron, which facilitates the redox cycling of iron between +2 and +3 oxidation states (Tachiev et al., 2000) and catalytic decomposition of H_2O_2. For example, the rate of reaction in Equation 4.54 is more than two orders of magnitudes faster than Equation 4.35 (Croft et al., 1992). The Fe(III)-polyhydroxy aromatics complexes are stable at pH ≤ 6.0 and work best in this pH range (Sun & Pignattelo, 1992, 1993b).

The activation of H_2O_2 with Fe(III)L at circumneutral pH has been used to degrade, for example, insecticides, herbicides, and endocrine disruptors (Huang et al., 2013; Sun & Pignatello, 1992, 1993a), diesel-contaminated soil (Watt & Dilly, 1996), and others. The rate of oxidation of contaminants by the $Fe(III)L/H_2O_2$ process is typically slow and becomes even slower when treating industrial wastewater, requiring a long HRT.

Table 4.7 presents a summary of the different iron-catalyzed decompositions of H_2O_2 that can be used to treat contaminated water, the treatment conditions, oxidizing species produced, and their advantages and disadvantages.

7.5 Application of the Fenton Process: Selected Examples

The Fenton process is ideal for treating or pretreating industrial wastewater with moderate to high concentrations COD. It has been used to destroy COD and TOC in textile (Lin and Peng, 1995), pharmaceuticals (Kulik et al., 2008), pulp and paper (Pérez et al., 2002), surfactants production (Lin et al., 1999), and refinery (Safarzadeh-Amiri, 2019) effluents. The process can also be used to pretreat highly contaminated industrial wastewaters that are toxic to bacteria, reduce their toxicity and organic loading, and make them amenable to biotreatment. The following examples demonstrate the power and versatility of the Fenton process for treating or pretreating effluents from three different industries: a nutrient manufacturer, a petroleum refinery, and a pharmaceutical production plant. These effluents are currently being treated or pretreated by the Fenton process.

7.5.1 Example 1: Nutrient Manufacturing Effluent

This effluent is contaminated with approximately 7000 mg/L of formaldehyde (FA) and more than 7000 mg/L of methanol (MeOH); the solution pH, alkalinity, and temperature are approximately 7.1, < 50 mg/L, and approximately 40 °C, respectively. The Fenton process is currently used to

TABLE 4.7 Summary of the iron-catalyzed decomposition of H_2O_2 processes that can be used to treat contaminated water, reagents, oxidizing species, treatment conditions, and interferences.

Oxidation Process	Fenton/Acidic pH	Fenton/Neutral pH	Fe-TAML/H_2O_2/Basic pH
Reagent	$Fe^{2+}(Fe^{3+})/H_2O_2$	Fe(III)-Chelate[a]/H_2O_{2a}	Fe-TAML/H_2O_2
Oxidizing Species	HO•/[b]$Fe^{IV}=O^+$	HO•/ [7]$Fe^{IV}=O^+$	$Fe^{IV}=O^+$ and/or $Fe^{V}=O^+$
Optimum pH range	2.0–3.0	3.0–7.0	9.0–13.0
Reduction Potential/V	2.4/1.4	2.3/1.4	1.4/2.0
Application	Treatment of industrial wastewater with moderate to high COD/TOC	In situ treatment of contaminated groundwater	Degradation of organic pollutants, such as some dyes, PCBs, etc.
Advantage	• Reasonable capital costs • Easy to operate • Ideal to treat contaminated groundwater and industrial wastewater	• Neutral pH • In-situ treatment • No need for pump and treat • Low capital costs; no sludge to manage	• Efficient at high pH • Ideal to treat pulp and Paper and some textile effluents • Useful for the oxidation of PCBs under basic pH
Disadvantages	• pH Adjustment, • Long residence time • High Fe and H_2O_2 dosages • Lots of sludge to manage/dispose	• Requires high Fe-HA complex	• Low efficiency at neutral pH • Selectivity • Limited ability to remove COD and TOC • Possibly high operating costs

[a]Chelator agent includes (a) small organic compounds (e.g., Gallic acid), (b) humic substances, or (c) a combination of the two;
[b]Ferryl ion ($Fe^{IV}=O^+$) is expected to be formed under special circumstances (see Section 4.0).

oxidize formaldehyde and reduce its concentration to approximately 200 mg/L. Figure 4.2 presents the rate of oxidation of formaldehyde in this effluent by the Fenton process.

7.5.2 Example 2: Refinery Wastewater

This wastewater is contaminated with sulfide, phenol, and saturated and unsaturated petroleum products; the solution COD and pH are approximately 25 000 mg/L and 12, respectively. The water is toxic and cannot be treated biologically without pretreatment to make in nontoxic. The bench-scale data presented in Figure 4.3 show that the water can be treated partially with the Fenton process using both the ferrous and ferric ions as catalysts, and treatment can be initiated at room temperature or at an elevated temperature depending on the concentration of iron; higher concentrations of iron are needed to initiate the process at room temperature.

The experimental data (not shown) show that the solution COD can be reduced by 50% in approximately 2 hours with $Fe(II)/H_2O_2$ (T = 60 °C, [Fe(II)] = 60 mg/L, and $[H_2O_2]$ = 17 000 mg/L); the same results were also achieved in approximately 3 hours with $Fe(III)/H_2O_2$ (T = 60 °C, [Fe(III)]

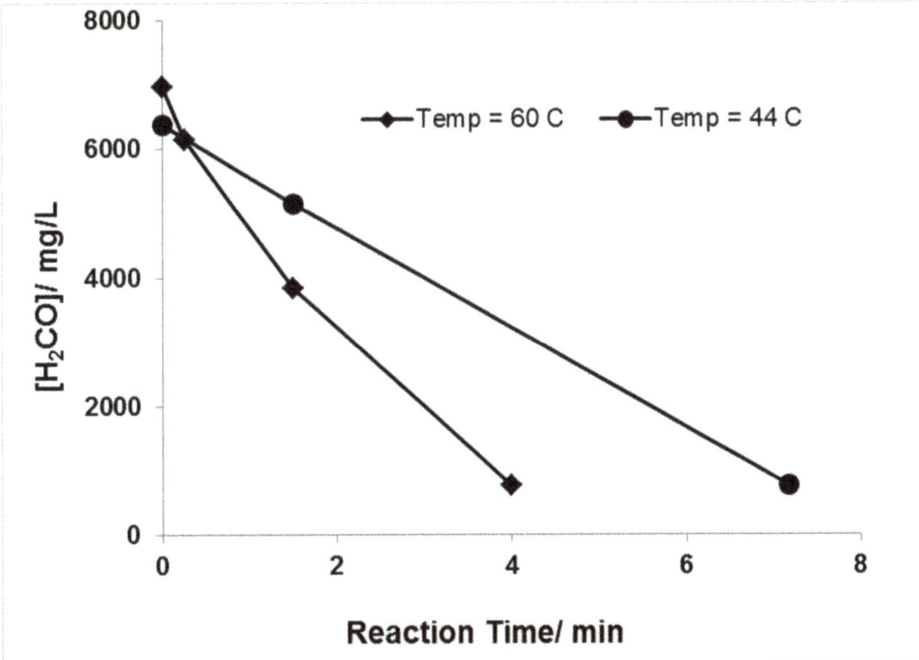

FIGURE 4.2 The rate of oxidation of formaldehyde in a process wastewater contaminated with approximately ≤400 mg/L of formaldehyde and ≥700 mg/L of methanol by the Fenton process; [Fe(II)] = 4500 mg/L, initial pH = 11.5 (Safarzadeh-Amiri, unpublished data).

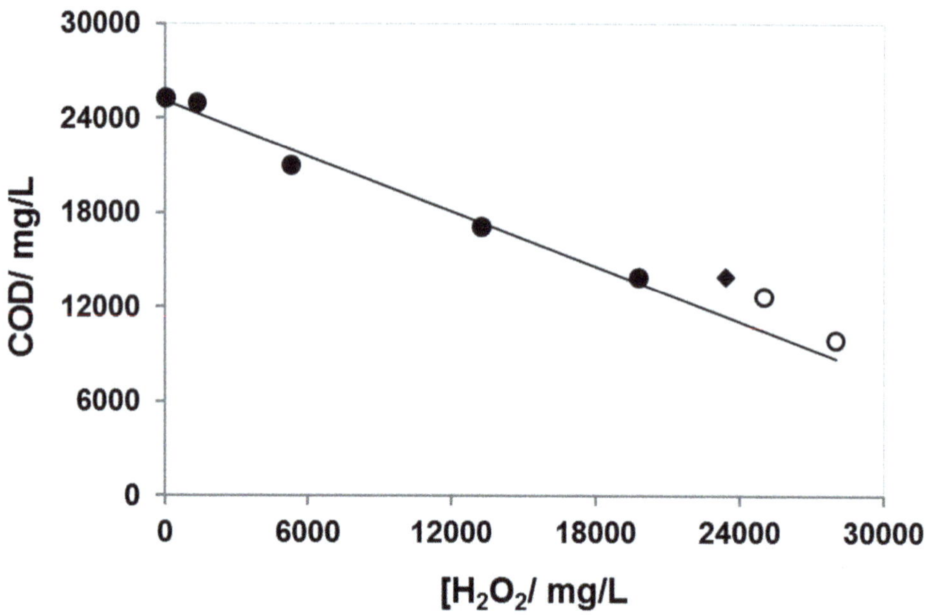

FIGURE 4.3 The rate of oxidation and removal COD versus H_2O_2 concentration of a refinery wastewater by the Fenton process; volume = 250 mL; ◆) [Fe(II)] = 60 mg/L, pH_i = 7, T_i = 60 °C, O) [Fe(II)] = 100 mg/L, pH_i = 7, T_i =RT, •) [Fe(III)] = 100 mg/L, pH_i ~ 5, T_i = 40 °C (Safarzadeh-Amiri, unpublished data).

= 100 mg/L, and [H_2O_2] = 24 000 mg/L). The wastewater is currently being pretreated by the Fenton process; the solution COD is reduced from 25 000 mg/L to approximately 17 000 mg/L and then polished by biological treatment before being released.

7.5.3 Example 3: Pharmaceutical Wastewaters

A pharmaceutical company generates three wastewater streams—S1, S2, and S3, with COD concentrations of approximately 12 350 mg/L 7650 mg/L, and 44 150 mg/L of COD, respectively. The S1 stream is slightly acidic (pH ~5.8), and the other two streams are slightly basic (pH ~9.5 to 9.9). All streams are toxic and cannot be treated biologically without pretreatment. The first stream is treated easily by the Fenton process without pH adjustment; the solution COD can be reduced from approximately 12 350 mg/L to less than 2500 mg/L in approximately 15 minutes using 200 mg/L of Fe(II) and 24 000 mg/L of H_2O_2 ([H_2O_2]/[COD] = 2.3).

The other streams were extremely recalcitrant and their COD was reduced only marginally after approximately 20 hours using 200 mg/L of Fe(II). However, these streams can be treated, at least partially, when diluted by 50% with S1. The diluted solution of S2 and S3 were treated partially using 200 mg/L of Fe(II) and 24 000 mg/L of H_2O_2 ($[H_2O_2]/[COD] = 3.6$), and 500 mg/L of Fe(II) and 5000 mg/L of H_2O_2 ($[H_2O_2]/[COD] = 2.8$), respectively.

8.0 THE SOLAR-FERRIOXALATE/HYDROGEN PEROXIDE PROCESS

Ferric ion forms stable complexes and/or associated ion pairs with carboxylates and polycarboxylates. The resulting complex exhibits a strong ligand-to-metal charge transfer (LMCT) absorption band in the UV region that tails into the visible region. The complex is typically photoreactive and generates Fe(II) when irradiated with UV-visible light. The quantum yield of formation Fe(II) ($\sigma_{Fe(II)}$) depends on the excitation wavelength, the concentration of Fe(III), and the structure and concentration of organic acid (Abrahamson, et al.,1994; Balzani & Caracceti, 1970). For example, the $\sigma_{Fe(II)}$ for Fe(III)-formate complex varies from 0.05 to 0.55 with varying concentrations of formic acid (Balzani & Caracceti, 1970). One of the best known and most studied examples of the Fe(III)-polycarboxylate complex is ferri-oxalate, $[Fe(C_2O_4)_3]^{3-}$, or Fe(III)-Ox with one of the highest $\sigma_{Fe(II)}$ values.

8.1 Photochemistry of Ferrioxalate

The photochemical properties of Fe(III)-Ox, perhaps one of the most popular compounds used as a chemical actinometer, have been studied extensively by Abrahamson et al. (1994), Cooper and De Graff (1972), Goldstein and Rabani (2008), Hatchard and Parker (1956), Mangiante et al. (2017), and Pozdnyakov et al. (2008). The complex exhibits a metal to solvent charge transfer (MSCT) band below approximately 250 nm and a LMCT band centered around 300 nm (Balzani & Caracceti, 1970; Goldstein & Rabani, 2008). Excitation of ferrioxalate into the LMCT band generates Fe(II)Ox—$[Fe(C_2O_4)_2]^{2-}$—and an extremely short-lived oxalate radical anion ($C_2O_4^{-\bullet}$) in the Fe(II) coordination sphere (Equation 4.55; Mangiante et al., 2017). The $C_2O_4^{-\bullet}$ dissociates rapidly into thermally excited CO_2 and $CO_2^{\bullet-}$, which thermally equilibrate and leave the Fe(II) coordination sphere within 100 ps and 10 ns, respectively. The $CO_2^{\bullet-}$ can react with ferrioxalate and generate Fe(II), as follows:

$$[Fe(C_2O_4)_3]^{4-} + h\nu \rightarrow [([Fe(II)(C_2O_4)_2]^{2-})C_2O_4^{-\bullet}] \quad (4.55)$$

$$[([Fe(II)(C_2O_4)_2]^{2-})..C_2O_4^{-\bullet}] \rightarrow [([Fe(II)(C_2O_4)_2]^{2-})..CO_2^{-\bullet}] + CO_2 \quad (4.56)$$

$$CO_2^{-\bullet} + [Fe(C_2O_4)_3]^{4-} \rightarrow CO_2 + Fe(II) + 3\ C_2O_4^{2-} \quad (4.57)$$

Goldstein and Rabani (2008) proposed that excitation of Fe(III)-Ox at the MSCT band generates $[Fe(C_2O_4)_2]^{2-}$ and a hydrated electron (e^-_{aq}) pair (Equation 4.58) in a solvent cage, which could recombine or escape from the solvent cage. The hydrated electron can react with $[Fe(C_2O_4)_3]^{3-}$ and reduce it to $[Fe(C_2O_4)_3]^{4-}$; the latter decomposes to $[Fe(C_2O_4)_2]^{-2}$ and $C_2O_4^{-\bullet}$, as follows:

$$[Fe(C_2O_4)_3]^{4-} + h\nu \rightarrow [Fe(C_2O_4)_3]^{2-} + e^-_{aq} \quad (4.58)$$

Theoretically, the quantum yield of formation of Fe(II), $\phi_{Fe(II)}$, could reach, according to Equations 4.55 to 4.57, a value of 2.0. However, the effective $\phi_{Fe(II)}$ is approximately 1.25 at 260 to 450 nm and 1.48 at 205 to 250 nm under actinometry conditions (i.e., [Fe(III)]/[Ox] = 1:3, $[H_2SO_4]$ = 0.05 M, [Fe(III)] = 4×10^{-4} M) (Cooper & De Graff, 1972; Goldstein & Rabani, 2008; Hatchard & Parker, 1956). This is because in air-saturated solution, oxygen competes with Fe(III)-Ox for $CO_2^{-\bullet}$ and generates superoxide ions ($O_2^{-\bullet}/HO_2^{\bullet}$), which in turn reacts with Fe(II) and generates hydrogen peroxide (Goldstein & Rabani, 2008; Zuo & Hoigne, 1992) according to the following reactions:

$$CO_2^{-\bullet} + O_2 \rightarrow CO_2 + O_2^{-\bullet} \quad (4.59)$$

$$O_2^{-\bullet} + H^+ \rightarrow HO_2^{\bullet} \quad pK_a = 4.8 \quad (4.60)$$

$$2\ HO_2^{\bullet} \rightarrow H_2O_2 + O_2 \quad (4.61)$$

$$Fe(II) + HO_2^{\bullet} + H^+ \rightarrow Fe(III) + H_2O_2 \quad (4.62)$$

Zuo and Hoigne (1992) have shown that the stoichiometry of the photodegradation reaction of Fe(III)-Ox in the absence of oxygen is in accord with Equations 4.55 through 4.57 and leads to generation of H_2O_2 when irradiated with sunlight in the presence of O_2.

Other Fe(III)-polycarboxylates exhibiting high photosensitivity are Fe(III) complexes of citric, tartaric, glyoxalic , and pryrovic acids (Abrahamson et al., 1994; Rodriguez et al., 1992; Zepp et al., 1992). The value of $\sigma_{Fe(II)}$ varies with varying ligand, solution pH, concentration of Fe(III), and [Fe(III)]/[polycarboxylate].

Irradiation of ferrioxalate in the presence of H_2O_2 generates •OH (Zepp et al., 1992) via the Fenton reaction (i.e., Equation 4.63), as follows:

$$[Fe(C_2O_4)_3]^{3-} + H_2O_2. \rightarrow [Fe(C_2O_4)_2]^- + OH^- + \bullet OH \quad (4.63)$$

Generation of •OH continues as long as oxalate is present in solution. Under ideal conditions and in the absence of OH-radical scavengers, the quantum yield of generation of •OH, $\sigma_{\bullet OH}$, is expected to be the same as $\sigma_{Fe(II)}$. In practice, however, $\sigma_{\bullet OH}$ could increase to as high as 3.65 in the presence of •OH scavengers, and this is attributable to redox recycling of iron by C-centered radicals (Hislop & Bolton, 1999). It has been shown more than two decades ago that, typically, photodegradation of organic pollutants mediated by ferrioxalate is a highly efficient process and remediation of contaminated water by this process is more efficient and less expensive than UV/H_2O_2 AOP (Safarzadeh-Amiri et al., 1997).

8.2 Solar Detoxification of Industrial Wastewater

The solar radiation (see Figure 4.4) is a source of both UV and visible lights; however, the terrestrial solar radiation drops sharply below approximately 500 nm and the light intensity drops to below the detection limits below approximately 300 nm. This is because almost all light below 300 nm is absorbed by the atmospheric ozone and does not reach the surface of the earth.

Approximately 4.5% of solar radiation is in the UV (300- to 400-nm) region and 18.5% is between the 300- to 500-nm region (see Figure 4.5). These limitations restrict the number of processes that can make use of the solar radiation. Titanium dioxide, as discussed previously, absorbs up to 388 nm and can be used as a catalyst for solar detoxification of wastewater.

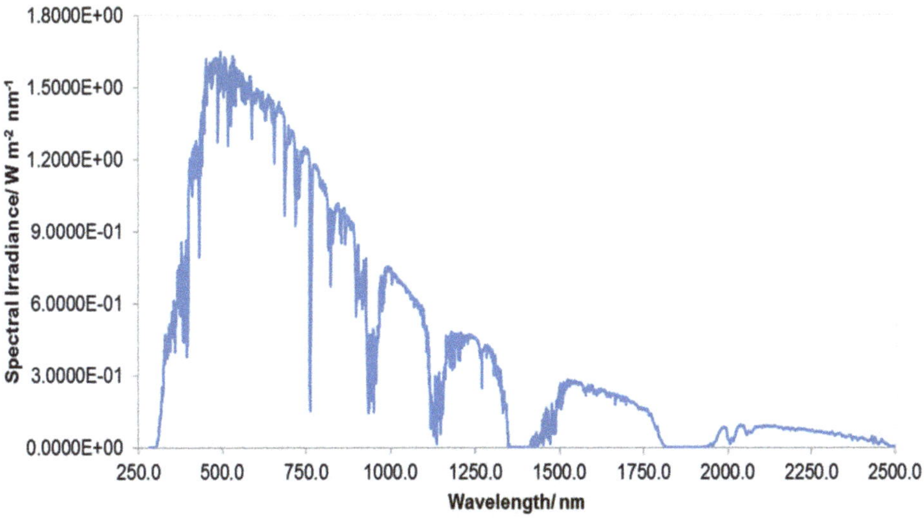

FIGURE 4.4 Solar spectral irradiance distribution for the standard AM 1.5: ASTM G-159 (from National Renewable Energy Laboratory http://www.nrel.gov/rredc/solar_data.html).

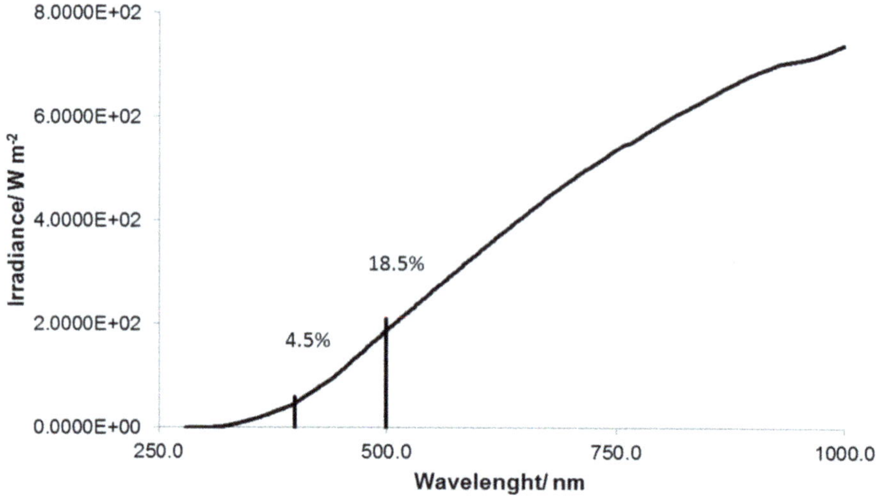

FIGURE 4.5 Integrated solar irradiance up to 1000 nm (using the solar spectrum given in Figure 4.4).

However, the fraction of solar light it can use is less than approximately 3.8% and insufficient for practical use considering the low quantum efficiency of the TiO_2 PWT process (Bolton et al., 1995).

Similarly, the ferric ion does not absorb and cannot use a significant fraction of the solar radiation (Safarzadeh-Amiri et al., 1996) either. Iron-polycarboxylate complexes (e.g., ferrioxalate) would be able to use up to approximately 18.5% solar light. The solar irradiation of Fe-polycarboxylates is believed to be responsible for the occurrence of H_2O_2 in atmospheric water, and has been proven experimentally (Zepp et al., 1992; Zuo & Hoigne, 1992).

Solar irradiation of Fe(III)-Ox/H_2O_2 (Solar-Fe(III)-OX) have been used to degrade organic pollutants in both the UPW and industrial wastewater in recent years. The mechanism of oxidation of an organic pollutant, RH, by Solar-Fe(III)-Ox AOP can be represented by Equations 4.61 to 4.63 and the following reactions:

$$[Fe(II)(C_2O_4)]_2^- + H_2O_2 \xrightarrow{k_9} [Fe(C_2O_4)_2]^- + OH^- + \bullet OH \quad (4.64)$$

$$RH + \bullet OH \rightarrow R^\bullet + H_2O \quad (4.65)$$

$$SH + \bullet OH \rightarrow S^\bullet + H_2O \quad (4.66)$$

$$R^\bullet + O_2 \rightarrow RO_2^\bullet \quad (4.67)$$

$$R^\bullet + [Fe(C_2O_4)_3]^{4-} \rightarrow [Fe(II)(C_2O_4)_2]^{2-} + C_2O_4^{2-} + R^- \quad (4.68)$$

$$S^{\bullet} + [Fe(C_2O_4)_3]^{4-} \rightarrow [Fe(II)(C_2O_4)_2]^{2-} + C_2O_4^{2-} + R^{-} \quad (4.69)$$

$$S^{\bullet} + O_2 \rightarrow SO_2^{\bullet} \quad (4.70)$$

where SH represents OH-radical scavengers, such as the background COD, bicarbonate, and so on, and other terms have been defined before. The rate of generation of Fe(II)—$R_{Fe(II)}$—in Equation 4.55 is given by $\phi_{Fe(II)} f I_0/V$ (M/s), where $\phi_{Fe(II)}$ is the quantum yield of Fe(II) formation, f is the fraction of light absorbed by Fe(III)-Ox, I_0 is the photon flux (einstein/s), and V is the irradiated volume (L).

The rate of degradation of RH at high (i.e., [RH]>>[SH]) and low (i.e., [SH]>>[RH]) concentrations and under steady-state conditions is given by Equations 4.71 and 4.72, respectively.

$$-\frac{d[RH]}{dt} = R_{Fe(II)} \times \left(\frac{k_{R,FeOx}[Fe(III)Ox]}{k_{R,O_2}[O_2]}\right) \quad (4.71)$$

$$-\frac{d[RH]}{dt} = \left(\frac{R_{Fe(II)}}{k_{OH,SH} \times [SH]}\right) \times \left(\frac{k_{R,FeOx}[Fe(III)-Ox]}{k_{OH,RH}[O_2]}\right) \quad (4.72)$$

where $k_{R,FeOx}$ and k_{R,O_2} = the rate constant of reaction of alkyl radical (R•) with ferrioxalate and molecular oxygen, respectively, and $k_{OH,RH}$ and $k_{OH,SH}$ = the rate constant of reaction of target (RH) and non-target (SH) contaminants with •OH, respectively. The validity of the aforementioned equations has been confirmed by Hislop and Bolton (1999).

8.3 Figures of Merit

As previously discussed, figures of merit are indicators of the energy efficiency of AOP. Their definition for the solar-driven process is based on the collector area required to achieve specific treatment (Bolton et al., 1996; 2001). These figures are "collector area per order" (CA/O) for the first-order kinetics and "collector area per mass" (CA/M) for the zero-order kinetics. Equations 4.73 and 4.75 can be used to calculate CA/O and CA/M, respectively, by replacing P (electrical power in kWh) with $A \times E/1000$, where A is the collector area and E (W/m^2) is the solar irradiance. Therefore, the figure of merit for the first-order kinetics is defined as the collector area (m^2) required to reduce the concentration of a pollutant by one order of magnitude (90%) in 1 m^3, or in 1 hour under "standard" solar conditions. The CA/O for the flow-through process can be calculated from the following equation:

$$CA/O = \frac{A \times E}{60 \times R \times \log(c_i - c_f)} \quad (4.73)$$

The corresponding equation for the batch process is given by the following:

$$CA/O = \frac{A \times E \times t}{60 \times V \times \log(c_i - c_f)} \tag{4.74}$$

In the aforementioned equations, A (m^2) is the actual collector area, R (L/min) is the flowrate, V (L) is the total batch volume, and t (min) is reaction time.

The figure of merit for the zero-order kinetic regime is defined as the collector area needed to oxidize 1 kg of pollutant in 1 m^3 (1000 L) of contaminated water in 1 hour under "standard" solar conditions. Assuming a "standardized" irradiance of 1000 W/m^2 (based on the AM 1.5 standard solar spectrum on a horizontal surface), the CA/M can be calculated using the following equation:

$$CA/M = \frac{A \times E \times t}{60 \times V \times (c_i - c_f)} \tag{4.75}$$

where M is the molecular weight (g/mol) and the other terms have been defined before.

Although the figures of merits for AOPs are now almost universally accepted and relevant values for a variety of pollutants have been reported in the literature, the corresponding figures for the solar-driven AOP are rare and seldom reported. The estimation of the CA/M and/or CA/O from the published data is not easy because it is not possible to access raw data. However, attempts have been made to estimate figures of merit where possible.

8.4 The Advantages and Disadvantages of the Solar-Ferrioxalate Advanced Oxidation Process

The advantages of Solar-FeOx AOP are similar to those of other AOPs, in addition to the fact that (a) solar light is free and renewable and (b) oxalate oxidizes to carbon dioxide and does not contribute organic carbon to the wastewater being remediated with process, whereas other ligands, such as citrate and tartrate, generate byproducts and add organic carbon to the water (Abrahamson et al., 1994). However, solar light is available for, at most, approximately 8 hours per day, and the solar irradiance depends on the altitude, latitude, and the cloud cover. These factors create obvious limitations to the practical applicability of this technology for industrial wastewater reuse applications.

8.5 Applications

The Solar-FeOx AOP has been used to degrade organic pollutants in UPW, tap water, and in industrial wastewater using either sunlight or a simulated solar light. The contaminants that are used in preparing synthetic wastewater include

1,4-D, benzene, toluene, xylenes, phenol and its derivatives, pharmaceuticals, herbicides and pesticides, and dyes, among others. The process efficacy and efficiency for the oxidation and removal of contaminants and mineralization of TOC and COD from both the synthetic and industrial wastewaters have been evaluated at bench- and pilot-scales; however, the technology is not known to have yet been applied at full-scale for industrial water reclamation.

The objectives of the majority investigations appear to have been to optimize mineralization of organic pollutants in synthetic and industrial wastewaters. As discussed previously, the Solar-FeOx process is efficient at oxidizing and degrading organic pollutants. However, its efficiency for mineralization of organic contaminants is not significantly better than, for example, the photo-Fenton process. This is because oxalate degrades rapidly and converts to CO_2 and Solar-FeOx process becomes solar-Fenton process unless it is continuously supplied with oxalate, and the addition of oxalate increases operating costs.

The efficacy and efficiency of Solar-FeOx process for the oxidation of contaminants and mineralization of TOC or COD in a synthetic wastewater is irrelevant to its performance for remediation of industrial wastewaters and will not be explored further. This is because synthetic wastewaters are free from organic and inorganic contaminants, dissolved solids, turbidity, and absorb weakly, if any, in the UV/vis. region. These factors can have profound influence on Solar-FeOx performance.

The oxidation and removal of TOC, COD, and/or specific contaminants from industrial wastewater by Solar-FeOx—as mentioned above—have been studied in recent years and Table 4.8 presents characterization data of a few of them. As can be seen, all wastewaters are characterized by high to very high TOC, COD, low to high turbidity (where the data is available), and varying pH. In addition to background contaminant, the pharmaceutical and beverage effluents are also characterized with approximately 2100 and 660 mg/L of acetate, respectively. Furthermore, pharmaceutical wastewater is also contaminated with high concentration (14 700 mg/L) of chloride ion and low concentration (0.39 mg/L) of iron, and refinery wastewater is contaminated with high concentration (1614 mg/L) of sulfate and moderate concentration (19 mg/L) of oil and grease. The BOD_5-to-COD—of three wastewaters—indicates that biodegradability of wastewaters varies from excellent (e.g., beverage effluent) to poor (e.g., textile effluent). The refinery wastewater is contaminated with approximately 66 mg/L of sulfide, a biocide; pharmaceutical wastewater may or may not contain biocides; and biotreatment can remove 90% to 95% of winery wastewater.

The pharmaceutical and winery wastewaters were diluted by approximately 33 and 40 times— to reduce their TOC to approximately 125 and 50 mg/L, respectively—before being treated, and refinery effluent was spiked with approximately 200 mg/L phenol and its oxidation was followed during

TABLE 4.8 Characterization data for a few industrial wastewaters.

Parameters (Unit)	Beverage	Latex	Pharmaceuticals	Refinery	Textile	Winery
pH	5.35	4.6	7.72	6.8	11.3	12.4
COD (mg/L)	6500	30 000	16 180	400	1239	—
TOC (mg/L)	1987	—	4111	—	408[a]	2674
BOD_5 (mg mg/L)	4400	10 867	—	—	198	—
BOD_5/COD	0.68	0.36	—	—	0.16	—
Turbidity (NTU)	1850	—	11.5	19		11.5
Total Solids (mg/L)	1300	2132[b]	34 000	78	902	12 060
Reference	Duran et al., 2015	Ashok et al., 2015	Estrada-Arriaga et al., 2016	Monteagudo et al., 2013	Manenti et al., 2015	Monteagudo et al., 2012

[a]Dissolved organic carbon
[b]Total suspended solids

solar irradiation. The beverage and winery wastewater were also pretreated to reduce their turbidity and TOC, respectively, before Solar-FeOx remediation. All tests, with the exception of those for refinery and latex wastewaters, were conducted using a pilot-scale unit. Table 4.9 presents the experimental parameters—i.e., treatment goal, catalyst, and H_2O_2 concentrations, % mineralization of TOC, COD, and/or oxidation of target contaminant removal, and CA/O or CA/M—for Solar-FeOx treatment of wastewater from Table 4.7.

The data presented in Table 4.8 show that Solar-FeOx AOP can mineralize TOC and COD and oxidize phenol. The data also show that:

1. Collector area required for mineralization of TOC in pharmaceutical wastewater is large and that Solar-FeOx may be impractical for this application.
2. Collector area needed for beverage and winery wastewater is also large.
3. The estimated operating costs for Solar-FeOx treatment of diluted pharmaceutical (\$1.87 m^{-3}; Monteagudo et al., 2013) and winery (\$0.90 m^{-3}) wastewaters are high.
4. The operating costs of beverage wastewater also appears to be high (≥\$5.6 m^{-3}).

The beverage wastewater is highly biodegradable; biological treatment can reduce winery wastewater COD by 90% to 95%. The pharmaceutical contains high concentration of acetate and could possibly be treated biologically with much lower operating costs. Could Solar-FeOx compete with biotreatment for these specific wastewaters? or, as Ashok et al. (2015) argue, would the simplicity and much shorter HRT of Solar-FeOx—compared to biotreatment process—make this process more favorable?

The Solar-FeOx appears to be efficient for mineralization of TOC/COD in latex and textile wastewaters and the collector area required both applications are reasonable. Furthermore, biodegradability of latex wastewater increases from 0.36 to approximately 0.5 after the oxidation and removal of approximately 33% of COD. Similarly, mineralization and removal of 65% of TOC from textile wastewater TOC increases its biodegradability significantly and the partially treated wastewater can be polished biologically to meet the discharge requirements. The estimated operating costs for latex and textile effluents are \$85 m^{-3} (Ashok et al., 2015) and \$1.76 m^{-3}, respectively. The operating costs for the former are high and as a result, Solar-FeOx cannot be viable treatment option in this case, providing that operation simplicity and much shorter HRT do not weight in it favor.

The refinery wastewater was treated with both the Solar-Fenton and Solar-FeOx processes. The rate oxidation of phenol (k = 0.22 min^{-1}) with Solar-Fenton was approximately one order of magnitude faster than

TABLE 4.9 Examples of industrial wastewaters that have been treated with Solar-FeOx process.

Wastewater	pH	Target	% Removal	Fe-Ox[a]/mg/L	H_2O_2/mg/L	CA/M/m^2 /kg	Reference
Beverage	2.98	TOC	70.6/98[b]	150	1000	88/109	Duran et al., 2015
Latex	4.0	COD	91.6	1300	82 500	1.81[c]	Ashok et al., 2015
Refinery	5.0	Phenol	100	20	300	—	Estrada-Arriaga et al., 2016
Pharmaceutical	7.72	TOC	84	120	5250	257[d]	Monteagudo et al., 2013
Textile	11.3	COD/TOC	65.2	36.2	1095	0.65[c]	Manenti et al., 2015
Winery	3.5	TOC	40	6[e]	260[f]	42[d]	Monteagudo et al., 2012

[a]As Fe and Fe-to-oxalate molar ratio of 1:3
[b]Both are TOC removal percentages
[c]CA/O (m^2 $Order^{-1}$ m^{-3})
[d]Assuming 35 L of wastewater was treated
[e]Fe was present in this wastewater and Fe-to-oxalate molar ratio was approximately 9.6
[f]The solution was spiked with approximately 1000 and 500 mg/L of H_2O_2 and oxalic acid, respectively, and more were injected continuously during treatment.

Solar-FeOx (k = 0.021 min^{-1}); approximately, 53% and 83% of COD was removed by the Solar-Fenton and Solar Fe-FeOX processes, respectively. Both processes used similar iron concentration (20 mg/L, however, H_2O_2 concentration (300 mg/L) for the former process was approximately 60% of the latter (500 mg/L). The treatment was carried in a beaker that was charged with approximately 1 L of wastewater, Fe(II) or FeOX, and H_2O_2 and placed in the sun. As a result, it is not possible to determine CA/O in this case.

9.0 SUMMARY

Advanced oxidation processes rely on generation of •OH, one of the most powerful oxidants. Once generated, •OH reacts fast (k = $10^8 - 10^{10}$ M^{-1} s^{-1}) with most organic contaminants and starts a cascade of reactions that can ultimately lead to their mineralization. The •OH can be generated by a number of AOPs, including UV/H_2O_2, UV/O_3, UV/H_2O_2/O_3, UV/TiO_2,

TABLE 4.10 Summary of the advantages and disadvantages of different AOPs.

AOP	Advantages	Disadvantages
UV/H_2O_2	The UV/H_2O_2 treatment process is fast, degrades the vast majority of organic contaminants at circumneutral pH, and is especially suited for treating drinking and reuse water.	The Op-Ex and Cap-Ex could be high for water with moderate background COD and alkalinity, residual H_2O_2 has to be destroyed in drinking water, and it is suited for treating industrial wastewater and contaminated groundwater.
O_3 and UV/O_3	Similar to UV/H_2O_2 AOP	High Cap-EX and Op-Ex because of low •OH yield, low O_3 solubility, not suitable for wastewater treatment, and is not offered commercially.
The Fenton Process	Ease of operation, low Cap-Ex and Op-Ex, and suitable for treating waters contaminated with moderate to relatively high concentration of contaminant(s)	Works best at acidic pH of approximately 2.8 and generates sludge.
The Solar FeOx AOP	Sunlight is free and renewable and oxalate converts to CO_2 and does increase water carbon load.	Limited volume water can be treated in a day because sunlight is available 8 hours a day; solar irradiance depends on location and cloud cover and sunlight collector area.

and the Fenton and photo Fenton processes. In the realm of industrial wastewater recycle, however, UV-based AOPs struggle as a result of poor UV transmissivity, which is a common trait for industrial wastewater. For this reason, non-UV technologies such as Fentons and O_3/H_2O_2 are almost always the AOP of choice for industrial wastewater. Table 4.10 summarizes the advantages, disadvantages, and application of each of these technologies.

10.0 REFERENCES

Abrahamson, H. B., Rezvani, A. B., & Brushmiller, J. G. (1994). Photochemical and spectroscopic studies of complexes of iron(III) citrate and other carboxylic acids, *Inorganica Chimica Acta, 226*, 117–127.

Amaral-Silva1, N., Martins, R. C., Castro-Silva, S., & Quinta-Ferreira, R. M. (2012). Utilization of ozone-based techniques for industrial effluents depuration and reuse. *Journal of Advanced Oxidation Technologies, 15*, 1–9.

Ashok, L., Adishkumar, S., Banu, J. R., & Yeom, I. T. (2015). Ferrioxalate-induced solar photo-Fenton treatment of natural rubber latex wastewaters. *Water Quality Research Journal of Canada, 50*, 349–358.

Azbar, N., Yonar, T., & Kestioglu, K. (2004). Comparison of various advanced oxidation processes and chemical treatment methods for COD and color removal from a polyester and acetate fiber dyeing effluent. *Chemosphere, 55*, 35–43.

Balzani, V., & Caracceti, V. (1970). Photochemistry of coordination compounds (Chapter 10, pp. 145–192). Academic Press.

Banerjee, D., Markley, A. L., Yano, T., Ghosh, A., Berget, P. B., Minkley, E. G. Jr., Khetan, S. K., & Collins, T. J. (2006). Green oxidation catalyst for rapid deactivation of bacteria spores. *Angewandte Chemie International Edition, 45*, 3974–3977.

Barb, W. G., Baxendale, J. H., George, P., & Hargrave, K. R. (1951). Reactions of ferrous and ferric ions with hydrogen peroxide. Part I—The ferrous ion reaction. *Transactions of the Faraday Society, 47*, 462–500, 591–616.

Bataineh, H., Pestovsky, O., & Bakac, A. (2012). pH-induced mechanistic change over from hydroxyl radicals to iron(IV) in the Fenton reaction. *Chemical Science, 3*, 1594–1599.

Baxendale, J. H., Hardy, H. R., & Sutcliffe, L. H. (1951). Kinetics and equilibria in the system ferrous iron + ferric iron + hydroquinone. *Transactions of the Faraday Society, 47*, 964–973.

Baxendale, J. H., & Wilson, J. A. (1957). Photolysis of hydrogen peroxide at high light intensities. *Transactions of the Faraday Society, 53*, 344–356.

Bedoui, A., Sindi, K., & Bensalah, N. (2008). Treatment of refractory organics contained in actual agro-industrial wastewaters by UV/H_2O_2. *Clean, 36*, 373–379.

Beltran-Heredia, J., Torregross, J., Dominguez, J. R., & Peres, A. (2001). Comparison of degradation of p-Hydroxybenzoic acid in aqueous solution by several oxidation processes. *Chemosphere, 45*, 351–359.

Benotti, M. J., Stanford, B. D., Wert, E. C., & Snyder, S. A. (2009). Evaluation of a photocatalytic reactor membrane pilot system for the removal of pharmaceuticals and endocrine disrupting compounds from water. *Water Research, 43*, 1513–1522.

Bishop, D. F., Stern, G., Fleischman, M., & Marshall, L. S. (1968). Hydrogen peroxide catalytic oxidation of refractory organics in municipal waste waters. *Industrial & Engineering Chemistry Product Research and Development, 7*, 110–117.

Boczkaj, G., Fernandes, A., & Makoś, P. (2017). Study of different advanced oxidation processes for wastewater treatment from petroleum bitumen production at basic pH. *Industrial & Engineering Chemistry Research, 56*, 8806–8814.

Bolton, J. R., Bircher, K. G., Tumas, W., & Tolman, C. A. (2001). Figures-of-merit for the technical development and application of advanced oxidation processes. *Pure Applied Chemistry, 73*, 627–637.

Bolton, J. R., Ravel, M., Cater, S. R., & Safarzadeh-Amiri, A. (1996). Homogeneous solar photodegradation of contaminants in water. *Proceedings of the ASME 1996 International Solar Energy Conference* (pp. 54–60), San Antonio, Texas. The American Society of Mechanical Engineering.

Bolton, J. R., Safarzadeh-Amiri, A., & Cater, S. R. (1995). The detoxification of wastewater streams using solar and artificial light. In F. S. Sterrett (Ed.), *Alternative fuels and the environment* (pp. 187–192). Lewis Publishers.

Bossman, S. H., Oliveros, E., Gob, S., Siegwart, S., Dahlen, P., Payawan, L. L., Straub, M., Worner, M., & Braun, A. M. (1998). New evidence against hydroxyl radicals as reactive intermediates in the thermal and photochemically enhanced Fenton reactions. *The Journal of Physical Chemistry A, 102, 5542–5550.*

Buxton, G. V., Greenstock, C. L., Helman, W. P., & Ross, A. B. (1988). Critical review of rate constants for reaction of hydrated electrons, hydrogen atoms and hydroxyl radicals (OH/O^-) in aqueous solution. *Journal of Physical and Chemical Reference Data, 17*, 513.

Chahbane, N., Popescu , D.-L., Mitchell, D. A.,Chanda, A., Lenoir, D., Ryabov, A. D., Schramm, K.-W., & Collins, T. J. (2007). Fe^{III}–TAML-catalyzed green oxidative degradation of the azo dye orange II *by* H_2O_2 *and* organic peroxides: Products, toxicity, kinetics, and mechanisms. *Green Chemistry, 9*, 49–57.

Chanda, A., Khetan, S. K., Banerjee, D., Ghosh, A., & Collins, T. J. (2006). Total Degradation of Fenitrothion and Other Organophosphorus Pesticides by Catalytic Oxidation Employing Fe-TAML Peroxide Activators. *Journal of the American Chemical Society, 128*, 12058–12059.

Chanda, A., Shan, X., Chakrabarti, M., Ellis, W. C., Popescu, D. L., & de Oliveira, F. T. (2008). (TAML)Fe^{IV}=O complex in aqueous solution: Synthesis and spectroscopic and computational characterization. *Inorganic Chemistry, 47*, 3669–3679.

Chen, R., & Pignatello, J. J. (1997). Role of quinone intermediates as electron shuttles in Fenton and photoassisted Fenton oxidations of aromatic compounds. *Environmental Science & Technology, 31*, 2399–2406.

Collins, T. J., & Ryabov, A. D. (2017). Targeting high-valent iron-TAML activators at hydrocarbons and beyond. *Chemical Reviews, 117*, 9140–9182.

Collins, T. J., & Walter, C. (2006). Little green molecules. *Scientific American*, 82–84.

Cooper, G. D., & De Graff, B. A. (1972). On the photochemistry of ferrioxalate system. *Journal of Physical Chemistry, 75*, 2897–2902.

Croft, S., Guilbert, B. C., Lindsay, J. R., & Stell, J. K. (1992). Mechanism of peroxide stabilization. An investigation of some reactions of hydrogen peroxide in the presence of aminophosphonic acids. *Journal of the Chemical Society, Perkin Transactions 2*, 154–160.

Duran, A., Monteagudo, J. M., Gil, J., Exposito, A. J., & San Martin, I. (2015). Solar photo-Fenton treatment of wastewater from the beverage industry: Intensification with Ferrioxalate. *Chemical Engineering Journal*, 270, 612–620.

Egorova, G. V., Voblikova, V. A., Sabitova, L. V., Tkachenko, I. S., Tkachenko, S. N., & Lunin, V. V. (2015). Ozone solubility in water. *Moscow University Chemistry Bulletin, 70*(5), 207–210.

Eisenhouer, H. R. (1964). Oxidation of phenolic wastes. *Journal of the Water Pollution Control Federation, 36*, 1116–1128.

Ensing, B., Francesco Buda, F., Blöchl, P. E., & Baerends, E. J. (2002). A Car–Parrinello study of the formation of oxidizing intermediates from Fenton's reagent in aqueous solution. *Physical Chemistry Chemical Physics, 4*, 3619–3627.

Estrada-Arriaga, E. B., Zepeda-Aviles, J. A., & García-Sánchez, L. (2016). Post-treatment of real oil refinery effluent with high concentrations of phenols using photo-ferrioxalate and Fenton's reactions with membrane process step. *Chemical Engineering Journal, 285*, 508–5016.

Fenton, H. J. H. (1894). Oxidation of tartaric acid in the presence of iron. *Journal of the Chemical Society, 65*, 899–910.

Getoff, N., & Schenck, G. O. (1968). Primary products of liquid water photolysis at 1236, 1470 and 1849 A. *Photochemistry and Photobiology, 8*, 167–178.

Ghosh, A., Mitchell, D. A., Chanda, A., Ryabov, A. D., Popescu, D. L., Upham, E. C., Collins, G. J., & Collins, T. J. (2008). Catalase-peroxidase activity of iron(III)-TAML activators of hydrogen peroxide. *Journal of the American Chemical Society, 130*, 15116–15126.

Giri, R. R., Ozaki, H., Okada, T., Takikita, S., Taniguchi, S., & Takanami, R. (1981). Water matrix effect on UV photodegradation of perfluorooctanoic acid. *Water Science & Technology, 64*, 1980–1986.

Glaze, W. H., Kang, J.-W., & Chapin, D. H. (1987). The chemistry of water treatment involving ozone, hydrogen peroxide and ultraviolet light. *Ozone: Science Engineering 9*, 335–352.

Goldstein, S., & Rabani, J. (2008). The ferrioxalate and iodide–iodate actinometers in the UV region. *Journal of Photochemistry and Photobiology A: Chemistry, 193*, 50–55.

Gupta, S. S., Stadler, M., Noser, C. A., Ghosh, A., Steinoff, B., Lenoir, D., Horwitz, C. P., Schramm, K. W., & Collins, T. J. (2002). Rapid total destruction of chlorophenols by activated hydrogen peroxide. *Science, 296*, 326–328.

Haber, F., & Weis, J. (1934). The catalytic decomposition of hydrogen peroxide by iron salts. *Proc. Royal Soc. A: Math., Phys. Eng. Science, 134*, 1332–251.

Hage, J. P., Lobel, A., & Sawyer, D. T. (1995). Aromatic hydroxylation by Fenton reagents {Reactive intermediate [$Lx^+Fe^{III}OOH(BH^+)$], not free hydroxyl radical (HO•)}. *Bioorganic & Medicinal Chemistry, 3*, 1383–1388.

Hatchard, C. G., & Parker, C. H. (1956). A new sensitive actinometer. II. Potassium ferrioxalate as a standard chemical actinometer. *Proceedings of the Royal Society A: Mathematical, Physical and Engineering Sciences, 235*, 518–538.

Heit, G., Neuner, A., Saugy, P.-Y., & Braun, A. M. (1998). Vacuum-UV (172 nm) actinometry. The quantum yield of the photolysis of water. *The Journal of Physical Chemistry A, 102*, 5551–5561.

Hislop, K. A., & Bolton, J. R. (1999). The photochemical generation of hydroxyl radicals in the UV-vis/ferrioxalate system/H_2O_2. *Environmental Science & Technology, 33*, 3119–3126.

Hoigne, J., & Bader, H. (1983a). Rate constant of reaction of ozone with organic and inorganic compounds in water. I. Non-dissociating organic compounds. *Water Research, 17*, 173–184.

Hoigne, J., & Bader, H. (1983b). Rate constant of reaction of ozone with organic and inorganic compounds in water. II. Dissociating organic compounds. *Water Research, 17*, 185–194.

Huang, W., Brigante, B., Wu, F., Mousty, C., Hanna, H., & Mailhot, M. (2013). Assessment of the Fe(III)–EDDS complex in Fenton-like processes: From the radical formation to the degradation of bisphenol A. *Environmental Science & Technology, 47*, 1952–1959.

Hunt, J. P., & Taube, H. (1952). The photochemical decomposition of hydrogen peroxide. Quantum yields, tracer and fractionation effects. *Journal of the American Chemical Society, 74*, 5999–6002.

Jacobson, F., Holcman, J., & Sehested, K. (1998). Reactions of the ferryl ion with some compounds found in cloud water. *International Journal of Chemical Kinetics, 30*, 215–221.

Kulik, N., Trapido, M., Goi, A., Veressinina, Y., & Munter, R. (2008). Combined chemical treatment of pharmaceutical effluents from medical ointment production. *Chemosphere, 70*, 1525–1531.

Laera, G., Cassano, D., Lopez, A., Pinto, A., Pollice A., Ricco, G., & Mascolo, G. (2012). Removal of organics and degradation products from industrial wastewater by a membrane bioreactor integrated with ozone or UV/H_2O_2 treatment. *Environmental Science & Technology, 46*, 1010–1018.

Legrini, O., Oliveros, E., & Braun, A. M. (1993). Photochemical processes for water treatment. *Chemical Reviews, 93*, 698.

Lin S. H., Lin, C. M., & Leu, H. G. (1999). Operating characteristics and kinetic studies of surfactant wastewater treatment by Fenton oxidation. *Water Research, 33*, 1735–1741.

Lin, S. H., & Peng, C. F. (1995). Treatment of textile wastewater by Fenton's reagent. *Journal of Environmental Science and Health, Part A: Environmental Science and Engineering and Toxicoloogy, 30*, 88–98.

Manenti, D. R., Soares, P. A., Modenes, A. N., Espinoza-Quinones, F. R., Boventura, R. A. R., Bergamasco, R., & Vilar, V. J. P. (2015). Insight into solar photo-Fenton process using iron(III)-organic ligand complexes applied to real textile wastewater treatment. *Chemical Engineering Journal, 266*, 204–212.

Mangiante, D. M., Schaller, R. D., Zarzycki, P., Banfield, J. F., & Gilbert, B. (2017). Mechanism of ferric oxalate photolysis. *ACS Earth and Space Chemistry, 1*, 270–276.

Martinez, N. S. S., Fernández, J. F., Segura, E. F., & Sánchez Ferrer, A. (2003). Pre-oxidation of an extremely polluted industrial wastewater by the Fenton's reagent. *Journal of Hazardous Materials, 101*, 315–322.

Menaseti, E., Pelizzeti, E., & Saini, G. (1976). Interactions of Fe(III) with adrenaline, L-dopa, and other catechol derivatives: Equilibria and kinetics of complex formation in acidic perchlorate media. *Journal of Inorganic and Nuclear Chemistry, 38*, 785–788.

Merenyi, G., Lind, J., Naumov, S., & von Sonntag, C. (2010a). The reaction of ozone with the hydroxide ion: Mechanistic considerations based on thermokinetic and quantum chemical calculations and the role of HO_4^- in superoxide dismutation. *Chemistry—A European Journal,* 16, 1372–1377.

Merenyi, G., Lind, J., Naumov, S., & von Sonntag, C. (2010b). Reaction of ozone with hydrogen peroxide (peroxone process): A revision of current mechanistic concepts based on thermokinetic and quantum-chemical considerations. *Environmental Science & Technology, 44,* 3505–3507.

Mohammadzadeh, J. J. S., Khoshfetrat, A. B., & Kaynezad, M. A. (2004). Pilot scale system for removal of phenol in phenolic wastewater of olefin plant. *Process Safety Environmental Protection, 82,* 365–370.

Monteagudo, J. M., Duran, A., Corral, J. M., Carnicer, A., Frades, J. M., & Alonson, M. A. (2012). Ferrioxalate-induced solar photo-Fenton system for the treatment of winery wastewater. *Chemical Engineering Journal, 181–182,* 281–288.

Monteagudo, J. M., Duran, A., Culebradas, R., San Martin, I., & Carnicer, A. (2013). Optimization of pharmaceutical wastewater treatment by solar/ferrioxalate photo-catalysis. *Journal of Environmental Management, 138,* 210–219.

Murphy, A. P., Boegli, W. J., Price, M. K., & Moody, C. D. (1989). A Fenton-like reaction to neutralize formaldehyde waste solution. *Environmental Science & Technology, 23,* 166–169.

Oppenlander, T., Walddorfer, C., Burgbacher, J., Kiermeier, M., & Lachner, K. (2005). Improved vacuum-UV (VUV)-initiated photomineralization of organic compounds in water with a xenon excimer flow-through photoreactor (Xe2* lamp, 172 nm) containing an axially centered ceramic oxygenator. *Chemosphere, 60,* 302–309.

Pérez, M., Torrades, F., Garcia-Hortal, J. A., Domènech, X., & Peral, P. (2002). Removal of organic contaminants in paper pulp treatment effluents under Fenton and photo-Fenton conditions. *Applied Catalysis BV: Environmental, 36,* 63–74.

Philippopoulos, C. J., & Poulopoulos, S. G. (2003). Photo-assisted oxidation of an oily wastewater using hydrogen peroxide. *Journal of Hazardous Materials, B98,* 201–210.

Pignatello, J. J. (1992). Dark and photoassisted Fe^{3+}-catalyzed degradation of chlorophenoxy herbicides by hydrogen peroxide. *Environmental Science & Technology, 26,* 944–951.

Pignatello, J. J., Oliveros, E., & MacKay, A. (2006). Advanced oxidation processes for organic contaminant destruction based on the Fenton reaction and related chemistry. *Critical Reviews in Environmental Science and Technology, 36,* 1–84.

Pozdnyakov, I. P., Kel, O. V., Plyusnin, V. F., Grivin, V. P., & Bazhin, N. M. (2008). New insight into photochemistry of ferrioxalate. *The Journal of Physical Chemistry A, 112*, 8316–8322.

Qin, W., Song, Y., Dai, Y., Qiu, G., Ren, M., & Zeng, P. (2015). Treatment of berberine hydrochloride pharmaceutical wastewater by $O_3/UV/H_2O_2$ advanced oxidation process. *Environmental Earth Sciences, 73*, 4939–4946.

Reisz, E., Schmidt, W., Schuchmann, H.-P., & von Sonntag, C. (2003). Photolysis of ozone in aqueous solution in the presence of tertiary butanol. *Environmental Science & Technology, 37*, 1941–1948.

Rodriguez, E. M., Nunez, B., Fernandez, G., & Beltran, F. J. (1992). Effects of some carboxylic acids on the Fe(III)/UVA photocatalytic oxidation of muconic acid in water. *Applied Catalysis B: Environmental, 26*, 314–319.

Safarzadeh-Amiri, A. (2019). [Unpublished data].

Safarzadeh-Amiri, A. (2001). O_3/H_2O_2 treatment of methyl-tert-butyl ether (MTBE) in contaminated water. *Water Research, 35*, 3706–3714.

Safarzadeh-Amiri, A., Bolton, J. R., & Cater, S. R. (1996). The use of iron in advanced oxidation processes. *Journal of Advanced Oxydation Technologies, 1*, 18–26.

Safarzadeh-Amiri, A., Bolton, J. R., & Cater, S. R. (1997). Ferrioxalate-mediated photodegradation of organic pollutants in contaminated water. *Water Research, 31*, 787–798.

Sawyer, D. T., & Valentine, J. S. (1981). How superoxide works? *Accounts of Chemical Research, 14*, 393–400.

Schneider, J., Matsuoka, M., Takeuchi, M., Zhang, J., Horiuchi, Y., Anpo, M., & Bahnemann, D. W. (2014). Understanding TiO_2 photocatalysis: Mechanisms and materials. *Chemical Reviews, 114*, 9919–9986.

Souza, B. M., Souza, B. S., Guimarães, T. M., Ribeiro, T. F. S., Cerqueira, A. C; Sant'Anna Jr., G. L., & Dezotti, M. (2016). Removal of recalcitrant organic matter content in wastewater by means of AOPs aiming industrial water reuse. *Environmental Science and Pollution Research, 23*, 22947–22956.

Staehalin, J., & Hoigne, J. (1982). Decomposition of ozone in water: Rate of initiation by hydroxide ions and hydrogen peroxide. *Environmental Science & Technology, 16*, 676–681.

Sun, L., & Bolton, J. R. (1996). Determination of the quantum yield for the photochemical generation of hydroxyl radicals in TiO_2 *suspensions. J. Phys. Chem. 1996, 100*, 4127–4134.

Sun, Y., & Pignatello, J. J. (1992). Chemical treatment of pesticides wastes: Evaluation of Fe(III) chelates for catalytic hydrogen peroxide oxidation of 2,4-D at circumneutral pH. *Journal of Agricultural and Food Chemistry, 40*, 322–327.

Sun, Y., & Pignatello, J. J. (1993a). Activation of hydrogen peroxide by iron(III) chelates for abiotic degradation of herbicides and insecticides in water. *Journal of Agricultural and Food Chemistry, 41*, 308–312.

Sun, Y., & Pignatello, J. J. (1993b). Photochemical reactions involved in the total mineralization of 2,4-D by Fe^{3+}/H_2O_2/UV. *Environmental Science & Technology, 27*, 304–310.

Tachiev G., Roth J. A., & Bowers A. R. (2000). Kinetics of hydrogen peroxide decomposition with complexed and "free" iron catalysts. *International Journal of Chemical Kinetics, 32*, 24–35.

Tekin, H., Bilkay, O., Ataberk, S. S., Balta, T. H., Ceribasi, I. H., & Sanin, F. D. (2006). Use of Fenton oxidation to improve the biodegradability of a pharmaceutical wastewater. *Journal of Hazardous Materials, 136*, 258–265.

Tong, A. Y. C., Braund, R., Warren, D. S., & Peake, B. M. (2012). TiO_2-assisted photodegradation of pharmaceuticals—A review. *Central European Journal of Chemistry, 10*, 989–1027.

Vogna, D., Marotta, R., Napolitano, A., Andreozzi, R., & d'Ischia, M. (2004). Advanced oxidation of the pharmaceutical drug diclofenac with UV/H_2O_2 and ozone. *Water Research, 38*, 414–422.

von Sonntag, C. (2008). Advanced oxidation processers: Mechanistic aspects. *Water Science & Technology, 58*, 1015–1021.

Walling, C. (1998). Intermediates in the reactions of Fenton type reagents. *Accounts of Chemical Research, 31*, 155–157.

Walling, C., & Goosen, A. (1973). Mechanism of the ferric ion catalyzed decomposition of hydrogen peroxide. Effect of organic substrates. *Journal of the American Chemical Society, 95*, 2987–2991.

Wardman, P. (1989). Reduction potential of one-electron couples involving free radicals in aqueous solution. *Journal of Physical and Chemical Reference Data, 18*, 1637–1755.

Watt, R. J., & Dilly, S. E. (1996). Evaluation of iron catalysts for the Fenton-like remediation of diesel-contaminated soil. *Journal of Hazardous Materials, 51*, 209–224.

Weeks, J. L., & Matheson, M. S. (1956). The primary yield of hydrogen peroxide decomposition. *Journal of the American Chemical Society, 78*, 1274–1278.

Weeks, J. L., Meaburn, G. M. A. C., & Gordon, S. (1963). Absorption coefficients of liquid water and aqueous solutions in the far ultraviolet. *Radiation Research, 19*, 559–567.

Wells, C. F., & Salam, M. A. (1968a). The effect of pH on the kinetics of the reaction of iron(II) with hydrogen peroxide in perchlorate media. *Journal of the Chemical Society A*, 24–29.

Wells, C. F., & Salam, M. A. (1968b). Complex formation between iron(II) and inorganic anions> part II. The effect of oxyanions on the reaction of iron(II) with hydrogen peroxide. *Journal of the Chemical Society A*, 308–315.

Wingate, K. G., Stuthridge, T. R., Wright, C. J., Horwitz, O. P., & Collins, T. J. (2004). Application of TAML catalysts to remove colour from pulp and paper mills. *Water Science & Technology, 49*, 255–260.

Wink, D. A., Nims, R. W., Saavedra, J. H., Utermahlen, W. E., & Ford, P. C. (1994). The Fenton oxidation mechanism: Reactivity of biologically relevant substrates with two oxidizing intermediates differ from those predicted for the hydroxyl radical. *Proceedings of the National Academy of Sciences, 91*, 6604–6608.

Wood, P. M. (1988). The potential diagram for oxygen at pH 7. *Biochemical Journal, 253*, 287–289.

Yasar, A., Ahmad, N., Chaudhry, V., Rehman, M. S. U., & Khan, A. A. A. (2006). Ozone for color and COD removal of raw and anaerobically biotreated combined industrial wastewater. *Polish Journal of Environmental Studies, 16*, 289–294.

Yoon, Y., Hwang, Y., Kwon, M., Jung, Y., Hwang, T.-M., & Kang, J.-W. (2014). Application of O_3 and O_3/H_2O_2 as post-treatment processes for color removal in swine wastewater from a membrane filtration system. *Journal of Industrial and Engineering Chemistry, 20*, 2801–2805.

Zangeneh, H., Zinatizadeh, A.A.L., & Feizy, M. (2014). A comparative study on the performance of different advanced oxidation processes (UV/O_3/H_2O_2) treating linear alkyl benzene (LAB) production plant's wastewater. *Journal of Industrial and Engineering Chemistry,* 20, 1453–1461.

Zepp, R. G., Faust, B. C., & Hoigne, J. (1992). Hydroxyl radical formation in aqueous reactions (pH4-8) of iron(II) with hydrogen peroxide: The photo-Fenton reaction. *Environmental Science & Technology, 26*, 314–319.

Zuo, Y., & Hoigne, J. (1992). Formation of hydrogen peroxide and depletion of oxalic acid in atmospheric water by photolysis iron(III)-oxalate complexes. *Environmental Science & Technology, 26*, 1014–1022.

11.0 SUGGESTED READINGS

Arslan, I., Balcioğlu, I. A., & Tuhkanen, T. (1999). Oxidative treatment of simulated dyehouse effluent by UV and near-UV light assisted Fenton's reagent. *Chemosphere, 39*, 2767–2683

Walling, C. (1975). Fenton's reagent revisited. *Accounts of Chemical Research, 8*, 125–131.

5

Desalination and Brine Management

Somnath Basu, Kushal Seth, Javier Casas, & Paul Sun

1.0 INTRODUCTION

Liquid streams containing high concentrations of dissolved inorganic solids, ranging from seawater concentrations of ~3.5% at the low end up to 25% or higher in water, is commonly termed "brine." These streams are produced as wastes from several industrial operations and—as with any industrial wastewater stream—they require safe management and disposal. In most cases, sodium chloride (NaCl) is the major constituent of total dissolved solids (TDS). However, industrial effluents also have other components such as calcium and sulfate ions, which often make brine management difficult. However, their high salinity and high scaling potential render these streams difficult to dispose.[1]

Additionally, many industrial effluents contain high concentrations of organic matter as well as high TDS. Although the most commonly practiced and economic means of managing high chemical oxygen demand (COD) wastewater streams is typically biological treatment, it is difficult to sustain microbial population beyond a certain level of salinity. This subject is discussed in further detail in Chapter 6. In some cases, industries resort to treatment of such streams by granular activated carbon, or other physicochemical processes to selectively remove COD first, and then manage disposal of high-salinity streams suitably. Alternatively, industries may be able to segregate streams that contribute high TDS levels (e.g., softener regeneration brine) and/or to blend a high TDS waste stream with other waste streams to reduce the TDS to a treatable level.

Leather, textile, seafood processing, and chlorinated organics production are among the industries generating high-salinity effluents. However,

[1]The authors gratefully acknowledge the help received from Mr. Malcolm Mann of Saltworks Technology, Richmond, British Columbia, Canada, in the preparation of the section on Electrodialysis Reversal (EDR) section. Mr. Saurabh Deshpande of Headworks International, Houston, Texas, conducted a thorough literature search and detailed editorial work in compiling the chapter.

a much larger and widespread challenge is the disposal of brine associated with flowback and produced water in oil and gas exploration and production operations throughout the United States. Technology can concentrate streams several fold, which helps minimize the volume to be disposed of and also to produce clean and highly pure water streams that can be used for beneficial purposes.

Desalination and brine concentration technologies are based on four major principles (see Figure 5.1): thermal (evaporation of water vapor); electrical; pressure (using a semipermeable, nonporous membrane to separate fresh water from saline solution); and chemical. Chemical processes, like ion exchange, can be used for selective constituent removal, but they are not suitable and/or practical for handling waters with high levels of TDS. Currently, thermal and membrane technologies are the most commonly used desalination techniques. Multistage flash distillation and reverse osmosis currently account for approximately two-thirds of the total installed seawater desalination capacity in the world (Henthorne, 2009).

Figure 5.2 provides a graphical view of commonly used desalination, brine treatment, and brine disposal alternatives. The X-axis presents the appropriate ranges of salinity for the various technologies. The Y-axis presents their relative costs in terms of barrels of water treated with respect to each other. (1 bbl equals 0.16 m^3, or 42 gal or 0.16 m^3.)

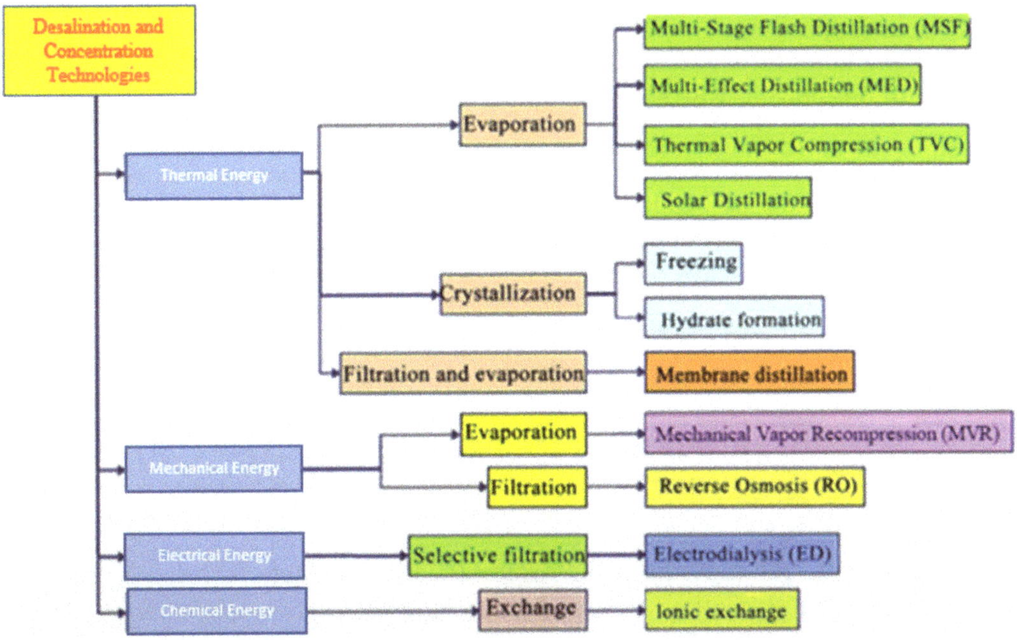

FIGURE 5.1 Different types of desalination technologies (adapted from Kabeel et al., 2013).

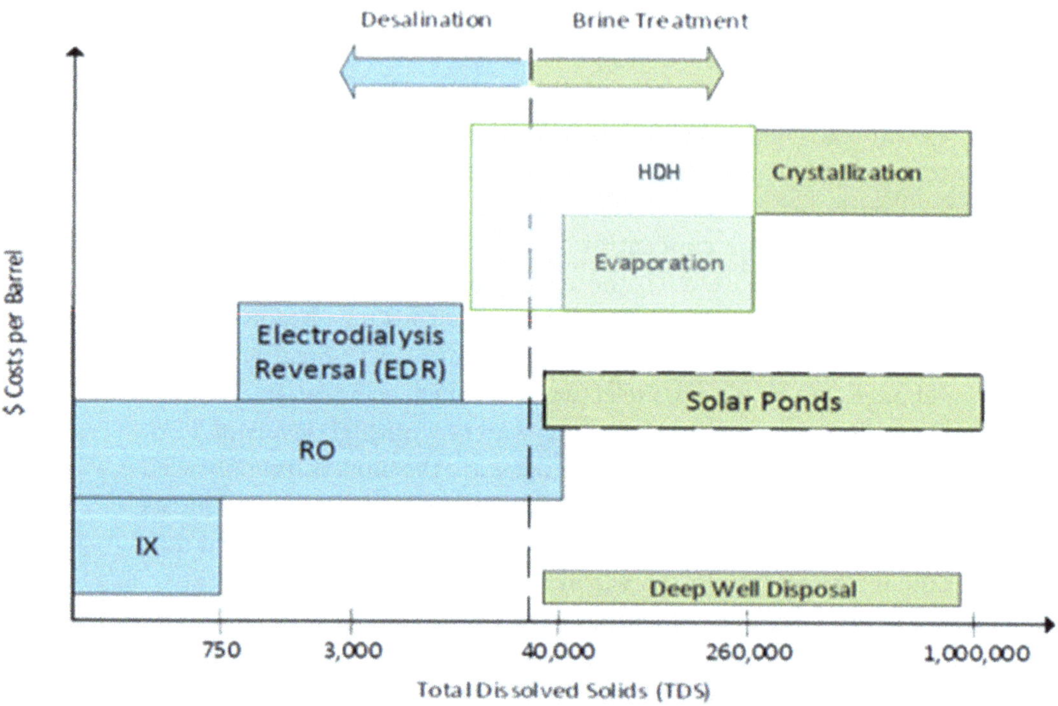

FIGURE 5.2 Desalination and brine treatment technologies and relative costs.

This chapter presents a detailed description of the technologies for concentration, disposal, and proper management of high-salinity streams. While keeping the discussion adequately broad based on the processing of high-salinity water from a range of industries, this chapter largely draws examples from two applications: (1) oil and gas exploration and production and (2) production of high-quality water from brackish water and seawater. The technologies are grouped into three broad categories based on their following functions:

- Desalination of brackish water and seawater
- Concentration of brine as a pretreatment step before disposal
- Disposal of brine concentrate and crystals

2.0 DESALINATION OF BRACKISH WATER AND SEAWATER

Water is the most precious resource on our planet, both for human consumption as well as industrial uses. However, the supply of fresh water is shrinking at an alarming rate and the gap between demand and supply has been ever increasing. This is more pronounced in many parts of the world. In certain regions of the United States, such as some parts of Florida, groundwater is highly saline because of seawater intrusion and is thus unsuitable for beneficial use. Availability of fresh water is also scarce in the Middle East because of climatic and geological conditions. Seawater and brackish water can be used in these regions to produce high-purity water for human consumption and industrial processes.

Reverse osmosis and water distillation by mechanical vapor recompression (MVR) are the two technologies that can be used in these regions for the production of high-quality water. The working principle and applications of reverse osmosis have been discussed in detail in Chapter 3. The ability of reverse osmosis to produce a consistent high-quality effluent makes it an important—a widely applied—technology for industrial wastewater recycle. However, management of the brine (reject) generated by reverse osmosis is often not given due attention, and management of this brine may create challenges. Later in this chapter, we will cover technologies that can be leveraged for appropriately managing such a brine stream. Electrodialysis reversal (EDR) is another membrane-based technology that is being increasingly used in brackish water desalination applications. This section presents a discussion of these technologies, including a section on freeze crystallization as an alternate process for the desalination of seawater.

2.1 Electrodialysis Reversal

Electrodialysis is a process in which separation of ions takes place with an application of electromotive force across ion-selective permeable membranes. Unlike reverse osmosis, this process does not require use of high pressure, and only ions pass through the membranes. In this process, under the electrical potential, positively charged ions (cations) migrate toward the cathode and negatively charged ions (anions) migrate toward the anode through ion-exchange membranes. The cation exchange membranes selectively transport the cations in the direction of the cathode and restrict the passage of anions through them. Similarly, the anion exchange membranes select only the anions to permeate through them in the direction of anode.

The membranes are arranged in parallel sets of alternating anion exchange membranes and cation exchange membranes, as shown in Figure 5.3. This figure shows the two streams entering into the EDR stack—a feed stream to be desalinated through a product compartment and a concentrate compartment with a salt-accepting stream (the reject stream).

Electrodialysis reversal is a special adaptation of electrodialysis. In this process, electrode reversal can take place during the process by alternating the polarity of electrodes at a regular interval. The benefit of polarity

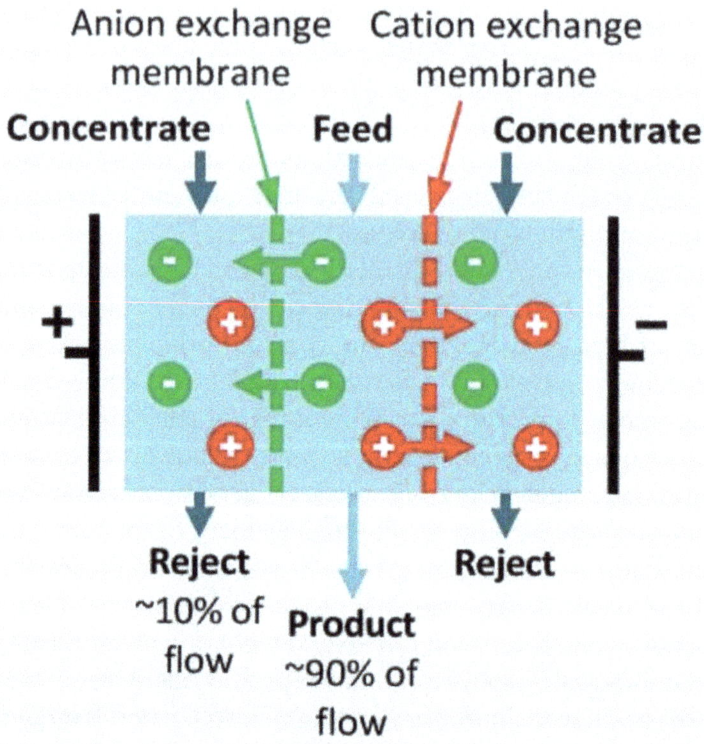

FIGURE 5.3 Operating principle of electrodialysis (reprinted with permission from Saltworks Technologies, Richmond, British Columbia, Canada).

reversal is that the anions and cations do not have an opportunity to accumulate, respectively, on the anion and cation exchange membrane surfaces, resulting in deposition of scales over time. Thus, the inorganic scaling and organic fouling can be avoided by EDR and efficiency maintained for a long time. This is in contrast to reverse osmosis, where the addition of antiscaling chemicals is necessary to maintain its performance; although, in some applications, using antiscalants in EDR is also beneficial.

Figure 5.4 presents a schematic view of EDR in forward and reverse polarity modes. As shown in the figure, the polarity of the electrodes and the hydraulic channels "reverse." These periodic reversals cause the ions to travel in an opposite direction, whereby the inorganic scale and foulants are "back flushed" off of the scaled side of the membrane surfaces. Moreover, EDR is more resilient to higher silt loading and organics because desalination is driven by applying electricity across ion-exchange membranes to separate ions by charge, rather than the use of a pressure difference across a membrane. This results in reduced pretreatment requirement and cost.

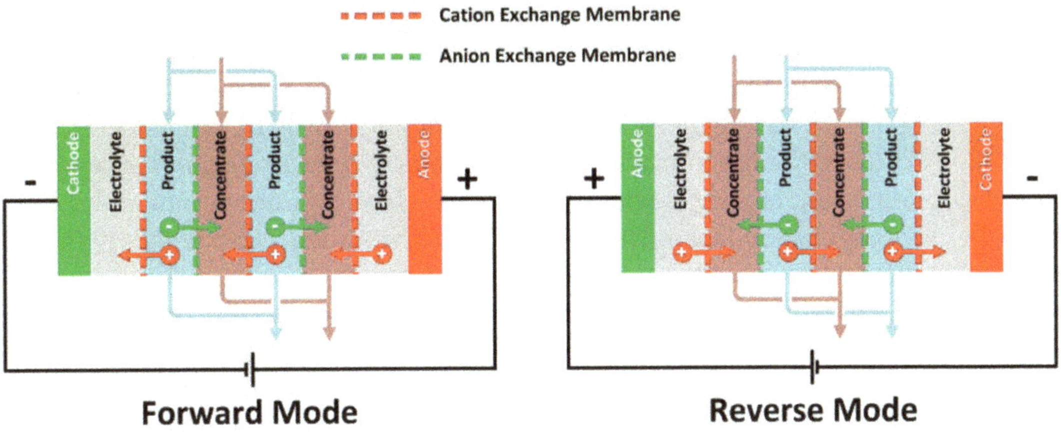

FIGURE 5.4 Electrodialysis reversal operation in forward and reverse polarity modes (reprinted with permission from Saltworks Technologies, Richmond, British Columbia, Canada).

The economics of EDR vary widely depending on many factors, especially the difference between TDS concentration in the feed and TDS concentration in the final desalted product water. The greater the difference in TDS, the more the EDR system has to desalt, resulting in larger membrane area and increased capital and power costs.

Because of the reduced potential for scaling and fouling, EDR technology can help reduce the brine volume by concentrating a saline water up to 180 000 mg/L TDS; more than that can be achieved by reverse osmosis. Depending on the TDS of the saline water, EDR recoveries of >95% can be achieved. However, the brine volume of EDR and reverse osmosis in industrial applications is often determined by the water chemistry (specifically the scaling compounds in the water). Hence, EDR and reverse osmosis often achieve the same volume reduction in industrial applications, in which the water has scaling potential and no upstream chemical softening is performed.

In a further development of EDR technology, anion exchange membranes, specifically for monovalent anions (e.g., chloride ions) are also available. These membranes selectively transport only the monovalent anions while restricting cations and multivalent anions (e.g., sulfate and carbonate ions, which are the most common scale-forming anions in water and wastewater from various sources). The action of monovalent EDR (mEDR) can be understood by considering the example of flue gas desulfurization (FGD) wastewater from coal fired power plants as explained below.

Flue gas desulfurization scrubber systems are used to remove sulfur dioxide (SO_2) air emissions with flue gas. They create waste streams often saturated with calcium sulfate while containing both metals and chlorides. These streams are recycled internally and then purged out of the system when chloride concentrations exceed a set level as high chloride concentrations inhibit SO_2 absorption from the flue gas and create corrosion concerns with wetted equipment. Chloride blowdown levels range with operations from 10 000 mg/L to as high as 30 000 mg/L. Monovalent EDR systems selectively remove chlorides from the FGD scrubber blowdown as a highly concentrated nonscaling brine of sodium and calcium chloride. Cations and multivalent anions (e.g., sulfate ions), are recycled in the scrubber and removed as gypsum (calcium sulfate) free from chloride ions.

Saltworks Technologies conducted an onsite pilot test program on an actual FGD wastewater with mEDR that demonstrated removing up to 85% of the chlorides to enable recycling of the treated FGD wastewater back to the scrubber. An economic analysis was completed to compare the cost of mEDR with (1) chemical softening followed by seawater reverse osmosis (SWRO) and ultrahigh pressure reverse osmosis (UHPRO) and (2) chemical softening followed by SWRO and an evaporator. This economic analysis was based on the assumption that capital costs depreciated over a 20-year period at 8% and including operating costs, such as energy, chemicals, and membrane replacements. Labor cost was excluded. The results showed that for a FGD stream of 300 m^3/d (~55 gpm) with 20 800 mg/L TDS and >11 000 mg/L chloride concentration, between 55% and 65% reduction in total cost of ownership could be achieved by mEDR (Saltworks Technologies, 2018). It is noted that the TDS of the chloride rich stream from mEDR is higher than what would be typically produced by SWRO or mechanical vapor recompression evaporation (discussed in detail in Section 3.1.1), but for some industrial reuse applications, such as recycling of the treated FGD wastewater with reduced chlorides back to the scrubber, this level of separation is adequate and higher purity may not be necessary.

2.2 Freeze Crystallization

2.2.1 Introduction

Freeze crystallization consists of freezing various brine streams under low temperatures to recover high-quality water. In freeze crystallization, process water is recovered as solid ice crystals by refrigeration and the crystals are melted by application of heat to produce liquid water. Dissolved salts leave the system as concentrated brine, which can be discharged to the sea.

Removal of dissolved solids by freeze crystallization can be achieved by using refrigeration available from vaporization of liquefied petroleum gas (LPG), liquefied natural gas (LNG), or another commercial refrigerant capable of attaining the temperature necessary for phase separation of water from liquid to solid phase. It may be an attractive alternative process for desalination of seawater and brackish and other low-salinity water.

2.2.2 Type and Configuration of Freeze Crystallization Processes

Freeze crystallization-based desalination processes can be classified according to (a) cooling loop integration and (b) type of concentration/crystallization process. There can be two types of cooling loops: single or dual. In a single-loop system, a refrigerant cools high TDS industrial wastewater through a single cooling loop. In a dual-loop system, a refrigerant (e.g., LPG, LNG) cools another refrigerant (e.g., glycol) through a primary loop, which, in turn, cools the industrial wastewater through a secondary loop. The two-refrigerant loop system is typically adopted by commercial systems.

Two types of freezing processes can be applied for freeze crystallization: (a) refrigeration under atmospheric pressure and (b) refrigeration under vacuum. In atmospheric freezing, process industrial wastewater is frozen under atmospheric pressure at (or just below) the freezing point (~273 to 269 °K or 0 to −4 °C) of water. Conceptually, cooling can proceed until the eutectic point (~243 to 253 °K or −30 to −20 °C, depending on salt concentration). However, for practical reasons, as described later in this chapter, it is not feasible to refrigerate the feed all the way to the eutectic point. Crystallization under vacuum freezing uses the principle of separation of ice crystals under three-phase equilibrium at the triple point of water.

Brine generated as the concentrate phase from both processes is then discharged. The following sections describe the working principles of both processes.

2.2.2.1 Eutectic Freeze Crystallization Under Atmospheric Pressure

Atmospheric freezing technology is based on the principle of separation by fractional crystallization (Rahman et al., 2006). In cases where the dissolved solids in a target wastewater are entirely made up of sodium chloride, with only insignificant amounts of other salts, the desalination process can be conceptually understood from the theoretical NaCl–water phase diagram (Figure 5.5) under atmospheric pressure. When seawater, containing 3.5% to 4% salt, is cooled, formation of ice crystals starts as the temperature reaches approximately −5 °C (23 °F). This can be observed from Figure 5.5 on the equilibrium line (on the left side of the diagram) between pure ice (solid phase) and liquid brine.

Further cooling results in further separation of ice crystals from increasing concentrations of brine as the temperature of the liquid gradually approaches the eutectic composition of approximately 23% NaCl in water at −21 °C (−6 °F). At the eutectic point, all NaCl in seawater will become bound with water in the form of a single-phase system consisting of a homogeneous solid solution of $NaCl{\cdot}2H_2O$. The remaining water, free from salt, can be recovered as ice crystals of pure water.

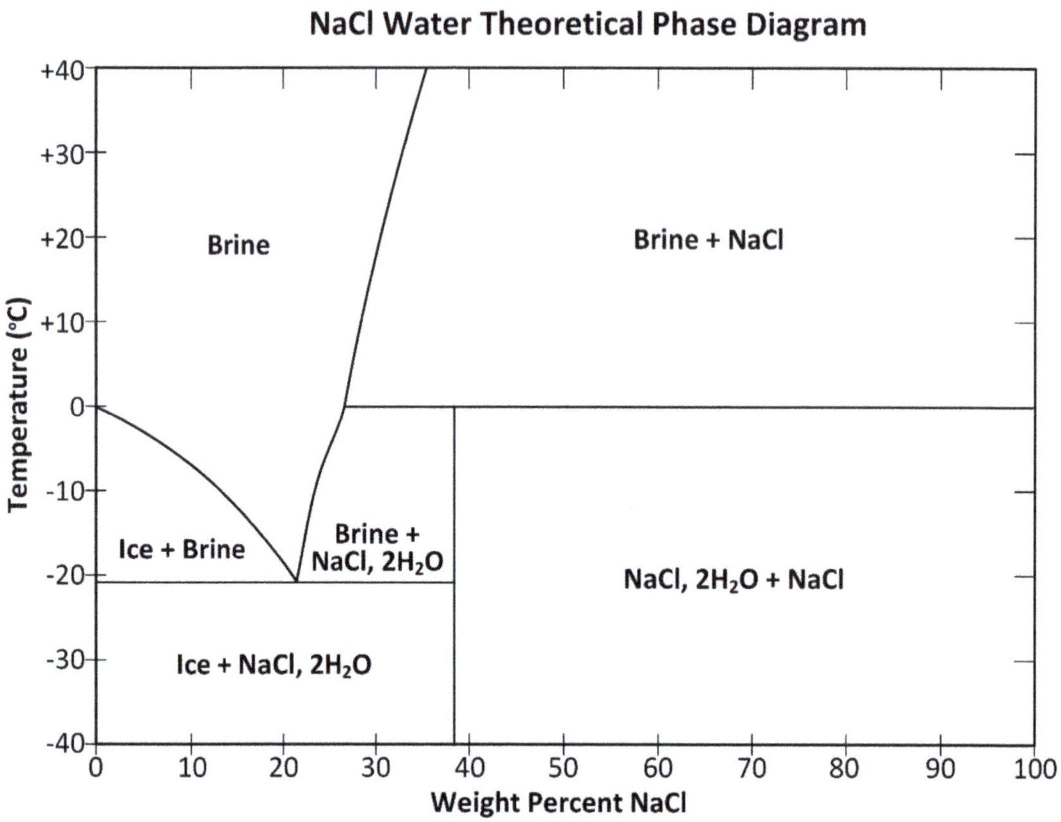

FIGURE 5.5 Sodium chloride water theoretical phase diagram (adapted from http://www.phasediagram.dk/binary/sodium_chloride.htm; re-created by Alan D. Ingham, Black & Veatch, Kansas City, Missouri).

2.2.2.2 Freeze Crystallization at Triple Point Under Vacuum

This technology is based on production and removal of ice at triple point of water (0 °C [32 °F] and 613 Pa). Figure 5.6 presents a phase diagram of water showing the location of the triple point. Under this condition, water coexists in solid–liquid–vapor forms in equilibrium. To maintain the triple-point condition, it is critical to remove the water vapor continuously as it is being formed. In industrial systems, this is achieved by using a compressor that removes the water vapor and compresses the vapor to a pressure and temperature at which it can be recondensed into liquid water.

2.2.3 Equipment Description

2.2.3.1 Equipment Based on Eutectic Crystallization

Eutectic freeze crystallization has been extensively studied in both laboratory and pilot (Verbeek, 2011) scales at Delft University, Netherlands, on synthetic potassium chloride solution using a 200-L crystallizer. GEA Niro PT B.V. of

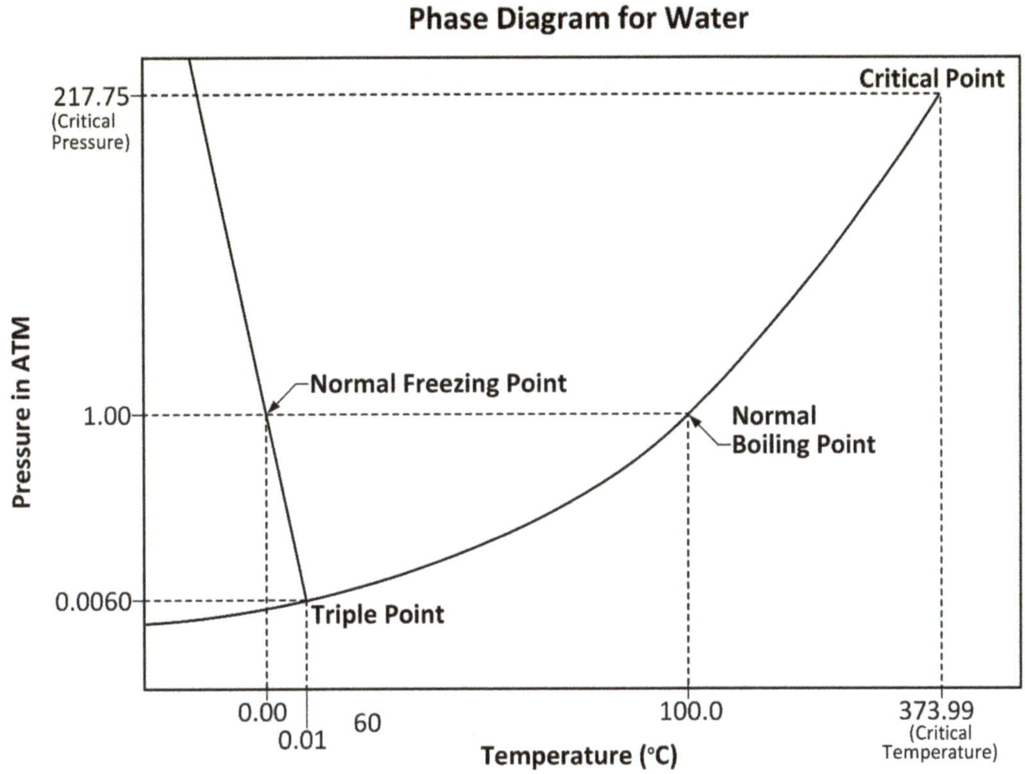

FIGURE 5.6 Phase diagram of water (adapted from http://en.wikipedia.org/wiki/Triple_point; re-created by Alan D. Ingham, Black & Veatch, Kansas City, Missouri).

the Netherlands used this principle to develop an equipment for recovery of water from seawater. Although it is theoretically possible to cool down the feed all the way to the eutectic point, which would allow recovery of 83% of pure water from the feed seawater, this process stops only at approximately 30% recovery because the concentrate becomes viscous and difficult to handle for pumping, and also deposition of the brine film on heat exchanger surfaces. The product is a slurry of ice crystals in a concentrated brine solution. A marketable product based on this technology would not be sustainable because of the high capital and operations and maintenance (O&M) cost.

2.2.3.2 Equipment Based on Crystallization at Triple Point of Water

This technology is based on production and recovery of ice from seawater at the triple point of water (0 °C and 613 Pa). IDE Technologies in Israel markets a vacuum ice maker freeze crystallization system using this principle for various low-salinity applications, such as district heating/cooling, thermal energy storage, and snow making (Tomer, 2011). In an AngloGold mine in South Africa, it has been used to treat mine drainage water of <1%

salinity. In the past, a seawater desalination facility based on this technology was used to provide drinking water for a community in southern Israel where drinking water was previously scarce. That facility, however, is no longer in service because of economic reasons. This process is capable of recovering 50% of feed as high-quality water.

Freeze crystallization offers the following distinct advantages over thermal crystallization:

- Energy duty is approximately 75% to 90% less because the latent heat of ice freezing and melting is 333 kJ/kg as opposed to that for water vaporization of 2500 kJ/kg.
- Equipment is not subject to an aggressive corrosive environment posed by high chloride concentration at boiling temperature.
- It is suitable for separation of chemicals and food (fruit juices, etc.) because it avoids degradation resulting from molecular breakdown; although this advantage is not specifically applicable for brine management systems.

Despite these advantages, application of freeze crystallization technology has not been commercially viable primarily because of the high capital and O&M costs. A recent survey conducted by El Kadi and Janajreh (2017) reported that "currently no commercial facility is available for desalination of seawater by freeze crystallization." At this time, it appears only feasible in the Middle East, where water is scarce in certain regions, but natural gas is plentiful. Evaporation of LNG in regasification terminals has the potential for application of freeze crystallization technology for the production of drinking water from seawater affordable on a large scale. However, this needs further technological and economic evaluation.

3.0 CONCENTRATION OF BRINE BEFORE DISPOSAL

Evaporation and humidification–dehumidification are the two main technologies under this category. These technologies are primarily used for treating and volume reduction of reverse osmosis concentrate streams because traditional reverse osmosis is limited, from technology and economic viewpoints, to handle liquids with TDS above 40 000 mg/L. At such a concentration the reverse osmosis would need to operate at 54 to 68 atm (800 to 1000 psi) as it further concentrates the solution by 50% to 100%. This pressure range is the upper limit of traditional desalination reverse osmosis units. It is noted that there are commercially available "high-pressure reverse osmosis" (HPRO) systems that can be operated up to approximately 115 to 136 atm

(1700 to 2000 psi), roughly doubling the upper TDS limit of traditional reverse osmosis units. It should also be noted that for industrial wastewater reclamation TDS is frequently not the limiting factor for operation of reverse osmosis. In most cases, the presence of scale-forming compounds become the limiting factor and are important to consider.

3.1 Evaporation

Evaporators are used for concentrating a solution by evaporation of the solvent using thermal energy or mechanical energy. This is a pretreatment step to minimize the disposal volume of liquid streams. In brine management application, the solvent is water.

In thermal evaporators, steam is used on the tube side as the source of heat, and the feed solution is heated to its boiling point on the shell side of a heat exchanger. The feed then enters the evaporator chamber, where a part of it flashes as vapor and the concentrated solution leaves the evaporator. The heating steam is condensed.

Thermal evaporators can have a single effect and multiple effects. In a single-effect evaporator, the water vapor from the boiling liquid is typically used to preheat the feed solution and the condensate is recovered for producing steam. A multiple-effect evaporator contains multiple evaporators in series. Vapor from the boiling liquid in the first evaporator serves as the heating steam for the second, which evaporates the concentrated liquid from the first. Similarly, vapor from the second becomes the heating steam for the third to evaporate the liquor exiting the second and so on. Forced circulation MVR evaporators are most widely used among several types available in the market.

3.1.1 Principle and Equipment Arrangement of Mechanical Vapor Recompression Evaporators

Mechanical vapor recompression is an energy-efficient process that separates the feed water into a large volume of (nearly) deionized water and a small volume of highly concentrated brine. It uses vapor recompression by a gas compressor as the main driver for water evaporation (see Figure 5.7). Evaporation by MVR is driven primarily by mechanical energy with some thermal input.

The system includes the evaporator/condenser tubes, feed preheaters, vapor compressor, ejector, and the pumping units. It takes advantage of the principle of using a heat pump to efficiently evaporate/condense the incoming brine. The driving energy is solely provided by a vapor compressor to lower the pressure at the feed water end, thus lowering its boiling temperature at the "boiling" brine side and delivering the compressed vapor to the super-heated steam side. The compressed steam is then condensed

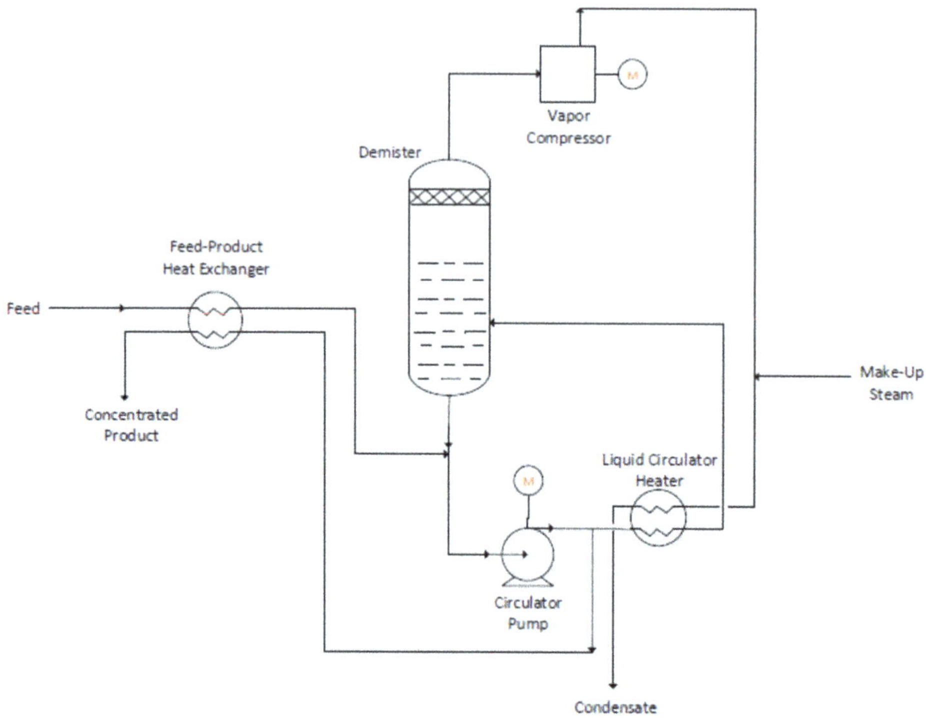

FIGURE 5.7 Schematic diagram of a mechanical vapor recompression evaporator.

in the shell side of the falling film heat exchanger at a higher condensation temperature, where the latent heat is transferred to the brine circulating through the internal tubes.

Readers are encouraged to study the principles of heat and mass balances provided in literature (Ettouney, 2006; Stickney & Fosberg, 1975). The actual energy required is estimated to be approximately 50 to 120 kWh per 3.785 m^3 (1000 gal) of water treated, depending on the design of the individual unit. This is energy efficient compared to other brine concentration techniques.

The vapor-liquid separation chamber is equipped with a demister to prevent carryover of concentrated liquid droplets with the vapor phase. The evaporator unit is typically constructed with corrosion-resistant alloys, such as 316L stainless steel or titanium. The feed/product heat exchangers in MVR systems are available both as shell and tube type and plate type. The compressor is typically of centrifugal type, or rotary-lobe positive displacement type. For most of the commercial systems, the compressor is electrically driven; however, gas-engine or steam-driven systems are also available.

The application range for a mechanical vapor recompression evaporator is for saline water from 20 000 to 200 000 mg/L based on NaCl concentration. Concentration of the solution leaving the MVR is typically at the high end of this range. Above the upper limit, NaCl salt will exceed its solubility and precipitate out of the brine solution. Evaporators are not designed to handle crystalizing solution because of the problem of scaling of the heat exchanger surfaces. A thermal crystallizer, on the other hand, is designed to precipitate out the salt crystals and recover more fresh water.

3.1.2 Limitations of Mechanical Vapor Recompression and Pretreatment Requirements

Mechanical vapor recompression evaporators are subject to the same scaling phenomenon as other thermal evaporation processes. When the divalent salts ($CaCO_3$, $CaSO_4$, $BaSO_4$, or silicates) reach their solubility limits at the elevated operating temperature, scaling can occur, especially in the feed/product heat exchanger. The calcium carbonate or calcium sulfate scaling can become a problem here. Figure 5.8 shows the solubility of various salts in water as a function of temperature. These types of scale-forming constituents are typically present in industrial wastewater. Water chemistry and the corresponding propensity for scale formation require careful attention when evaluating this and other brine management technologies.

The noncondensable dissolved air and carbon dioxide (CO_2) are removed in the feed deaerator, which is essentially a steam stripper. In certain cases,

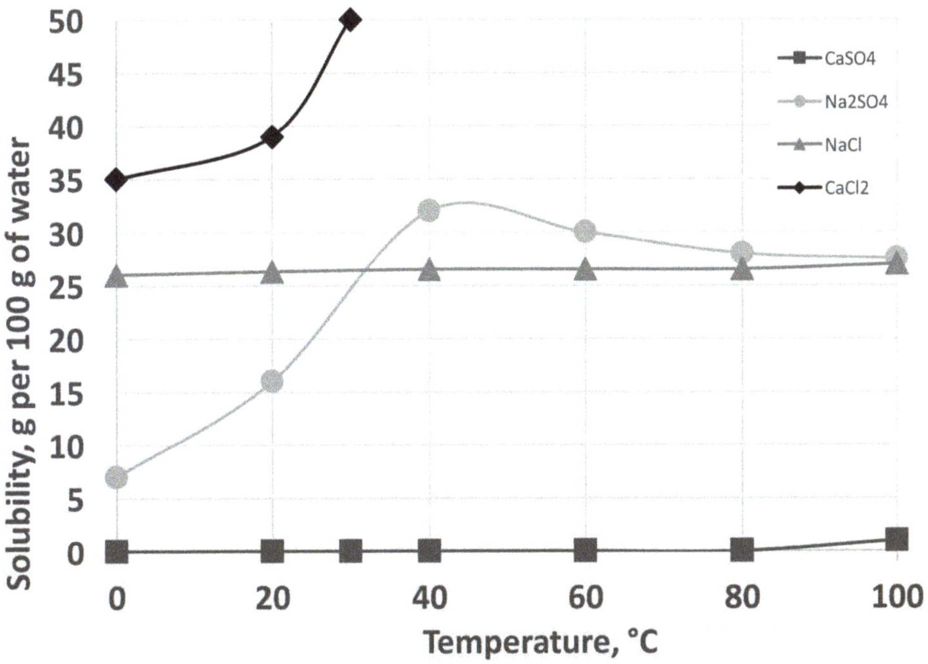

FIGURE 5.8 Solubility of some common salts in heated water.

acid is added to lower the pH to approximately 6 or 6.5 to strip out the carbonate. This will prevent the formation of $CaCO_3$ and $Mg(OH)_2$ scale inside the evaporator (Jenkins et al., 2011). Furthermore, the internal falling film heat exchanger can be protected by using the seed slurry circulating system where crystals (e.g., $CaSO_4$) circulate in the brine stream (Stickney & Fosberg, 1975). There, the crystal growth would preferentially occur on the crystals instead of on the exchanger surfaces. Silicate can also be co-precipitated with calcium sulfate, but within certain limits.

Other techniques to overcome the scaling problem are, for example, using pretreatment techniques to remove the precipitating elements or specially design condensate/influent heat exchangers to increase the scouring to clean up the scaling on the surface. Antiscaling chemicals are sometimes used to control precipitation. These improvements can increase the operating range of the NaCl precipitation conditions or operating MVR at the brine concentrations of ~250 000 mg/L. However, elevated boiling point and corrosion rates at higher recirculation salt concentrations limit the MVR evaporation system to use as the brine concentrator not a precipitator/crystallizer.

Although the pretreatment requirements for the MVR evaporator are not as stringent as those for the reverse osmosis system, there are other areas of concern besides scaling and corrosion limitation previously mentioned. These are as follows:

- **Boiling point elevation:** The increase in the boiling point of the brine can be unpredictable. For an efficient MVR evaporator operation, the boiling point elevation is limited to 3.6 °C (6.5 °F). Some of the compounds in the wastewater, such as some organic compounds or salts (e.g., $CaCl_2$, $MgCl_2$) may cause significant brine boiling point elevation. As a result, they may need special handling.
- **Foaming possibilities:** Controlling the influent concentrations of organic compounds is important to prevent foaming which can create significant operational issues. Manufacturer and vendor experiences are invaluable in assessing the foaming potential. The following levels of organics are listed by some manufacturers to be restricted in the evaporator for foam prevention. Addition of antifoaming agents for removal of organics at pretreatment are the possible prevention steps:
    - Total organic carbon (TOC) < 1500 mg/L
    - Oil and grease < 50 mg/L
- **Volatile compound carryover:** Volatile compounds, such as ammonia or methanol in the feed, will be boiled over to the condensate stream. Their removal from the produced clean water may be required.

Mechanical vapor recompression is suited for smaller flow rates of 2 to 3 m^3/min (<500 to 800 gpm) because of the limitation of finding a large and efficient compressor. So, for larger applications, other evaporation techniques may be more suitable. However, it is simple, energy efficient, robust, and easier to operate than other evaporation or distillation processes.

When considering use of this technique for brine concentration, the appropriate equipment manufacturers should be consulted as early as possible.

The MVR evaporation principle can be used further in crystallization of saturated brine to precipitate salt crystals (Ertel et al., 2012). However, because of its restriction on limited boiling point elevation, application is restricted to produce crystals of low-solubility NaCl and other sodium salts. If other highly soluble salts, such as calcium chloride or magnesium salts, are present, as is normally the case with industrial wastewaters—other types of thermal crystallizers or dryers are needed to remove the last bit of water before achieving the target of zero liquid discharge (ZLD).

3.1.3 Application Cases

Mechanical vapor recompression systems have been successfully used for decades in seawater desalination (Ettouney, 2006), power plant cooling tower blowdown brine concentration (Stickney & Fosberg, 1975), salt removal from industrial wastewater, oil- and gas-produced water treatment, and treatment of wastewater from oil sand extractions (Jenkins et al., 2011). A special case study is described here to illustrate its application in the treatment of high-salt wastewater.

A portable MVR evaporator system has been used in the reclaiming of shale gas flowback and produced water at Devon Energy's Maggie Spain Water Recycling Facility in Decatur, Texas. The whole system was powered by on-site natural gas-powered electricity generators, and the MVR compressor system was powered by gas engines. A 60-day system evaluation study was conducted in 2011. During this period, the incoming produced water flow ranged from 640 to 950 m^3/d (170 000 to 250 000 gpd). It was treated in the system characterized by the process flow diagram depicted in Figure 5.9.

The received water was first stored in a ~5.7 ML (1.5 mil. gal) reservoir and a steady flow was delivered to the rapid mix tank where caustic soda solution was added to increase the water pH to approximately 10. The purpose here is to flocculate the suspended solids and oil and use the iron in the water as a coagulant. An organic polymer was also added to assist the floc buildup. The flocculated slurry was then separated in a gravity clarifier. This pretreatment step was primarily for oil, suspended solids, and iron cleanup

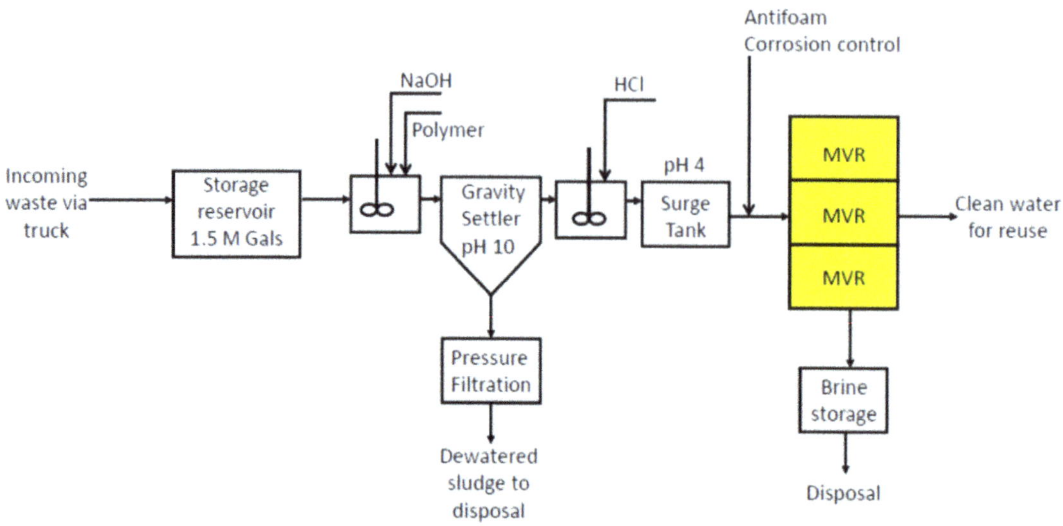

FIGURE 5.9 Shale gas flowback and produced water treatment flow sheet, Maggie Spain Facility.

from the incoming water to prepare for the evaporator system. The settled sludge was dewatered in a pressure filtration system.

The dewatered sludge, now at a solid concentration of 20% to 25%, was disposed of offsite. The clarified water was acidified by HCl to a pH of 4 and put into a surge tank to be processed by the MVR system. There were three mobile MVR evaporators, each with 2 m^3/h (80-gpm) capacity. The feed water was separated via evaporation/condensation into the clean water and concentrated brine streams. The average clean water recovery during this period was 72.5%, and the rest became concentrated brine to be disposed of or reused.

Table 5.1 shows the measured pertinent components of different streams along the treatment train. First, the pretreatment system only removed total suspended solids (TSS), oil (expressed in terms of total petroleum hydrocarbon, or TPH), and iron; no other dissolved ions were affected. The MVR concentrated the total dissolved salts almost completely into the brine stream, which averaged 27.5% of the produced/flowback water feed (a concentration factor of 3.64). Almost all inorganic ions were captured in the brine stream, with final TDS of approximately 170 000 mg/L.

Some of the volatile components, such as ammonia and some organics, were carried over to the distillate side. These volatiles in the produced clean water may require further cleanup depending on the reuse criteria. Some of the divalent cations (Ba^{+2}, Ca^{+2}, Sr^{+2}) may have exceeded their solubility limits when combined with the sulfate ions. However, no scaling problem was reported because of the addition of antiscaling agent. Therefore, the use of antiscaling agents is also important in most cases.

TABLE 5.1 Water analysis across the treatment train.

Parameters	Raw Wastewater	Pretreated Wastewater	MVR Distillate	MVR Concentrated Brine
Flow distribution	100%	100%	72.50%	27.50%
pH, s.u.	9	4	11	7
TDS, mg/L	49 500	49 133	171	168 465
TSS, mg/L	1272	140	9	617
Alkalinity, as $CaCO_3$, mg/L	405	12	263	162
TOC, as C, mg/L	42	10	22	12
TPH, mg/L	38	5	4	4
NH_3, mg/L	84	81	68	113
Sulfate, mg/L	309	221	6	887
Phosphate, mg/L	3	2	0.1	7
Boron (B), mg/L	17	16	0.4	63
Sodium (Na), mg/L	10 741	12 400	14	41 302
Potassium (K), mg/L	484	504	0.5	2028
Lithium (Li), mg/L	12	12	0.1	42
Calcium (Ca), mg/L	2916	2876	3	9699
Magnesium (Mg), mg/L	316	319	0.4	1132
Barium (Ba), mg/L	15	13	0.1	27
Strontium (Sr), mg/L	505	528	0.5	1739
Total Iron (Fe), mg/L	28	3	0.1	4

The energy used was estimated to be approximately 151 kWh per 3785 L (1000 gal) of clean water produced. This is higher than normal; however, considering that the system consisted of movable equipment with gas-powered generators, the energy consumption is considered acceptable. The overall cost of processing was recorded at \$7.2/1000 gal (~\$2/1000 L) of clean water generated or \$5.22/1000 gal (~\$1.4/1000 L) of water processed.

3.2 Multieffect Distillation

This is a special application of MVR evaporators for desalination of seawater or brackish water primarily as a source water for industrial use. All the heat exchangers are plate type. As shown in the simple flow schematic (see Figure 5.10), the feed first recovers heat from the outgoing distillate and brine streams in a feed-effluent heat exchanger. The heated feed is next evaporated in two stages in series. Steam from a boiler is used as the heating medium in the first stage. Vapor from stage 1 is recompressed and used as the heating medium for the second stage. The concentrate from the second stage leaves as brine and is then disposed of. Condensate from both stages are collected in a condensate header and stored in a tank as product. Vapor from the second stage contains a large fraction of noncondensable gases that are vented.

The recovered product is high quality, with TDS ~300 mg/L, and it is used for many purposes. On ships and offshore oil and gas exploration platforms, for example, this water is used as the source for drinking, cooling, boiler feed, and service water for cleaning. Remineralization of this water is conducted by adding certain minerals to make it suitable for use as potable water. In order for it to be used as boiler feed water, further treatment by ion exchange is done. For use as service water, however, no further treatment is necessary.

The feed is pretreated to protect the evaporator internals from scaling and fouling. Screens are installed to arrest the passage of large solids in the equipment. Antiscalant chemicals are added to minimize the scaling potential of inorganic salts. Chlorine is added to prevent biofouling, followed by dosing of sodium bisulfite for dichlorination of any residual chlorine.

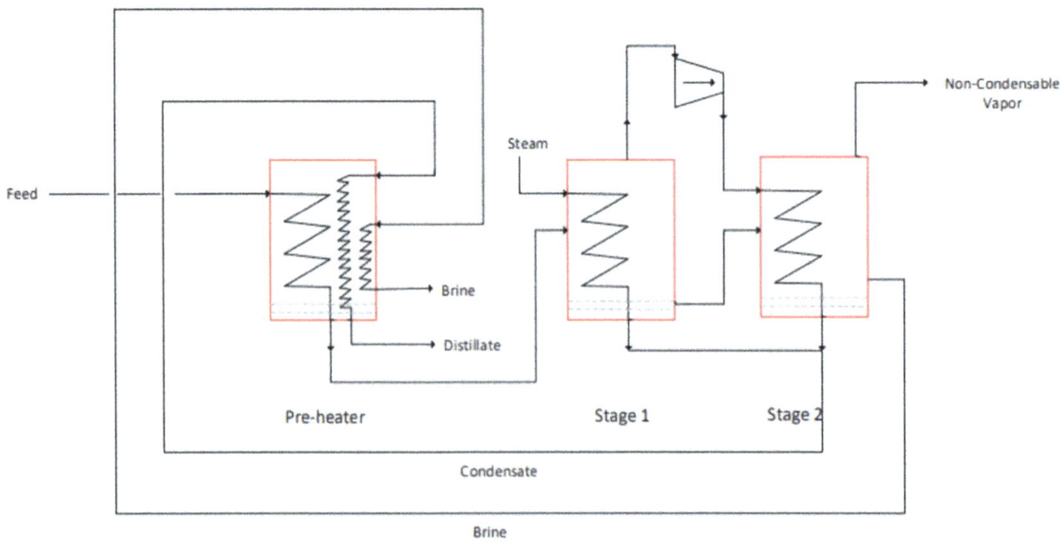

FIGURE 5.10 Flow schematic of multieffect distillation equipment.

3.3 Dynamic Vapor Recovery by Humidification–Dehumidification

Humidification–dehumidification (HDH) is a competing technology of evaporation for concentration of brine. It is a carrier gas-based thermal technique similar to nature's rain cycle. The typical components of an HDH process include a humidifier, a dehumidifier, and a heater, which can all be set up to run at atmospheric pressure. The system can use low-grade waste heat and renewable energies as heat sources to drive the process.

3.3.1 Theory and Principle of Operation

Humidification–dehumidification is used to produce fresh water from streams of high salinity. It is particularly effective at processing waters of salinity greater than what traditional membrane technologies can tolerate and is capable of desalinating to the limit of saturation for NaCl. Although there are various embodiments of HDH technologies, they all operate based on the same principles and use the same unit processes.

Humidification–dehumidification uses a gas, referred to as "carrier gas," to recover fresh water through direct contact heat and mass transfer with a brine stream. This is accomplished by first heating either the gas stream or the brine stream before they enter the humidifier. The carrier gas flows counter to the flow of the brine and extracts water from the brine until it is fully humidified. This process is typically carried out at elevated temperatures up to 90 °C (194 °F) because the water-carrying capacity of air, the most commonly used carrier gas, is significantly higher at higher temperatures than it is at lower temperatures. Heat is simultaneously transferred from one stream to the other as the carrier gas and brine come into contact. The brine leaving the humidifier is either sent to the next process in a treatment train or recirculated for additional desalination.

The water vapor leaving the humidifier is carried through a dehumidifier where it is cooled, condensed, and collected as fresh water via either direct or indirect contact with a lower temperature fluid. The cooling fluid can be the cool feed brine or even fresh water. Depending on the system design, the carrier gas is either recovered for rehumidification in a closed-loop fashion or released to the environment.

Humidification–dehumidification systems can incorporate heat recovery devices to reduce thermal energy consumption and improve energy efficiency. Existing technologies have recovered heat either through carrier gas extractions from the humidifier to the dehumidifier or from the heat carried by the condensed water vapor in the dehumidifier.

3.3.2 Effect of Inlet Carrier Gas Quality

As mentioned in the previous section, the temperature of the carrier gas has a large effect on the water-carrying capacity of the gas. Figure 5.11 depicts the effect of inlet temperature of the carrier gas on the humidity ratio using air as the carrier gas. Modeling of properties for brine and humid air was done using the Cool Prop Excel add-in (Bell et al., 2014), which is based on a publication by Herrmann et al. (2009) that was used to establish saturation property tables for the 2009 ASHRAE Handbook.

The model used to develop Figure 5.11 assumes that air enters the HDH system fully saturated—a worst case scenario. Humidity ratio is defined as the mass of water that can be held by a given mass of dry air. From this analysis, it can be concluded that the mass of water carried by humid air through the humidifier increases exponentially with temperature. Examining the figure, dry air carries 52 times more water at 90 °C (194 °F) than at 30° C (86 °F), which explains why HDH systems are operated at approximately 90° C (194 °F). Moreover, operating at these temperatures allows the facility to be built out of mostly noncorrosive, nonscaling engineered plastics.

With ambient temperatures rarely exceeding 50° C (122 °F) in the United States, it is safe to draw the conclusion that high ambient temperature of the inlet air does not have a significant effect on the amount of evaporation performed in the humidifier. At temperatures greater than 50° C (122 °F),

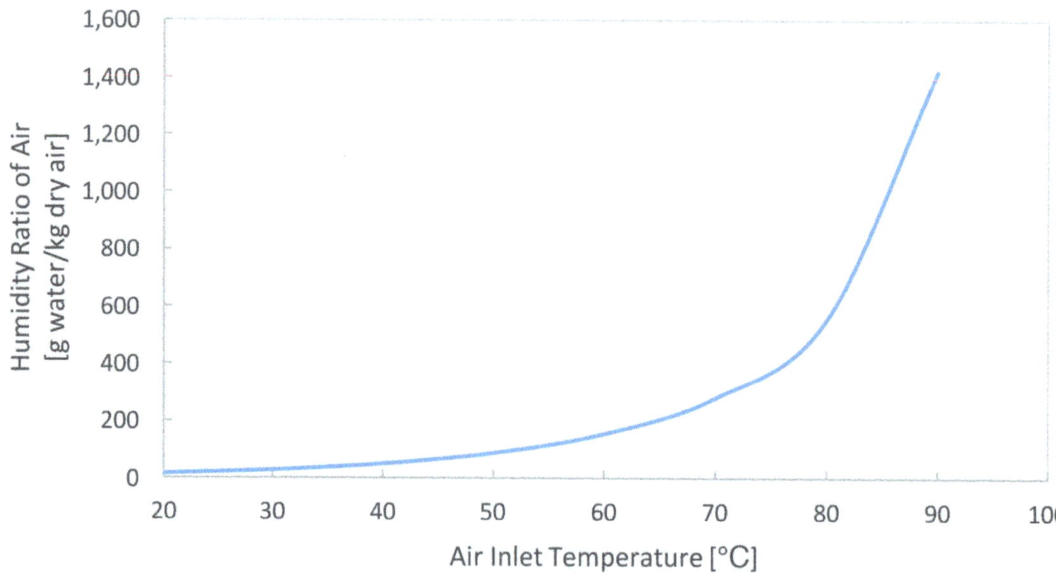

FIGURE 5.11 Water-carrying capacity of inlet air at 100% relative humidity as a function of temperature.

however, there is a sharp decrease in the ability for air to take on more moisture (again, assuming the inlet air is fully saturated).

The aforementioned case is a worst-case scenario that does not hold true when the inlet air is not fully saturated with water before it enters the humidifier. This makes it important to also consider the effect of relative humidity of the inlet air on the performance of HDH systems. As Figure 5.12 shows, inlet air with low relative humidity makes a large difference in the humidity ratio with the saturated air at the outlet of the humidifier. In other words, the inlet air maintains a large capacity for carrying water vapor at lower relative humidity.

At higher relative humidity of air entering the humidifier, there is a slight decrease in humidity ratio when the temperature is below 50 °C (122 °F), but it does not significantly affect the amount of fresh water recovered (<6% difference). It is only when both relative humidity and temperature of the inlet air are high (>0.3 and >50 °C, respectively) that a significant effect is observed on the amount of fresh water recovered.

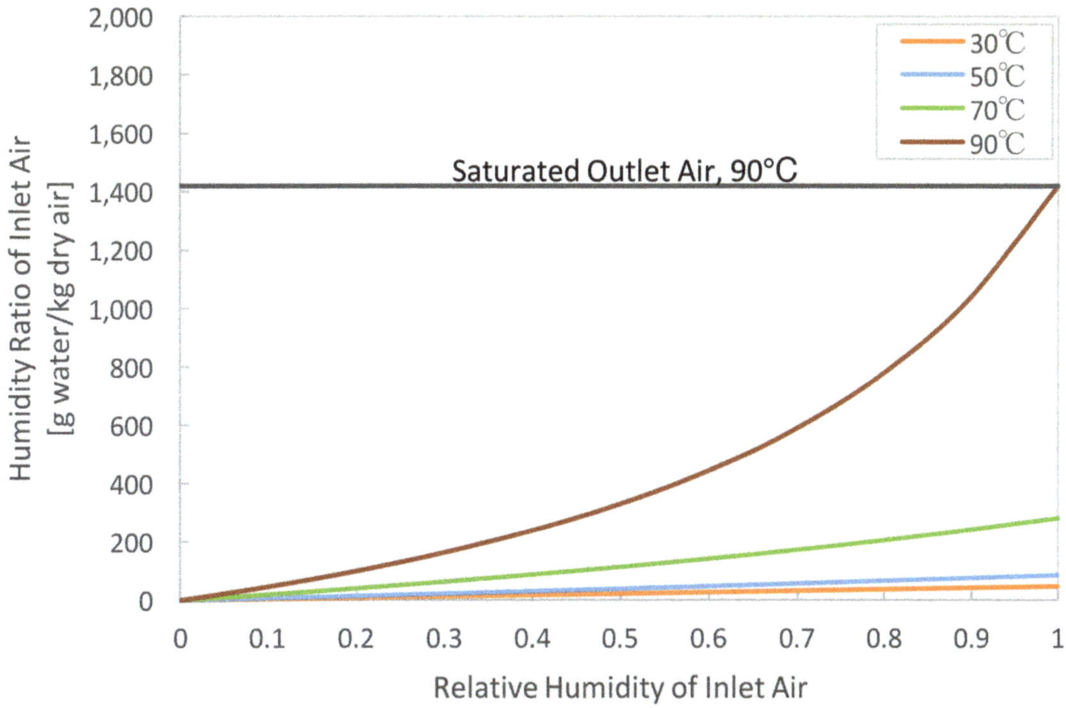

FIGURE 5.12 Humidity ratio of inlet air as a function of relative humidity at different temperatures and compared to humidity ratio of saturated air at 90 °C.

3.3.3 Evaluation of Humidification–Dehumidification Systems

Assessing performance of thermal desalination technologies such as HDH systems requires understanding of two key parameters: gained output ratio (GOR) and recovery ratio.

The GOR is a dimensionless number, defined in Equation 5.1, as the ratio of latent heat of vaporization (h_v) for the total mass of fresh water produced ($\dot{m}_p$) to the heat put into the system ($\dot{Q}_{in}$) and expressed as follows:

$$\text{GOR} = \frac{\dot{m}_p \cdot h_v}{\dot{Q}_{in}} \tag{5.1}$$

This is an important parameter for evaluating performance of HDH technologies because it is a measure of the energy efficiency to produce fresh water in a thermal process. A GOR of 1 is equivalent to achieving evaporation of water with energy input equivalent to the heat of vaporization of water alone. In other words, it indicates that water was evaporated using only the energy required to boil water, no more or no less.

In a real system, however, there are inefficiencies in unit operations that prevent achievement of a GOR value of 1, which necessitates use of heat recovery devices for improved performance. Through implementation of heat recovery devices, GOR can be enhanced and may even exceed a value of 1, indicating that evaporation in excess of the theoretical maximum has been achieved without increasing the thermal energy input. This is attributable to the reuse of the recovered heat to evaporate additional water.

The second parameter of importance is the recovery ratio. The recovery ratio is the mass of fresh water produced ($\dot{m}_p$) per mass of feed water ($\dot{m}_f$) in the system, as follows:

$$\text{Recovery ratio} = \frac{\dot{m}_p}{\dot{m}_f} \tag{5.2}$$

Closed water designs of HDH systems allow for improved recovery ratio as the brine can be recirculated in a semicontinuous process. It is possible to achieve up to 90% recovery of fresh water depending on the quality of the inlet brine. Salinity and concentration of scaling elements such as calcium and magnesium are the biggest determinants of the maximum recovery ratio for a given feed stream.

3.3.4 Humidification–Dehumidification Designs

3.3.4.1 Humidifier

Humidifiers are heat and mass transfer devices used for the humidification of a carrier gas through direct contact with a feed brine. They have the dual function of reducing the volume of a feed brine as well as removing high-quality water as vapor that can be recovered in a dehumidifier.

There are two types considered when designing a humidifier: packed bed and bubble column. Packed bed humidifiers feature packing material, typically made of plastic, with geometries designed to evenly distribute the hot brine entering the humidifier to maximize the contact surface area with the carrier gas. This design is robust and performs well in mitigating issues related to scaling. Most of the scaling that occurs from cooling of the brine can be induced onto the packing material. However, replacing the packing material is inexpensive and does not meaningfully affect the economics of HDH.

Humidifiers of the bubble column design operate by flowing brine across trays with perforations through which the carrier gas bubbles through. More efficient heat and mass transfer is achieved than the packed bed design because of the increased contact area between the brine and carrier gas. The improved contact area allows for design of a smaller humidifier and a decrease in capital cost compared to a packed bed unit. The size of the bubbles formed, velocity, and properties of the carrier gas all affect the rate of humidification that can be achieved.

3.3.4.2 Dehumidifier

In a review of dehumidifier designs by Dehghani, et al. (2018), two classes of dehumidifier are identified: indirect contact and direct contact. There are many examples of these designs in literature that the authors covered in depth.

Direct contact dehumidifiers function similar to humidifiers as heat and mass are exchanged simultaneously between a humid carrier gas and fresh water. One type of direct contact dehumidifier is the bubble column dehumidifier. As with humidifiers of the same design, the humid carrier gas is bubbled through a pool of fresh water to achieve condensation of water vapor. This design maximizes the contact surface area between the two fluids, resulting in more efficient heat and mass transfer and eliminating any possible issues related to corrosion. Additionally, the footprint is reduced compared to an indirect contact dehumidifier to achieve the same amount of dehumidification.

The most common form of indirect contact dehumidifiers uses a coil as the heat exchange surface for dehumidification. A coolant, typically the feed brine, passes through the coil and functions as a heat sink for the humid carrier gas as it passes across the outside of the coil. As the heat is transferred from the carrier gas through the coil to the coolant, water vapor condenses on the outside of the coil. The material selection of the coil is important as it should be conducive to heat transfer and be capable of withstanding corrosion from the feed brine.

3.3.5 Open Versus Closed Cycles

There are three primary configurations used for HDH system design: closed-air open-water (CAOW) water heated, closed-water open-air (CWOA) water heated, and CAOW air heated. There are also alternative configurations, such as open-air open-water (OAOW), that are less commonly used in commercial applications. Air emissions control is important for OAOW, especially in oil and gas industry applications. The word "open" refers to a system in which a fluid makes a single pass along a process path before exiting the system. The opposite is referred to as a closed system. In a closed system, a fluid is circulated in a continuous fashion. In the case of a closed water system, as the feed is evaporated it becomes necessary to remove the concentrated liquid and add new feed to prevent issues with salting out. Water typically refers to the feed brine stream, whereas air refers to the carrier gas. Figure 5.13 illustrates a few examples of HDH cycle configurations.

In a CAOW water-heated system, feed brine is heated using a waste heat source before contacting a carrier gas. The energy from the hot brine is used to heat the carrier gas, which results in humidification of the gas. The concentrated brine exits the humidifier. The humid carrier gas is then fed through a dehumidifier to cool and condense the water vapor it is carrying. The fluid used to cool the humid carrier gas is the feed brine that, after further heating, is fed into the humidifier. The dehumidified carrier gas is recirculated through the humidifier to repeat the process in a closed loop, as illustrated in Figure 5.13(a).

A CAOW water-heated HDH system can carry out the dehumidification process through either direct or indirect contact dehumidification (covered in the previous section). The design in Figure 5.13(a) uses indirect contact dehumidification by passing feed brine through a coil inside of the dehumidifier. As the humid, hot carrier gas passes along the outside of the coil, heat is transferred to the cooler feed brine causing the water vapor from the gas to condense on the outside of the coil. Figure 5.13(b) depicts a case of direct contact cooling in the dehumidifier. This design uses a pool of fresh water in the dehumidifier as the cooling fluid. As the humid carrier gas passes through the pool of fresh water, heat is transferred from the carrier gas to the pool. The decrease in temperature of the carrier gas allows the water vapor to condense directly into the freshwater pool.

Closed-air open-water air-heated systems operate similar to the water-heated alternative, but instead of heating the feed brine, the carrier gas is

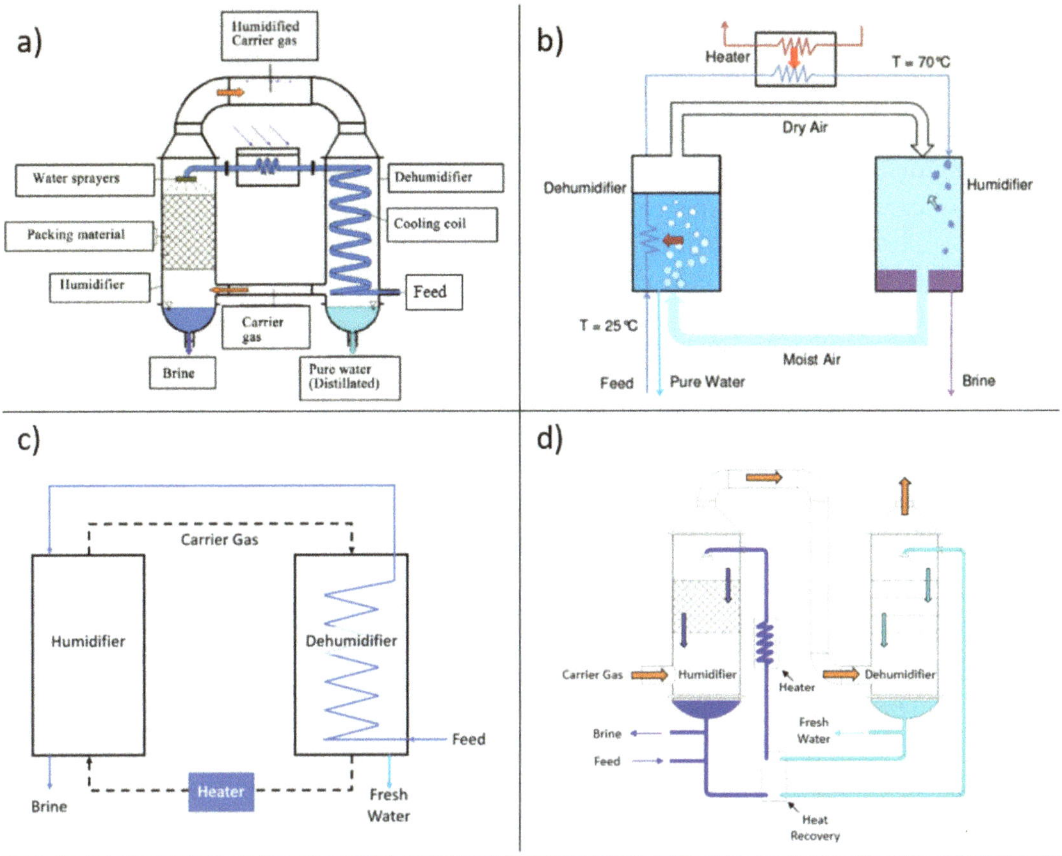

FIGURE 5.13 Various embodiments of HDH technologies: (a) CAOW water heated with indirect contact dehumidification (adapted from Kabeel et al., 2013); (b) CAOW water heated with direct contact dehumidification (Warsinger et al., 2015); (c) CAOW air heated with indirect contact dehumidification; and (d) CWOA water heated with direct contact dehumidification.

heated directly to the operating temperature. As depicted in Figure 5.13(c), the hot carrier gas passes through the humidifier where it is saturated and cooled simultaneously by the brine. The carrier gas then passes through the dehumidifier, where it is cooled further to condense the water vapor. The dry carrier gas is recovered and heated to begin the cycle again. This configuration is not optimal when maximum freshwater recovery is desired because the lower carrier gas temperature leaving the humidifier limits the absolute humidity that the carrier gas can reach, thus limiting freshwater recovery from the feed brine.

Closed-water open-air water-heated systems begin with heating of the feed brine to operating temperature before entering the humidifier, as depicted in Figure 5.13(d). Once heated, the brine enters the humidifier where it heats and humidifies the carrier gas. The humid carrier gas is then cooled in the dehumidifier to condense the water vapor from the carrier gas and the dry gas is released to the atmosphere. As observed by Kabeel et al. (2013), the benefit of a CWOA water-heated configuration comes from the dehumidification step which, at optimal humidifier conditions, can cool the carrier gas to temperatures below ambient, thus improving water recovery. However, if optimal conditions are not achieved, the reverse effect may be observed and the carrier gas may not be cooled sufficiently, resulting in loss of water recovery.

Another feature of a closed water system is the recirculation of brine, which enables a much higher recovery ratio than an open water system. Recovery ratio is severely limited through a single pass of an HDH system. Through recirculation of the brine, however, the recovery ratio is limited only by the saturation point of the brine and the efficiency of the dehumidifier. It is possible to achieve up to 90% recovery in commercial applications at optimal conditions.

3.3.6 Carrier Gas

The carrier gas selected for HDH is a substantial design factor for determining the cost of desalination. There are three thermodynamic properties identified by Narayan (2012) as being particularly important for selection of the carrier gas: thermal conductivity k, specific heat capacity c_p, and density ρ. Table 5.2 summarizes these properties for different carrier gases.

The thermal conductivity of the carrier gas indicates how efficiently heat will transfer from the hot brine to the gas. This has implications not only for cost but also for the maximum rate of mass transfer one can achieve. Table 5.2 shows that helium has a thermal conductivity nearly 6 times

TABLE 5.2 Thermophysical properties of different carrier gases at standard temperature and pressure (Narayan, 2012).

Carrier Gas	Thermal Conductivity, k [W/m·K]	Specific Heat, c_p [kJ/kg·K]	Density, ρ [kg/m^3]
Air	0.02551	1.005	1.169
He	0.1502	5.193	0.1615
N_2	0.02568	1.038	1.13
Ar	0.01796	0.5203	1.611
CO_2	0.01657	0.8415	1.775

greater than that of air. Ideally, this would be the carrier gas of choice as it is heated most readily and thus can be humidified most easily. However, helium and other low-density carrier gas can present operational challenges on the commercial scale. More work needs to be performed to fully evaluate operational feasibility of low-density gases.

The specific heat capacity of the carrier gas determines the mass ratio of dry carrier gas to hot brine that is required for efficient operation of the humidifier. A high specific heat capacity (such as that of helium) requires less carrier gas introduced per unit of hot brine to achieve the same quantity of heat transfer as a carrier gas with lower specific heat capacity. The optimal ratio of carrier gas to brine varies based on operating conditions. Another implication in the use of a carrier gas with high specific heat is the possibility of using a smaller dehumidifier.

Selection of the carrier gas is of utmost importance for open-air HDH systems, with cost and availability being key considerations. Because the gas is not recovered, it is necessary to have a ready and ample supply on hand for steady operation. This presents an operational challenge for using helium as a carrier gas. Helium generators are prohibitively expensive and storing them onsite increases footprint. Another option is to use nitrogen. Nitrogen does not have the same issues of cost for onsite generation, but its thermophysical properties are similar to air, and hence this does not justify the added operational expense.

On the other hand, air is readily available and practically free. There is no need for storage, which greatly minimizes the operational footprint. Upon weighing cost and availability against the thermodynamic properties of the carrier gas, it becomes apparent that ambient air is the optimal choice for most open-air HDH applications.

However, there are certain drawbacks with air as a carrier gas for HDH systems. Ideally, the carrier gas used for HDH operations should be inert to avoid any unwanted side reactions. High oxygen content in air can cause oxidation of the feed wastewater and create undesirable byproducts. Additionally, passing air through the brine increases the amount of dissolved oxygen in the brine. For some applications, minimizing the concentration of dissolved oxygen in the product streams is of interest for long-term prevention of oxygen-induced corrosion.

3.3.7 Energy Source

Humidification–dehumidification systems can use gas-powered boilers as the main source of energy input for heating brine. In applications where there is ample supply of natural gas, such as in unconventional oil and gas developments in the United States, this is a major advantage. However, one significant advantage of HDH technologies is the potential to use alternative low-grade waste heat sources for operation. Though not yet implemented commercially, novel processes exist for using renewable energy sources such as geothermal energy coproduced during oil production and solar energy in regions with an abundance of solar radiation throughout the year (Li, 2009).

3.3.8 Commercial Applications and Operations Snapshot

Carrier gas extraction (CGE), is a commercial version of the thermal HDH technology. It is a modified CWOA water-heated HDH system that uses air as the carrier gas. As noted before, its major advantage is in desalinating highly saline feed water. Typically, membrane processes are not effective in these high ranges of feed salinity because of lower tolerance to scale-forming compounds and high osmotic pressure limits. Carrier gas extraction is being applied in oil and gas exploration and production, textiles and tanneries, mining, power generation, food and beverage, chemicals, and refining sectors.

Particularly in the oil and gas industry, HDH-based CGE technology has an important application. The recent and rapid development of horizontal drilling and fracturing has greatly increased the overall production in the unconventional oil and gas. Multifold increases in water consumption and subsequent disposal issues are prompting exploration and production operators to look for more economical and environmentally sustainable options for hydraulic fracturing. In areas that have limited access to fresh water and saturated brine disposal, CGE can be used as a total water solution technology.

Carrier gas extraction reduces the high transportation costs of produced water by treating the wastewater onsite, producing fresh water and saturated brine and thereby reducing any negative environmental impact. The technology can produce a consistent effluent stream despite varying influent water quality, which is typically the case in upstream produced water. The heavy brine generated as during the process, can be used for drilling, workover, and completion application. The fresh water generated can have TDS content of even less than 500 mg/L, which can be either reused in hydraulic fracturing and drilling or surface water discharge. (See Table 5.4.)

The HDH-based CGE technology was applied as a part of a turnkey ZLD facility in Permian Basin in West Texas (see Figure 5.14). Permian Basin is one of the most prolific shale reservoirs in the world.

During the operation of the facility, the freshwater recovery ratio was approximately 60% when the TDS concentration in the feed water was 100 000 mg/L, and approximately 40% when the TDS in the feed water was 150 000 mg/L. From this perspective, CGE is a competing technology of evaporation.

FIGURE 5.14 HDH-based CGE system in operation in Permian Basin, West Texas (reprinted with permission from Gradiant Energy Services, Houston, Texas).

3.3.9 Economic and Technological Advantages

To evaluate the HDH-based CGE system and compare the performance of various thermal and electricity-driven desalination technologies, GOR and equivalent electricity consumption are used. The GOR is a thermal energy-based performance parameter, as discussed earlier (see Equation 5.1).

In the CGE system, there is a large percentage of air (60% to 90% by mass), which is noncondensable in the condensing stream. To counteract the low heat and mass transfer rates in a conventional heat exchanger, a bubble column is used to perform dehumidification in a column of cold liquid rather than on a cold surface. The bubble column dehumidifier design enables the CGE system to have a smaller footprint.

Additionally, in CGE, the valuable heat transfer surface is decoupled from the sacrificial separation surface. The heat input to the system takes place through plate and frame heat exchangers, whereas the mass transfer occurs through the direct contact of air bubbles and hot brine droplets. Formation of solids during cooling of the brine as heat is transferred to the air directly affects performance of packing material, which is inexpensive compared to the heat exchanger tubes made of stainless steel or titanium used in evaporators. Decoupling of valuable heat transfer surface from the sacrificial separation surface reduces pretreatment requirements and saves cost.

TABLE 5.3 Constituents of feed and effluent for the CGE facility.

Constituent	Feed Brine	Fresh Water	Units
Barium	4	ND	mg/L
Bromide	4620	3.1	mg/L
Calcium	2330	0.6	mg/L
Chloride	67 600	45	mg/L
Iron	14.7	ND	mg/L
Magnesium	369	ND	mg/L
Oil and grease	9.8	ND	mg/L
Sodium	37 400	29	mg/L
Strontium	708	0.2	mg/L
Benzene	2320	ND	mg/L
Toluene	1740	ND	mg/L
pH	6.8	7 to 9	s.u.
Total hardness	7660	3.3	mg/L as $CaCO_3$
TDS	117 000	113	mg/L
Total suspended solids	277	ND	mg/L

ND = Nondetectable as the measurement was below detection limit

Elimination of metallic evaporators and condensers significantly reduces the capital cost. Carrier gas extraction also operates at much lower temperatures, which allows for use of waste heat to eliminate thermal energy costs. Desalination through CGE provides a compact and effective technology for brine management, which can meet the water challenges for tomorrow. Also, it may be noted that HDH technology covers a much larger range of TDS than evaporation as shown in Figure 5.2.

4.0 PROCESSING AND DISPOSAL OF BRINE CONCENTRATE

Final disposal of brine can follow one of the following four methods: crystallization, deep well injection, solar pond evaporation, and landfilling. To minimize cost and complexity of operation, every disposal method seeks to handle as little volume as possible. As such, it is preferable for deep wells to accept liquids in high concentrations, and solar ponds can accept both concentrated liquids and solid–liquid slurries. However, those methods have the flexibility to also receive dilute liquids, which will require larger capacity and incur higher costs. On the other hand, crystallizers are meant to process high concentration liquids only, typically the concentrated brine from evaporators and HDH systems. Reverse osmosis concentrates can bypass evaporators and be fed directly to a crystallizer, but that will require a larger capacity crystallizer, leading to higher capital and O&M costs.

4.1 Crystallization

Crystallization is a process for separation of dissolved solids (solute) by forming solid particles from a saturated solution. In addition to handling brine, it is also an important separation process for production of different varieties of chemicals.

In a solution, the solid and liquid phases coexist in thermodynamic equilibrium under the prevailing conditions of temperature, pressure, composition, pH, and so on. Crystallizers are the equipment carrying out solid–liquid separation using the property of crystallization under solid–liquid equilibrium. Crystals can be made up of a single phase, as in the case of freezing ice from water by cooling. Also, crystallization is often used as a technique for separating and purifying one or more components from a liquid solution. In such cases, crystals of one phase or multiple phases can form from a liquid solution, depending on the temperature and composition. Several substances can crystallize from solutions forming different phases of solid crystals consisting of different amounts of water of hydration.

Crystals are made up of units, called "crystal lattices," that are geometrically organized in definite shapes in a three-dimensional space. A crystal lattice represents the arrangement of its constituent particles, which may be atoms, molecules, or ions.

In brine management applications, separation of salt from water is achieved at high temperatures under boiling conditions of water, whereas freeze crystallization conducts desalination of seawater, brackish water, and mine drainage water at low temperature under freezing conditions of water. Thermal crystallizers separate solute (dissolved solids in brine) as crystals under boiling conditions, whereas freeze crystallizers remove the solvent (water) as solid crystals (ice) under freezing conditions. These form the basis of thermal crystallization and freeze crystallization processes, respectively. A brief introduction of freeze crystallization was provided earlier in this chapter. Thermal crystallization is discussed in detail in the following sections.

4.1.1 Thermal Crystallizers

Thermal crystallizers remove water as vapor from a concentrated (TDS ~250 000 mg/L) brine solution by application of heat. The water vapor is condensed and the concentrated brine leaves as a slurry of solid crystals. The liquid phase, from which the solid crystals are formed, is called the "mother liquor." The solid–liquid slurry is called the "magma."

Water produced by thermal crystallizers is in highly pure state relative to the brine from which is was produced, with a TDS concentration of <500 mg/L and can be used for beneficial reuse. The concentrated stream is disposed of by one or multiple means (e.g., landfilling or discharge into a solar pond). However, it may require further processing to pass a paint filter test (U.S. EPA, 2004) for the landfill to accept.

4.1.2 Definitions and Technology Description

Crystallization is a two-step process that involves nucleation and crystal growth. "Nucleation" refers to the beginning of crystallization from the mother liquor. The solid–liquid equilibrium of a solution is depicted in the form of a composition versus temperature diagram, as shown in Figure 5.15. The solute can exist in a stable, unstable, or metastable zone. A solution with a temperature and composition above the solubility curve is unsaturated and stable. Below the solubility line, the solution is supersaturated and unstable. Between the stable and unstable zones is a transition or metastable zone, where the solution is marginally supersaturated and suitable for formation of small nuclei and their growth into larger crystals. Many crystallizers operate within the metastable region.

In thermal crystallization processes, a solution is supersaturated by evaporating an amount of solvent (which is water in case of brines) and increasing the concentration of the dissolved solids to levels higher than saturation. In cases of brine crystallization, because the NaCl solubility is independent of temperature, the driving force for supersaturation of the mother liquor is concentration by driving off water by evaporation. When a solution is heated to the liquid–vapor equilibrium point, the solvent becomes saturated and starts vaporizing. It must be noted that, in the context industrial wastes, brine streams contain many more dissolved species than just sodium chloride. Therefore, the crystallization phenomenon is not dictated by the sodium chloride solubility alone and the saturation point is dependent on each of the components present in the matrix. Figure 5.8 depicts the variation of solubilities of a few types of dissolved solids present in typical brine streams. It is always advisable to conduct a thorough analysis of brine samples and also to run a pilot crystallizer tests ahead of undertaking a full-scale system design.

Nucleation of Crystals from Supersaturated Solution in Metastable Zone

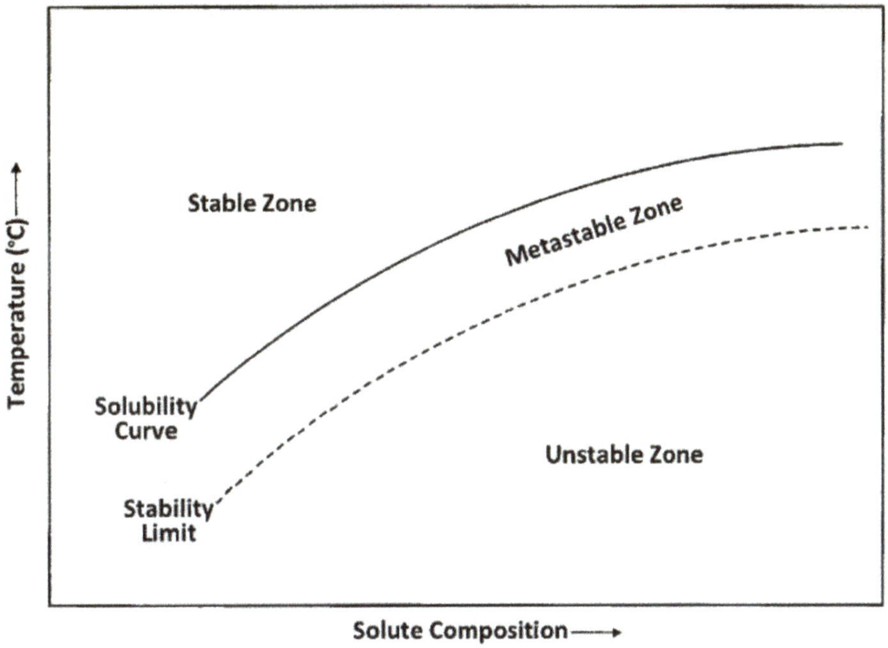

FIGURE 5.15 Nucleation of crystals from supersaturated solution in metastable zone (Samant & Young, 2006; reprinted with permission from Chemical Engineering Progress).

Nucleation can start spontaneously, without the aid of any foreign particles, or "seed." However, in reality, unaided nucleation rarely happens in a crystallization process, except for precipitation reactions (e.g., precipitation of metal hydroxides from aqueous solutions as a result of elevation of pH). Nucleation in almost all the practical applications of a crystallization process starts with the help of particles added from an external source, which makes it possible within finite time. The rate of nucleation is defined as the number of new particles formed per unit volume of magma or solids-free mother liquor.

The nucleation step is followed by crystal growth. Once nuclei of crystals are formed in a supersaturated solution, growth of the particles into larger-size crystals takes place surrounding each nucleus. New atoms/molecules/ions of the solute transfer through the liquid phase on the solid surfaces of the nucleus and deposit, enlarging the crystal size. This process is controlled by the rate of transport of the solute through the liquid phase. Control of nucleation to a minimum and growth of crystal size to the maximum are important considerations to avoid formation of excessive fine-size crystals, resulting in difficulty of separation of crystals based on their size.

For effective handling and disposal of solid crystals, the crystal size is an important parameter. The crystal size distribution is carefully controlled in the operation of a crystallizer. Crystallizers are designed to meet specific needs of crystal size distribution, crystal growth rate, and so on.

Crystals are often separated from the mother liquor by filtration or centrifugation to recover the solids free from liquid. When the crystal growth is slow, then a significant amount of time is required to reach equilibrium. In such situations, the final mother liquor may retain appreciable supersaturation, and the actual yield will be lower than that calculated by the solubility curve.

4.1.3 Crystallizer Design Considerations

Before undertaking the design of a crystallizer, a number of process and mechanical engineering considerations are taken into account, depending on the service. These include flow, the type and characteristics of solute and solvent, and the desired goal. For example, in cases of production of chemicals like sodium sulfate, potassium nitrate, and so on, the crystals are actually the products of value. However, in other cases, the environmental regulations dictate the treatment goals and the crystallizer design. This includes removal of certain types of waste materials, including radioactive wastes, from industrial discharges. In the brine management applications for water and wastewater treatment, crystallization of salt fulfills the role of a ZLD technology if a production site is required to totally eliminate liquid discharges.

A crystallizer is an evaporator that can tolerate scaling, so water is evaporated from an undersaturated brine, making it supersaturated. Therefore, the presence of solids in the feed should not be a problem for a mixed-salt crystallizer. There is no requirement for the removal of precipitating salt from a crystallizer because the unit itself is a scaling unit.

4.1.4 Crystallizer Description

Brine crystallizers use evaporation of the mother liquor to create a supersaturated liquid phase as the first step. Figure 5.16 demonstrates the configuration of a vacuum crystallizer in its simplified form. Here, the feed is introduced in the downpipe, where it blends with the circulating mother liquor. It enters the tube side of a vertical shell and tube exchanger, where it is heated up to the boiling point by steam on the shell side. The condensate may be reused for raising fresh steam in a boiler. A pump helps circulation of the feed liquid through the heater to the crystallizer body. The point of introduction of the feed to the crystallizer is approximately 0.3 m (1 ft) or more below the liquid (magma) surface such that the feed is under a

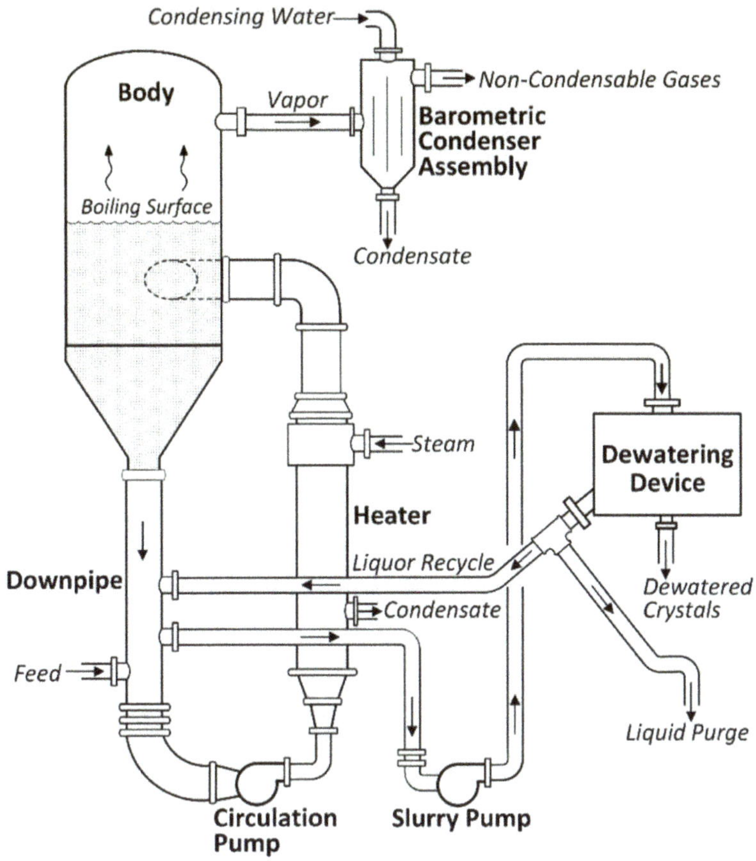

FIGURE 5.16 Forced circulation vacuum crystallizer (adapted from Unit Operations in Chemical Engineering by McCabe et al., 1993; re-created by Alan D. Ingham, Black & Veatch, Kansas City, Missouri).

hydrostatic head and it does not flash under vacuum as it enters the body. This arrangement helps control nucleation and growth of crystals.

Vaporization occurs from the top surface of magma. Vapor disengages from the liquid and leaves through a connecting line under vacuum. Relieved of the vapor, the liquid becomes supersaturated and flows down vertical leg of the downpipe. As the nuclei are formed from supersaturated magma and crystals grow, these act as seeds for formation of new crystals. Saturated liquid is combined with fresh feed and recirculated through the heater. A slurry of crystals and mother liquor is continuously withdrawn by a pump for feeding into a centrifuge to separate the crystals from mother liquor. The separated liquid is returned to the crystallizer for further reprocessing. A small part of the liquid may be bled out intermittently. The withdrawal rate of crystals is carefully controlled to properly balance the feed rate to maintain steady-state operation.

Several types of crystallizers are available on the market. Although they differ in the internal arrangements and their construction, the basic principle is the same as previously described. In brine management applications, another commonly used piece of equipment is the draft tube baffle crystallizer, as depicted in Figure 5.17. In addition to the basic features of a forced circulation type, these are equipped with a vertical draft tube, an internal cylindrical baffle, a central propeller, and an elutriation leg. These additional features help in the segregation of larger and optimum-size crystals that can be withdrawn (directly or through a centrifuge) as product from fine and undergrown ones.

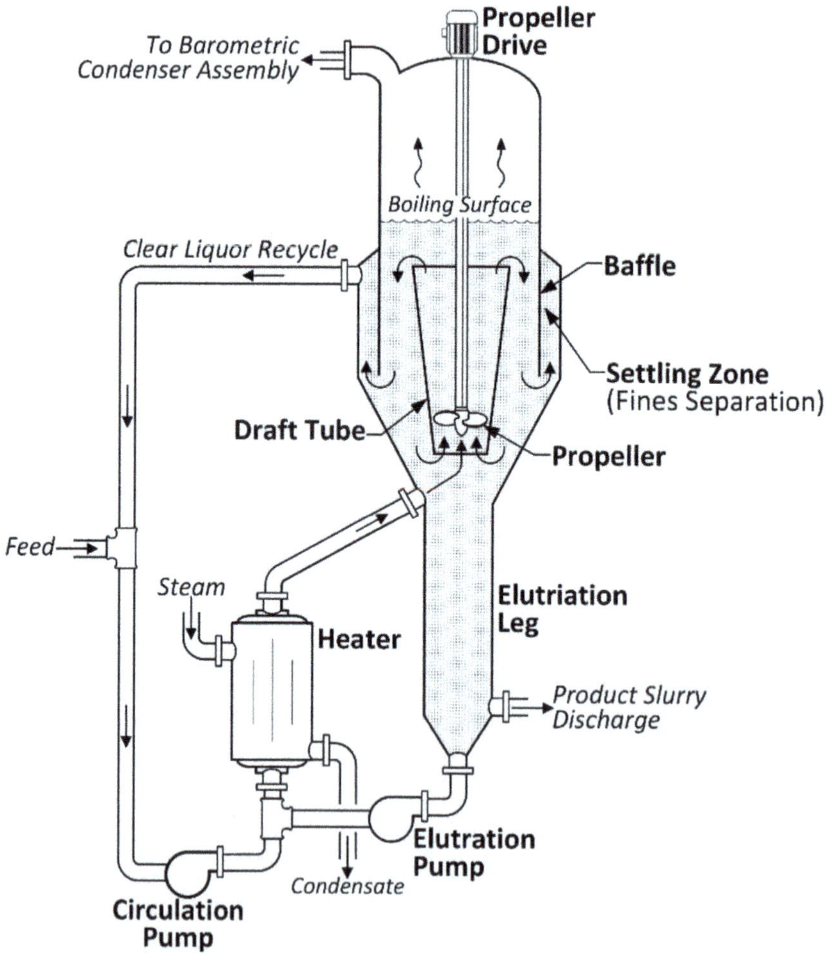

FIGURE 5.17 Draft tube baffle crystallizer (adapted from Unit Operations in Chemical Engineering by McCabe et al., 1993; re-created by Alan D. Ingham, Black & Veatch, Kansas City, Missouri).

The flow pattern created by the propeller and the elutriation system induces a vertical motion of the liquid through the draft tube and carries over the fine and light crystals through a baffle following an underflow-overflow path to the recirculation line. Feed is blended with this stream and introduced to the crystallizer via a heat exchanger. Draft tube baffle crystallizers produce much denser crystals than the product from the basic forced circulation type and may be suitable for disposal without any further centrifugation step.

As in the case of evaporators, the crystallizers are also available as multistage units, in which the vapor produced from the flash chamber of one stage can be used as the heating steam of the next stage.

Steam is the source of energy for crystallizers. Approximately one pound of steam is used for evaporation per pound of water. Thermal recompression type is one of the most energy-efficient systems in which the energy is supplied by a vapor compressor driven by electricity or by a gas or steam turbine.

Some of the key strengths of both evaporation and crystallization technologies include the following:

- Easily adaptable to high variability of brine quality and quantity
- Able to treat water with a high TDS concentration compared to other nonthermal technologies such as reverse osmosis
- Smooth and quick recovery from process variation and upsets with minimal operator attention

The limitations of MVR evaporators were outlined earlier in this chapter (see Section 3.1.2). Crystallizers are also subject to all of those encountered by evaporators. However, crystallizers encounter a few additional difficulties in properly controlling nucleation, crystal classification, removal of excess nuclei, and removal of small crystals.

As in the case of evaporators, crystallizer designs are also dependent on manufacturers. Before making a selection for a specific application, the manufacturers should be consulted.

4.1.5 Application Case

Eureka Resources operates an MVR crystallizer as a part of the overall treatment of high-TDS produced water at Standing Stone, Pennsylvania. The treatment train consists of the following steps:

1. Collection in storage tanks
2. Primary clarification to remove TSS
3. Chemical addition and pH adjustment

4. Secondary clarification to remove hardness, alkalinity, and scaling ions as sludge
5. MVR evaporator (NOMAD Evaporator from Fountain Quail)

Brine from the MVR distiller is crystallized by a single-stage MVR crystallizer (SEP, Switzerland) to produce solid NaCl crystals for sale in the market. The crystallizer capacity is 80 m^3/d (21 000 gpd or 5000 bbl/day) with a 50% turndown capability. The product salt after centrifuge and drying is of 99.7% purity and is marketable for industrial applications.

Distillate from the crystallizer and evaporator contain methanol, which is added during the exploration and production process as antifreeze and carried over with the produced water. This stream is treated by membrane bioreactor and reverse osmosis to produce effluent that fulfills the Pennsylvania Department of Environmental Protection (PADEP) permit limits for beneficial reuse or discharge. Table 5.4 presents a profile of raw produced water characteristics, distillate, and PADEP limits for treated effluent.

TABLE 5.4 Qualities for raw produced water, crystallizer distillate, and PADEP limits (Ertel et al., 2012).

	Raw Produced Water	Crystallizer Distillate	PADEP Permit (WMGER 123) Limits
Ammonia, mg/L	NA	50–135	2
Barium, mg/L	100–40 000	0.37–3.0	2
COD, mg/L	NA	1300–2700	15
Ethylene Glycol, μg/L	NA	1600–50 000	13
Gross Alpha, pCi/L	NA	3	15
Gross Beta, pCi/L	NA	4	1000
Radium 226, pCi/L Radium 228, pCi/L	NA	2.6–20	5
Iron, mg/L	100–200	0.03–0.08	0.03
Magnesium, mg/L	100–200	0.04–2.0	10
Sodium, mg/L	10 000–32 000	3–33	25
Sulfate, mg/L	100–1000	0.17–0.82	25
TDS, mg/L	1000–365 000	1–220	500

NA = not analyzed

Chapter 6 provides a detailed discussion of the aforementioned principles of brine pretreatment and distillate posttreatment by physicochemical and biological processes; therefore, they and are not separately considered in this chapter.

4.2 Deep-Well Injection

Injection wells are widely used for disposal of liquid waste in oil and gas exploration and production activities. Two main applications of deep-well injection are

- a means of disposal of produced water with no or minimum treatment, and
- a means to achieve the goal of ZLD after treatment and concentration and significant volume reduction by reverse osmosis or evaporation.

Brine is injected to porous subsurface rock formations. The rock structures receiving this waste need to have the ability to contain the waste.

It is necessary that waste from the wells never migrates out of the injection zone. According to Strycker and Collins (1987), there are at least five ways a waste material may migrate and contaminate potable groundwater, as waste may

- escape through the injection well because of insufficient well casing or because of corrosion or extensive injection pressure;
- escape vertically outside of the well into an underground source of drinking water;
- escape through confining beds that may have high primary permeability, solution channels, joints, faults, or induced fractures;
- travel vertically upward through nearby wells, which are inadequately cemented, plugged, or have leaky or faulty casing; and
- cause groundwater contamination because of lateral movement of liquid waste from a region of saline water to a region of fresh water in the same aquifer.

These wells are typically between 304.8 to 2438.4 m (1000 to 8000 ft) deep. Depending on the desired rate of throughput, brine is typically injected at a velocity of 3 m/s (10 ft/s) through pipes that are 0.08 to 0.1 m (2 to 24 in.) in diameter. The injection well drilling requires several materials, including cement, mud, and drill bits, although significant drilling costs are attributed to labor and drill rig rental.

4.2.1 Site Selection

Site selection is one of the most important design considerations. It is dependent on geologic and hydrogeologic conditions. Rock formations such as sandstones are highly porous and are able to take in large volumes of liquid. Permeability of these types of formations can potentially give rise to migration of disposed liquids by one or a combination of the mechanisms listed above. Other rock formations like shales and clay are impermeable and act as defining layers for the injection well to provide a barrier against migration. Another critical consideration in site selection is the potential for rock movement. This involves careful site survey and extensive seismic three-dimensional modeling. Recent seismic activities resulting from deep-well injection in parts of Ohio and Oklahoma call attention to an extensive study before the actual selection of an injection site.

4.2.2 Construction

According to the Safe Drinking Water Act Underground Injection Control program under U.S. EPA requires all Class 1 wells be cased and cemented to prevent movement of fluids into or between underground sources of drinking water. A typical injection well is constructed in successive stages of drilling (or reaming), casing, and cementing until a well of the required depth and diameter is completed.

4.2.3 Well Disposal Systems

Two types of deep-well disposal systems are widely used—slurry fracture injection (SFI) and cuttings reinjection. However, SFI wells are primarily used for brine injection. Slurry injection involves processing solid materials to make particles of suitable size and blending them with a fluid (often water based) to create a slurry. The slurry is injected to a suitable confined formation at a pressure high enough to continuously fracture the rock structure receiving the slurry. The following are some important features of SFI:

- high-pressure injection
- injection of large volumes of waste (as high as 10 000 m^3/mo)
- multiple waste streams
- wells dedicated to disposal of produced water and brine

4.3 Solar Evaporation Ponds

Evaporation ponds represent an effective and reliable method for removal of water from a concentrated solution. As in the case of deep-well injection, in brine management application, evaporation ponds are used to achieve the goal of ZLD. These ponds rely on solar energy to evaporate water and leave behind precipitated salts, which are periodically collected and disposed of in landfills.

As a means of ultimate disposal or brine evaporation, evaporation ponds offer the following advantages:

- easy to build and operate
- require no major moving operating energy, except pumping the liquid
- low operating cost

However, the following disadvantages are also associated with evaporation ponds:

- require large land area
- application limited to dry and arid regions
- liner integrity is a major concern to avoid seepage into the ground

Evaporation ponds often draw criticism because the water lost by evaporation cannot be recovered. However, that is not accurate because nature returns that water to the ground in the form of rain.

Major considerations when developing solar ponds include the following:

- concentrate volume
- geographic location (evaporation rate and rainfall)
- land cost
- storage requirements
- disposal costs to landfill

Evaporation ponds are most appropriate for regions having a relatively warm, dry climate with high evaporation rates, level terrain, and low land costs.

4.3.1 Sizing

The surface area of the pond depends primarily on the evaporation rate at the location. The depth should be calculated after proper consideration has been given to surge capacity, water storage capacity, and storage capacity for precipitated salts. Studies have indicated that pond depths ranging from 0.3 to 0.45 m (12 to 18 in.) are optimal for maximizing evaporation rate. A freeboard of a minimum of 0.61 m (24 in.) is provided to accommodate level fluctuations because of rainstorms and waves caused by wind.

Evaporation ponds are land intensive and require large surface area. However, pond size can be dramatically reduced with inclusion of technologies such as Droplet Spray Evaporators and Wind Aided Intensive Evaporation (WAIV). In cases where land is expensive or limited area on existing property is available, these technologies can be extremely useful.

4.3.2 Evaporation Rate

The evaporation rate is high in arid climates. An annual evaporation rate is calculated for an evaporation pond by using a Class A pan coefficient. Class A pan coefficient is compiled in a database for areas throughout the United States for periods of 10 years or more. The pans are fabricated to standard dimensions to represent a natural body of water as closely as possible. A standard evaporation pond is referred to as Class A pan. Multiplying the pan evaporation rate by the pan coefficient results in a mean annual lake evaporation rate for a specific region.

Major cost parameters considered while developing an evaporation pond are as follows:

- Land costs
  - The cost of land varies not only from city to city, but also from location to location within a given city.
  - Land costs can easily vary by a factor of 10 or more, depending on the exact location near a city.
- Earth work
  - Earth work costs depend on the whether the terrain is flat or hilly, rocky or sandy, and so on.
  - Earth work involves land clearing; building dikes; and developing perimeter area for road, fence, dike, and dike covers.
- Lining
  - Lining is an important instrument in preventing salts and other contaminants from leaching, migrating, and contaminating groundwater.
  - Typical types of liners include polyvinyl chloride (PVC), high density polyethylene (HDPE), butyl rubber, etc.
- Miscellaneous costs
  - The costs depend on the specific project site.
  - Some possible costs include fencing, maintenance roadways, disposal costs, seepage monitoring, and contaminated ground cleanup.

- Operation and maintenance
  - Operating costs include security and damage inspection.
  - The ponds operate mostly maintenance free.

As mentioned earlier, evaporation enhancing technologies help large reduction in volume of evaporation ponds. The fundamental principle of these technologies is increase in mass transfer rate by—(a) increasing the air–water contact surface area, and (b) turbulence (Reynolds Number). In case of droplet spray technology, brine is pumped and sprayed through a sprinkler system on the top of evaporator surfaces. As the droplets travel vertically up before falling on to evaporator pond water evaporates from their surface. An overall increase in evaporation rate from 30% to 35% has been reported. Higher evaporation rates can be achieved by decreasing droplet size by using fine spray nozzles, but the downside is the potential of choking the nozzles by solid precipitate formation.

In case of WAIV, high air–water contact surface is created by pumping a brine stream at an elevation higher than the evaporation pond and allowing it to slowly flow down the surface of a geotextile fabric. The fabric offers large surface area and helps enhance the evaporation rate. A 13-fold evaporation rate has been reported from a small-scale study with TDS concentrations up to 18 000 mg/L. A comprehensive review of these technologies and the technical and cost comparison between several evaporation enhancing technologies have been presented by Hoque et al. (2010).

The aforementioned sections provide introductory-level discussions on deep-well injection and solar evaporation ponds. For detailed discussions on their design and cost estimations, readers are referred to the Mickley & Associates (2009) report, *Reclamation: Managing water in the West.*

4.4 Landfills

Landfilling is the ultimate means of disposal of brine. Brine solutions cannot be directly disposed of in landfills and need to be solidified before release to landfill. The nature and source of solids dictate the classification of landfill suitable for disposal of solids. The nature of solids, hazardous or not, is determined from Toxicity Characteristic Leaching Procedure leaching tests (U.S. EPA, 1992). Also, the material needs to pass a paint filter test (U.S. EPA, 2004) to demonstrate absence of free liquids in the solids.

Crystallization followed by landfilling are two means of achieving ZLD goals for brine management. The chemical composition of the crystals determines the type of landfill for their disposal.

Brine may be disposed of in hazardous waste landfills. These landfills are constructed as a double-liner system to protect groundwater from contamination by leachate. The liner material should be a flexible membrane lining like HDPE, ethylene propylene diene monomer, PVC, and so on. The landfills are required to monitor odor and water quality on a regular basis. Landfills are supposed to have monitoring facilities placed upgradient and downgradient from the site.

5.0 SUMMARY

Mechanical vapor recompression evaporators have been used in various desalination processes for decades. In industrial wastewater reuse areas, it occupies an application niche as brine concentrator for treatment of reverse osmosis concentrate to reduce its flow to the near-ZLD level. It is a matured and cost-effective technology.

Considering the highly corrosive environment posed by concentrated brine at the high operating temperatures, thermal evaporators, including MVR, are expensive from both capital and O&M viewpoints. Competition from newer technologies (e.g., HDH is growing).

For seawater desalination, reverse osmosis is the dominant technology throughout the world. Electrodialysis reversal is another membrane-based technology that largely overcomes the scaling potential on reverse osmosis membrane surfaces. However, because of the relatively low energy prices in the Middle East, thermal desalination processes (e.g., multieffect distillation) are quite common. This scenario is changing rapidly because the old and aging thermal desalination facilities are planned to be replaced in the region by membrane-based processes as a result of the abundance of inexpensive renewable energy available from solar source. Abu Dhabi is planning to build "a huge new desalination facility every year for the next three years at least—even though demand for water in the emirate is relatively stable—as the tumbling cost of desal makes it efficient to shutter old thermal facilities in favor of new membrane based facilities" (Global Water Intelligence, 2019).

Like other areas of water and wastewater treatment, minimum or ZLD will become the guiding principle of new projects that also involve desalination and brine management. Innovations and new applications of existing technologies are drivers for sustainable processes. In this regard, a new collaboration between the Technical University of Delft, Netherlands, and various industries in Europe, called Zero Brine (www.zerobrine.eu), is worth mentioning. It is a 4-year innovation action plan funded by the European Commission with the goal of creating a circular economy. One of the demonstration projects undertaken by the Zero Brine program is to recover resources from spent brine solutions from industries near Rotterdam port.

This is one of the largest industrial complexes in the world, and spent brine streams are produced from a large number of industries (e.g., power plants, petroleum refineries, chemical and petrochemical plants). Cooling tower blowdown, ion exchange spent regeneration liquid, and reverse osmosis reject streams are the most common sources of brine from each of these industries. The goal of this project is to demonstrate the efficacy of a combined treatment by nanofiltration, evaporation, and eutectic freeze crystallization for the recovery of high-purity (>90%) magnesium, commercial-grade sodium chloride, and sulfate suitable for reuse in industrial applications. The project is aiming for approximately 70% resource recovery from spent brine solutions generated in this complex to minimize disposal to the environment.

Selection of desalination and brine management technologies in the coming decades will be dictated by the following major drivers:

- Innovations for achieving minimum liquid discharge and maximizing resource recovery
- Capital and long-term O&M costs
- Ability to be operated by renewable energy
- Low effect on carbon footprint
- Ease of operation by maximizing the use of smart technology

6.0 REFERENCES

Aqueous Solutions Aps. (2020). *Phase diagrams.* http://www.phasediagram.dk/binary/sodium_chloride.htm

Bell, I. H., Wronski, J., Quoilin, S., & Lemort, V. (2014). Pure and pseudo-pure fluid thermophysical property evaluation and the open-source thermophysical property library CoolProp. *Industrial & Engineering Chemistry Research*, *53*, 2498–2508.

Dehghani, S., Date, A., & Akbarzadeh, A. (2018). Humidification–dehumidification desalination cycle. In *Emerging technologies for sustainable desalination handbook*.

El Kadi, K., & Janajreh, I. (2017). Desalination by freeze crystallization: An overview. *International Journal of Thermal & Environmental Engineering*, *15*(2), 103–110.

Ertel, E., McManus, K., Butters, T., & Bogdan, J. (2012). Wastewater treatment challenges associated with nonconventional oil and gas activity in Pennsylvania. *Proceedings of the 85th Annual Water Environment Federation Technical Exhibition and Conference*, New Orleans, Louisiana, Sept 29–Oct 3, p. 137.

Ettouney, H. (2006). Design of single-effect mechanical vapor compression. *Desalination, 190*, 1–15.

Global Water Intelligence. (2019). Abu Dhabi build out to replace thermal desal. Global Water Intelligence, p. 13.

Henthorne, L. (2009). The current state of desalination. *Proceedings of the IDA World Conference*, Dubai, United Arab Emirates.

Herrmann, S., Kretzschmar, H., & Gatley, D. P. (2009). Thermodynamic properties of real moist air, dry air, steam, water, and ice (RP-1485). *HVAC&R Research*, *15*, 961–986.

Hoque, S., Alexander, T., & Gurian, P. L. (2010). Innovative technologies increase evaporation pond efficiency. *IDA Journal of Desalination and Water Reuse, 2* (72–78).

Kabeel, A. E.; Hamed, M. H., Omara, A. M., & Sharshir, S. W. (2013). Water desalination using a humidification–dehumidification technique—A detailed review. *Natural Resources Journal*, *4*, 286–305.

Jenkins, K., Higgins, T., Mavis, J., Sandy, T., Reid, L., & Martins, K. (2011). Conceptual design and evaluation of zero liquid discharge for management of industrial wastewater. *Proceedings of the 84th Annual Water Environment Federation Technical Exhibition and Conference*, Oct 16–19, Los Angeles, California, p. 1753.

Li, X. (2009). Experimental analysis of produced water desalination by a humidification–dehumidification process [PhD Thesis, New Mexico Institute of Mining and Technology].

McCabe, W. L., Smith, J. C., & Harriott, P. (1993). *Unit operations in chemical engineering*. McGraw Hill.

Mickley & Associates. (2009). Reclamation: Managing water in the West. Report submitted to the U.S. Department of the Interior, Bureau of Reclamation, Agreement No. 04-FC-81-1050.

Narayan, G. P. (2012). Thermal design of humidification dehumidification systems for affordable and small-scale desalination [PhD thesis, Massachusetts Institute of Technology].

Rahman, A. S., Ahmed, M., & Dong Chen, X. (2006). Freezing melting process and desalination: Review of the state of the art. *Separation and Purification Reviews*, *35*, 59–96

Saltworks Technologies. (2018). *How to improve FGD economics with chloride removal using selective EDR*. https://www.saltworkstech.com/articles/how-to-improve-fgd-economics-with-chloride-removal-using-selective-edr/

Samant, D. S., & O'Young, L. (2006). Understanding crystallization and crystallizers. *Chemical Engineering Progress*, *102*(10), 28–33.

Stickney, W. W., & Fosberg, T. M. (1975). Chemical waste treatment by evaporation. 80th National Meeting, AIChE, Boston, Massachusetts.

Strycker, A., & Collins, A. G. (1987). *State of the art report: Injection of hazardous wastes into deep wells*. Robert S. Kerr Environmental Research Laboratory.

Tomer, E. (2011). *Utilizing available "coldness" from liquefied natural gas (LNG) regasification process for seawater desalination*. IDE World Congress, Perth, Western Australia.

Triple point. (2020, March 31). In *Wikipedia*. https://en.wikipedia.org/w/index.php?title=Triple_point&action=history

U.S. Environmental Protection Agency. (1992). *SW-846 Test method 1311: Toxicity characteristic leaching procedure (TCLP)*. U.S. Environmental Protection Agency.

U.S. Environmental Protection Agency. (2004). *SW-846 Test method 9095B: Paint filter liquids test*. U.S. Environmental Protection Agency.

Verbeek, B. (2011). *Eutectic freeze crystallization on sodium chloride: Analysis of a full experimental cycle* [MS thesis, TU-Delft, Document No. 2461, 1–88].

Warsinger, D. M., Mistry, K. H., Nayar, K. G., Chung, H. W., & Lienhard V., J. H. (2015). Entropy generation of desalination powered by variable temperature waste heat. *Entropy*, *17*, 7530–7566.

Zero Brine. Zero Brine Home Page. http://www.zerobrine.eu

6

Minimum Liquid Discharge Technology Applications—Food and Beverage

Brent Sutter; Christopher Lewis; Elyse Dumas; Gennaro Dicataldo, PhD, PE; Derek Stevens; Somnath Basu; Dylan Christenson; Brian Moore, PhD; Patrick Regan; & Walt Kozlowski

1.0 MINIMUM LIQUID DISCHARGE TECHNOLOGY APPLICATIONS—FOOD AND BEVERAGE

The food and beverage industry is a great candidate for water reuse based on the sheer amount of water the industry uses. Some of this water is used as an ingredient during production and some is used during washdown, general cleaning, and cleaning of equipment at the facility. However, most of this water not added directly to the product can typically be tied to various processes during production that require water. Common contenders for high water use include equipment used for heating and cooling products during production, product washdown (such as carcass washing at meat processing facilities), and clean-in-place (CIP) processes for production and storage equipment. Pasteurizers, for example, can require tens of thousands of gallons per day of water, and wastewater from bottle washing comprises 50% of all wastewater produced by the beverage industry (Barbera & Gurnari, 2018). Water for utility systems such as noncontact cooling water or steam makeup are also a big component of the industry's overall water use.

1.1 Why Minimum Liquid Discharge for Food and Beverage?

There are many reasons a food and beverage company would want to minimize liquid discharge. This subsection explores two drivers for water reuse: the economics of water reuse and corporate social responsibility.

1.1.1 Company Bottom Line

The U.S. food and beverage industry spends an estimated $5.1 billion on water and wastewater treatment equipment annually, compared to $160 million and $1.3 billion for pharmaceuticals and mining, respectively (Bluefield Research, 2016). Categorical effluent guidelines and industrial wastewater pretreatment requirements drive a large portion of this spending; however, as the cost of water supply and wastewater discharge and treatment increases, industrial water reuse is becoming more common.

By recycling treated industrial wastewater, a food and beverage facility can reduce both water supply fees and fees for offsite wastewater treatment and disposal. The savings associated with these fees can justify capital and operational expenditures for additional onsite treatment and reuse, although the short return on investment favored by food and beverage companies makes a tough business case. To boost a company's bottom line, water reuse systems can be designed with other opportunities for project payback in mind beyond the traditional water and wastewater fees, such as biogas production, utilization, and food waste processing.

1.1.2 Corporate Responsibility

In an age of social media, increased environmental awareness, and consumer choice, certain food and beverage producers have set themselves apart by establishing themselves as environmental stewards. More than a quarter of consumers (29%) claim to base their purchasing decisions on concerns regarding environmental sustainability, and 19% of consumers claim to have increased their sustainable purchasing in the last year (The Hartman Group, Inc., 2017). Investing in water reuse systems to reduce the strain on municipal water supplies can better align a food and beverage producer's environmental footprint with potential consumer preferences.

1.2 Water Reuse Regulatory Landscape

Because of the health risks associated with food and beverage consumption, producers must follow strict regulations for food safety. Although there are no national regulations for water reuse, two governing agencies provide oversight for food and beverage production within the United States: The U.S. Department of Agriculture (USDA) and the U.S. Food and Drug Administration (FDA). Both agencies provide oversight and regulation for food safety during production; although, the stringency of these requirements differs between agencies when it comes to water quality requirements.

The USDA, which regulates meat and poultry products, publishes a list of production activities for which recycled water can be used, along with relevant corresponding water quality requirements. Such activities include water used to chill or cook ready-to-eat products, chill or wash raw products, and select steps associated with meat and poultry processing per USDA's part in the *Code of Federal Regulations* (Establishment Grounds and Facilities, 2016). Any activities not included on this list of activities must comply with U.S. National Primary Drinking Water Regulations and follow the requirements published by USDA. Conversely, the FDA, which regulates the remaining food and beverage products not governed by USDA, only states that water may be reused if it does not increase the level of contamination of the food (Good Manufacturing Practice, 2018), leaving the water quality requirements, monitoring, and detailed implementation requirements up to the producer.

The ambiguity of these regulations can make it hard to define, approve, and implement high-quality water reuse systems within the food and beverage industry, although there are many reasons a facility would invest in water reuse systems (either traditional utility systems or high-quality systems unique to food and beverage production), as described in the following subsections.

1.3 Food and Beverage Industry Water Use

A common metric used to rate a food and beverage facility's water use efficiency is the ratio of cubic meters (gallons) of water used per equivalent unit of product produced." The more efficient the facility is with the water it takes in, or with the amount of water that can be recycled back to the head of the facility or for specific uses within production, the lower this ratio will be. Table 6.1 presents examples of water use ratios for common industries within the food and beverage sector.

Many things can affect a food and beverage facility's water use efficiency ratio, but the age of the facility and its equipment is often a key variable that

TABLE 6.1 Water use ratios for key food and beverage sectors (Bluefield Research, 2016).

Industry	Water Use Ratio
Alcohol distilleries	15 m^3/ton of product
Seafood	11 m^3/ton of product
Beer	6.5 m^3/ton of product
Meat	6.5 m^3/ton of product
Dairy	5 m^3/ton of product
Wine	4 m^3/ton of product
Fruits and vegetables	3 m^3/ton of product
Sugar	1 m^3/ton of product

contributes to inefficient water use. Facilities that have added noncontact cooling water recycling or upgraded equipment such as spray nozzles with new water-efficient nozzles have been able to better reduce this ratio when compared to their dated counterparts.

1.4 Generated Wastewater

Because food production is often a batch process, either because of the ways the food production facility is staffed or the nature of the production itself, wastewater in the food and beverage industry has a unique flow pattern compared to other manufacturing industries. Even for facilities that are staffed 24 hours a day, production lines frequently operate for a week or two before shutting down for cleaning, maintenance, or product changeovers. In addition to the variations in flow, wastewater characteristics can vary wildly depending on what product is being made and whether the facility is running CIP processes between production runs.

Regardless of the product, overarching trends and characteristics can be applied to wastewater in the food and beverage industry (see Table 6.2). Clean-in-place systems, which are systems that circulate a sequence of chemicals through production equipment so that it can be cleaned without disassembling the system, contribute to large swings in wastewater pH and temperature. Depending on what is being produced, wastewater often has high organic loads and oil and grease (O&G) content that must be pretreated onsite to ensure local water resource recovery facilities (WRRFs) are not overwhelmed.

TABLE 6.2 Food and beverage wastewater characteristics.

Food and Beverage Sector/Activity	Defining Wastewater Characteristics
Slaughterhouses	High organic content (BOD and COD), nitrogen, and phosphorus Complex composition of fats, proteins, and fibers
Beverage/Soft drink	High TDS, COD
Distilleries	Dark color Acidic, high organic matter Odorous
Winery	High BOD/COD
Dairy	High organic content (BOD and COD), nitrogen and phosphorus
Bakery	High organic content (BOD and COD), O&G, starches

Note. BOD = biochemical oxygen demand; COD = chemical oxygen demand; TDS = total dissolved solids.

It is important to note that reducing a facility's water footprint by reusing water for utility systems and other low-load activities can negatively affect final wastewater effluent that is permitted on a concentration basis. Although the load of organics and other wastewater constituents from process and production wastewater does not change (assuming production is held constant over this process improvement), reusing water for cooling removes the dilution effect this waste stream previously had when mixing with other high-strength waste. Eliminating the dilution effect can lead to larger pH swings and new, higher constituent concentrations in wastewater effluent.

1.5 Matching Treatment Technology with Water Quality Requirements

While the need to conserve and reuse water has already been established in previous sections, water quality requirements may vary substantially throughout different applications at a typical food and beverage facility. Depending on the type of water or wastewater treatment available, not all treated water may be suitable for reuse in a given application. Examples of different water quality requirements may include disinfected water, demineralized water, deionized water, pH controlled water, or any combination of these quality requirements and more.

As such, a food and beverage facility must determine minimum water quality requirements for each unique water use application before matching recycled water streams and their treatment processes to potential reuse applications. Intuitively, water use processes that require higher quality water must be matched with higher quality reuse water. For example, it would not be appropriate to use rinse water directly from slaughterhouse processing in any direct contact applications with the final product. In contrast, it may prove too costly, and even counterproductive, to use ultrapure water for irrigation. Processes with relaxed water quality requirements may be more cost-effective candidates for water reuse than ones that require potable water quality or better. Examples of applications requiring less stringent treatment methods may include cooling tower, toilet flushing, irrigation, and general facility washdown.

In general, high-quality water reuse applications can be broken down into three categories: indirect contact with products, direct contact, and direct addition into the product. Indirect contact involves applications that contact surfaces that directly handle food and beverage products, such as washing tanks, pipes, machinery, and so on. Direct contact includes water used for washing the product or other similar processes and, lastly, direct addition into the product would qualify as anytime water is used as an ingredient.

Water reuse for indirect contact applications is the easiest to use out of these three high-quality categories, but limited installations and regulatory uncertainty regarding the approval process can make this difficult to implement. Additionally, treatment systems for indirect contact may be cost-prohibitive at small scales. If a facility only intends to reuse tens of thousands of cubic meters (gallons) each day for indirect contact applications, it can be difficult to justify the capital and operational costs of the required water treatment system.

There are few successful examples of water reuse for direct contact with products because these applications are difficult to define and difficult to approve given the risk associated with water quality and food safety. Reuse water as an ingredient and direct addition into the product has the highest risk of all. Because this notion is so dependent on the regulatory climate, water sources, and cultural norms, there are few references detailing large-scale direct addition reuse applications within the food and beverage industry.

Facilities with onsite wastewater treatment are better positioned to capitalize on water reuse than those that directly discharge to local WRRFs for wastewater management. Existing onsite wastewater treatment systems, either for pretreatment before WRRFs or direct discharge to surface water-bodies, may be retrofitted or expanded to create water reuse opportunities if effluent quality does not already meet the requirements set forth by the intended application of reuse.

1.6 Achieving Minimum Liquid Discharge

Detailed understanding of water use at a production facility is critical to determining the most cost-effective course of action to achieving minimum liquid discharge (MLD). Facility employees are experts in knowing how, when, why, and where water is used in their facility, as well as knowing the quality of water needed for each process in the facility; however, they may not be experienced or aware of reuse technologies and typically outsource to experts (consultants, or sometimes equipment providers). Water efficiency audits can be performed internally, although a portion of opportunities are typically missed as a result of lack of water conservation expertise as well as lack of fresh set of eyes Because self-evaluation may not reveal well-worn routines of water consumption, professional auditors and industry-developed guidelines (Brewers Association, 2013) may help to achieve an accurate account of water use and disposal. Typically, a facility water audit would involve the tasks outlined in the following subsections.

1.6.1 Facility Water Audit

An audit should begin with a complete account of all water use points in the facility. Usage points may include

- direct addition to product (water going to the consumer);
- direct contact sources (such as product rinse water);
- indirect contact sources (such as equipment CIP, washdowns, floor wash);
- domestic use (toilets, faucets, showers);
- makeup water for facilities' mechanical systems; and
- landscape irrigation.

Source water should also be identified, whether it is coming from the local municipality or private well water. Note any fees that are associated with both incoming source water and onsite wastewater treatment and/or wastewater fees associated with disposal to a WRRF. This topic is described in more detail in Chapter 2.

1.6.1.1 Water Accounting

One of the more challenging (and enlightening) tasks of an audit may be identifying and understanding unexplained water flow. Piping systems, particularly in older facilities, may have been modified over the years, leading to inadvertent water consumption through decommissioned equipment or forgotten rinse lines left open.

In addition to identifying sources and usage points, ongoing water accounting should include water temperatures, special treatment before use (filtration, pH adjustment, dechlorination, etc.), and how/where spent water is diverted after use. Be sure to note any flow restrictors or flow control devices.

A complete accounting effort should include breakdowns of each process or equipment subset and will provide a baseline of water use for the facility and each of its processes. Because many food and beverage facilities operate on a per batch basis, water use may be tracked similarly, thus allowing a baseline to be generated/compared between process batches, noting that different product types may involve different water consumption requirements. Readings taken between batch processes, during CIP, or while the facility is shut down (if that occasion exists) may help discover leaks or unexplained losses.

Unfortunately, most facilities have only a single master water meter that measures incoming water to the facility. As such, determining water use will probably require the use of portable or temporary meters (for measuring multiple flows simultaneously) and temporary manual observation sessions of process runs. This is known as "submetering." Submetering is a critical (if tedious) process, particularly in older or complex facilities. Individual zones and large-use equipment, like boil kettles or filtration skids, may have individual submeters.

All submeter readings should be logged and reviewed on a regular basis. Ideally, each meter shall be read at least monthly. More frequent readings can be taken to narrow down water consumption during highlighted process steps. Although more frequent readings may incur additional costs, the additional data collected more quickly helps to identify unexplained losses. It is easier to cull too much data (and revise collection frequency) than to handle insufficient data collection.

1.6.1.2 Data Review and Evaluation

After an established metering period or selected number of batch processes, responsible parties should review the data, explicitly looking for unusual or unexplained readings, comparing those with any past data (if it exists) or expected water consumption values. Data review is particularly important because it may point to aspects of water use that, intuitively, may not seem relevant.

1.6.2 Taking Action

Once the collected data have been evaluated, an action plan may be established for water conservation. The plan should set specific water use reduction goals for the overall facility and also each subprocess that is measurable and within reach. The plan should identify responsible participants and corresponding detailed actions to minimize water discharge. Regular review of the ongoing implementation versus stated goals is critical to success.

1.6.3 Risk Assessment

In the United States, incoming water from a regulated public water system is routinely audited to ensure regulations are met under both local specifications and those defined by law in the 1974 Safe Drinking Water Act. As such, when considering the reuse of water within a private manufacturing setting, such as a food and beverage production facility, it is critical to understand the targeted process water requirements, reuse sources, water treatment strategies, and associated risks. This responsibility is incumbent on the processer to ensure, via regular audits, measurements, or some combination thereof, that defined requirements are met. This level of water management requires a systematic preventive approach to food safety in production processes. This approach is known as "Hazard Analysis and Critical Control Points" (HACCP). The HACCP approach is recognized by the National Advisory Committee on Microbiological Criteria for Foods as an effective and rational means of assuring food safety from harvest to consumption per FDA. This includes a principal focus on problem prevention to the fundamental system. Under the HACCP system, documented steps shall be taken to reestablish control in the event of an observed deviation to ensure potentially hazardous products do not reach the consumer. To achieve this, HAACP uses seven principles:

1. Conduct hazard analysis
2. Perform critical control point identification
3. Establish critical limits
4. Establish monitoring procedures
5. Establish corrective actions
6. Establish verification procedures
7. Establish recordkeeping and documentation procedures

In recognition of the wide-ranging variety of products and processes in the marketplace, the HACCP system must be created by each processor to meet the needs of its unique product and processes, from raw material acquisition to distribution and consumer consumption. In cooperation with the efforts of the HACCP Alliance, FDA recommends implementing HACCP systems for the food industry. Commitment and enforcement by facility management are critical to the success of an HACCP system.

1.6.4 Reusing Untreated Process Water

It may be possible to reuse spent water in other areas of the facility. For example, water spent on final bottle rinse (before product fill) is anticipated to have minimal pollutants and, as such, may be applicable to first bottle or initial equipment rinse, or any application where a secondary or tertiary cleaning will ensure specified final clean. In the case of bottle rinse, high-quality water used during final bottle rinse is effectively used for two different processes. Although screening may be required to prevent the introduction of large particles, by reusing spent final rinse water in an earlier process step, the facility has effectively reduced its water consumption.

1.7 Treatment Technologies for Reuse Applications

Often, multiple technologies must be used in coordination with one another to produce a desired water quality for treated wastewater. Common treatment technologies (like those listed here) may be applied independently or complementary with one another. For many treatment technologies, there are dependencies on upstream pretreatment systems such as screening, coagulation/flocculation, and pH adjustment. Depending on industry characteristics, usage, or treatment objectives, some pretreatment steps may be more relevant than others to meet specific objectives. For more detailed descriptions of each of these technologies and their fundamental operation, please see Chapter 3 of this publication.

1.7.1 Screening

Screening technologies involve a physical barrier to separate solids from liquid in the wastewater stream. Depending on the application, screens may contain perforations or gaps from 0.5 to 25 mm (or larger). Similar to membrane and media filtration, captured screened solids must be periodically removed, resulting in a solids disposal requirement. Depending on the screened product value, such as spent grain, screened solids may be considered a value-added product.

1.7.2 pH Adjust/Neutralization and Equalization

Chemical treatment technologies may include pH adjustment or neutralization and are often executed via a pH adjust tank, sump, or diversion pit within the factory. Typically, adjusting pH is a relatively simple process and involves varying acidic or alkaline chemical dosing to the wastewater stream to meet a desired pH. However, it should be noted that pH adjustment chemicals must be maintained and managed. Additionally, if pH dosing is performed before downstream treatment, additional considerations may be required to account for feed strategy, solids management (for pumping and piping), probe calibration, and materials of construction.

Flow equalization, or balancing, may be used in conjunction with neutralization or as a unique process step. Equalization smooths surging flow and loads to the treatment system, often serving to dilute delivered batches or slugs of chemicals, pH, solids, and more. Equalization requires volumetric storage (e.g., up to 24 hours of hydraulic residence time), and thus increases the overall system footprint of the treatment system.

1.7.3 Dissolved Air Flotation

Dissolved air flotation (DAF) systems often combine chemical dosing with flotation technology in a single tank, using fine-bubble technology to lift colloidal particles to the surface for removal and disposal while allowing effluent to continue. Dissolved air flotation systems may often be used as primary treatment processes, intended to remove larger particles before additional treatment downstream, but may be used as a secondary clarification step or even as a complete system treatment, depending on the required effluent water quality. Common primary applications within the food and beverage industry include the poultry and beef industries for capturing high-value protein tissue or tallow for value-added rendering processes. Because dissolved material is not removed from a DAF system, additional processing may be required after DAF before reuse.

1.7.4 Gravity Separation Technologies

Like DAF systems, sedimentation technologies, including lamella plates and clarifiers, are designed to separate colloidal solids from liquid. Also similar to DAF systems, sedimentation will not remove dissolved solids from the effluent stream. These technologies use longer retention times (clarifiers) or surface contact (lamella plates) to separate solids from liquids. The longer retention times of sedimentation technologies may require larger footprints when compared to other separation technologies; this is a critical consideration when evaluating solids separation technologies.

1.7.5 Filtration

Filtration technologies come in a variety of form factors, including membrane filtration (e.g., microfiltration, ultrafiltration, nanofiltration, and reverse osmosis), and media filtration (e.g., diatomaceous earth, sand, and activated carbon), with widely varying feed requirements and capture rates. Filtration is primarily used in the separation of solids and liquid, but may also be combined with biological treatment as part of a holistic approach to removing dissolved substances and also solids in certain applications, like membrane bioreactors (MBRs). Because of inherent high solids capture efficiency, filtration technologies often require additional attention and cleaning compared to settling or flotation solids–liquid separation technologies.

1.7.6 Disinfection/Advanced Oxidation Process

Disinfection is intended to destroy or render ineffective any bacterial or viral constituents in the wastewater. This is frequently performed through chlorination; however, this is not the only option available. The advanced oxidation process (AOP) is intended to remove organic material by oxidation through reactions with hydroxyl radicals, specifically to a subset of such chemical processes that use combinations of ozone (O_3), hydrogen peroxide (H_2O_2), and ultraviolet (UV) radiation.

1.7.7 Additional Nonbiological Separation Technologies

Although covered in more detail in Chapter 3, additional separation technologies include chemical precipitation, centrifugal separation, adsorption onto activated carbon, and chemical adsorption or ion exchange for removal of dissolved constituents.

1.7.8 Biological Treatment

Biological treatment involves using microbiological organisms to remove pollutants (and possibly nutrients as well) from wastewater. Through the process of degrading organics, the mass of microbes will increase, and a portion of this biomass must be disposed of to maintain a steady amount of biomass in the system, resulting in necessary biosolids management. Biological treatment may be used as an aerobic or anaerobic treatment step and typically requires experienced and skilled operators to ensure best treatment capability. Although Chapter 3 goes into greater detail on biological processes, they are addressed here briefly because they may figure into food and beverage processes.

1.7.8.1 Anaerobic

Anaerobic digestion uses microorganisms to break down biodegradable material in the absence of oxygen. Biogas is produced as a beneficial byproduct of anaerobic digestion. For food and beverage applications, anaerobic systems, such as upflow anaerobic sludge blanket technologies, may provide a process benefit that includes energy generation/savings. Water reuse with anaerobic systems often involves an additional process afterward to refine the ammonia-containing effluent into a more reusable water source.

1.7.8.2 Aerobic

The aerobic treatment process uses oxygen and microorganisms (aerobic bacteria) to treat wastewater. Typical aerobic applications for wastewater treatment include conventional activated sludge (CAS) systems, MBRs, oxidation ditches, trickling filters, moving bed biofilm reactors (MBBRs), and sequencing batch reactors (SBRs). In all cases, aerobic treatment may reduce biochemical oxygen demand (BOD), total suspended solids (TSS), and nitrogen and phosphorus compounds; however, each of these systems has limitations and requires varying footprints, energy consumption, and labor to properly operate. Unlike anaerobic processes, aerobic systems do not generate biogas.

1.7.8.3 Membrane Bioreactor Technology

Because the focus of this publication is on minimizing water use, it is important to note that MBR is recognized as the best available technology for achieving the highest quality effluent with the fewest number of process steps (Iglesias et al., 2017). Like all biological treatment systems, aerobic MBRs, and less common anaerobic MBRs, require biological processes that must be monitored and maintained. This can prove challenging when treating batch production effluent or during periods of facility shutdown, in which case there is no source of energy to maintain the biological treatment itself. Like all biological systems, poorly maintained biology can result in costly WRRF upsets, membrane fouling, and inadequate wastewater treatment.

1.8 Treatment Technology Assessment

To determine the most appropriate treatment solution for water reuse for a given application, it is important to evaluate all related technology parameters and how they will combine with the reuse needs of the facility.

1.8.1 Footprint

In many cases, older factories, or those residing on limited real estate, may have limited technology choices based on the physical size of treatment equipment. Additionally, it is critical that some treatment systems reside indoors, protected from challenging outdoor environments. These types of logistical requirements can drive solutions that may otherwise not be considered.

1.8.2 Operational Risks and Operability

Operational requirements may also drive an end user in one particular direction, especially if a given technology requires manual input or labor that is currently not part of the staff payroll. Unless new personnel will be hired to attend to wastewater treatment, then some technologies could be ruled out. This may be especially true for biological versus nonbiological solutions.

1.8.3 Life-Cycle Analysis, Capital Expense, and Operating Expenses

Evaluating treatment costs versus benefits over the life of the WRRF is common for municipal treatment systems; however, for industrial facilities, it is also important to evaluate the life of the product mix or planned production life for a given facility and the effect of that production on water use. When evaluating treatment technology and how it may function to reduce overall factory water consumption, it is important to consider the projected period for the factory's planned product mix before any adjustments or expansions to the manufacturing process, which could affect the effectiveness of the chosen treatment technology.

Operating expenses (OpEx) include labor, energy and chemical consumption, and additional disposal costs to properly run a treatment system. Although some treatment technologies require less operator attendance, treating water to higher quality specifications commonly requires more attention and, in some cases, additional energy and chemical costs. Further, maintaining strict operational targets in a 24 hour/7 days a week operating environment can drive higher labor costs. More complex controls, remote access, and the integration of more advanced automation and predictive analytics can help to reduce these costs. The as-a-service (XaaS) business model is relatively new to the wastewater treatment industry. With the XaaS model, treatment equipment is effectively leased and operated by someone other than the end user to ensure proper operation and quality effluent (Cambrian Innovation, 2019).

Table 6.3 lists a few common treatment technologies and how they vary, relatively, in terms of footprint, effluent quality, and operational cost. Note that pretreatment and tertiary treatment (not included) may be applied to almost any treatment system; this essentially represents another step to further remove pollutants to varying degrees of quality, including drinking water standards.

1.9 Case Study: Tuna Canning Facility

A commercial canned tuna process used by Grupo PINSA resulted in wastewater disposal that was costly and wasteful for the company. Critical parameters include flows of 1140 m^3/d (0.30 mgd) with constituents of elevated nitrogen and organic matter. To reduce overall water consumption and associated costs, Grupo PINSA decided to enhance the existing DAF-only process in 2012 by adding an aerobic MBR to the process flow (membranes by Microdyn-Nadir, BIO-CEL® modules, see Figure 6.1). The MBR process is located downstream of the existing DAF, thus treating the high BOD and nitrogen levels of the DAF effluent. Tertiary treatments via ozone oxidation,

TABLE 6.3 Relative footprint, CapEx, OpEx, effluent quality amongst common treatment technologies for food and beverage effluent treatment (Guo et al., 2014; Young et al., 2013).

Technology	Footprint	CapEx	OpEx	Treatment Capability for Reuse (TSS, BOD, Nutrient removal)
CAS with clarifier	Large	Low	Medium	Medium
Oxidation ditch	Large	Low	Low	Medium
MBBR with DAF	Medium	Medium	Medium	Medium
Trickling filter	Large	Low	Low	Low
MBR	Small	High	High	High
Anaerobic system[a]	Small	High	High[a]	Low
SBR	Medium	Medium	Medium	Medium
pH + DAF (nonbiological)	Small	Low	Low	Low

Note. CapEx = capital expenditures;

[a]Anaerobic system assume biogas production and, if used, must be evaluated for OpEx.

activated carbon, and reverse osmosis were also added. With the goals of both water reuse and lower discharge costs, the facility is now capable of meeting local regulations for potable water requirements. And, after nine years of continuous operation and a 30% increase in flow and membrane capacity (2018), the MBR treatment system has delivered excellent results in terms of quality and flow.

1.9.1 Membrane Bioreactor System Operation

Challenge 1: Biological fouling. Maintaining healthy and stable biology is critical to stable performance in an MBR, particularly regarding minimizing membrane fouling potential. The tuna canning MBR encountered membrane fouling effects resulting from periodic high/low loading conditions (not uncommon for batch-type processing for many food and beverage facilities), extreme swings in incoming temperature and pH, and biosolids management.

Challenge 2: Inorganic fouling. A more unique constituent with the incoming wastewater (a result of the canning process) is a high salt concentration. It was noted that high concentrations of dissolved salt would accumulate on the membranes, resulting in lower membrane permeability. A correlation was found between sludge age, mixed liquor suspended solids (MLSS) concentration, and salt concentration in the mixed liquor. As such, it was necessary to control MLSS to concentrations below 12 000 ppm to minimize accumulation of salt on the membranes to reduce membrane cleaning frequency.

To meet these unique treatment challenges, the WRRF adopted a number of successful measures to control both biological and inorganic-based membrane fouling. These included the following:

- Increase the contact time in the equalization tank to balance high/low slug loads.
- Increase aeration during high load conditions.
- Implement automatic pH control in the biological reactor for healthier sludge.
- Maintain MLSS concentration below 12 000 ppm.
- Improve control of biosolids removal.

1.9.2 Results

Table 6.4 presents typical system operations data after several years in operation.

TABLE 6.4 Typical influent concentrations and removal rates from Grupo PINSA facility.

Sampling	BOD	COD	TN	TP	TSS	Turbidity
Raw water	5790 mg/L	12 310 mg/L	785 mg/L	243 mg/L	2580 mg/L	2134 NTU
DAF effluent	2100 mg/L	3235 mg/L	472 mg/L	25 mg/L	87 mg/L	75 NTU
MBR effluent	25 mg/L	147 mg/L	15 mg/L	5 mg/L	2 mg/L	0.35 NTU
MBR % removal	98.8%	95.5%	96.8%	80.0%	97.7%	99.5%
Tertiary filtration + disinfection effluent	0	0	1 mg/L	0	0	0
Tertiary filtration + disinfection % removal	100%	100%	93.3%	100%	100%	100%

Note. TN = total nitrogen; TP = total phosphorus.

1.9.3 Conclusions and Reuse Strategy

As a result of the successful treatment of the tuna canning wastewater, the operator realized its objective of using 380 m^3/d (100 000 gpd) of treated water in the following tuna processing facility areas: washing docks, washing boats, preparation of DAF chemicals and membrane backwash, general facility washdown, facility cooling towers, and employee washrooms. Finally, by reusing treated water in the aforementioned areas listed, annual cost savings of $80 000 (U.S. dollars) now contribute to the company's bottom line.

2.0 MINIMUM LIQUID DISCHARGE TECHNOLOGY APPLICATIONS—POWER

2.1 Overview of Power Generation Processes

Electricity demand is growing at a robust pace around the world, with generation capacity expected to double by 2030, exceeding 30 trillion kWh of output. Asia represents a large portion of the growth over that period, with additional strong growth in capacity in Europe and North America.

FIGURE 6.1 PINSA MBR.

Considering that the cost of fuel is typically the largest cost for a power plant, the emergence of the United States as a leader in natural gas production has had a significant effect on the cost and has driven a shift in the U.S. power generation capacity to take advantage of this lower-cost—and cleaner-burning—fuel. However, the type of power generation plant cannot be described by its fuel type alone. There are various types of loading that can affect capital and operations and maintenance (O&M) costs. These types of loading range from base load to peaking load. The capital and operating expense associated with various power generation technologies and fuel types play a big role in defining what type of loading the plant is supporting (see Table 6.5).

In terms of total cost of ownership, a typical base loading power generation plant requires a large upfront capital cost, but operates with relatively lower O&M and fuel costs. This makes sense considering this has historically consisted of large coal and nuclear plants. An intermediate loading facility will have lower capital costs, but slightly higher O&M costs. This has traditionally been the role of the older coal-fired facilities with lower thermal efficiency. However, with an increased supply of natural gas, more power companies are shifting their base load capacity to include more combined cycle

TABLE 6.5 Typical CapEx and OpEx for various types of power plants (Penn State University, Department of Energy and Mineral Engineering).

Technology	CapEx ($/kW)	OpEx ($/kWh)
Coal-fired combustion turbine	500–1000	0.20–0.04
Natural gas combustion turbine	400–800	0.04–0.10
Coal gasification combined-cycle (IGCC)	1000–1500	0.04–0.08
Natural gas combined-cycle	600–1200	0.04–0.10
Wind turbine (includes offshore wind)	1200–5000	Less than 0.01
Nuclear	1200–5000	0.02–0.05
Photovoltaic solar	4500 and up	Less than 0.01
Hydroelectric	1200–5000	Less than 0.01

generation capacity to take advantage of the inexpensive fuel source. This shift is further accelerated by growing regulations on coal-fired power generation, which will be touched on in more detail in the following subsections.

The third loading capacity is commonly referred to as a "peaking facility." These can range from a small combined cycle facility to a combustion engine. Peaking facilities are relatively inexpensive to build, but they require a higher O&M budget. Thus, they typically sit idle until the demand from the grid is enough to justify their operation.

In the case of a grid dispatch model, the electric utility or grid operator will define what role various facilities will take. The grid operator would commit or schedule a set of base load facilities, taking advantage of low operating costs. These facilities would run 24 hours a day. While another set of facilities would be scheduled to meet daytime or "shoulder" demand. Lastly, a set of facilities would be on standby able to be turned on in minutes if necessary.

To better understand the operation of a power plant, it is important to understand what the drivers behind the plant's income statement are. The first category is the operating revenue. This is a product of the production, measured in megawatts hours (MWh) multiplied by the price of electricity ($/kWh). As much as this may be a simple way to understand operating revenue, there are a range of factors that ultimately affect the availability or uptime of the plant. Critical failures, such as boiler and condenser tube failures, can cause costly unplanned or forced outages. Similarly, scheduled maintenance and repairs during planned outage can cost valuable uptime, especially when unexpected issues occur.

On the other side of operating revenue, there are OpEx. These can be summarized into fuel or feedstock cost and O&M. The performance on a condenser or on the fireside of a boiler can significantly affect the net heat rate and ability to generate heat from the feedstock. Under O&M, there are expected OpEx associated with chemicals and other consumables, labor, and water (influent/effluent). For many coal-fired power plants, compliance costs under O&M are significant as plants work to meet increasingly more stringent effluent limitation guidelines (ELGs). The waste streams in these plants can originate from multiple sources (Figure 6.2). In many cases, this has forced coal-fired power plants to shut down given the significant cost required to meet ELG regulations.

At a power plant it is best to consider a typical water balance. Because water consumption can be high at a power plant, understanding where water is being used inefficiently represents a significant opportunity to

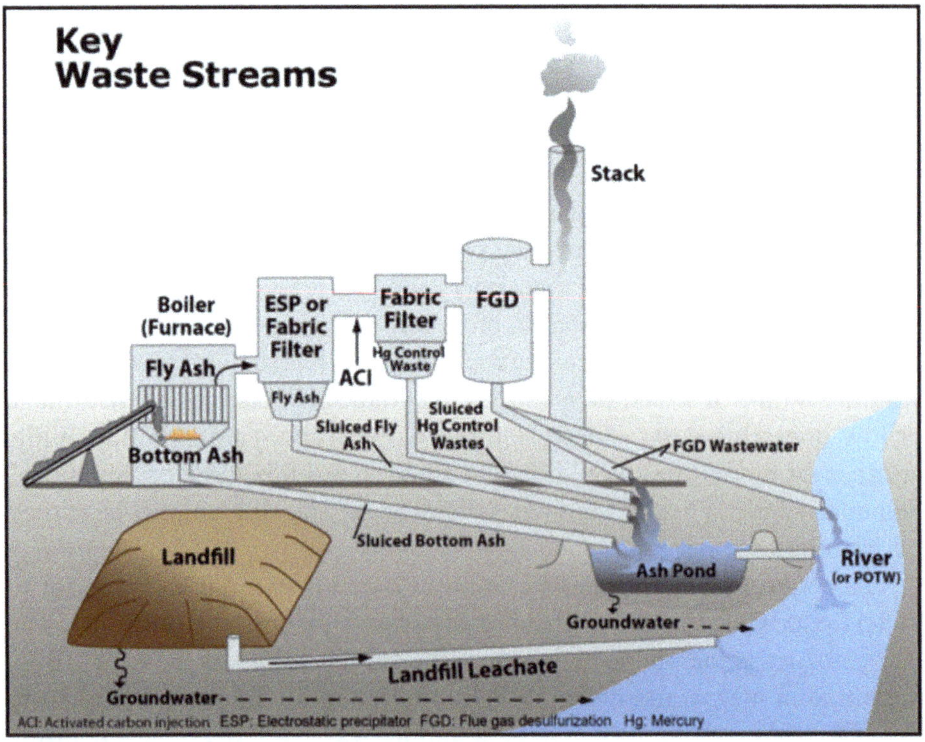

FIGURE 6.2 Key waste streams under ELG guidelines for coal-fired power plants (ESP = electrostatic precipitator, ACI = activated carbon injection, FGD = flue gas desulphfurization) (U.S. EPA, 2015).

reduce O&M costs. The consumption of water can be directly tied to the influent or withdrawal, the unrecoverable evaporative loss, and amount that is discharged to surface water or, in the most inefficient cases, to deep-well injection. The influent or withdrawal can be thought of as the amount required primarily by process cooling and emissions reduction requirements, and steam requirements for driving power generation turbines. The unrecoverable evaporative loss is directly correlated to the amount of cooling required. The discharge amount is subject to regulatory compliance. More importantly, and relevant to the subject of MLD, the cost or availability of influent water typically drives the decision to recycle or reuse that water after treatment.

If we consider the average water balance for a 650-MW capacity operation, it requires roughly m^3/s (10 000 gpm) of high-quality boiler feedwater (BFW) to produce the 630 kg/s (5 000 000 lb/hr) of steam required to drive the turbine. It also requires 59 052 m^3/h (260 000 gpm) of cool water to operate the steam condenser. There is an additional 1590 m^3/h (7000 gpm) of makeup water required for the cooling tower to compensate for the roughly 1476 m^3/h (6500) gpm of water loss through evaporation and non-productive water loss through water droplets (drift) as well as the 114 m^3/h (500 gpm) of blowdown required to maintain the optimal chemistry to prevent the formation of scale and deposits on critical heat-exchanger surfaces.

More simply put, a 650-MW capacity power plant typically uses approximately 59 042 m^3/h (260 000 gpm) of water (or 400 gpm/MWh). This number is much higher in those power plants where once-through cooling is used versus recirculating cooling.

In summary, there is a broad set of challenges that the power industry is currently facing. These challenges range from shifting power demand to a shift in fleet or type of plants to align with trends in fuel sources (or feedstock). Regulatory pressure represents a significant challenge recently as coal-fired power plants work to meet effluent quality requirements. This is coupled with an aging workforce and infrastructure.

2.2 Wet Processes and Wastewater-Generating Processes

Figure 6.3 takes into consideration the unit processes required to meet water quality needs across a typical power plant.

The treatment of source water into a power plant depends on the source itself. In the case of surface water, intake screens are required to keep everything out from logs to frogs. This requirement is specified in the U.S. Environmental Protection Agency (U.S. EPA) 316b regulation, which stipulates that the design and construction of the intake structure minimizes adverse

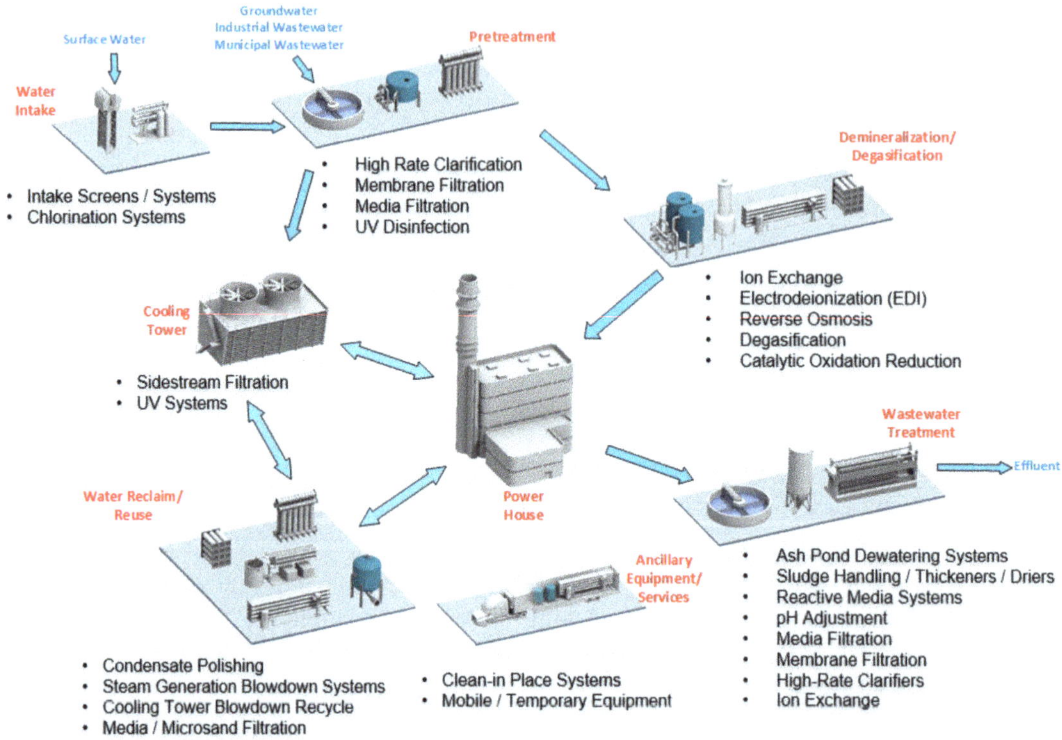

FIGURE 6.3 Typical unit processes to address water requirements of a power generation facility.

effects to aquatic life. For a plant to be 316b compliant, special considerations are required to minimize the risk of fish and other aquatic life from being killed or negatively affected by the intake operations. This is typically managed with some type of a traveling water screen. In the case of groundwater sources, as well as treated industrial and municipal wastewater, a range of pretreatment processes may be required, including clarification (conventional or high rate), ultrafiltration membrane, media filtration, and/or UV disinfection.

2.3 Case Study—Zero Liquid Discharge Example at a Combined Cycle Gas Power Plant

The growing trend to limit the use of fresh water while minimizing, if not eliminating, the discharge of wastewater in the power industry has made the need to provide an integrated approach a critical factor. An example project in the southwestern United States presents how a WRRF design that uses precipitation with microfiltration combined with offsite service was the solution for a gas turbine combined cycle power plant. The solution primarily incorporates membrane technology and operates at an overall recovery of greater than 95% while discharging a stream of wastewater to evaporation ponds and trucking solids offsite.

The feedwater to the plant is sourced from three different wells, which vary in constituent characteristics, whereas the wastewater is a combination of reverse osmosis concentrate, steam generator blowdown, evaporative cooler blowdown, wet surface air cooler blowdown, multimedia backwash water, oil/water separator wastes, and other service water wastes. This design overcomes many of the shortcomings of a conventional approach or a thermal design, which would have much higher capital and operating costs.

2.3.1 Background

The Harry Allen Generating Station is a natural gas-fueled power plant located in Southern Nevada approximately 25 miles northeast of Las Vegas. The plant was originally built in 1995 as a simple cycle facility with a 72-MW combustion turbine. A second, similarly sized combustion turbine was installed in 2006, doubling the plant capacity to 144 MW. The final buildout of the plant was completed in 2011 with the addition of two heat recovery steam generators that capture waste heat from the combustion turbines, which then drive a single steam turbine generator to produce an additional capacity of 484 MW, raising the total plant capacity to 628 MW. This modification resulted in the transformation of the plant from a peaking facility to a base-loaded combined cycle facility.

One of the design objectives of the plant was to minimize the quantity of water required while eliminating any water discharge offsite. The plant consumes a mere 121 L (32 gal) of water for each megawatt of power produced, whereas a conventional plant may use more than 2.3 m^3/MW (600 gal/MW) of energy produced. The reduction in water demand was principally addressed by the use of air-cooled condenser cells and evaporative coolers.

The combined wastewater from the plant is discharged to a pair of refurbished 10 117 m^2 (2.5 acre) geo-textile, riprap-lined evaporation ponds. The solids produced from the wastewater are recovered by a filter press and sent offsite for disposal to a sanitary landfill. The objective of the plant's water and wastewater treatment systems was to achieve zero liquid discharge (ZLD).

2.3.2 Design Requirements

The water and wastewater requirements of the plant can be summarized as follows:

- Potable water treatment system
- Makeup water treatment system
- Wastewater treatment system

The source of raw water for the plant is from three local wells. The system design is based on the worst possible blend of any of the three wells; this was critical in the design criteria to ensure that the solution would effectively handle the water characteristics during all periods. This water quality was used for the design of both the potable water and the makeup water treatment systems (in this case, the potable water system is of insignificant capacity).

2.3.2.1 Makeup Water System

The makeup water treatment system (MWTS) in Figure 6.4 consists of three 50% trains. The system is divided into a filtration system followed by a polishing system. This system is capable of treating 68 m^3/h (300 gpm) of well water, with the flexibility to also treat a portion of the recovered wastewater. The treated water from this roughing system is channeled to a blend storage tank that also has the ability to store recovered water from the wastewater treatment system. This partially treated and blended water is used for various plant demands, including fire protection. The balance of this stored water is fed to a polishing makeup demineralizer system. The multimedia filter backwash/rinse and reverse osmosis concentrate is sent to the WRRF.

The polishing system consists of three 40-gpm feed second-pass reverse osmosis trains and offsite regenerated mixed bed ion-exchange polishers. The polished water is stored in a demineralized water storage tank. The second-pass reverse osmosis concentrate is returned to the filtered/blend water storage tank, so the polishing system does not produce any wastewater during normal operation.

Because the raw water source was well water, the design of the pretreatment system uses well-proven technologies. The well water analysis reports suspended solids in the range of 3 to 5 mg/L, dictating the need for filtration to reduce the silt density index (SDI) to an acceptable value for reverse osmosis feed. The alternatives for filtration were either multimedia filtration (MMF) or ultrafiltration. Although ultrafiltration would provide a more positive barrier and produce a superior filtrate water quality, it does have some drawbacks given the fact that its wastewaters would require recovery to minimize the volume of waste that would have to be sent to the wastewater treatment system.

The backwash water for the MMF is less than 1%, whereas ultrafiltration would require a backwash water volume of approximately 5%. In addition, the MMF does not require periodic chemical cleanings that would further complicate the wastewater recovery system. Newer high-efficiency media filtration technologies, such as those using microsand, present the potential for even less water use than MMF. Microsand filtration should be considered during future ZLD/MLD projects.

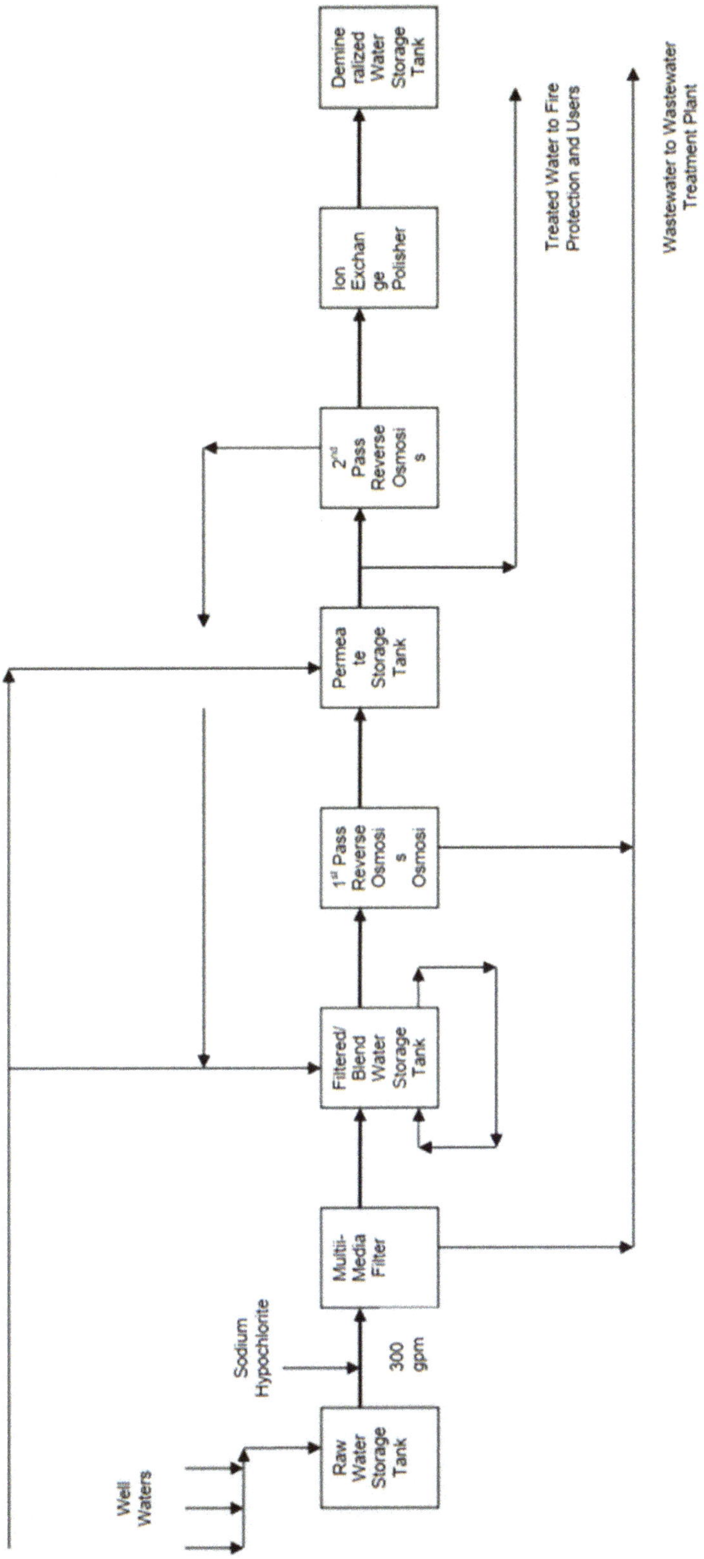

FIGURE 6.4 Makeup water treatment system.

The well water is dosed with sodium hypochlorite to ensure the complete oxidation of any soluble iron. The filtrate from the multimedia filters is stored in a filtered water/blend water storage tank. In addition to the filtered water, this tank will receive concentrate from the downstream makeup second-pass reverse osmosis, permeate from the wastewater recovery reverse osmosis, and potentially some of the wastewater recovery reverse osmosis concentrate.

The filtered water is dosed with an antiscalant, a reducing agent (sodium bisulfite), and processed through 5-μm (nominal) cartridge filters before being sent to the first-pass, roughing reverse osmosis trains. The selection of a two-pass reverse osmosis was chosen to minimize the loading on the downstream polishing demineralizer. Although the pH of the well water would indicate the minimal presence of carbon dioxide, provisions are included to eliminate the presence of any carbon dioxide by the addition of interstage sodium hydroxide.

An interpass permeate storage tank is included to collect and blend the first-pass reverse osmosis permeate and the reclaimed wastewater reverse osmosis permeate. A portion of the permeate stored in this tank is used for various facility uses, including first-pass reverse osmosis flushing and firewater. The balance of the permeate is treated by the second-pass reverse osmosis and then by the offsite regenerated mixed-bed polishers to produce demineralized makeup water.

The concentrate from the first-pass reverse osmosis is sent to the WRRF for treatment. This pass operates at a comfortable recovery of approximately 75%. The relatively low recovery ensures simple, reliable, and trouble-free operation without producing excessive wastewater while its concentrate stream will be recovered by the wastewater treatment system. The second pass operates at a recovery of approximately 85%, with its concentrate being sent to the filtered water/blend tank. This second-pass concentrate will be blended with the filtrate from the multimedia filters and permeate from the wastewater reverse osmosis.

The second-pass reverse osmosis permeate could have been polished by either electrodeionization (EDI) or conventional mixed-bed ion exchange. Electrodeionization has the advantage of eliminating any waste stream; although, this advantage comes with the need to consider the possibility of periodic cleaning. Its concentrate stream could be returned to one of the upstream reverse osmosis permeate storage tanks or the filtrate/blend storage tank. The ion-exchange approach would require offsite regeneration to avoid the production of spent regenerant wastewater. Offsite regeneration also shifts the capital investment and the operation and maintenance costs to the service provider. Finally, the mixed-bed ion exchange simplifies the recycling of the stored, demineralized water to ensure the highest possible water quality is always available. The final decision was to use a mixed-bed ion-exchange solution with offsite regeneration in lieu of EDI.

2.3.2.2 Wastewater Treatment System

The wastewater treatment system (WWTS) in Figure 6.5 is designed to treat the heat recovery steam generator blowdown, the MWTS multimedia filter backwash water, the MWTS first-pass reverse osmosis concentrate, the combustion turbine generator evaporative cooler blowdown, the wet surface air cooler blowdown, the existing service water wastes, the discharge from the oil/water separator, and recycle water from the WWTS after these streams are collected in a wastewater equalization tank. The design capacity of the WWTS is 68 m^3/h (300 gpm). The water processed by this system is either sent to the evaporation ponds or returned to the makeup treatment system for further processing and recovery. The solids produced are dewatered and hauled offsite.

The wastewater that must be treated comes from a multitude of sources, which increases the complexity of any wastewater treatment process that would be used to reduce the volume of wastes and its discharge to the evaporation ponds. The first option for treatment would have been a thermal process, or evaporator. This process has been used in many facilities throughout the western United States, but, because of its typically high cost for both installation and operation, it was not an ideal fit in this facility.

The second option was a design based on the use of reverse osmosis membrane technology to concentrate the wastewater into an acceptable stream that could be sent to the evaporation ponds. The key to the membrane approach is to manage the water chemistry and avoid exceeding the solubility of any inorganic contaminants that could be possible. In addition, it is necessary to reduce particulates, remove undesirable organic constituents, and manage the biofouling that can result in excessive membrane fouling. Finally, it is important to minimize the cleaning frequency of the reverse osmosis that would place an unacceptable burden on the facility's operators. It should be noted although a full CIP membrane cleaning system was provided, the current facility operation uses offsite membrane cleaning. The reverse osmosis membrane approach was selected for this facility as the most cost-effective design and one that could be operated and maintained without placing a burden on the operating personnel.

As with all reverse osmosis membranes found in power plants, the performance of the reverse osmosis system is a function of the pretreatment system's ability to minimize membrane fouling. The variety and the variability of the wastewaters for this application would have been a challenge for any pretreatment system. Two options were evaluated for the reverse osmosis pretreatment system. The first option consisted of a conventional solids contact clarifier for the reduction in sparingly soluble salts followed by media filtration. The use of a solids contact clarifier is commonly used in

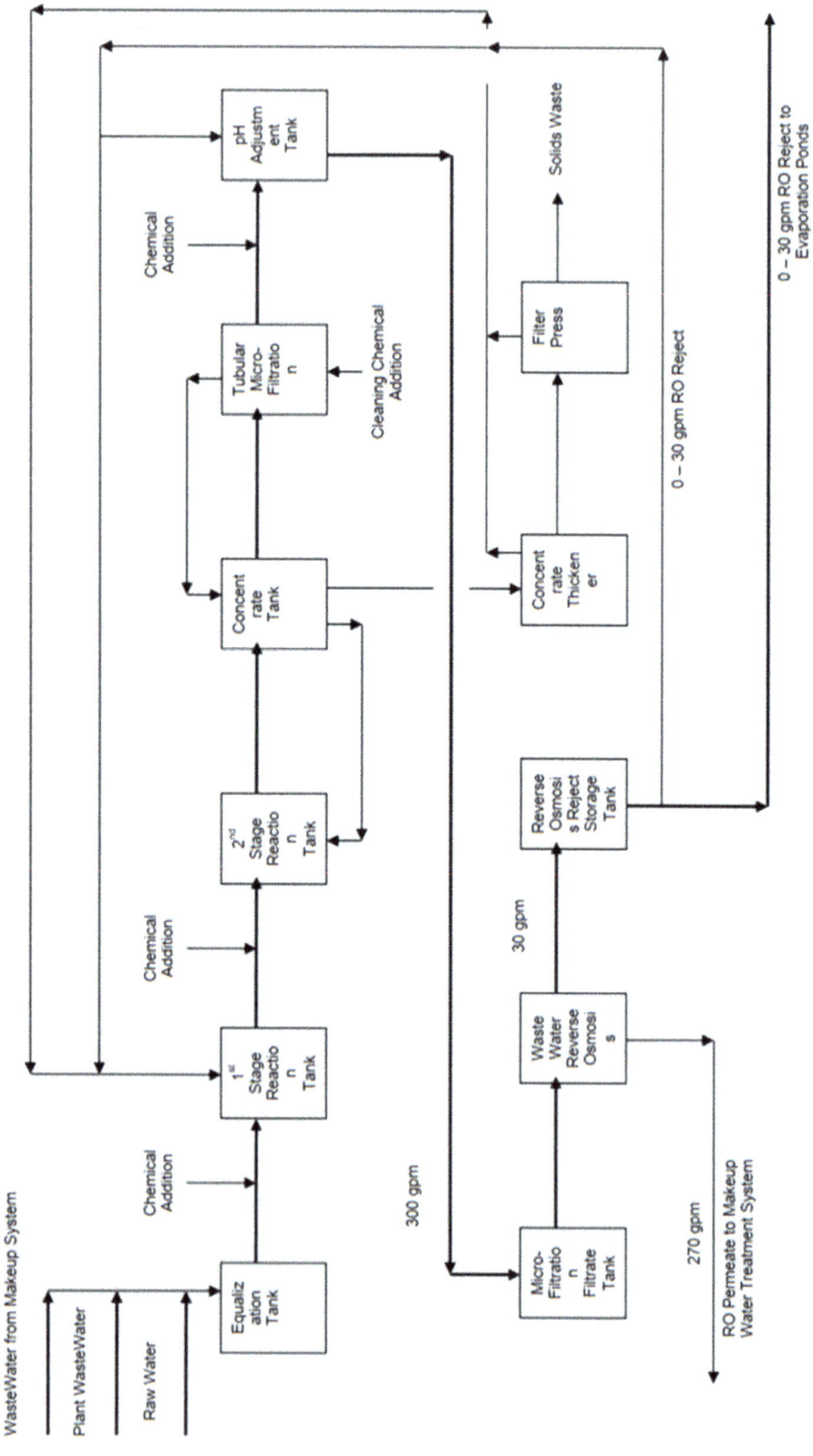

FIGURE 6.5 Wastewater treatment system.

power plants as a standalone cooling tower blowdown minimization system or in combination with media filters. It can also be used as pretreatment in a thermal ZLD system to reduce the size of the evaporator.

The second option involved chemical precipitation in combination with a crossflow tubular microfiltration system. As with a solids contact clarifier, this design uses chemical precipitation to reduce the salts and concentration of chemical species that can limit the recovery of the wastewater reverse osmosis. The unique feature of this approach is the use of a crossflow tubular microfiltration system to enhance the reduction in the unwanted salts while providing a physical barrier to reduce the level of suspended solids that cannot be achieved with a media filter.

In essence, the slurry crossflow tubular microfiltration system replaces the function of a typical precipitator/clarifier/filter system into a single unit operation. The chemically pretreated wastewater flows through the tubular modules at a high crossflow velocity sufficient to maintain turbulent flow. This design is a pressure-driven process, with typical operating pressures in the range of 40 to 60 psi (2.7 to 4.1 atm). The fluid is forced through the membrane with a pore size in the range of 0.1 to 0.3 μm that is cast onto the inside of a tubular porous material. The turbulent flow prevents the buildup of particles on the inner surface of the modules while the filtrate is forced through the pores of the membrane. The design incorporates a periodic back-pulse with air to prevent solids accumulation on the surface of the membrane and extend the service cycle before a chemical CIP is required. This turbulent crossflow and tubular design also eliminates the need for prefiltration with the ability to handle solids levels up to 5% by weight.

Experience indicates this type of design is particularly beneficial when treating a mixture of waters. Chemical dispersants/soaps routinely found in cooling water interfere with critical coagulation, flocculation, and settling processes in a conventional precipitator/clarifier. Consistency of both the precipitation and solids removal processes is more effective when both processes occur within the microfiltration process.

Perhaps the key consideration in the decision to select the crossflow tubular microfiltration system over a clarifier was its ability to deal with the variability of the wastewater that was to be treated. Although some of the streams to the WRRF are relatively constant in flow and contaminants, and an equalization tank is used to moderate the variability, it would be impossible to maintain a constant flow and feedwater analysis. One of the traits of a solids contact clarifier is the fact it is a constant rate process. Variations in temperature, flowrate, and salinity can result in an upset in the sludge blanket and the settling velocity of the solids unless the changes are gradual and controlled, which, in this instance, was not possible.

In addition, there can be carryover of polymers that are typically used to enhance the performance of the clarifier. Most polymers will result in severe membrane fouling of either low-pressure filtration or high-pressure reverse osmosis membranes. Operation of the crossflow microfiltration, however, does not require the use of a polymer. The use of an equalization tank attempts to level out such variations in operating parameters, but it is not possible to provide a tank sized large enough to minimize the effect on the performance of a clarifier. Such variations will result in excessive solids carryover from the solids contact clarifier because the suspended matter will not be able to settle. This carryover can produce excessive loading on the downstream filtration system and the resultant short service runs.

Variations in the raw water analysis require corresponding modifications in chemical addition to ensure the required precipitation of the undesirable salts. This includes salts that are not only precipitated, but those that are adsorbed on precipitated material, such as silica. With the elimination of sparingly insoluble salts from the process, the limiting factor of the downstream reverse osmosis system becomes the level of total dissolved solids (TDS) in the concentrate stream. The use of seawater membranes in the reverse osmosis system will permit a reduction in the wastewater volume corresponding to the limitation of the osmotic pressure, which would be a concentrate steam that contains TDS in the range of approximately 80 000 mg/L. This would equate to an overall recovery of approximately 98% to 99%, assuming a feedwater TDS of approximately 1000 mg/L.

The WWTS begins with an equalization tank that blends the multimedia filter backwash water, the first-pass reverse osmosis reject from the MWTS, the raw water, and the various facility waste streams. The effluent from the equalization tank is pumped to the two-stage reaction tanks, where polymer addition occurs. The polymer is specific to the precipitation/adsorption process. This process relies on hardness reduction by precipitation and silica reduction by the adsorption of silica on magnesium hydroxide precipitate. Although the design feedwater contained ample magnesium for this purpose, the decision was made to incorporate the ability to add a magnesium salt to ensure the required silica reduction could be produced.

Sodium hypochlorite and hydrochloric acid are added to the first reaction tank for pH control and to oxidize any incoming organic matter to prevent potential membrane fouling. Provisions are also included to feed ferric chloride to act as a coagulant, which can often be helpful in minimizing the frequency of membrane cleaning. The adjusted water flows to the second-stage reaction tank where soda ash and caustic soda are added to precipitate calcium carbonate and magnesium as magnesium hydroxide.

The effluent from the two-stage reaction tanks flows by gravity to a microfiltration feed/concentration tank. This tank acts as a pump suction tank, a level control device, and a slurry recirculation tank for the downstream microfiltration. By increasing the slurry concentration of the precipitated materials, it enhances the reduction of the silica by increasing the concentrations of the solids that are used to remove certain species such as silica. The concentration of this slurry is in the range of 2% to 5% solids. Excess sludge is intermittently bled from the system by pumping some of the sludge from the concentrate tank to a thickener.

Each crossflow tubular microfiltration train consists of a feed/recirculation pump, membrane modules, a cleaning pump, and a chemical cleaning tank and water rinse tanks. The flow is controlled to each module by the level in the concentrate tank. If this level drops below a given setpoint, the flow to the first-stage reaction tank is increased. When the flow to a microfiltration train drops below the design rate that was established to process all the wastewater, the train is removed from service and cleaned. There are four 33% microfiltration trains. The trains use 10 tubular polyvinylidene fluoride membranes with a pore size in the range of 0.1 μm (nominal). Each module is approximately 3 cm (1 in.) diameter and 1.8 m (72 in.) long, and there are four modules per housing. There are a total of 36 20-cm (8-in.) diameter and 3-m (10-ft) long polyvinyl chloride housings per train. The microfiltration membranes are typically rated to operate at a flux of approximately 510 $L/m^2 \cdot h$ (300 gal/sq ft/d) for this application. The actual operating flux will depend on the nature of the precipitated materials, the particulates, and other impurities combined with the effectiveness of the cleaning regime.

A redundant microfiltration train is provided to ensure the design flowrate is maintained when a train is taken offline for module cleaning. The automated design cleaning frequency for each microfiltration train is approximately 3 days. When the microfiltration is cleaned, the feed/recirculation pump is shut down and the microfiltration is flushed to displace the slurry back to the feed tank. For this application, cleaning is conducted with hydrochloric acid and occasionally sodium hypochlorite. This is accomplished by recirculating the cleaning solution back to the cleaning tank for approximately 30 to 60 minutes. After the cleaning cycle is completed, the cleaning chemicals will be reclaimed by feeding them back to the concentrate feed tank.

The solids that are sent to the sludge thickener are processed by a filter press that produces a cake of approximately 50% solids. The filtrate from the press along with the overflow from the thickener is returned to the first reaction tank for reprocessing. The filtrate from the microfiltration is pH adjusted with hydrochloric acid in a pH-adjustment tank for optimum pH for the downstream reverse osmosis and then flow by gravity to a microfiltration filtrate tank. The filtrate from this tank is pumped to the reverse osmosis units. Chemical addition ahead of reverse osmosis consists of sodium bisulfite for the removal of any residual free chlorine that could be present in the filtrate as a result of the feed to the reaction tanks or the microfiltration cleaning process. A nonoxidizing biocide can also be added to retard biological activity. It should be noted that an antiscalant is not required or included because the sparingly insoluble salts are all kept well below their saturation levels by the upstream process.

The reverse osmosis concentrate system consists of five three-stage trains designed to treat a feedwater capacity of 68 m^3/h (300 gpm). The system is designed to operate at an overall recovery of 90%, with 61 m^3/h (270 gpm) of permeate being reclaimed by returning this stream to the makeup treatment system to either the filtrate tank for further reprocessing or to the first-pass permeate storage tank for facility use, including the wet surface air cooler. As noted, the design recovery of reverse osmosis is 90%. However, because a portion of the concentrate stream can be returned upstream to either the reaction tanks or the pH correction tanks for reprocessing, it is possible to operate the reverse osmosis at a comfortable recovery of 75%, producing a total concentrate flowrate of 17 m^3/h (75 gpm) while still maintaining the overall system recovery at a minimum of 90%.

The 17 m^3/h (75 gpm) of concentrate is routed to a reverse osmosis reject holding tank where it can be split into three streams. The reverse osmosis concentrate is either directed to the evaporation ponds, the first-stage reaction tank, and/or the pH adjustment tank. The exact ratios depend on the actual concentrate water quality. There are conditions under which the reverse osmosis reject's water quality is such that most of the reject can be recycled to the makeup facility, increasing the recovery above the 90% value. To date, a recovery of up to 98% has been achieved under certain operating conditions.

2.3.3 Operation and Performance

A well-trained and competent operator is critical to the reliable and efficient performance of any water or WWTS. This is especially true when a wastewater stream varies in composition, temperature, and flowrate, as is the case for a reverse osmosis system that is designed to optimize recovery. Although the various unit operations are primarily pressure-driven membrane processes, the correct chemistry must be maintained to ensure the necessary operating parameters are maintained and that the required membrane cleanings are performed when required. Based on the design data presented, the key constituents that must be monitored and controlled to permit the required reverse osmosis recovery at a minimum of 90% are total hardness and silica. The microfiltration system's primary function is to reduce these impurities and any other constituents that could limit recovery if they are concentrated in the reverse osmosis and permitted to result in scaling. Naturally, the microfiltration system must also ensure the SDI of the reverse osmosis feedwater is below 3.0.

To ensure the system is in chemical balance, operators monitor the following key parameters of the microfiltration filtrate to determine if any adjustments to the chemical addition are required:

- silica,
- total hardness,
- alkalinity, and
- pH.

Because the raw water can contain a wide variety of contaminants because of the various wastewater streams that are fed to the equalization tank, provisions were incorporated to the design to use a wide variety of chemicals to precipitate the hardness, silica, and any other constituents before microfiltration. It has not been necessary, however, to feed each of the available chemicals at all times. The filtrate is then pH adjusted and dosed with a biocide to further condition the filtrate for feed to the reverse osmosis. The chemicals required for the microfiltration system include

- sodium hypochlorite or hydrochloric acid,
- sodium hydroxide,
- magnesium sulfate or sodium carbonate, and
- ferric chloride or calcium chloride.

In addition to the preceding chemicals, the following additional chemicals are used for operation of microfiltration and reverse osmosis: sodium hypochlorite, sodium bisulfate, antiscalant, biocide. It should be noted that the reverse osmosis elements are cleaned offsite in a service center while a portion of the microfiltration cleaning chemicals are sent to the evaporation ponds for final disposal.

Based on the composition of the wastewater that has been treated to date, chemical use has primarily consisted of sodium hydroxide for hardness reduction and hydrochloric acid for pH adjustment. Given this operating condition, the actual operating flux of the membrane is substantially higher than the design flux, permitting fewer online units during normal operation and extending the period of time between microfiltration membrane cleanings.

The operator has a great deal of flexibility in determining the timing for microfiltration system cleaning. Unlike other membrane processes that rely on transmembrane pressure loss or flux decline to initiate a cleaning, the cleaning of the microfiltration system is based on the ability of the microfiltration system to process a specific wastewater flowrate. If the level in the equalization tank begins to rise to an unacceptable level because of a limit in the downstream microfiltration processing capacity, a microfiltration membrane system is taken offline and replaced with a standby microfiltration membrane system, thereby increasing the wastewater treatment processing rate.

Because the maximum operating flux for easy-to-treat wastewaters can be can be up to 1020 L/m^2·h (600 GFD), the operator has a greater flexibility in deciding when to perform a cleaning of a microfiltration system. For example, although each of the microfiltration trains were designed to operate at a capacity of 23 m^3/h (100 gpm), the trains have been operated at flowrates up to approximately 57 m^3/h (250 gpm). The operator may also find it convenient to schedule microfiltration cleanings based on the total production capacity or use the calendar for initiation. To date, an acid cleaning using hydrochloric acid has been effective in returning the membrane to its required flux, which accounts for the higher than anticipated flowrates.

As is the case for all reverse osmosis systems, the performance of the wastewater reverse osmosis concentrator is a function of the quality of the feedwater produced by the pretreatment system, which, in this case, is the microfiltration membrane system. Since being placed into operation around February 2011, the wastewater reverse osmosis system has been cleaned only one time. Reverse osmosis is the governing unit process that determines if the required overall recovery of 90% can be achieved. The system has consistently exceeded this objective by returning a portion of its concentrate/ reject stream to the first-stage reaction tank.

After the initial commissioning of the system, the average recovery of the system has been approximately 98%. The high system recovery has been possible using seawater reverse osmosis membranes in the wastewater reverse osmosis, which operate at pressures up to approximately atm 34 atm (500 psig). The wastewater reverse osmosis produces a permeate water quality with a TDS concentration of approximately 25 mg/L. Because this relatively low TDS water quality is returned to the makeup demineralizer system and further polished by the makeup reverse osmosis, the reduced loading on the service ion-exchange demineralizers has extended their service cycle. Currently, the regeneration frequency for the ion-exchange units is in the range of 6 to 8 weeks. This also reduces the raw water demand on the well water that supports the facility overall.

2.3.4 Conclusions

Integrated water and wastewater systems have been operating for several years. After a normal commissioning and startup period, the required objective of producing the required makeup water and treating a variety of wastewaters, the facility is continuously exceeding the primary goal of minimizing the wastewater discharge to the evaporation ponds by a maximum of 90% for design. In fact, the discharge is often consistently in the range of

5% of design. Although the wastewater to date is less demanding than the original design analysis, it should be recognized that the demands on the system because of start-stop operation and the variable feedwater create a difficult operating environment. This would be especially true if the system had included a conventional softener/clarifier and filtration process in lieu of the installed slurry crossflow tubular microfiltration system. The guaranteed performance by the equipment vendor has been validated not only by the minimal discharge to the evaporation ponds, but also the trouble-free performance of the reverse osmosis wastewater concentrator.

The successful performance of this integrated membrane makeup and ZLD system is based on the following considerations:

- A complete profile of the various raw water and wastewater streams that must be treated. This information must not only include the design basis for the system, but also the actual water quality variations that can occur during the facility's operation.
- An understanding of the water and wastewater treatment chemistries required to produce the desired results. The overall recovery of the system is ultimately based on the control of the solubility of the potential compounds that can be produced within the system.
- An understanding of the basic unit processes required to achieve the specified objectives. The system should be flexible enough to deal with reasonable changes in the design criteria, which will occur in all water facilities and WRRFs.
- A trained operating staff committed to operating and maintaining the system. Their success will ultimately be a function of the data collection, analysis, and ability to implement periodic changes in operation of the system. This also includes their ability to address periodic performance shortcomings as a result of unanticipated changes.

3.0 MINIMUM LIQUID DISCHARGE TECHNOLOGY APPLICATIONS—OIL AND GAS EXPLORATION AND PRODUCTION

3.1 Background

Energy is needed daily to sustain modern human civilization. The primary sources of energy in the United States are from nonrenewable sources such as petroleum and coal with limited contribution from renewable energy sources (e.g., solar, wind). In addition to being a decreasing resource, energy production from nonrenewable sources pollutes the environment and affects the U.S. carbon footprint significantly. Nuclear energy has limited environmental impact, but it can result in catastrophic disasters. The incidents at Three Mile Island (United States), Chernobyl (Russia), and, most recently, Fukushima (Japan), have exposed the inherent safety limitations of nuclear power generation plants and created doubts in the minds of policymakers about the future of nuclear energy for new generation power plants. Although the probability of such incidents is limited, their effects are catastrophic, long lasting, and widespread.

Some renewable energy resources are under development and some forms (i.e., solar, wind, geothermal, biomass, etc.) are fully developed now, both technologically and commercially; however, their share of contribution is small with respect to the total demand for energy because of their efficiency, cost, and maintenance. Moreover, some of these renewable energy resources, especially solar and wind energy, need backup systems to offset and provide baseload because they are heavily dependent on weather conditions.

Against this backdrop, a quiet revolution is taking place in the arena of energy resources in the United States. Advances in hydrofracturing technology have allowed the production of oil and natural gas from tight shale plays that were previously inaccessible to the conventional methods used to produce from porous formations.

The recent surge in unconventional production by hydrofracturing has produced a shift in the potential oil and gas resources available as well as the water management challenges inherent in production. Unconventional production has multiple benefits, including the following:

- It produces clean burning fuels, with reduced greenhouse gas emissions per unit of energy production relative to coal.
- It helps control energy prices because of the abundant supply in the United States.
- It helps the local economy and job creation.
- It also helps the United States by decreasing the need for foreign oil dependence; thus, the United States has become a net exporter of natural gas as opposed to being an importer for a long time.

Conventional or traditional oil and gas production involves a vertical well drilled into a porous formation from which hydrocarbons are extracted and brought to the surface. The fluid that is produced from the well contains a mixture of hydrocarbons, water, and other constituents (i.e., sand, drilling fluids, etc.). This water is termed "produced water." The produced water from conventional production is often reused onsite.

Unconventional production involves horizontal directional drilling with the technique of hydrofracturing, also called hydrofracturing or "fracking" used to fracture the tight shale formation to extract the hydrocarbons. Unconventional production also involves water, although in two distinct categories: flowback water and produced water, which are discussed in more detail in Section 3.6. The location of the oil and gas wells also affects the water management approaches applied. Offshore operations are able to discharge produced water directly back into the waterbody in which they are operating, whereas onshore operations have more complex management and discharge challenges. This section primarily focuses on onshore unconventional production because (a) the industry is moving toward unconventional production, and (b) there is a greater magnitude of challenges and subsequent opportunities associated with the produced water from these operations.

3.2 Production Technology—Unconventional Hydrofracturing and Horizontal Drilling

Conventional reserves of oil and gas in the United States are dwindling. The recent surge of oil and gas production has been possible because of the availability of light tight oil and shale gas from various unconventional sources. Tight gas and shale gas reservoirs are much deeper than coal bed methane and conventional oil and gas reservoirs, as Figure 6.6 shows (U.S. Energy Information Administration [EIA], 2011). Low permeability and large depths of these formations historically made these resources rather inaccessible.

However, exploration in difficult terrains at depths in the range of 1.5 to 5.5 km (5000 to 18 000 ft) has been made possible by hydrofracturing technology. In principle, this involves application of high-pressurized water to fracture the rocks to release the oil and gas trapped in its tiny pores. Hydrofracturing, in combination with horizontal drilling, has enabled one single well to extend horizontally to a radius greater than 4.5 km (15 000 ft) in the proximity of the production well and has made drilling for hydrofracturing a cost-effective operation. Contribution of shale oil and gas from unconventional resources with respect to the total U.S. production has risen steeply over the last decade A large part is driven by the shale plays in West Texas. Figure 6.7 shows the distribution of shale oil and gas basins throughout the United States (U.S. EIA, 2016).

3.3 Water Streams in Hydrofracturing Operation

Injection of water at high pressure and high flowrates is critical to the success of hydrofracturing operations. High pressure is essential to overcome

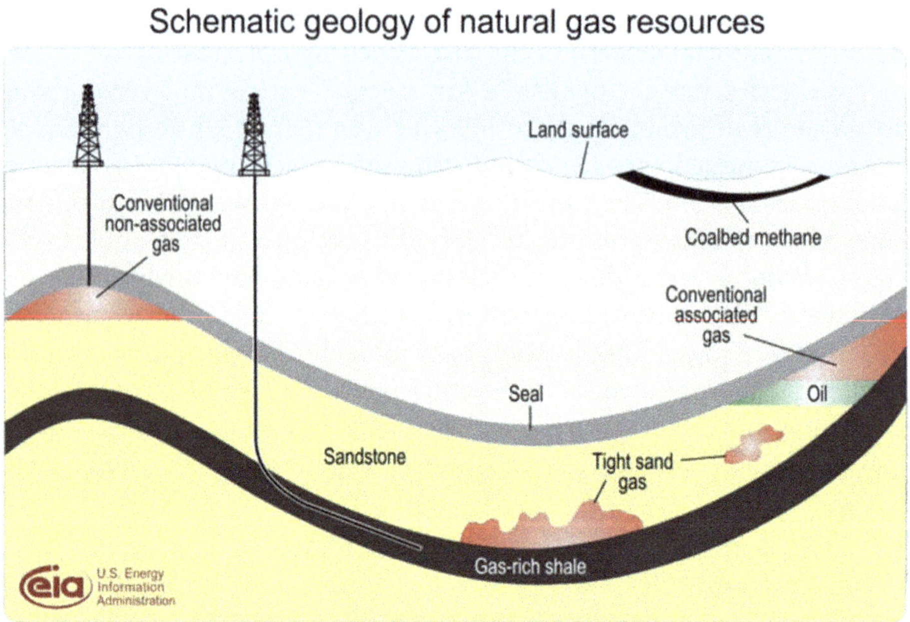

FIGURE 6.6 Schematic geology of natural gas resources (U.S. EIA, 2011).

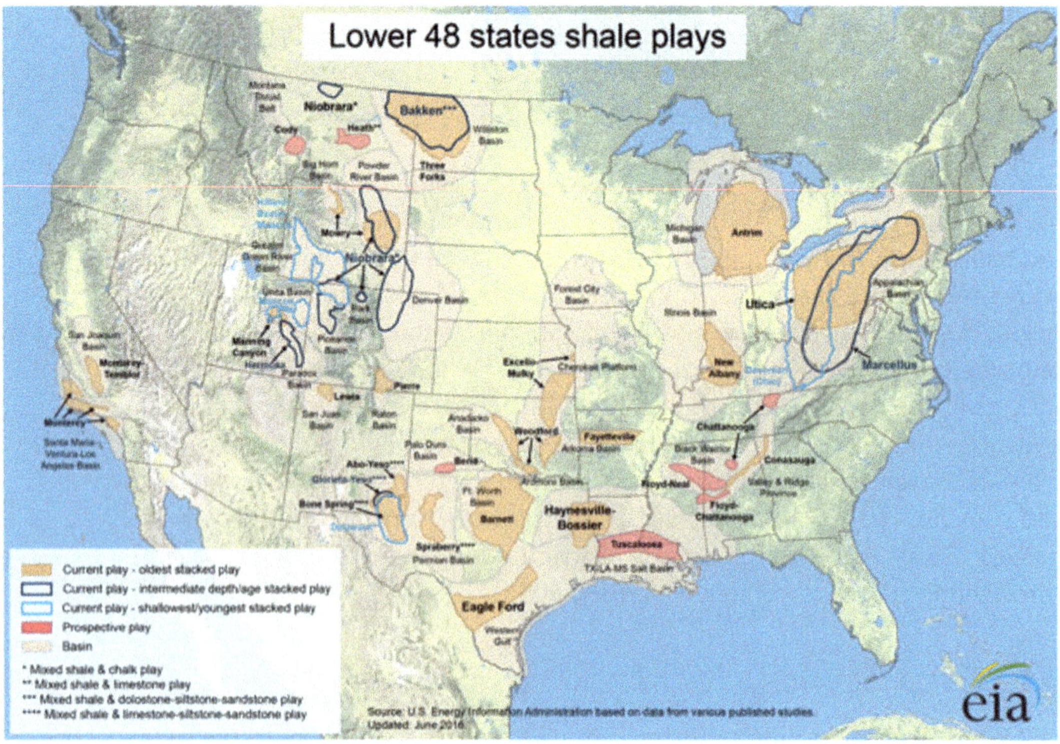

FIGURE 6.7 Shale plays in the lower 48 states (U.S. EIA, 2016).

the bottom hole pressure, whereas high flow is required to fracture the rocks by shearing mechanism that opens up the pores.

Typically, hydrofracturing operations on a single well create a water demand of approximately 50 000 bbl, or 8000 m^3 (2 mil. gal) per week. Each well requires 2 to 3 weeks of fracturing to become fully productive, which amounts to a water demand anywhere between 100 000 and 150 000 bbl/well (15 000 to 23 000 m^3, or 4 to 6 mil. gal/well). However, once the initial hydrofracturing is completed, often in multiple stages, the well remains productive for at least 5 to 7 years until the production declines significantly. Typically, a refracturing operation is conducted after the significant decline in oil production.

Oil and gas explorations are always associated with production of water. In cases of exploration from conventional sources, water is produced continuously during the production process. This produced water is reinjected to the porous formation as part of the water flooding technique used to push the hydrocarbons toward the well head to maximize extraction to the surface. However, in cases of unconventional oil and gas exploration, initially a part of the hydrofracturing water returns with the gas as flowback water.

Typically, water flows back only at the rate of 25% to 35% of the water introduced as fracturing water, which translates into anywhere between 3800 and 15 000 m^3 (1 and 4 mil. gal) of water/well for the first few weeks of operation. This is followed by a continuous production of water through the lifetime of the well, which is called "produced water." Produced water originates from the shale formation layers. Thus, flowback water is generated during the work of the well completion team, which typically lasts a few days to weeks. On the other hand, produced water flows when the production team takes over the wells. Figure 6.8 shows the typical flow profiles of hydrofractured water injection and flowback and produced water production (Sun et al., 2018).

3.4 Hydrofracturing Water Resources and Produced Water Disposal

Hydrofracturing operation is conducted by introducing "Frac" water at high pressure and high rates underground to shear the shale structures and free up the oil and gas trapped in the fine pores. Production starts at a shale play with only a few wells, but it develops into several thousands of wells in the course of its development, typically in only 2 to 3 years. Because hydrofracturing of each well requires large amounts of water, sourcing of water is a significant concern. Traditionally, operators have used surface water and groundwater from within or outside the regions of exploration. However, there are two significant disadvantages to these practices. First, bringing water from outside sources requires trucking. Not only is this an

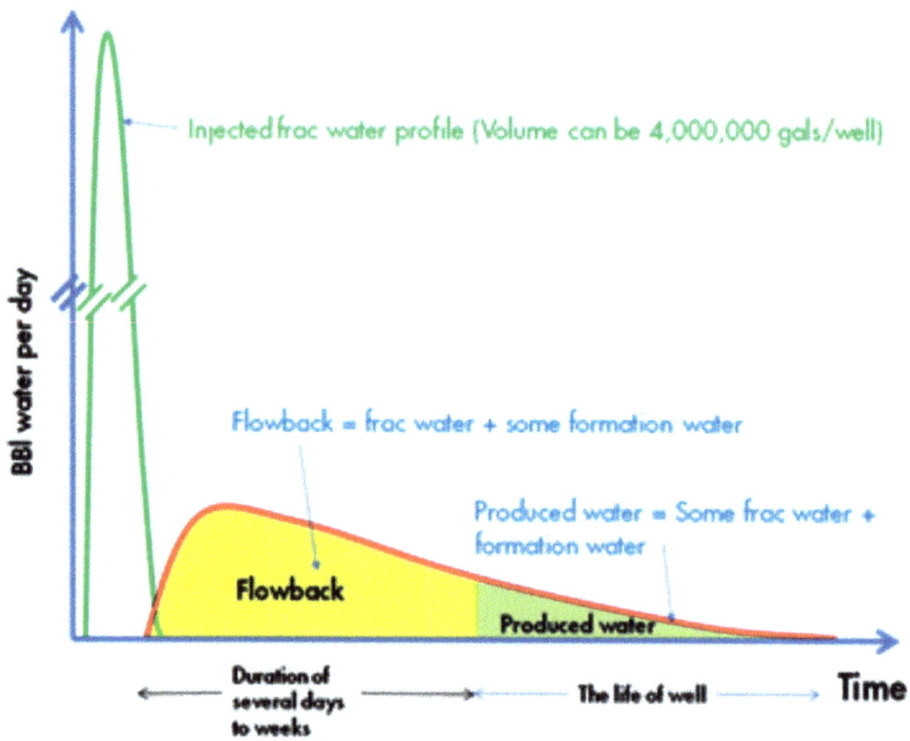

FIGURE 6.8 Profiles of hydrofractured water, flowback water, and produced water flows with time.

expensive operation, but it may also create several other problems over time. When the number of wells grows to the full capacity of the basin, this would significantly increase the truck traffic. The resulting high volume of traffic for long periods of time creates a heavy burden on the existing infrastructure of all the neighboring communities. Secondly, sustainability of truck transportation of fresh water in and produced water out is also questionable from an environmental standpoint as it relates to air emissions of particulate matters and volatile organic carbon.

Availability of trucks and drivers is another important consideration during the planning stages of undertaking exploratory activities in a shale play. Several basins throughout North America are undergoing rapid development simultaneously. Given this high level of activity, it is becoming difficult for the operators to receive a continuous supply of trucks and drivers for prolonged periods of time. Secondly, even when surface or groundwater sources are available within a region of exploration, their use for hydrofracturing is often not sustainable for a long time because of the water demand versus rapidly depleting water resources that may be needed for current and future human consumption or agricultural purposes. This raises a concern for using fresh water for industrial activities, including hydrofracturing. Moreover, produced water also needs disposal and is often transported offsite to saltwater disposal wells to be injected to deep, brackish formations as produced water contains high concentrations of dissolved salts and other chemical constituents.

Trucking large quantities of fresh water for hydrofracturing and produced water to disposal sites not only creates logistics, health and safety, and environmental problems, but it is also expensive. Thus, recycling of produced water is receiving increasing national attention and many states are considering mandatory recycling for the development of existing and new exploration sites. However, the quality and quantity of produced water may not be adequate for reuse. Therefore, this water may need proper treatment to render it fit for reuse. A good understanding of the chemistry and characteristics of produced water is a prerequisite for making a sound decision on its effective treatment.

3.5 Chemical Characteristics of Hydrofractured Water, Flowback, and Produced Water

Several chemicals are added to water to make it suitable for hydrofracturing operation. Table 6.6 (U.S. Department of Energy [U.S. DOE], 2009) presents a typical composition of some of the more commonly used chemicals, their compositions and functions. The total amount of additives ranges from 0.5% to 2.0% on a volume/volume basis.

There are two types of hydrofracturing fluids widely used: gel-based and slick water-based. The composition given in Table 6.6 represents that for a gel-based hydrofracturing fluid, the main ingredient is cross-linked guar gum. Here, calcium or magnesium borate salts are used to cross-link large polysaccharide chains to form large polymers of molecular weight as high as several thousand kilodaltons (kDa). This creates a high-viscosity fluid that supports proppant transport and ensures uniform distribution of the proppant in the rock structure. However, the downside of a gel-based hydrofractured fluid is the high energy cost of pumping caused by a high pressure drop in the transmission line.

The current trend is to use slick water systems-based hydrofractured fluids with significant reduction in friction by the addition of the chemicals to reduce the energy for pumping large quantities of water into the shale formation. However, these chemicals and the dosage rates are changing with time as increasingly benign and "green" chemicals are being introduced to the market to address any potential health hazards that could be associated with hydrofracturing operation. Also, both gel-based and slick water-based hydrofractured fluids have their own applications depending on the rock structure and specific site.

TABLE 6.6 Hydrofracturing water additives (U.S. Department of Energy, 2009).

Main Compound(s)	Volumetric Composition (%)	Additive Type	Purpose
Hydrochloric acid	0.123	Diluted acid (15%)	Dissolves minerals to initiate fractures
Glutaraldehyde	0.001	Biocide	Controls bacteria (corrosive)
Ammonium persulfate	0.010	Emulsion breaker	Allows a delayed breakdown of gel polymer
N,N-dimethyl formamide	0.002	Corrosion inhibitor	Prevents pipe corrosion
Borate salts	0.007	Cross-linker for gel	Increases molecular weights of gel polymers by cross-linking, which increases elasticity and suspension properties
Polyacrylamide/mineral oil	0.088	Friction reducer	Minimizes friction between fluid, pipe, and formation
Guar gum, hydroxyethyl cellulose	0.056	Gel polymers	Thickens the water to suspend the proppant, typically sand
Citric acid	0.004	Iron control	Helps control precipitation of metals
Potassium chloride	0.060	KCl Brine carrier fluid	Stabilizes clays in the shale formation (reduce fines)
Ammonium bisulfite	0.002	Oxygen scavenger	Removes oxygen to help prevent bacterial corrosion
Sodium, or potassium carbonate	0.011	pH adjusting agent	Maintains effectiveness of materials such as cross-linkers
Silica, quartz sand	0.500	Proppant	Remains in fractures to prop open providing gas pathway
Ethylene glycol	0.043	Scale inhibitor	Helps prevent scale deposition and a lubricant
Isopropanol	0.085	Cosurfactant	Helps viscosity fluid
Lauryl sulfate	0.001	Surfactants	Helps carry proppant, enhances water injection rate, cleans fractures, reduces friction, and breaks oil water emulsions
Water	99.007	Aqueous base	Carrier fluid

As flowback water is generated from hydrofracturing operation, it may contain some of the components contained in that hydrofractured water, although much of it is adsorbed in the formation rock underground. A part of the proppant, which is most commonly sand particles, return and exhibit as TSS in flowback water. The characteristics of produced water at a shale play vary considerably with time, and even from well to well, depending on the geologic formation of the reservoir and its mineral matter content that the water has been in contact with. Typically, produced water contains high levels of TDS in the forms of hardness; cations of various metals, including high concentrations of sodium ions; and various anions, including high concentrations of chloride and sulfate ions.

Apart from sodium, the other commonly encountered cations are those of calcium, magnesium, iron, boron, barium, strontium, arsenic, etc. Typically, produced water also contains dissolved hydrogen sulfide. Wells that produce both gas and liquid products will have hydrocarbon liquids associated with produced water. Blauch et al. (2009) conducted a comprehensive study thorough chemical analyses of flowback water from various locations in the Marcellus Shale in southwestern Pennsylvania as summarized in Table 6.7.

The concentrations of all constituent ions, except for those of sulfate, increase as the flowback water volume increases. This may be attributable to two possible reasons: (a) capture of sulfate by calcium ions and (b) biogenic reduction of sulfate to sulfide ions under the anaerobic condition prevailing at the large depths of shale gas reserves. Part of the sulfide ions may also be immobilized in place by ferrous ions generated under the reducing conditions. The ratio of ferrous iron to total iron increased with progress of flowback water volume (see Table 6.7), which confirms the gradual conversion of iron to the ferrous ions, followed by precipitation in situ as ferrous sulfide (FeS).

The pH and alkalinity show a decreasing trend because of the increasing acidity caused by dissolution of carbon dioxide produced from anaerobic biological activity in situ. This gradual drop of pH in turn results in increased dissolution of ionic species in the flowback water. Sodium and calcium ions are dominant in all samples. The steep rise in concentration of strontium ions raises the possibility of the formation of strontium sulfate, which forms a hard scale on pipe and equipment surfaces that is difficult to remove.

At Marcellus, Antrim, and several other shale plays, naturally occurring radioactive materials (NORM; e.g., radium and uranium ions) are commonly associated with produced water as a result of the geologic formation of the basin. The most common radioactive constituents are Radium 226 and Radium 228. These are present in the form of Ra^{2+} cations. Radium

TABLE 6.7 Typical flowback water composition at Marcellus Shale (Blauch et al., 2009).

	Flowback water volume (bbl)			
Concentration	**12 000**	**13 000**	**14 000**	**16 000**
pH	6.22	6.08	5.98	5.88
Alkalinity (bicarbonate only) as mg/L of $CaCO_3$	280	240	200	160
Total hardness as mg/L of $CaCO_3$	19 300	21 100	24 200	27 600
TSS, mg/L	144	175	498	502
Chloride, mg/L	54 000	59 000	62 000	67 800
Sulfate, mg/L	31	20	20	24
Sodium, mg/L	26 220	28 630	31 810	35 350
Potassium, mg/L	1119	1201	1350	1480
Calcium, mg/L	7160	7680	8880	9720
Magnesium, mg/L	341	463	488	805
Barium, mg/L	28.9	43.3	99.6	175.7
Strontium, mg/L	1110	1305	1,513	1837
Iron (ferrous), mg/L	0.4	0.9	1.1	3.3
Iron (total), mg/L	63	66	72	78
Specific gravity	1.065	1.068	1.077	1.087
Conductivity, μmhos	133 100	141 500	157 600	173 200
Microbial content	Low	Low	Low	Low
Langelier Saturation Index	1.02	0.84	0.72	0.55

ions are mobile in water and present in concentrations from 0 to 1000 pico curie/gram (pCi/gm). This is not enough to be regulated. However, if a produced water is scale forming, NORM partitions into barite (barium sulfate) scale by co-precipitation in the form of a complex salt, as follows:

$$Ba^{2+} + Ra^{2+} + 2\ SO_4^{2-} \rightarrow Ba\ (Ra)(SO_4)_2 \qquad (6.1)$$

As opposed to calcium carbonate, the solubility of which increases with decreasing temperature, the solubility of barium sulfate decreases with decreasing temperature. The significant drop in pressure and temperature during the gas production process facilitates scaling potential of barium sulfate over calcium carbonate or sulfate in the course of its travel from the reservoir to the surface. The NORM accumulation rate in produced water scale in the United States is estimated at 300 000 to 1 000 000 tons/year. For scales containing radiation levels >2000 pCi/gm, the rate drops to 15 000 to 50 000 tons/year (Tomson et al., 2003). Barite has low solubility and is hard to remove without cutting. Handling and disposal of materials with radioactive scale is a problem that requires careful consideration. It is important to note that the complexities and challenges of produced water with NORM accumulation eliminate it for consideration for potential treatment and reuse.

Bank (2011) reported the presence of uranium in the concentration range of 8 to 53 mg/L in produced water samples in Marcellus Shale. The samples were observed to be rich in barium, uranium, chromium, and zinc. Uranium was observed to be closely associated with total organic carbon (TOC) of the samples. This may be explained by two possible mechanisms: (a) dissolution of uranium ions in water resulting from the acidic condition in the shale as a result of anaerobic activity and (b) formation of water-soluble complex ions in association with organic compounds responsible for elevated TOC, which ranged from 2.4% to 13.7% in the water samples.

At several shale plays (e.g., at Bakken and in parts of Eagle Ford and Permian Basin), gas is associated with liquid products. Produced water at those locations contains organic liquids that contribute to high levels of TOC. The typical range of TOC of produced water is 100 to 2000 mg/L (EFX Systems, Inc., 2000).

High pressure associated with low temperature (particularly in offshore fields) can cause formation of slugs of gas hydrate in the transmission line. These are gas bubbles trapped within crystals of ice. Slugs deposit on the internal surfaces of transmission pipes and cause obstruction to flow. In such cases, monoethylene glycol (MEG) and methanol are dosed in the well and pipes to prevent slugging in the pipe and ensure the flow of gas. The function of these alcohols is to depress the freezing point of water so that it remains in liquid form and does not create slugs, even under low temperatures. In the gas–liquid separation process aboveground, these chemicals remain dissolved in water, raising the TOC levels of produced water.

Methanol is dosed only intermittently during production and at startup. Dosed methanol in produced water will raise the COD levels, with the potential for exceeding several thousand milligrams per liter. Tomson et al.

(2003) demonstrated that hydrate inhibitors, MEG and methanol, are detrimental from the standpoint of scaling because both promote scaling by accelerating nucleation of barite. Therefore, addition of both of these chemicals needs to be optimized to avoid barite scaling.

Knowledge of the characteristics of water, as previously described, is essential to devise a treatment process for suitable disposal and reuse. A general outline of the overall treatment process is presented in the following section.

3.6 Produced Water Treatment

Selection of a process train and the individual equipment required depend on the influent and specified effluent characteristics. The effluent characteristics are based on the intended means of disposal (e.g., surface discharge to sewer system or a body of water, deep well injection) or reuse (e.g., recycling for hydrofracturing, land application, cooling water, industrial reuse, and use in a power plant).

3.6.1 Treatment Logistics

Onsite treatment is a viable alternative to truck-transportation and offsite disposal/management of water. There can be different spatial configurations of the treatment system, as follows:

- **Dispersed:** These typically refer to truck-mounted packaged treatment systems; they are mobile and moved from region to region to serve clusters of wells.
- **Stationary–Centralized:** This refers to a relatively large stationary and centralized treatment system that draws in the produced water from wells by a piping network and also distributes the treated water to new well locations by another piping network. The piping systems are made flexible to accommodate the changing locations of wells.
- **Stationary–Decentralized:** The concept is similar to centralized treatment system, but the main difference is to set up a number (3 to 4) of treatment systems to serve wells at various regions within a given shale area. This may be more applicable for large shale plays to avoid long and complex piping networks to and from one single treatment system.

Selection of the logistics will depend on cost and social and environmental effect. These are discussed in detail in section 3.9.

3.6.2 Treatment Process

Treatment process steps are determined on a case by case basis depending on the types of contaminants present in the water and the required quality of treated effluent, which, in turn, is determined by the ultimate destination for the treated water—reuse for hydrofracturing, discharge to a receiving body of water, or deep well injection. Sections 3.6.2.1 through 3.6.2.9 discuss unit operations and processes that are commonly used for produced water treatment. However, not all the treatment steps are necessary at a given site because those are determined by the water characteristics at that specific site.

3.6.2.1 Equalization

Produced water characteristics vary not only from region to region, but also within the same geography. They vary widely from time to time, and even from well to well. Given this high variability, it is necessary to collect the water as it comes out of the ground into an equalization tank, allowing between 6 and 12 hours of retention time depending on the level of fluctuation.

3.6.2.2 Oxidation

The environment is highly reducing in shale formation. There is also the presence of anaerobic bacterial population. Any iron and manganese in the geologic formation is present in a reduced state as dissolved ferrous (Fe^{2+}) and manganous (Mn^{2+}) ions. Once above ground, as water comes in contact with air, these are oxidized into insoluble ferric hydroxide and manganese oxide, which settle in quiescent locations throughout the treatment process train and interfere with equipment performance.

The presence of gypsum (calcium sulfate) is common in geologic structures. In such shales, sulfate is reduced to sulfide ions by anaerobic bacteria. Produced water streams from such sites contain large concentrations of dissolved hydrogen sulfide, which is corrosive to equipment in the treatment process and poses a human health hazard and safety risk. At such sites, an upfront chemical oxidation step is included after equalization. Most widely, chlorine dioxide (ClO_2) is used as the oxidant of choice because it does not produce chlorinated organic residuals that have been identified as carcinogens. Additionally, chemical oxidants also act as biocides to kill the anaerobic bacteria transported from underground with produced water. This is particularly important to prevent biochemical reactions if the treated water is reused for hydrofracturing.

3.6.2.3 Primary Solids and Oil Removal

Total suspended solids (TSS) are associated with produced water. These are carried over from the breakage of rocks during fracturing operation, return of some proppant fines, ferric hydroxide and manganese oxide particles formed by chemical oxidation, and some ferric sulfide formed by the reaction of ferric ions and hydrogen sulfide in water. Also, some amount of liquid hydrocarbon is associated with the produced water streams. These need to be removed before the downstream treatment processes.

The simplest process is gravity separation, such as an American Petroleum Institute (API) separator, whereby solids settle down and oil floats up in a clarifier, with oil being skimmed off from the top surface and solids removed as sludge. Some other technologies that can be applied are corrugated plate separators and hydrocyclones; although, these technologies are seen less frequently in produced water applications because of challenges with operation and maintenance. A more advanced process is electrocoagulation. This process consists of an electrochemical cell using iron or aluminum plates as sacrificial anodes. Under the influence of an electric potential applied from an external source, the anode plates dissolve as aluminum (Al^{3+}) or ferric (Fe^{3+}) ions, which act as fresh source coagulants for solids particles to facilitate settling.

On the cathode plate, water molecules are electrolyzed to form hydrogen gas. The rising gas bubbles help coalesce fine oil droplets into larger drops and rise to the top surface of water. In addition, the electrolysis of water produces some amount of high energy hydroxyl radicals (•OH), which carry out breakdown of dissolved COD by advanced oxidation (Basu & Monroe, 2012). For further details on the principle of advanced oxidation, see Chapter 4.

3.6.2.4 Secondary Solids and Oil Removal

It is assumed that most of the free-phase oil is removed in the primary removal step. The final traces of residual solids and emulsified oil are removed by induced gas flotation (IGF) or dissolved gas flotation (DGF). Natural gas, as opposed to air, is used as the means of flotation because it is available in abundance at the exploration sites. The coalescence of fine and dispersed oil droplets is facilitated by the addition of suitable flocculant and demulsifying polymers.

Typically, IGF is favored over DGF in produced water applications. Dissolved gas flotation works on the principle of dissolution of gas (produced in the field) in water under high pressure followed by release of pressure in the tank that liberates fine microbubbles of gas (10- to 100-μm range), which attach the fine oil droplets and TSS particles and float those up onto the surface as a blanket that is separated by skimming. For further details on this unit operation, see Chapter 3.

Induced gas flotation also takes advantage of gas bubbles for oil and TSS removal, but those are generated by dispersing gas into water by mechanical means (e.g., educator, impellers) that create fine-size bubbles (100- to 500-μm size range) by shearing, which help attach the fine oil droplets and particulate matter to float up and separate. These are less efficient than DGF because of the larger size of gas bubbles, which directly relate to separation efficiency of oil and TSS from water. However, these are still favored because of the following two reasons:

- Solubility of gas in water is largely diminished because of high TDS in produced water, adversely affecting DGF performance.
- Operation of IGF is simpler and more appropriate for field application than DGF.

Despite these disadvantages, many sites still use DGF where salt concentration is not extremely high. Dissolved gas flotation performs at a superior level if it is properly attended to from operation and maintenance points of view.

3.6.2.5 Divalent Cations Removal

The presence of high concentrations of divalent cations, especially calcium, magnesium, barium, and strontium, present the challenges of scaling in the equipment, piping, tanks, and vessels of treatment systems. Because of this, removal of divalent cations is required to maintain stable operation. If the treated water is used for refracturing, then there is risk of scaling the pipe, resulting in high pressure drop and pumping energy consumption. Scales can also block the pores in the shale structure. Based on the relative amounts of cations and anions present in water and their solubility products, calcium can form scales as calcium carbonate and sulfate, magnesium as magnesium hydroxide, and barium and strontium as their respective sulfate salts.

Removal of divalent ions is achieved by precipitation as carbonate (under alkaline conditions) and/or sulfate solids in reactor tanks, followed by settling and separation in a gravity clarifier. To enhance the removal process, sulfate precipitation is sometimes followed by carbonate precipitation in two reactors in series with a common gravity clarifier to remove the solids. The reaction is conducted by dosing sodium sulfate solution as a source of sulfate ions and/or sodium bicarbonate (soda ash) solution as the source of carbonate ions into the reactor, as follows:

$$Ca^{2+} + HCO_3^- + OH^- \rightarrow CaCO_3\ (s)\downarrow + H_2O$$

$$Mg^{2+} + 2\ OH^- \rightarrow Mg\ (OH)_2\ (s)\downarrow$$

$$Ba^{2+} + SO_4^{2-} \rightarrow BaSO_4\ (s)\downarrow$$

$$Sr^{2+} + SO_4^{2-} \rightarrow SrSO_4\ (s)\downarrow$$

3.6.2.6 Biological Treatment

Biological treatment is used for reducing the COD concentration of produced water. Chemical oxygen demand results from the carryover of hydrocarbons associated with produced water. A good part of free-phase and emulsified oil is removed in primary and secondary solids and oil removal steps. However, a part of hydrocarbon remains in the aqueous phase as dissolved COD. Moreover, if MEG and methanol are used as antifreeze during drilling and exploration, then the produced water will contain high concentrations of readily degradable COD.

Biological treatment is not universally used for produced water treatment. It is currently infrequently practiced because of high and variable TDS present in the produced water, significantly affecting any biological processes. However, advancements in understandings of biological treatment processes and different economic and feasibility drivers provide opportunities where biological treatment may be integrated as part of an overall treatment system, particularly for shale plays producing lower TDS produced and flowback waters. It is case specific. In certain reuse applications, the hydrofractured water can be COD tolerant to some extent. In those cases, biological treatment may not be required to create a water suitable for reuse and may be bypassed.

Biotreatment processes in both full scale and pilot scale have been reported in the literature using different variants of activated sludge (including MBR) as well as attached growth processes. The main challenge with biological treatment of produced water is its high salinity. The following are three significant effects of high salinity:

- high osmotic pressure across cell walls, potentially leading to cell rupture;
- interference of biological flocculation and settling in the secondary clarifier by adversely affecting biopolymer formation mechanism on cell walls; and
- adverse effects of settling of solids in the secondary clarifier because the difference in specific gravities between high saline water and biomass in the mixed liquor is insignificant.

Although the first effect can be overcome effectively by allowing long acclimation periods, the second and the third effects, which both involve solid–liquid separation by gravity settling, are difficult. From this point of view, an attached growth process is more effective because the separation of biomass from water is not dependent on bioflocculation and settling sedimentation. Biological active filters (BAFs) with pre- and post-sedimentation tanks have been recommended (Guerra et al., 2011) for effective treatment of high saline water by BAFs. Freedman et al. (2017) also demonstrated effective biological treatment of produced water through treatment by BAFs.

The Gas Technology Institute conducted extensive laboratory and field on Appalachian Shale water and produced excellent results, achieving more than 90% removal of TOC, O&G, and benzene, toluene, ethylbenzene, and xylene (BTEX) for TDS concentrations up to 18% (EFX Systems, Inc., 2000). This study used a granular activated carbon (GAC) fluidized bed reactor. Special features of this setup include the following:

- It contains features similar to an MBBR.
- It has biomass attached on GAC particles.
- Its bed expands during operation and operates as a fluidized bed reactor.
- The advantage of GAC is that it first adsorbs TOC within its pores, which provides the site for growth and attachment of biomass within the larger-diameter pores of the GAC.
- It offers high biomass concentration and high solids retention time.
- It eliminates channeling, which is quite common to fixed-bed reactors.

Nichols (2018) demonstrated the efficacy of produced water treatment by SBR, a variant of the suspended growth process. The tests were conducted in bench and pilot scales on samples collected from the Denver-Julesburg Basin. Total COD, soluble COD, and ammonia removals have been reported to range from 10% to 20%, 15% to 30%, and 10% to 20%, respectively. Presence of refractory COD in significant amounts accounted for low COD removal rates.

Lawrence Berkeley National Laboratory (Camarillo & Stringfellow, 2018) reported a comprehensive review of 59 biological treatment studies globally on produced water that were published in the literature from 1979 to 2018. Some of the highlights of this report are as follows:

- Forty-one of these studies were conducted in bench scale.
- Sixteen studies were conducted in pilot scale.

- In 73% of these studies, oil field-produced water samples were used, but others used synthetic water for testing.
- The majority of the tests involved MBR or fixed-film static bed processes.
- When using actual produced water samples, the threshold TDS was 50 000 mg/L, with a COD reduction of 73% or better.

Few full-scale operating biological treatment systems have been reported for produced water. One full-scale example is the produced water treatment facility in the Pinedale Anticline field in southwestern Wyoming. The treatment train is a multistep process consisting of oil/water separation, anaerobic treatment, aerobic treatment, secondary clarification, and a final polishing filtration. Water in excess of that needed to meet the fracturing and other ancillary demands is further processed to meet the discharge standards to a river by a MBR, followed by a reverse osmosis system and a boron-removing ion-exchange system in series. The improvements in the final quality of the discharge water over the recycle water are reduction of TDS from 8000 mg/L to 100 mg/L, boron from 15 to 30 mg/L to below 0.75 mg/L, and TOC to below the detection limit (Boschee, 2012).

There is an increasing interest in biological treatment of produced water because of the following reasons:

- Environmental concerns and regulatory requirements
- Low-cost option for moderate- and high-COD waters compared to physicochemical processes
- Limited deep well injection sites and public resistance to injection
- Important pretreatment step before advanced treatment for reuse

The drivers for reuse include an increased emphasis on environmental stewardship and the pragmatic need for reuse because trucking fresh water and hauling produced water are expensive. Individual bioreactor configurations are described in detail in Chapter 3. The reader is directed to the reference materials for a better understanding of the biological treatment case studies.

3.6.2.7 Tertiary Treatment for Salt Removal

Produced water contains high salinity. As shown in Table 6.7, the ionic concentrations of water at Marcellus Shale are high, which accounts for TDS concentrations more than 100 000 mg/L. For reuse applications, it is necessary to remove salt to an acceptable level.

The upstream treatment processes previously described do not remove TDS. Reverse osmosis and evaporation are two commonly practiced processes to remove dissolved salt from water. However, both processes are quite energy intensive. In the case of reverse osmosis, the recovery of treated water gradually diminishes with the increase of feed TDS. Beyond a TDS of 40 000 mg/L, the recovery becomes too low to practically justify use of reverse osmosis as a process for salt removal from produced waters. Additionally, the presence of any salts in significant concentrations of divalent cations in the feed has the potential to deposit and scale on the membrane surface further limiting the applicability of reverse osmosis for produced waters.

Therefore, it will be necessary to remove divalent cations by the processes previously described before treatment by reverse osmosis. Chapter 3 describes the working principle of reverse osmosis and its advantages, and challenges in detail.

For high TDS concentrations, evaporation is the more suitable process. It is much less limited than reverse osmosis by the TDS of the feed. The evaporation process is energy intensive, requiring high operating costs. Moreover, at feed TDS above 300 000 mg/L, there is a high potential for crystal formation and scaling on the heat exchanger tube surfaces, which also leads to high maintenance costs. As with reverse osmosis, scaling by salts of divalent cations is a concern here also, and prior removal of such cations is strongly recommended. Capital costs are also high because of the cost of tube material operating under a corrosive environment resulting from handling of salt at high temperatures. Chapter 5 discusses brine evaporators in detail.

Ion exchange is another process to remove certain components of TDS and produces high-quality effluent. However, its application is limited to an upper limit of a few thousand milligrams per liter of TDS. Ion exchange also requires pretreatment to remove TSS and dissolved organic compounds to protect the active sites on the surfaces of ion-exchange resins. As a further challenge, regeneration of the resin produces a highly saline brine that must be managed. Chapter 3 discusses the ion-exchange process in detail.

A few other processes are also used for TDS control. These include forward osmosis, membrane distillation, capacitance deionization, mechanical vapor recompression (MVR), and humidification–dehumidification. Please reference Chapter 5 for further discussion of these technologies.

3.6.2.8 Treatment to Achieve Zero Liquid Discharge

This is the last step in the produced water treatment train if ZLD is desired or mandated. There are three avenues to achieve ZLD: (a) producing and disposing solid crystals of salt by thermal crystallizers, (b) disposal in a solar evaporation pond, and (c) deep well injection.

Drying high-concentration salt from reverse osmosis reject stream or evaporator blowdown stream in a solar evaporation pond is a viable alternative, but it has a few downsides. These include potential seepage of high-concentration salts into the ground, with the possibility of groundwater contamination. This becomes even more concerning if the concentrated waste stream contains NORM. The other concern is its potential to contaminate surrounding water resources in case of overflow during rainstorm events. Also, it is best suited for warmer and dry climates. Please refer to Chapter 5 for a detailed discussion on this subject.

Thermal crystallization is frequently the preferred ultimate step to remove and recover the last drop of water in concentrated brine from reverse osmosis reject and evaporator blowdown streams. In principle, feed at TDS $> 400\ 000$ mg/L is heated up at high pressure by steam in heat exchangers. This stream is then led to a chamber where it is released under atmospheric pressure. As a result, water flashes into vapor, leaving the solids behind as crystals. The solid–liquid slurry is withdrawn continuously and dewatered, the separated solids are disposed of, and the liquid is recycled to crystallizer. Chapter 5 describes the working principle of crystallizers in detail.

3.6.2.9 Summary of Produced Water Treatment Processes

For most purposes, removal of TSS, O&G, dissolved hydrogen sulfide, divalent cations, and TDS (primarily arising from sodium chloride) to moderate levels should be adequate. Technologies to achieve this level of treatment are oxidation (for dissolved hydrogen sulfide, iron and manganese ions, odors, color, and organic matter); electro-coagulation and/or flotation (for TSS and oil removal); chemical addition; setting sedimentation; pressure filtration; and nanofiltration/ultrafiltration.

Produced waters with large amounts of organic compounds from condensates (e.g., simple alcohols and organic acids) and aromatic compounds (e.g., BTEX, phenols, and derivatives) contain high levels of TOC.

Figure 6.9 presents a comprehensive block flow diagram of a produced water treatment train. Not all the unit operations and processes are applicable in every situation. Some of these may be taken out depending on the feed composition and the treated water quality requirements. Each application must be reviewed on a case by case basis during the planning and design stage.

3.7 Reuse Opportunities

Water is a precious resource. Reuse is an important consideration given the water scarcity in many highly active regions of the United States in upstream oil and gas exploration and production, especially West Texas, New Mexico, Oklahoma, and California. There are two significant avenues of produced water recycling in this industry: use of treated produced water

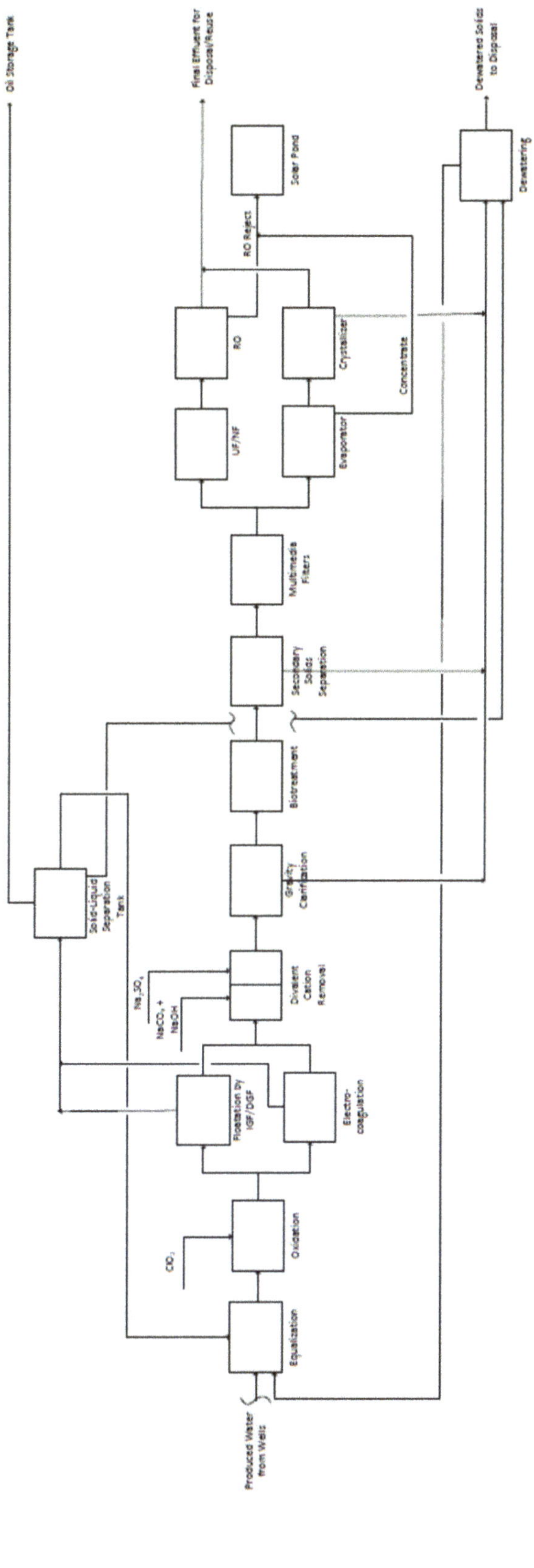

FIGURE 6.9 Block flow diagram of a comprehensive treatment train for flowback/produced water.

for refracking and use of treated produced water for land application in irrigation, landscaping, and agriculture.

Use of treated produced water for refracking has become more common at exploration sites. This helps to significantly reduce the demand for fresh water, disposal of produced water, and trucking water/wastewater through communities. Chemistry for hydrofracturing formulations has advanced significantly so that extensive levels of treatment, particularly TDS removal, is no longer necessary to the extent that it was in the past. This has made reuse of treated produced water much more feasible and economical at the present time than ever before.

In some cases, the produced water is of sufficient quality that it can be reused directly without treatment. It is important to note that the decisions around reuse of produced water are complex because of the shifting water availability market in which other potential sources of water such as brackish water, local landowner water supplies, and even municipal wastewater effluent may be competing water supply options. The availability and effectiveness of the treatment technology to provide the water quality required is just one of the factors. Although the technologies exist for treatment, as discussed in this section, a significant portion of produced water is still disposed of in saltwater disposal wells.

As the unconventional production industry has increased, midstream water management companies have grown to provide comprehensive water solutions encompassing both water supply and produced water disposal for production sites. Reuse of produced water provides an additional opportunity for these midstream companies to use a water resource and maximize profit. It is likely the shift of control in the water market from individual operators to larger midstream companies that handle the water logistics for many sites simultaneously will also result in the growth of produced water reuse and reduction of saltwater disposal.

For certain industrial reuse and power plant applications, extremely high-quality water is necessary. For such applications, advanced treatments consisting of one or a combination of a few others (e.g., ion exchange, reverse osmosis, electrodialysis, vapor recompression evaporation) may be necessary.

Land application of produced water is also being considered, especially in drought-prone regions of the United States. It should be noted that, for high TDS (>100 000 mg/L) water reverse osmosis will not be effective in bringing the TDS down significantly. The permeate water may be suitable for reuse in hydrofracturing, but may not be of adequate quality for land application or agriculture because of the potential of high sodium adsorption ratio (SAR), expressed as follows:

$$\text{SAR} = (\text{Na}^{+})\ \text{meq/L}/\{\sqrt{[(\text{Ca}^{2+}+\text{Mg}^{2+})\ \text{meq/L}]/2}\} \qquad (6.2)$$

A SAR value in the range of 12 to 15 is the upper limit to protect soil from severe deterioration of vegetation (Colorado School of Mines, 2009). In such a case, the permeate stream may need to be blended with some amount of fresh water to readjust the SAR value by introducing calcium and magnesium ions. As water scarcity challenges continue to shape the economy of water, it is likely that further investigations and advancements will allow for an even better understanding of and the ability to reuse produced water in applications outside of oil and gas production.

3.8 Residuals Disposal

Every treatment process generates a residual stream that is typically unusable and needs suitable disposal without creating any environmental or health hazards. Although upstream oil and gas activities are exempt from regulation under the Resource Conservation and Recovery Act, there are still certain requirements for disposal that must be met. In cases of produced water treatment, the sedimentation processes will generate sludge that needs thickening and dewatering before disposal. Dewatered sludge from hydrofracturing water treatment facilities needs to be disposed of in hazardous waste landfills because of the high concentration of heavy metals.

3.9 Environmental and Social Impacts and Regulatory Considerations

The health, safety, and environmental concerns associated with hydrofracturing operation are as follows:

- potential for groundwater aquifer contamination by hydrofracturing chemicals;
- potential for groundwater contamination by NORM;
- health and safety concerns for workers exposed to radiation hazard; and
- environmental concerns associated with disposal of large amounts of dissolved solids (TDS) and radioactive materials.

Several regulations are in place to address these concerns. For offshore oil and gas production activities in the United States, requirements are covered under U.S. EPA offshore effluent guidelines (1993) for the Oil & Gas Subcategory. The Primary Drinking Water Regulations for radionuclides (40 C.F.R. Part 141.15, 1976) established the limits of radioactivity in water. No federal regulations specific to hydrofracturing activities have been established yet. However, U.S. EPA initiated several technical workshops in 2011 that served as brainstorming sessions to gain a deep understanding of various aspects of the technology.

At the state level, regulations vary from state to state. Regulations are new in some states and developing in others. The Commonwealth of Pennsylvania (Pennsylvania Department of Environmental Protection [PA DEP]) imposed a limit of 500 mg/L of TDS from all industrial discharges that is applicable to Marcellus Shale. Additionally, PA DEP mandates that disposal of the concentrate can only be done by evaporation/crystallization, which will produce <363 metric tons/d (<400 tons/d) of salt waste per 4 ML/d (1 mgd) of concentrate at Marcellus Shale.

The U.S. state of Louisiana regulates radiation to a level of 50 µrems/h, while the Occupational Safety and Health Administration (OSHA) limit for worker exposure is 2000 µrems/h. In New York, among other regulations, the Draft Permit at Condition 31 [b] addresses the issue of dissolved radionuclides in liquid wastes for landfilling, but the same problem with solid wastes has not been considered. Other states that are currently enforcing or developing regulations include Pennsylvania, West Virginia, Louisiana, Oklahoma, and Wyoming. More states are expected to join in the near future.

Cross-contamination of groundwater is a potential concern, but is highly improbable because ground aquifers are at much shallower depths compared to the hydrofracturing operation depths, and the possibility of vertical upward transport of hydrofractured water to groundwater level is remote as these are separated by impervious rock layers that are hard to penetrate. A more likely contamination point for groundwater is by leakage through drill pipes if the pipe joints are not constructed properly and well secured.

Social impact is an important consideration because the heavy traffic caused by trucking in fresh water for hydrofracturing and trucking out produced water for treatment and disposal has led to a number of fatal accidents and injuries to residents. Even transportation by pipelines also may encounter social conflicts as some of the shale plays are located in native reservation areas and infringe upon the land rights of those communities. These will need careful consideration.

Although not a preferred option, exploration sites continue to dispose flowback and produced water by deep well injection in saltwater disposal wells. This practice has faced public and regulatory resistance because of several seismic disturbances in the recent past.

Current innovations have advanced the reuse of produced water in production operations at the wellhead, with the level and type of treatment guided by the reuse goal. It is likely that with continued external factors impacting water availability for the broader population as well as for oil and gas production, there will be future investigations and advancements in the area of produced water reuse for nonproduction.

4.0 MINIMUM LIQUID DISCHARGE TECHNOLOGY APPLICATIONS—REFINING

4.1 Refinery Processes and Water Requirements

Petroleum refineries are complex systems that require energy and water to refine crude oil to valuable products. Petroleum refining includes five basic classes of operations (Gary et al., 2007; U.S. EPA, 2019):

- Separation
- Impurity removal
- Conversion
- Upgrading
- Blending and other operations

4.1.1 Separation

Figure 6.10 shows a typical refinery schematic. Separation is the first step in crude oil processing. First, salts and other impurities present in the crude are removed in the desalter unit. This operation requires large amounts of water to be mixed with the crude to "wash out" salts and solids. The crude and water are separated typically by electrical desalting using high voltage electrostatic charges to coalesce and settle suspended water droplets. Caustic and acid may be added to the water wash to adjust the pH, and ammonia to reduce corrosion. Surfactants or demulsifiers are typically injected into the crude when it contains elevated amounts of suspended solids (Eckenfelder et al., 2009). Desalted crude oil is then routed to the atmospheric and vacuum distillation columns, where it is separated into fractions based on their boiling points. Water in the form of steam is used in the atmospheric crude, side-cut strippers, and vacuum distillation tower to strip oil and reduce coke formation in the furnace. Also, wash water and scale inhibitors are used in the crude overhead to minimize corrosion. Steam can be injected to the distillation column to drive separation in place of reboilers. Oily wastewater that results from washing the crude is sent to the WWTS for further processing.

4.1.2 Impurity Removal

Impurity removal is a process in which substances such as sulfur, nitrogen, oxygen, halides, and trace metals are eliminated from products. This is typically accomplished in naphtha, gas oil, gasoline, and distillate hydrotreaters. Hydrotreating is a process that removes impurities by catalytically reacting them with hydrogen (Gary et al., 2007). The hydrogen is mixed with the oil

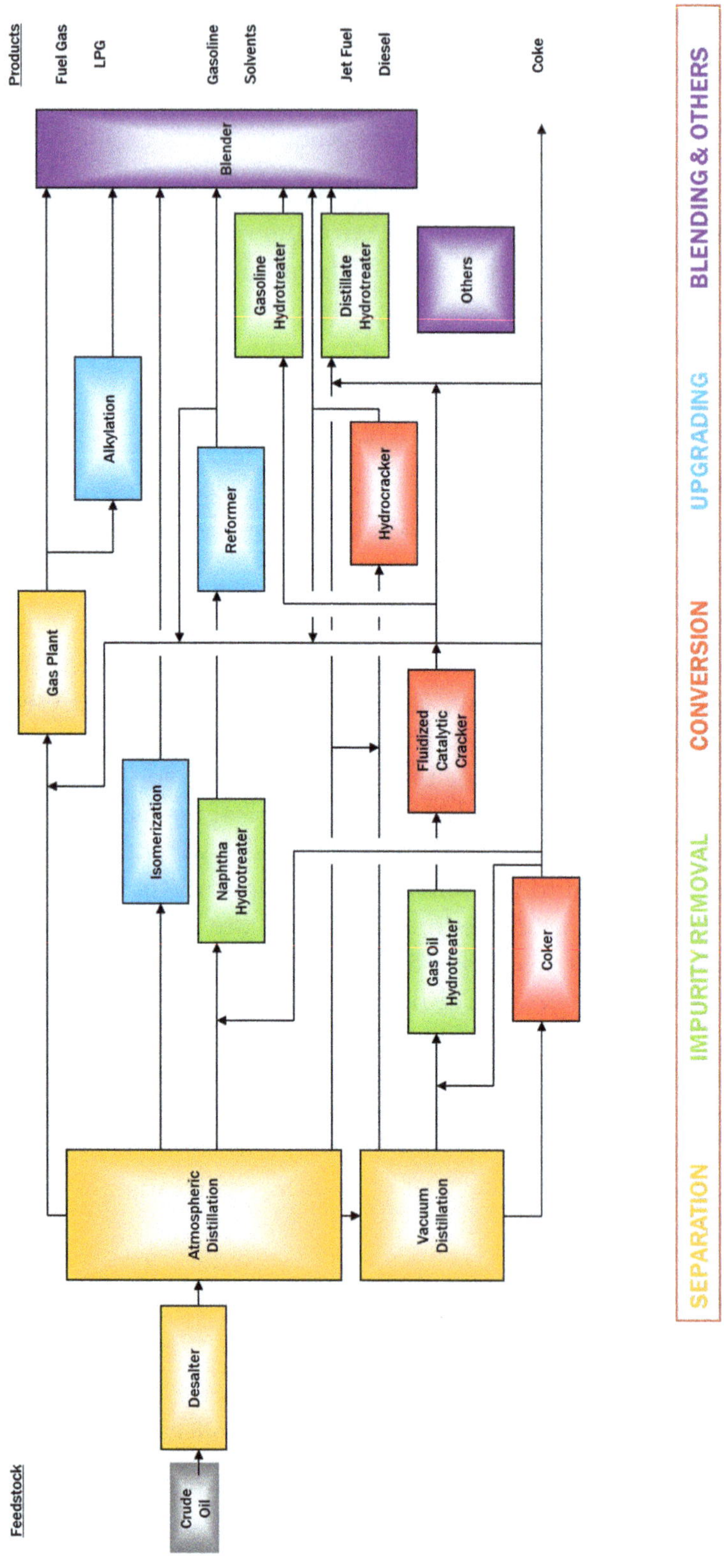

FIGURE 6.10 Simplified refinery process diagram (Johnson, personal communication, June 3, 2019).

feed through a fixed-bed reactor at high pressure. In the presence of a metal-sulfide catalyst, the hydrogen reacts with oil to produce hydrogen sulfide, ammonia, saturated hydrocarbons, and free metals. The contaminant metals attach to the surface of the catalyst, while the other products exit the reactor with the oil–hydrogen stream. In the product cooling section, the hydrogen sulfide and ammonia can react to form ammonium bisulfide. To reduce corrosion and fouling, this salt is removed from the system by injecting liquid water as a water wash (Johnson, personal communication, June 3, 2019).

4.1.3 Conversion

Conversion is the process by which the molecular size or structure of the hydrocarbon molecules is reduced by decomposition. Large molecules are broken down into smaller molecules through cracking and related processes. This is typically accomplished in a coker, fluid catalytic cracker (FCC), and hydrocracker.

The coker unit converts heavy feedstocks into a solid coke and lower-boiling hydrocarbons. A typical delayed coking unit uses steam for heating and gas oil stripping. Quench and cutting water and caustic are also used during various steps in the coking process. The FCC converts heavy oils into more valuable gasoline, diesel, and lighter products. The hot oil feed is placed in contact with the catalyst and molecules are broken down while coke forms on the catalyst. Steam is used to strip the catalyst of coke before the catalyst is regenerated. Steam is also used to remove impurities in other related processes. Hydrocracking reduces molecular size in a process similar to an FCC, except that the catalytic cracking includes a hydrogenation of aromatic molecules. Steam is also needed during hydrocracking unit operation.

4.1.4 Upgrading

Upgrading is the process by which the structure of the hydrocarbon molecules is changed via unification, reforming, and isomerization. Unification causes larger molecules to be formed from smaller ones through alkylation and related processes. Reforming is the process by which molecules are rearranged into different geometric structures via isomerization, catalytic reforming, and other similar processes. Alkylation adds an alkyl group to a compound; in refinery terms, it is the reaction of low-molecular-weight olefins with an isoparaffin to form higher molecular weight isoparraffins by using hydrofluoric or sulfuric acid as a catalyst. During isomerization and reforming, hydrocarbon molecules are rearranged to form higher octane molecules. Steam and liquid water are used in the various upgrading units (Gary et al., 2007).

4.1.5 Blending and Other Operations

Blending is the process that combines various hydrocarbon fractions, additives, and other components to produce specific products. Other operations include light-ends and sulfur recovery; sour-water stripping; wastewater treatment and solid waste handling; process water treatment and cooling; acid and tail gas treatment; hydrogen production, storage, and handling; and product movement.

During the refinery's processes and operations previously described, water, steam, natural gas, and fuel are needed to produce salable products (Gary et al., 2007). Refinery byproducts, such as wastewater and solids, carry impurities (e.g., hydrocarbons, metals) that need to be treated and properly disposed of to meet regulatory requirements for water returned to the environment.

4.1.6 Water and Steam Requirements in Refineries

Typical water and steam utilities required in refineries are as follows (Gary et al., 2007; U.S. DOE, 2016):

- Cooling water
- Boiler feed water for steam production (steam is typically 3.4 atm [50 psig] to 47.6 atm [700 psig])
- Process water
- Potable water
- Utility water
- Fire water

Cooling water is used in evaporative, closed-loop, and once-through cooling water systems to remove heat. One example is the separation or distillation process. During this process, crude oil is heated and vaporized to produce fuel oil and refinery fuel gases. The various fuel fractions are separated by cooling and condensing products that are withdrawn from the crude tower. Heat is removed from the distillation system by using heat exchangers, which reject heat by using cooling water and air coolers. Heat is also transferred via heat exchange to waste heat boilers that produce steam or to process streams that need to be preheated. Air coolers are typically used in conjunction with water-cooled heat exchangers to reduce water use.

Evaporative cooling water systems exchange heat by recirculating cooling water and rejecting heat by spraying the hot water against a stream of air. The water that is not lost by evaporation is recirculated back to the cooling system. As the water is recirculated and evaporated through the cooling system, the concentration of salts in the water increases, and fresh makeup water must be introduced to lower the salt concentration because salt can cause corrosion and scale formation.

The efficiency of cooling water systems is based on the number of cycles of concentration (COC), which depends, in large part, on the quality of the makeup water used. Typical COCs in refinery cooling systems range between 4 and 10 (International Petroleum Industry Environmental Conservation [IPIECA], 2010). Best-in-class cooling systems using good quality water (i.e., low in hardness and conductivity) coupled with an effective chemical program can achieve a COC of 7 or higher and significantly reduce makeup water consumption (see Section 4.1.8, "Water Quality Requirements").

Closed-loop water systems circulate water in a closed-loop piping system that provides cooling and heating without evaporation or air contact. Heat absorbed by the water in the closed-loop system is rejected using a heat exchanger in a once-through cooling system. Once-through cooling systems withdraw water from a large waterbody (e.g., lake, river, ocean), exchange heat, and typically return it to its source. Typically, cooling water represents a large amount of the water withdrawn and consumed in refineries (see Section 4.2.2).

Boiler feedwater is the high-purity water required for steam production. It is nearly free of organics, salts, and particulate matter (see Section 4.1.8). The higher the steam pressure required (3.4 atm [50 psig] to 47.6 atm [700 psig]), the higher the purity of the BFW required. Steam is used for heating, cooling, crude processing (e.g., atmospheric crude unit furnace, towers, and side-cut strippers), running turbines, and fluid catalytic cracking, among other processes (Gary et al., 2007).

Process water is typically the water that comes in close contact with hydrocarbons for various refining purposes (e.g., desalter wash water to remove salts and impurities from crude, hydroprocessing water wash for preventing salt deposition). To avoid corrosion or scaling, this water is typically softened by removing, for example, calcium and magnesium ions.

Potable water is used mostly in bathrooms, kitchens, wash areas, and eyewash and safety showers, where human contact is likely. Other uses for potable water include special chemical operations (e.g., as a diluent) or as the feed to reverse osmosis units.

Utility water is used primarily for cleaning and washing concrete slabs and process unit areas. Fire water is primarily used for firefighting purposes; however, it may occasionally be used in process units (e.g., as a cooling agent for WRRFs) or as the utility water.

Fire water needs to be available in large quantities. Typically, refineries collect stormwater, treated wastewater, or other relatively clean water streams that do not contact hydrocarbons and store it in large reservoirs that are dedicated to the fire system (IPIECA, 2010).

4.1.7 Quantitative Water Requirements

Typically, refineries withdraw water from one or more sources, for example, surface water (rivers, lakes, etc.), groundwater, city or municipal water, or seawater. Water withdrawn into a refinery may be consumed (e.g., in steam methane reformers, flare systems, evaporated in cooling towers, contained in crude products), returned to the original source, or discharged to the environment. A typical petroleum refinery currently requires between 0.05 m^3 (0.3 bbl) and 0.30 m^3 (1.9 bbl) of water per 0.16 m^3 (1.0 bbl) of crude oil refined with an average of 0.24 m^3 (1.5 bbl) of water per 0.16 m^3 (1.0 bbl) of crude oil refined (Sun et al., 2018; Wu & Xu, 2018). However, water requirements depend heavily on the complexity of the refinery (i.e., based on the Nelson Index) and the types of processes and equipment present (e.g., cooling system) (U.S. DOE, 2016). Because of yield gain during crude processing (i.e., 0.16 m^3 [1.0 bbl] of crude generates 0.17 m^3 [1.06 bbl] of refined product), consumptive water use may be expressed as 0.22 m^3 (1.40 bbl) of water/m^3 (bbl) of refined product. Water consumption may range from 0.08 m^3 (0.50 bbl) to 0.40 m^3 (2.50 bbl) of water/m^3 (bbl) of refined product depending on the refining process (Wu & Xu, 2018).

4.1.8 Water Quality Requirements

Water quality requirements depend greatly on the complexity of the refinery. No two refineries are exactly alike, and every process requires various amounts and grades of steam, process water, cooling water, and so on. The more steam that is needed by the various refinery processes, the more BFW or high-quality water must be provided compared to other types of water. Table 6.8 lists the requirements for BFW and typical water quality parameters for makeup water for a best-in-class cooling water system. Typically, BFW is treated by using cold/warm lime or soda ash softening to remove hardness (i.e., calcium, magnesium, silicon, etc.), pressure/gravity filtration (e.g., zeolite filters) to remove solids, and ion exchange and decarbonation to remove undesirable cations and anions.

In some refineries, reverse osmosis is used as a treatment step. Hardness and dissolved solids (primarily silica) were found to be the most common water constituents that limit COC in both cooling towers and boiler operations. Refinery sites with lower hardness and silica concentrations in cooling tower and BFW makeup water achieve above-average COCs.

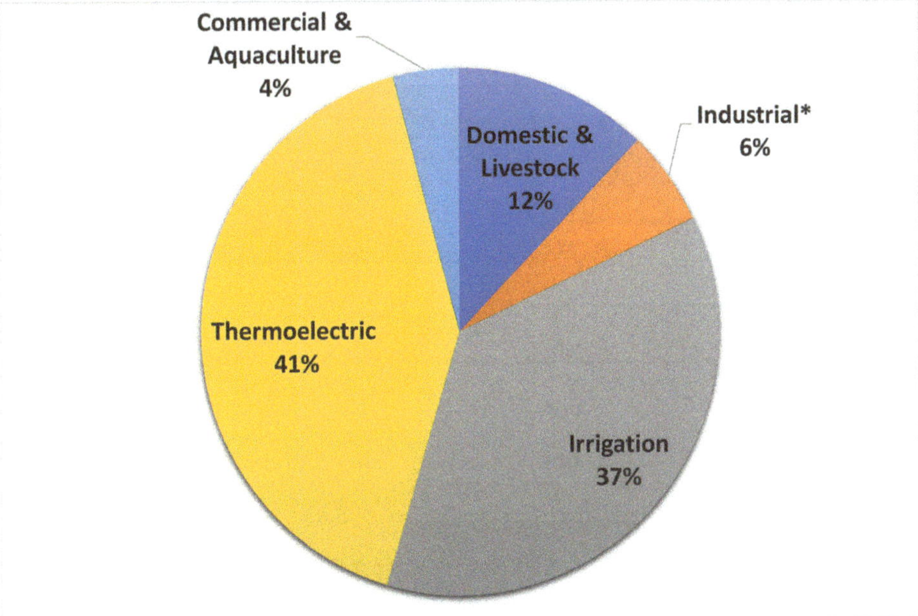

FIGURE 6.11 Total U.S. freshwater use among various sectors (USGS, 2015). Industrial includes petroleum refining, chemical, food, mining, paper, and steel industries.

4.1.9 Factors Affecting Water Use

Several factors affect water use in refineries, including the following:

- Regional availability based on wet or dry conditions and/or competing demand for the same water resources
- Seasonality (summer vs. winter) and demand (e.g., high demand for cooling water in summer and for steam in winter)
- Quality and physical characteristics of water (TDS, organics, temperature, etc.)
- Refinery size and complexity

4.1.10 Trends in Water Use

Historically, U.S. petroleum refineries have been large water users. Water use surveys conducted from 1954 to 1955 of 61 refineries showed, on average, a water requirement of 1.7 m^3 (11 bbl) water/m^3 (bbl) of crude oil refined (Otts, 1963). The water was mostly used in contact and noncontact cooling (once-through cooling systems), steam production, and process units (e.g., desalting). Since 1972, the petroleum refining industry has steadily reduced the amount of water used and consequently discharged (U.S. EPA, 1996).

TABLE 6.8 Water quality characteristics for refinery boiler feed water and cooling system makeup water.

Parameter	Unit	Water Quality	
		Boiler Feed Water[a]	Cooling Tower Makeup Water[b]
pH	—	8.3–10.0[c]	7.0–9.0
Dissolved oxygen	ppb	<7[c]	NS
Conductivity	µS/cm	5–11	280
Iron	ppm	≤0.1[c]	N/A
Copper	ppb	≤0.05[c]	N/A
Silica	ppm	<0.05	14
Total hardness	ppm as $CaCO_3$	≤0.3	96
Chloride	ppm	<0.05	14
TDS	ppm	<0.5	168
Sodium	ppm	<0.05	N/A
Sulfate	ppm	<0.1	37
O&G	ppm	<0.1	ND
TOC	pm	<1[c]	<10
Turbidity	NTU	ND	<5.0

Note. NS = Not specified; ND = not detected; N/A = not available
[a]Typical requirements (after demineralization) of U.S. Gulf Coast and Pacific Coast refineries.
[b]Water quality data for best-in-class cooling tower makeup water with an efficient chemical program in a Pacific Coast refinery.
[c]American Society of Mechanical Engineers Guidelines for ≤900 psi boilers (ASME, 1994).

In the early 1970s, barometric condensers were commonly used, but they required large amounts of water and generated contaminated wastewater. The barometric condenser was used to remove air, exhaust steam, and other vapors from vacuum equipment (Schutte & Koerting, n.d.). However, to conserve water and minimize liquid discharge during the mid-1970s, barometric condensers were mostly replaced with surface condensers (U.S. EPA, 1996). Once-through cooling systems were also replaced with closed-loop and evaporative cooling water systems, and this change drastically reduced the water intake to refineries. Figure 6.12 shows the trend in refinery water use, in terms of barrels of water/barrels of crude oil refined, from 1955 to 2018.

The water use trend in refineries over the past 64 years shows a significant decrease from 1.7 m^3 (11 bbl) water/m^3 (bbl) of crude (including water withdrawn in once-through cooling systems) in 1954 to an estimated average 0.2 m^3 (1.5 bbl) water/m^3 (bbl) of crude in 2018 (Wu & Xu, 2018). Water conservation, internal recycle/reuse, and use of air coolers (e.g., fin-fans) or oil coolers (instead of cooling water systems), in addition to cooling system process advancements, have resulted in significant water use reduction in the U.S. crude oil refining industry in the past several decades (Eble & Feathers, 1992a; Ellis et al., 2001).

4.2 Current Water Management in the Refining Industry

4.2.1 Sources of Water

4.2.1.1 Surface Water

Surface water can be supplied to a refinery from various sources, including rivers and lakes. Surface water typically contains dissolved solids (e.g., calcium, silica), suspended solids, and organics that need to be removed

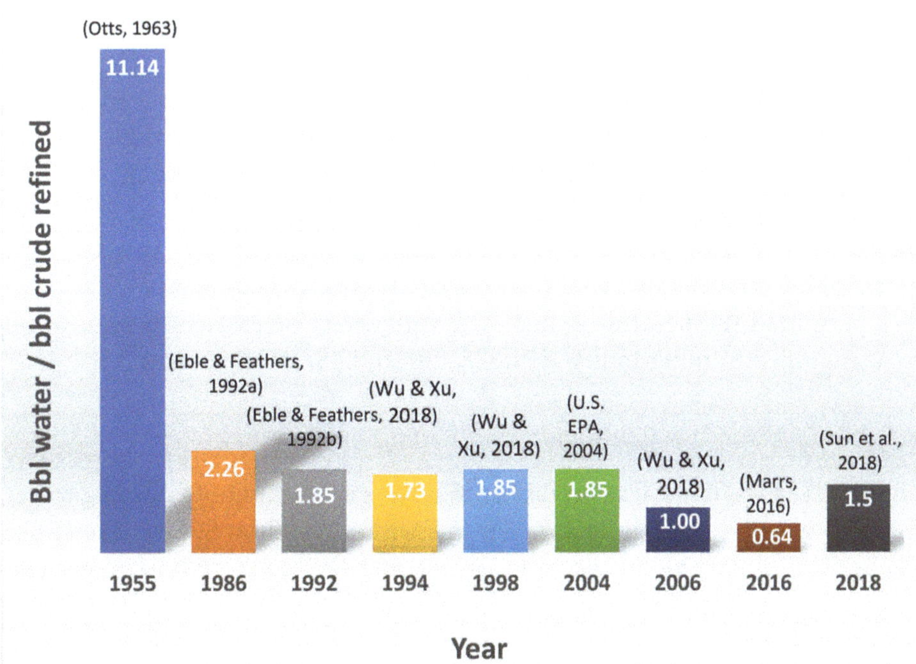

FIGURE 6.12 Trends in water use in refineries from 1955 to 2018.

before it can be used as cooling tower and BFW makeup water (IPIECA, 2010; U.S. DOE, 2016).

4.2.1.2 Purchased Water

Purchased city or municipal water is typically the cleanest but also the most expensive source of water. This water is typically used as cooling tower makeup water without any treatment. The temperature of this water may vary significantly depending on its source (e.g., surface water tends to be warmer than groundwater). In the past decade, treated effluent water (gray water) from municipal WRRFs has been used to reduce the demand stress on city potable water (IPIECA, 2010; U.S. DOE, 2016).

4.2.1.3 Groundwater

Groundwater is commonly used in inland or more arid areas or to augment surface water. The quality of this water may vary based on the aquifer characteristics. In some cases, groundwater can be hard (having a high mineral content), thus making it more expensive to treat to BFW quality. An advantage of using water from an aquifer is that the temperature is typically cooler and more consistent than that of surface water. However, a significant disadvantage could be the finite volumes that may be available as well as the desire of local and state water authorities to conserve this freshwater resource for other uses (IPIECA, 2010; U.S. DOE, 2016).

4.2.1.4 Seawater

Seawater is typically used in coastal refineries for once-through cooling systems. It requires upgraded metallurgies for pumping, transfer, and heat exchangers because of the high TDS content (3.5% [wt]) and corrosive nature of saltwater. The water is typically used untreated and must be returned uncontaminated to the same location from which it was withdrawn at a permitted temperature range. The water may be treated to remove oil by using oil-skimming tanks. (IPIECA, 2010; U.S. DOE, 2016).

4.2.1.5 Water Use Consumed and Exiting the Refinery

Figure 6.13a shows typical water use in refineries. Commonly, cooling tower makeup water is the significant user (52%); followed by BFW makeup use (42%); and then process, potable, utility, and fire water (6%) (Ostenberg, 2009). Typical water consumption in refineries (see Figure 6.13b) includes water loss through evaporation in cooling towers (38%), water vented as steam (11%), water consumed in processes (5%) (i.e., steam methane

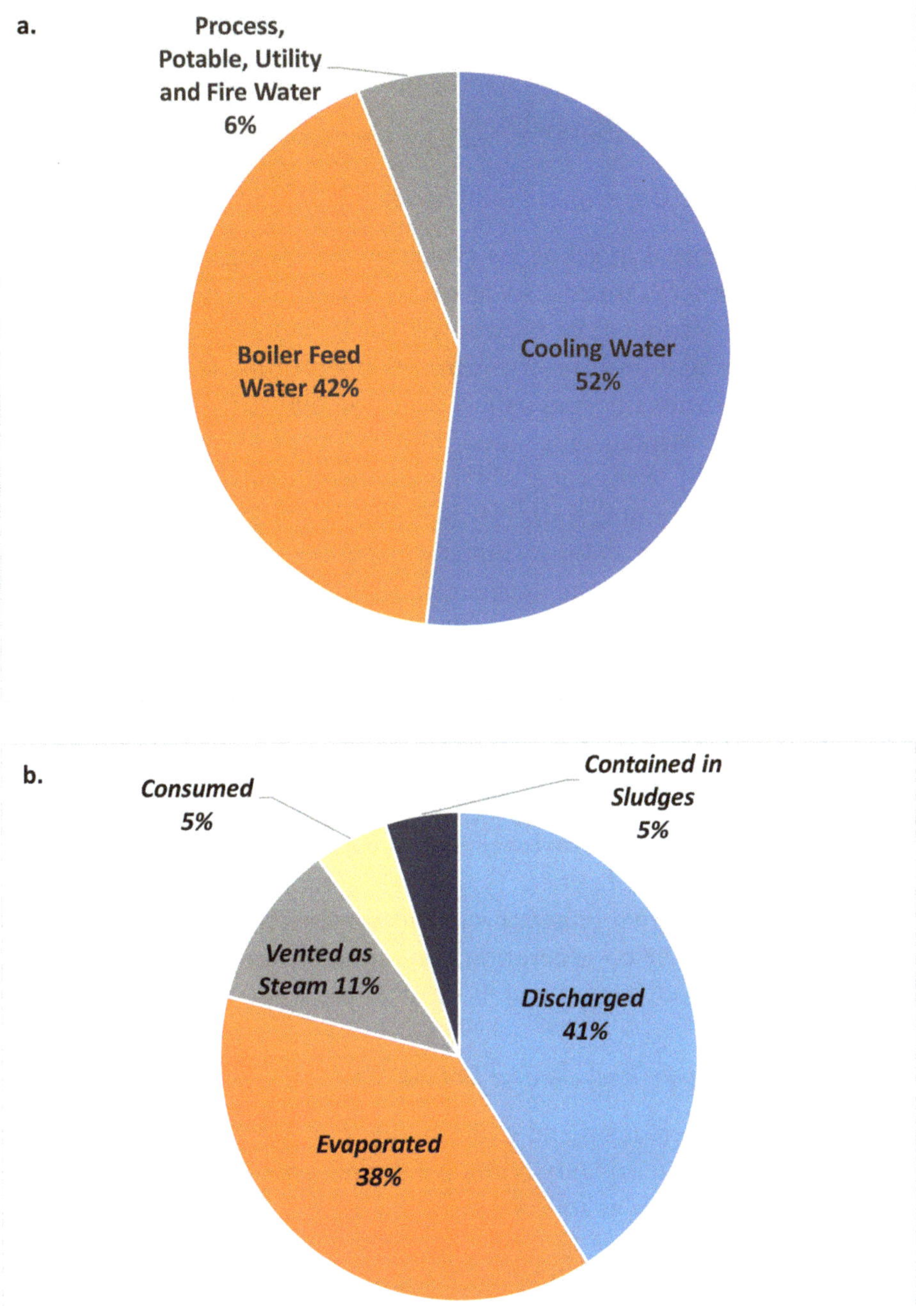

FIGURE 6.13 (a) Typical water use in refineries and (b) water consumed by and exiting the refinery (adapted from Ostenberg, 2009).

reformers, flare systems, etc.), and water contained in primary and secondary sludges (5%). The water exiting the refinery as discharge treated wastewater effluent is typically less than half of the water withdrawn (41%) from outside sources (Ostenberg, 2009).

4.2.2 Wastewater Generated in Refineries

Wastewater generated in refineries typically includes desalter water, sour water, various process waters, spent caustic, tank bottoms water, cooling tower blowdown, condensate blowdown, boiler blowdown, ion exchange backwash associated with reverse osmosis operations, effluent from wet gas scrubbers, stormwater, and ballast water. Significant generators of sour water include the distillation process unit, which produces almost half of the total volume of wastewater (44%), followed by the FCC unit (26%) and the catalytic reforming units (10%). The thermal cracking/visbreaking, catalytic hydrocracking, coking, catalytic hydrotreating, isomerization, and sulfur removal/Claus units also generate sour water (~12%). Alkylation generates wastewater containing spent potassium hydroxide (4%) and crude desalting desalted wastewater (4%) (see Table 6.9).

4.3 General Review of Refinery Water Conservation, Recycle, and Reuse Practices

Water conservation and reuse/recycle best practices are critical for reducing the water footprint in refineries. Typical water reduction best practices in refineries include internal reuse/recycle and conservation, steam leak elimination, refinery power co-generation, downgraded water quality use, and water use elimination, as outlined in the next sections.

4.3.1 Internal Reuse/Recycle and Conservation

Internal water conservation and reuse/recycle opportunities are made possible by matching water quality with the requirements of refinery processes. These opportunities are as follows:

- Stripped sour water (SSW) is relatively free of salts and is preferentially recycled as wash water for desalters (National Association of Corrosion Engineers International, 2009). Desalters require, on average, 5% (vol) to 6% (vol) water per crude oil volume for proper desalting (Gary et al., 2007). Also, SSW can be used for wet gas scrubbers and hydrocracker operation. Because of the presence of selenium in the SSW, this practice may not be suitable for all refineries and it needs to be evaluated on a case by case basis.

TABLE 6.9 Typical wastewater generated in various refinery processes at petroleum refineries (U.S. EPA, 2004).

Process	Description	Volume (bbl water/bbl of crude)	Percent of Total Volume
Distillation	Sour water	0.619	44%
Fluid catalytic cracking	Sour water	0.357	26%
Catalytic reforming	Sour water	0.143	10%
Alkylation	Spent KOH	0.062	4%
Crude desalting	Desalted wastewater	0.050	4%
Visbreaking/thermal cracking	Sour water	0.048	3%
Hydrocracking	Sour water	0.048	3%
Coking	Sour water	0.024	2%
Isomerization	Sour water and caustic wash water	0.024	2%
Additive production: ethers manufacture	Pretreatment wash water	<0.024	Negligible
Catalytic hydrotreating	Sour water	0.024	2%
Chemical treating: sweetening/merox process	Negligible	Negligible	Negligible
Sulfur removal/Claus process	Sour water	<0.024	Negligible
Lubricating oil manufacture	Steam stripping and solvent recovery wastewater	<0.024	Negligible
Total		**1.405**	**100%**

- Vacuum tower overhead condensate, atmospheric tower, and preflash tower overhead sour water, BFW, cooling tower and boiler blowdown, and water recovered from slop may be used as wash water for desalters depending on their quality (IPIECA, 2010).
- Steam condensate (clean condensate, not process condensate) can be recycled and used as BFW makeup water or cooling tower makeup water (Langer, 1983).

- Boiler blowdown can be recycled as cooling tower makeup water; however, TDS may be too high.
- Cooling tower blowdown can be reused as fire water.
- Ultrafiltration backwash water can be used as cooling tower makeup water after removal of solids. Also, reverse osmosis systems reject for reverse osmosis units pretreated by ion exchange, to remove calcium and magnesium, are suitable to be used as cooling tower makeup water.
- Treated remediation groundwater and SSW can be used, in some cases, as cooling tower makeup water, fire water, coker unit cutting and quench water, and dust control. This water needs to be low in suspended solids, ammonia, nitrates, and phosphates to be suitable as cooling tower makeup water.
- Treated wastewater effluent can also be used as fire water and fuel-grade coker quench water.
- High-pressure steam is reused to BFW by pressure letdown (Venkatesan, 2012).
- Waste activated sludge can be used for coke dust control.
- Spent caustic water may be reused for the Claus process or sulfur recovery facility or for pH control in the wastewater treatment system.
- Stormwater can be collected and used for fire water and as feed to a reverse osmosis system.

An example of internal reuse at a Gulf Coast refinery in the United States shows a significant volume of water (~7629 m^3/d [~1400 gpm] or ~7.42 ML/d [~1.96 mgd]) that is conserved internally every day by matching the cooling tower, boiler blowdown, and SSW to the fire water, cooling tower makeup, and desalter mud wash water quality and demand, respectively.

4.3.2 Steam Leak Elimination

Steam losses in refineries represent a high-energy and high-quality water waste. Steam lines in refineries may develop leaks because of the corrosive nature of the steam. Steam leak identification and elimination are best practices that can conserve significant water and energy (Langer, 1983; Venkatesan, 2012).

4.3.3 Power Co-generation

A common practice for oil refineries is to use co-generation where excess steam or heat can be used to create additional electrical power. This practice increases efficiency and results in water conservation because water requirements for co-generation often are lower than for the same power generated by coal-fired boilers or steam-condensing turbines. For example, a 23-ML/d (6-mgd) water intake at a refinery allows co-generation of 525 MW using gas turbines versus the more than 53 ML/d (14 mgd) needed to generate 525 MW with a coal-fired boiler using a condensing turbine in a commercial power plant (American Petroleum Institute, n.d.).

4.3.4 Downgraded Water Quality Use

To reduce the demand on potable, city, or municipal water, use of treated municipal effluent, or gray water, for feed to a filtration and reverse osmosis system or cooling tower makeup is becoming more widespread (Francis et al., 1996; Puckorius et al., 1998). Some examples include the Chevron Richmond and El Segundo Refineries in California. Richmond Refinery uses 13 ML/d (3.5 mgd) of high-quality recycled water from the East Bay Municipal Utility District (Towey & Baldwin, 2015). CHS McPherson Refinery in Kansas uses city gray water (CHS Inc., 2019). Also, untreated saltwater is used in some coastal refinery process units (in addition to once-through cooling).

4.3.5 Elimination of Water Use

As previously mentioned, air-cooled (e.g., fin-fans) or oil cooling systems are used in refinery process units to eliminate water use and consumption.

4.3.6 Case Studies

4.3.6.1 Ponca City Refinery (Phillips 66), Oklahoma, United States

Phillips 66 is a diversified energy manufacturing and logistics company. With a portfolio of midstream, chemicals, refining, and marketing and specialty businesses, the company processes, transports, stores, and markets fuels and products globally. Phillips 66 owns 13 refineries worldwide, including Ponca City Refinery (PCR), in Ponca City, Oklahoma, which processes a mixture of light, medium, and heavy crude oils with a total throughput of 34 967 m^3/d (220 000 bbl/d). The refinery is a high-conversion facility that produces a full range of products, including gasoline, diesel, and aviation fuels; liquefied petroleum gas (LPG); and anode-grade petroleum coke. Its facilities include two FCC units, as well as alkylation, delayed coking, naphtha reforming, and hydrodesulfurization units (Phillips 66, n.d.). Ponca City Refinery's water supply includes city water, two refinery-operated groundwater wells, treated groundwater from a series of remediation recovery wells, and stormwater that is stored in two large surface impoundments (see Table 6.10).

TABLE 6.10 Water supply sources for Phillips 66 Ponca City Refinery, Ponca City, Oklahoma (Meyers et al., 2008).

Water Source	Flow m³/d (gpm)	Percent of Total Water Withdrawn
Refinery groundwater wells	13 079 (2400)	50%
Purchased city water	10 354 (1900)	39%
Remediation wells	1962 (360)	7%
Stormwater runoff	1035 (190)	4%
Total	26 430 (4850)	100%

Because of seasonal and yearly variability in precipitation, seasonal demand for fresh water, and refinery water demand variability (large demand for steam during cold winters and cooling systems makeup water during hot summers), PCR has implemented several in-place recycle and efficiency practices for water conservation and, in turn, MLD (Meyers et al., 2008). Ponca City Refinery has nine cooling towers that are large consumers of water because of evaporative loss from cooling towers. Through optimization efforts and a targeted chemical program, PCR has been able to operate the cooling systems at optimal COC (>4 to 7) and significantly minimize water and chemical consumption (Meyers et al., 2008).

A small amount of the SSW, one of the largest wastewater streams in PCR 218 m^3/d (40 gpm) or 218 016 L/d (57 600 gpd), is currently being recycled to one of the wet gas scrubbers. Fresh water is being conserved by using SSW instead. Also, remediation water from recovery wells (1226 m^3/d (225 gpm) or 1 226 340 L/d (324 000 gpd) is treated and then used internally as fire water, coker unit cutting water, service water, and dust control. Other minor streams are also reused. For example, some spent caustic is reused by the sulfur facility (Claus unit). Studies conducted at PCR showed that water conservation, TDS, and whole-effluent toxicity (WET) testing are correlated. Water losses of up to 50% to 60% across the refinery, SO_x reduction regulations requiring two new wet gas scrubbers (which added 1000 ppm of TDS in the wastewater), other refinery water recycling and conservation efforts, and the conservative nature of salts all led to a doubling in concentration of TDS (from 2000 ppm to 3500 to 5000 ppm) from the freshwater source to the wastewater effluent. Because of the potential for TDS toxicity, WET bioassays become a de facto limiting concentration standard for the amount of possible water conservation in a refinery facility (Fort et al., 2013; Meyers et al., 2008).

4.3.6.2 Toledo Refining Company (PBF Energy), Ohio, United States

PBF Energy (PBF) is one of the largest independent petroleum refiners and suppliers of unbranded transportation fuels, heating oil, petrochemical feedstocks, lubricants, and other petroleum products in the United States. PBF currently owns and operates five domestic oil refineries, including the Toledo Refinery in Oregon, Ohio, which has throughput capacity of approximately 27 027 m^3/d (170 000 bbl/d). Toledo processes a slate of light, sweet crudes and produces a high volume of finished products, including gasoline, ultra-low sulfur diesel, and a variety of high-value petrochemicals including nonene, xylene, tetramer, and toluene (Water Tech Online, October 25, 2013).

To minimize liquid discharge and reduce demand to its local water supply, Toledo Refinery reclaimed 50% of its wastewater effluent as cooling tower makeup water. Refinery wastewater is currently pretreated before it is discharged to the local, publicly owned treatment works. The pretreatment consists of an API separator and a dissolved nitrogen flotation system to remove oil, grease, and solids followed by a secondary treatment—an activated sludge unit to remove residual dissolved organics and inorganics—and a two-train clarification system.

The quality of the effluent is high enough to split the effluent stream to send a portion to tertiary treatment (other than the WRRF) for polishing using five sand filters. The polished water is then recycled as cooling tower makeup water. By recycling 50% of this water, the refinery has significantly reduced city water purchases, thereby saving money while also conserving water resources (Industrial Water World, 2013).

4.3.6.3 Federated Co-Operatives Limited's Co-Op Refinery Complex, Saskatchewan, Canada

Federated Co-Operatives Limited's Co-Op Refinery Complex (CRC) in Regina, Saskatchewan, is one of Canada's largest refineries, with 20 668 m^3/d (130 000 bbl/d) of refining capacity. The refinery requires an average of 8719 m^3/d (1600 gpm or 54 860 bbl/d) of water for steam generation. The steam is used for heating, hydrogen production, power generation, and cooling towers. The refinery was one of the largest freshwater consumers in Regina, drawing water from groundwater wells and the city's municipal system during high water demand.

When CRC increased production from 15 898 m^3/d (100 000 bbl/d) to 20 668 m^3/d (130 000 bbl/d), it needed additional water and steam to support this growth in refinery capacity. The provincial body responsible for water supply, the Water Security Agency, mandated that the refinery find a new source of water to support its expansion. Regina's arid climate and the need for water conservation spurred CRC to invest $200 million in Canadian dollars in a wastewater improvement project (WIP) to enable the refinery to reuse most of its wastewater.

To meet the ambitious recycling target and the high-quality requirement to generate steam, the WIP required a treatment system capable of removing suspended solids, O&G, phenols, nitrogen compounds, organics, and other unwanted compounds. The WIP was built in mid-2015 and has been in operation since November 2015.

The WWTS includes the following five treatment stages:

1. Equalization tanks and oily water separators
2. Dissolved gas flotation unit
3. MBR
4. Centrifuge
5. Demineralization facility

Details of the five stage treatment train technologies used are as follows:

1. **Stage 1:** Refinery wastewater is processed for oil and solids through equalization tanks to skim floating oil by gravity from the top of the wastewater. The floating oil is subsequently reprocessed within the refinery. Wastewater is then processed through an oily water separator where sediment and additional oil are separated.
2. **Stage 2:** The wastewater from stage 1 is routed to a DGF Clarifier (Poseidon Model #S400TCE10-E) unit for further removal of oil by using microscopic nitrogen bubbles that attach to the oil droplets and float them to the surface, where they are skimmed off. This DGF uses nitrogen instead of oxygen as the blanketing/recirculation gas to mitigate air emissions and prevent a potentially explosive environment in the DGF headspace. This stage uses PolyFloc (AE1701) chemistry as a flocculant and ferric sulfate as coagulant. To optimize removal of the finely dispersed oil droplets and suspended solids, flocculant, coagulant, and caustic (to adjust pH) are injected at the inlet of the DGF.
3. **Stage 3:** Following DGF, the de-oiled wastewater is combined with recycled mixed liquor from the MBR tanks and equally distributed to two parallel bioreactor tanks. The tanks are designed with an anoxic zone inlet that is 20% of the tank volume for denitrification. The water from the inlet zone then flows under a baffle wall to the aerated portion of the tank. Effluent from the bioreactors is combined in a header that feeds three aerated MBR trains. The MBRs remove organics and residual ammonia by using biological treatment. The water is then filtered by using ZeeWeed 500 ultrafiltration hollow-fiber

membranes. The permeate is then sent to a demineralization facility (stage 5) to generate BFW. Stage 3 uses BioPlus BA2970 and BA3900 supplemental bacteria for biological seeding and bioaugmentation in the biological reactor and FoamTrol AF3031 for foam control.

4. **Stage 4:** In this stage, a centrifuge dewaters waste activated sludge and returns a portion, as needed, back to the MBR system. A small amount of water is also extracted from the waste activated sludge and reprocessed. A flocculent (Novus CE2655) is used to generate biosolids, which are disposed of safely.
5. **Stage 5:** The permeate from stage 3 undergoes further treatment in a demineralization facility to make BFW. The facility includes a high-pH reverse osmosis system, which consists of ion-exchange units for hardness and TDS reduction and a decarbonator to remove alkalinity and carbon dioxide. The permeate is further treated by another reverse osmosis unit to polish the water. Finally, a mixed-bed ion exchanger removes residual ions to produce ultrapure water suitable for steam production.

An asset performance management system monitors the performance and long-term health of the MBR membranes; tracks the operation, chemistry, and overall product quality between stages; and closely evaluates the quality of water leaving stage 3 to ensure it is on spec to be reused during stage 5. CRC currently recovers 7.6 ML/d (2.0 mgd) of wastewater, the majority of which is used in its steam facility. The WIP has cut the refinery's freshwater needs by 28%, which is equivalent to the annual water use of approximately 3100 Regina households (Elliott, 2018; Hilts et al., 2018).

4.3.6.4 Singapore Refining Company Private Limited (Chevron), Jurong Island, Singapore

Singapore Refining Company Private Limited (SRC) is a joint venture between Singapore Petroleum Company Limited and Chevron (viz., Chevron Singapore Pte. Ltd. and Chevron Trading Pte. Ltd.). Located on Jurong Island, Singapore, SRC operates a refinery that processes 46 105 m^3/d (290 000 bbl/d) of crude oil. SRC refinery produces primarily fuel products that are distributed locally and exported overseas. On the basis of a water audit conducted to identify streams for recycling, SRC decided to build an onsite effluent treatment recycling facility (ETRP). The ETRP uses ceramic microfiltration and reverse osmosis to remove suspended solids, oil, grease, and other contaminants from the treated wastewater. The ETRP currently treats 2000 m^3/d (367 gpm) and can meet approximately 20% of SRC's water demand.

Other water conservation initiatives implemented by SRC include the following:

- Maximizing the reuse of treated effluent water internally (up to 40% from wastewater effluent)
- Maximizing the use of seawater and fin-fan coolers for cooling
- Using closed-circulation cooling water or tempered coolers
- Steam trap surveillance and close monitoring of water consumption (PUB, 2018)

4.3.6.5 British Petroleum Kwinana Refinery, Australia

British Petroleum (BP) Kwinana Refinery is located approximately 35 km south of Perth, adjacent to the Cockburn Sound, an intensively used marine bay on the coast of Australia. Kwinana Refinery has been in operation since 1955 and has a refining capacity of 22 735 m^3/d (143 000 bbl/d). Because of increased water withdrawal for agricultural, industrial, and public uses, and decreased annual rainfall, seasonal water restrictions have been implemented since 1997. Kwinana Refinery potable water use in 1996 was ~6150 m^3/d (~1129 gpm), which was provided by the Water Corporation of Western Australia (WCWA) from groundwater and surface sources.

Water supply shortages and cost pressure risks spurred BP Kwinana to start a water-minimization program with the following key purposes:

- Significantly reduce potable water use by using lower quality water sources
- Increase the refinery's water recycle and conservation
- Achieve ZLD of process wastewater to the environmentally sensitive marine bay

To minimize potable water use, a five-step program was implemented as follows:

- A refinery-wide cross-functional team was formed and tasked with developing a holistic view of water use among all refinery units.
- A detailed water balance was created to understand water use and the water quality needs of various refinery processes. The water quality changes that occurred when moving from one process to another were also evaluated. This effort identified possible water conservation opportunities that could match water quality needs to process demands.

- Cost analyses of various options were completed and key savings potentials from water conservation efforts were identified.
- Water-use targets were established and employees were encouraged to reach those targets.
- A detailed analysis of each refining process was completed to better understand opportunities for water conservation, recycle, and replacement activities.

Kwinana Refinery water conservation, recycle, and replacement efforts resulted in significant reduction in use of potable water from WCWA. The use of lower quality groundwater realized an estimated savings of 2000 m^3/d (367 gpm) of potable water. Internal refinery reuse of process water (where quality matches the requirements of process units), condensate recycle from boilers and steam generators, and increased maintenance and housekeeping realized a water savings of 690 m^3/d (127 gpm). Over a 10-year period, Kwinana Refinery's water demand was reduced by an estimated 48% (IPIECA, 2015).

4.3.6.5 ExxonMobil, Integrated Manufacturing Complex, Singapore

The Singapore Refinery is fully integrated with the Singapore Chemical Plant (SCP) to form ExxonMobil's largest integrated manufacturing complex worldwide. The Singapore Refinery produces a range of fuel products, including gasoline, diesel, jet fuel, LPG, asphalt, and lubricant base stock. It also provides the chemical plant with feedstock. The SCP, in turn, produces ethylene, propylene, polymers, and specialty chemicals. These are the basic building blocks for high-performance products and consumer goods such as diapers, food packaging, solvents, cosmetics, and automotive parts.

ExxonMobil uses alternative water sources and constantly seeks opportunities to replace, reduce, reuse, and recycle water, which they use primarily for steam generation and process cooling. As part of the 2013 SCP expansion, ExxonMobil implemented a MBR technology to treat wastewater so that it could be reused for cooling in addition to seawater and fin-fan cooling. Currently, more than 55% of MBR-treated water is recycled as cooling water (ExxonMobil, n.d.). The MBR system is designed to treat an average wastewater flow close to 10 800 m^3/d (1982 gpm). Treatment occurs in two parallel biological treatment trains and four membrane operating system basins.

Carbonaceous biological oxygen demand, biological nitrogen removal reduction, and biomass and other suspended solids removal are also performed in this system. Together with the upstream primary oil/water removal units such as corrugated plate interceptors and DAF units, MBR is capable of treating wastewater from process units to effluent that not only meets the cooling tower makeup requirements, but also the National Environment Agency of Singapore regulatory specifications for discharge into the sea (PUB, 2018).

5.0 MINIMUM LIQUID DISCHARGE TECHNOLOGY APPLICATIONS—TEXTILES

5.1 India Textile Industry—Background

The global textile market was valued at approximately \$870 billion in 2017. India contributed \$150 billion in U.S. dollars in 2017, with projected growth to \$223 billion by 2021 (India Brand Equity Foundation [IBEF], 2019). Textiles represent a key market segment for the country. From 2017 to 2018, it contributed to 2% of India's gross domestic product, employed 45 million people, and contributed to 15% of the country's export earnings (IBEF, 2018).

On the downside, the textile industry is water intensive and competes with needs for domestic and agricultural water use. This competition is especially intense in India because of the lack of available water resources (KPMG, 2010). India has 16% of the world's people but only 4% of the world's water resources. By 2030, it is forecasted that India will face a 50% deficiency of water (Rajakumar, 2019). In response, Indian state governments are placing more stringent regulations on textile manufacturing facilities to incentivize more sustainable water management and to enable continued growth of this important market segment, but with reduced effect on the environment.

The discharge of water and high salt concentration into an ecosystem with sparse rainfall, such as in southern India, is not tolerated, and causes the natural salinity levels of local fresh surface waters to increase. Raising freshwater salinity levels not only affects the ecology of the water inhabitants, but also translates to yield loss for crops being irrigated by the water and to higher treatment costs for municipalities to convert the water into safe drinking water.

The textile industry is both water and salt intensive. Therefore, India's Central Pollution Control Board set effluent standards for wastewater management for the textile industry to incentivize both water conservation and reduce salt discharge. Amendments to the 1986 Environment Protection Act restrict the TDS level in textile discharge wastewater to be no more than 2100 mg/L. In addition to TDS, limits for pH, TSS, color, COD, BOD, chlorides, and sulfates are also established. The typical combined TDS level of a textile facility's wastewater is between 3000 and 10 000 mg/L; therefore, to comply with these regulation, they are forced to practice ZLD.

Although having documented regulations is important, enforcing them is equally important. In the textile industry in India, auditing facilities is a challenge because of the thousands of small facilities generating and discharging wastewater throughout the country. Consequently, many textile facilities continued to discharge outside regulated limits because of the significant expense of the treatment process required to achieve the discharge limits.

In 2011, the Farmers Association near Tirupur, a city in southern India and a significant textile hub, filed a petition claiming that textile facilities were unlawfully discharging contaminants into the Noyyal River, causing it to become toxic. Because the river was also used to irrigate local farmlands and provide water for livestock, the Farmers Association claimed that the pollution resulted in sterile fields and sick and dying livestock. This petition spurred local regulatory enforcement action and more than 700 textile dyeing facilities were forced to shut down by the Madras High Court (Narayanan, 2012). The closure of the textile facilities severely affected the economy in Tirupur, but it sent a clear message to the local textile mill owners that, to stay in business, they must find a way to comply with the discharge limits.

Today, all textile facilities and common effluent treatment facilities in Tirupur have installed ZLD technology to manage their wastewaters. Because of the constricted margins within a textile factory, controlling the operating cost of ZLD treatment was paramount to staying in business. Minimum liquid discharge using pressure-driven membrane technology was adopted as a cost-effective solution to minimize the volume of water requiring expensive thermal dewatering treatment. The successful adoption of MLD to lower the cost of ZLD wastewater treatment in India's textile industry is now being used as an example for other wastewater-producing industries in India, such as paper and pulp and distilleries.

This section presents an overview of the water-intensive wet processing steps of the textile manufacturing process and the general composition and volume of wastewater generated. Approaches used to treat and recycle the wastewater for reuse and meet stringent discharge requirements of ZLD are also reviewed.

5.2 Textile Wet Processing Steps

Several processing steps are used to produce fiber, yarn, and fabric for further processing into apparel, home furnishings, and industrial goods. Textiles typically go through four stages of production that may include (1) yarn formation, (2) fabric formation, (3) wet processing, and (4) textile finishing. Among these stages, the wet processing step is the largest water and chemical consuming stage, and it also generates a significant volume of wastewater. The main wet processing steps are summarized as follows:

- **Sizing/Slashing:** This process involves sizing of yarn with starch or polyvinyl alcohol (PVA) or carboxymethyl cellulose (CMC) to give the necessary tensile strength and smoothness required for weaving. The water required for sizing varies from 0.5 to 8.2 L/kg of yarn, with an average of 4.35 L/kg.
- **Desizing:** The sizing agents, which are made water soluble through treatment with acid (sulfuric acid) or enzymes, are removed from the cloth to make it suitable for dyeing and further processing. The water required for this process varies from 2.5 to 21 L/kg of material, with an average of 11.75 L/kg.
- **Scouring/Kiering:** The desized cloth is subjected to scouring to remove natural impurities such as greases, waxes, fats, and other impurities. The conventional method for doing this is through a batch kier boiling process; however, modern techniques using a "J" box provide continuous scouring. The water used in this process is alkaline, containing approximately 1% caustic soda, soda ash (sodium carbonate), sodium silicate, and sodium peroxide, with a small amount of detergent. The water required in this process ranges from 20 to 45 L/kg of material, with an average of 32.5 L/kg.
- **Bleaching:** The natural coloring of the fabric is removed through a bleaching process to render the material white. The bleaching agent is alkaline hypochlorite or chlorine. For high-quality fibers, peroxide is used. The water and chemicals used and the effluent generated in this process vary based on the type of operation and the material being processed. Bleaching yarn through the hypochlorite or hydrogen peroxide methods requires similar amounts of water (i.e., 24 to 32 L/kg of material). Cloth bleaching, however, requires more reagent (i.e., 40 to 48 L/kg of material).
- **Mercerising:** Mercerisation of fabrics provides luster, strength, dye affinity, and abrasion resistance, and it is applied to cotton fabrics to increase dyeing efficiency. The process is carried out using cold caustic soda (~3% to 5%) followed by washing with water several times. The water required for this process ranges from 17 to 32 L/kg of material, with an average of 24.5 L/kg.
- **Dyeing:** The dyeing process is the most complex step in the wet processing sequence. It is carried out at the fiber stage, as a yarn or as

a fabric. The dyeing process can use hundreds of dyes and auxiliary chemicals. Salt, either sodium chloride or sodium sulfate, is used to block the fabric's negative charge and allow for dye penetration into the fabric and fixation. Sodium chloride historically has been used, but the industry is shifting toward the use of sodium sulfate to avoid issues related to chloride-induced corrosion of processing equipment. Regardless of the type of salt used, the quality is given careful consideration because salts with high hardness can cause the precipitation of soaps or some dyes, which affects the dye uptake and appearance of the fabric. The grades of salt sourced for textile dyeing range in hardness from 0 to 180 mg/L. The amount of salt used depends on the dye being used, but can be as high as equal weight of salt to fabric. This is a salt concentration as high as 100 g/L. The water requirement in this step can range from 36 to 176 L/kg of material, with the average being 106 L/kg.

- **Rinsing and washing:** Fabric rinsing and washing is the largest wastewater-generating process, accounting for approximately 40% of the wastewater generated. The best quality water is required for the final stages/batches of washing, whereas lower-grade water can be used for the first washing steps. Therefore, to help minimize the water required, the washing effluents from the final washings are collected and used as the rinsing water for the earlier stages/batches of washing without treatment. The water use in this step can range from 250 to 700 L/kg of material.

The wastewaters from each of these processing steps are commonly combined to form a complex composite wastewater. Table 6.11 provides a summary of the composite composition range of the wet processing wastewater (Chakraborty, 2010; Ciabatti et al., 2010; Khandegar & Saroha, 2013; Majcen Le Marechal et al., 2012).

5.3 Textile Wet Processing Wastewater Treatment

Although segregating each wet processing waste stream could simplify wastewater treatment, the volumes of each are small, making it more efficient to combine and treat them as a composite in a single wastewater treatment process. The exception being that some facilities will treat the dye wastewater separate for immediate recycle of dyes and/or salts in the dye process. Because managing a wastewater treatment facility can be onerous, it is also common for smaller textile facilities to send their wastewater offsite and pay to have it treated at common effluent treatment facilities.

For each volume of wastewater delivered to the common effluent treatment facility, a fraction of clean water is provided in return. The textile facility must pay per volume of wastewater delivered and only gets returned fresh water based on the water recovery practice of the common effluent facility. Wastewater treatment in the common effluent treatment facility can be especially complex because each textile mill contributing wastewater may use different textile wet processing chemistry. Water recovery for reuse may not be as high for these facilities as it is for an effluent treatment facility within a textile factory because of the complex composition of the combined wastewater.

5.3.1 In-Facility Dye Wastewater Reuse

Exhausted dye baths have a high COD (up to 2000 mg/L) and high salt concentration (up to 100 g/L). For dye processes using sodium chloride as the salt type, in-process nanofiltration of the dye bath offers a mechanism to directly recover and reuse the dye until it becomes exhausted and/or to reuse the sodium chloride solution and concentrate the exhausted dye materials before sending them to the WRRF's common wastewater treatment process (see general schematic in this section).

The type of nanofiltration used in this process must be effective at passing monovalent salts through the membrane, but rejecting larger dye molecules. Because of the high concentration of dye components and the exposure to fabric fines, fouling of the nanofiltration membrane is common. Ultrafiltration is often applied before nanofiltration to remove the fines and reduce fouling; however, the organic dye components still pose a fouling risk and can result in a decline in water flow through the membrane over time. Typically, the nanofiltration membranes will last 1.5 years in this process. The investment in this treatment process is quickly recovered through dye recycle, reduced salt consumption, reduced demand on fresh water, and reduced water treated by the WRRF. The following is a general schematic of in-process dye bath treatment for reuse:

Color removal → ultrafiltration → nanofiltration (concentrate contains dyestuffs; permeate contains sodium chloride)

The in-process dye and salt recovery process is more limited when sodium sulfate is used in the dye baths. The sulfate ion is large and primarily rejected by commonly available nanofiltration membranes. Nanofiltration membranes capable of allowing sulfate-ion passage are often less selective at rejecting lower molecular weight dyes. In these instances, the dye bath may simply be treated for decolorization and the salts are reused in the dye process until the COD levels become too high, at which time it is sent to the common WRRF.

5.3.2 Primary and Secondary Treatment

Because the wastewater streams are not segregated, the composition is complex, with many kinds of organic and inorganic materials, and typically requires more than one treatment process. The discharge and reuse objective of the effluent treatment facility's water is used to define the levels of treatment used. The use of only primary and secondary treatment processes before discharging is typically used in regions that have moderate discharge restrictions with tolerance for high TDS levels and where fresh water is readily available. The focus of the primary and secondary treatment processes is to reduce the COD, BOD, total nitrogen, and total phosphorus, but these steps will not reduce the TDS of the water.

A combination of physicochemical and biochemical treatment is used. Physicochemical treatments are applied to reduce suspended solids, color, and some part of the organics and nonbiodegradable dyestuffs and chemicals. The most popular physiochemical process is coagulation, using common coagulants such as polyaluminum chloride and polyacrylamide followed by flocculation.

Secondary treatment using a biochemical process includes aeration and sometimes anaerobic methods. Because of the high level of dyestuffs and special auxiliaries in the wastewater, the BOD/COD of the mixed textile wastewater is quite low, at less than 0.2 (0.5 is considered easy to treat using biological processes), and thus the anaerobic process (e.g., hydrolysis-acidification process) may be necessary to break down the high molecular weight nonbiodegradable organic substances into smaller molecules. However, even with anaerobic treatment, the subsequent aerobic CAS process is not always able to achieve the full extent of COD reduction required to meet the COD discharge requirements. The raw wastewater from a textile facility has approximately 2000 mg/L COD and, after secondary biological treatment, it is typically reduced to 150 to 200 mg/L. If lower COD levels are required for discharging, AOPs may be necessary. The following is a primary and secondary treatment general schematic:

Equalization → Neutralization → Flocculation & Coagulation → Primary Clarifier → Biotreatment → Secondary Clarifier → discharged or sent to tertiary/advanced treatment

5.3.3 Tertiary Treatment

Tertiary treatment and advanced treatment are used when primary and secondary treatments do not meet the facility's permitted discharge requirements or water reuse objectives. For example, the TDS of the combined wastewater is typically 3000 to 10 000 mg/L; therefore, textile facilities in southern India, which are under the strict regulation to only discharge waters with less than 2100 mg/L TDS, need to revert to ZLD methods.

To reduce the economic impact of the ZLD process on the textile mill's overall cost of manufacturing, both water recovery and recycle and salt recovery and recycle are targeted to offset the operational costs of ZLD. The goals of the ZLD process are as follows: (a) produce water that is suitable for general reuse, less than 500 mg/L TDS; (b) optimize the amount of salt that can be reused by the textile mill dyeing process and minimize waste salt sent for disposal to landfills; and (c) minimize the overall operational cost of the ZLD process. Membrane-based MLD followed by evaporation/crystallization has been adopted to achieve these objectives (also referred to as "MLD/ZLD").

Depending on the composition of the TDS, there are two common MLD/ZLD schemes that are adopted. Scheme I is common for any composition of TDS, whereas Scheme II is adopted by textile mills having wastewaters with TDS comprising primarily sodium chloride. The general composite wastewater treatment Scheme I is as follows:

Primary Treatment → Secondary Treatment → decolorization → softening (ion exchange) → suspended solids removal (ultrafiltration) → pure water recovery (RO1, RO2, RO3) → Evaporation (MEE, MVR) → Crystallization (salt recovery) → evaporation (MEE) → solar pond

The general composite wastewater treatment Scheme II is as follows:

Secondary Treatment → decolorization → softening (IEX) → suspended solids removal (ultrafiltration) → pure water recovery (RO1, RO2, RO3) → Brine recovery (nanofiltration) → Evaporation (MEE, MVR) → solar pond

In both schemes, the pretreatment and the reverse osmosis systems are similar; it is the downstream processes for salt recovery that differ. The following subsections (Sections 5.3.3.1 to 5.3.3.4) discuss the subcategories of reverse osmosis pretreatment, reverse osmosis system, and salt recovery of tertiary and advanced treatment in more detail.

5.3.3.1 Reverse Osmosis Pretreatment

Before entering the reverse osmosis system, the secondary treated water is further treated for suspended solids removal using ultrafiltration modules or a sand filtration system, color reduction using a chemical process or decolorization resin, and hardness removal using lime softening and/or ion-exchange resins. The outcome of the pretreatment process provides a reverse osmosis feed comprised of 3000 to 10 000 mg/L TDS, 150 to 200 mg/L

COD, <20 ppm hardness, and 300 to 400 Pt/Co. Although this is significantly cleaner water than the raw wastewater, the high organic contamination level (COD, color) presents a fouling challenge for the reverse osmosis system, especially because it is concentrated by 10 times through the pure water recovery process.

5.3.3.2 Reverse Osmosis System

The reverse osmosis system is designed with multiple stages (i.e., RO1, RO2, RO3, etc). The concentrate of each stage becomes the feed to the subsequent stage to recover more water and concentrate the reverse osmosis reject stream, and to minimize water sent to the evaporator system. Most systems are designed with break tanks between reverse osmosis stages to enable independent operation of each and the ability to clean a reverse osmosis stage without taking the entire system out of service. However, to conserve energy and capital, some water treatment system designers have enough confidence in the pretreatment process and the performance reliability of each reverse osmosis stage to fully integrate them using only booster pumps between each stage.

With initial reverse osmosis feed TDS between 3000 and 10 000 mg/L, fouling-resistant elements with maximum operating pressures of 600 psi are used in the first reverse osmosis stage (RO1) to recover up to 75% of the water as pure water for reuse. The concentrate produced in RO1 has a TDS level in the range of 15 000 to 20 000 mg/L, and this water becomes the feed to RO2. In the second and third reverse osmosis stages (i.e., reverse osmosis and RO3), fouling-resistant elements with maximum operating pressure limits up to 1200 psi are used.

Few elements are on the market that that are designed for the challenging fouling conditions in this pressure rating. Seawater-type elements fit the pressure rating requirement, but are not designed to withstand the high organic contaminant (COD) levels of the water and, therefore, require frequent acid and base cleaning of the systems is required to remove the foulants. This can lead to high labor demand, high chemical use, and short element lifetimes.

To aid in reducing fouling, reverse osmosis elements installed in the later stages in the concentration process are often operated a low operating flux (6 to 10 $L/m^2{\cdot}h$) and achieve only 25% to 35% water recovery in each stage. The necessary operating flux to afford manageable fouling rates is highly dependent on the composition and fouling nature of the COD present in the water, with some COD blocking the membranes faster than others. When the systems are designed, it is difficult to predict what operating flux will be required to manage the fouling; therefore, additional reverse osmosis stages are often added to the design after the system is operational to squeeze out more water if low fluxes are needed to control fouling.

The final reject concentration of the multistage reverse osmosis system is limited by the upper pressure limit of the final stage of the reverse osmosis system, typically approximately 69 atm (1015 psi or 70 bar). The pressure limit is typically 15% lower than the maximum allowed operating pressure of the reverse osmosis element or reverse osmosis system, which is typically approximately 82 atm (1200 psi or 82 bar). On average, reverse osmosis–MLD systems in India's textile market recover between 88% and 91% of the water and concentrate salts in the water up to 60 000 mg/L TDS, while operating up to 69 atm (1015 psi or 70 bar).

Recently, specialty reverse osmosis elements have been developed to operate at pressures up to 118 atm (1740 psi or 120 bar), enabling the recovery of up to 50% more water and achieving a wastewater TDS level up to 120 000 mg/L. Textile wastewater treatment facilities are currently in the process of evaluating the use of these elements to further reduce the volume of water sent to the final, expensive thermal dewatering process in the ZLD sequence.

5.3.3.3 Salt Recovery (Scheme 1)

In southern India, many textile facilities are converting to the use of sodium sulfate in place of sodium chloride in the dye process. When sodium sulfate constitutes a large fraction of TDS in the wastewater, the most common salt recovery process used after reverse osmosis concentration is the use of a multieffect evaporator (MEE) or mechanical vapor recompression (MVR) to further dewater the reverse osmosis reject stream to near the saturation level of sodium sulfate followed by crystallization. A TDS concentration of at least 160 g/L is needed when the salt composition is primarily sodium sulfate to operate the crystallizer within a practical temperature range to recrystallize sodium sulfate (Figure 6.14). Achieving higher concentrations is desirable to minimize the cooling capacity of the crystallizer.

In the crystallization process, the sodium chloride and organic contaminants stay in solution and the sodium sulfate precipitates and is removed as a purified solid for reuse. The supernatant is typically contaminated with a mixture of sodium chloride, hardness, and organics, making it difficult to reuse; thus, it is dehydrated with a second evaporator sequence to recover more water before being sent to a solar pond for final evaporation. This process can also be applied if the TDS in the reverse osmosis concentrate stream is primarily sodium chloride and contaminated by hardness and organics; however, much higher concentrations are needed to achieve the saturation point of sodium chloride (see Figure 6.14).

5.3.3.4 Salt Recovery (Scheme 2)

When the textile wastewater TDS are primarily composed of sodium chloride, a membrane option is available to facilitate salt recovery without the need for recrystallization. The reverse osmosis concentrate stream is directed to a nanofiltration system equipped with ion-selective membranes that selectively pass monovalent ions (e.g., sodium and chloride) while rejecting divalent ions (e.g., sulfate and calcium) and COD. Because the monovalent ions pass into the permeate side of the membrane, the osmotic pressure of the permeate counteracts the osmotic pressure on the feed side of the membrane, resulting in low applied feed pressures to produce permeate and little energy requirement.

A simulation of the separation performance of a membrane of this type is provided in Table 6.12. In this example, the nanofiltration system is treating a feedwater with a total TDS of 58 239 mg/L, comprising a 50:50 mixture of sodium chloride and sodium sulfate, and is operated to achieve 50% recovery. Because of the high chloride passage, an applied feed pressure of only 18.03 atm (265 psi or 18.27 bar) is required to achieve 50% recovery. Note: If this same high osmotic strength feedwater was treated using a reverse osmosis element, an ultra-high pressure system operating at greater than 98.7 atm (1450 psi or 100 bar) would be needed. In this simulated example, the permeate from the nanofiltration system is a 96.6% pure sodium chloride solution, which can be recycled back to the dyeing process as a solution or can be sent to an evaporator/crystallizer system to produce solid purified salt. The concentrate stream is reduced in volume by 50% and is at a higher TDS then the starting feedwater, resulting in less evaporation before sending to the solar pond. A second-stage nanofiltration system could be used to capture more purified sodium chloride and reduce the volume evaporated even more.

5.3.4 Evaporator/Crystallizer

The evaporator process is the most expensive to build and operate. The materials must be constructed using corrosion-resistant alloys resulting in high capital cost. The heat capacity of water is one of the highest of common solvents used in industry; thus, significant energy is required to vaporize water driving the operating costs. Because of these expenses, all previous wastewater treatment steps are optimized to minimize the volume of water requiring treatment in this step.

The configuration of the evaporator system depends on the type of salt recovery process being used. In either salt recovery option mentioned in Section 5.3.3.3 and 5.3.3.4, the resulting water being sent to the evaporator is typically between 50 000 to 65 000 mg/L, unless an ultra-high-pressure reverse osmosis stage is used to concentrate it further to 120 000 mg/L. In either case, an MEE or MVR is used to concentrate the water to the saturation point (i.e., ~160 000 mg/L for sodium sulfate and ~300 000 mg/L for sodium chloride). The condensate collected from the evaporators can achieve the 500-mg/L TDS pure water requirement for reuse in the textile mill and, when combined with the reverse osmosis permeate collected at the facilities, the overall water recovery of the wastewater treatment process can reach 99%.

The final saturated salt solution from the evaporator is cooled in a crystallizer to precipitate the purified solid salt for reuse. The purity requirement for reusing these salts in the dye process is not stringent, with the exception of requiring low hardness and low color. Finally, the supernatant from the crystallizer containing the waste organic contaminants and residual salt mixture is dewatered in an evaporator pan or solar pond for final waste solid collection and disposal.

Typically, the reverse osmosis concentrated water is only minimally treated before being sent to the evaporator/crystallization process. However, some operations may require additives to be dosed to control foaming and scaling in the evaporator. Minimizing additives is important to avoid negatively affecting the crystallization process. To reduce the use of additives, some operations may install an ultrafiltration step to remove suspended solids and/or lime softening to reduce scaling. Because of the challenging water composition, it is not uncommon that an evaporator used to dewater textile wastewater is cleaned at least once a month, if not weekly, to remove scaling and fouling.

5.3.5 General Minimum Liquid Discharge/Zero Liquid Discharge System Reliability

In general, one-third of the operating costs of the wastewater treatment process is primary, secondary, and tertiary treatment; one-third is water recovery by reverse osmosis; and one-third is final dewatering and salt management. The negative economic impact of the MLD/ZLD system on the textile mill is reduced as more water is recovered by reverse osmosis for reuse, less water is recovered by thermal evaporation, less salt is landfilled, and more salt is reused in the process. The MLD/ZLD systems treating textile wastewater have successfully operated in southern India for more than 8 years, achieving an overall water recovery of 98% to 99%. The biggest challenges these systems face is high operation costs and operation reliability; thus, further innovations to alleviate these issues are welcomed.

The type of fabric being processed at the facility or the type of dye being used in the process can significantly affect the reliability of the wastewater treatment process. In general, yarning wastewaters are considered the easiest, followed by knitting. Woolen wastewaters are the most challenging. It is believed that the PVA used in the woolen process is removed during sizing and does not break down well in the secondary wastewater treatment steps, leaving a high fouling, high molecular weight contaminant in the wastewater entering the reverse osmosis system. These high molecular weight contaminants will block the reverse osmosis membrane and cause flow loss. The other leading foulant causing membrane blockage is dyes. It is typically recognized that darker dyes present the most significant wastewater treatment challenge.

The bottleneck of the water treatment system is typically reverse osmosis and not pretreatment. To manage the fouling and scaling that occurs in reverse osmosis systems, facility operators in southern India have adopted routine cleaning practices. Because Sundays are routinely slower days at the textile mills, the water facility uses this weekly downtime to take the membranes out of service to perform a thorough cleaning using caustic followed by acid.

In addition to these weekly cleanings, some facility operators have also adopted daily reverse osmosis permeate flushing routines and/or mini-CIP practices to maintain performance in the later reverse osmosis stages (RO2, RO3, etc.) in the system. These practices are considered acceptable when taking into account the poor quality of the feedwater entering the reverse osmosis system and high system water recovery. Regardless, the operators welcome more robust solutions to provide more assurance of stable operation. If the routine cleaning practices do not work to manage the fouling, flow through the membrane is reduced and more water is sent to the evaporator, driving the treatment costs up significantly.

5.3.6 Textile Minimum Liquid Discharge/Zero Liquid Discharge Process Conclusion

When ZLD was first required by the textile mills in southern India, it was assumed to be cost prohibitive. However, with the effective implementation of membrane-based MLD followed by thermal evaporation, textile mills have found that they are able to meet the wastewater discharge requirements and still maintain a profitable business. The successful implementation of this process enables both a reduced water and salt environmental footprint. Continued membrane innovations in select ion-separation and

TABLE 6.11 Composite raw wastewater from wet processing process.

Volume (L/kg of product)	pH	TDS (mg/L)	TSS (mg/L)	Color (Pt/Co)	BOD (mg/L)	COD (mg/L)
100–1000	Alkaline	3000–10 000	100–700	>14 000	50–550	250–8000

Additives: PVA, CMC, lubricants, surfactants, detergents, dyestuff, $NaHSO_3$, Na_2SiO_3, NaClO, H_2O_2, $NaClO_2$, NaCl, Na_2SO_4, NaOH, etc.

TABLE 6.12 Specific ion and corresponding sodium salt separation of a high TDS feed composition using a commercially available ion-separation membrane operated at 50% recovery.

Solute	Feed	Permeate	Reject
Na^+ (mg/L)	20 894	13 963	27 828
Cl^- (mg/L)	17 745	21 082	14 406
SO_4^- (mg/L)	19 600	609	38 595
NaCl (mg/L)	28 992	34 444	23 537
Na_2SO_4 (mg/L)	28 992	901	57 088
Total TDS (mg/L)	58 239	35 654	80 829

Note. Results from modeling the performance of FILMTEC FORTILIFE™ XC-N elements using FILMTEC's ROSA software.

ultra-high-pressure operation will further lower the cost of treatment as will innovations to improve the fouling resistance and durability of reverse osmosis membranes. The example MLD/ZLD application developed in southern India can be used to drive similar treatment practices for textile industries in other parts of the world as well as other industries looking to reduce their environmental footprint.

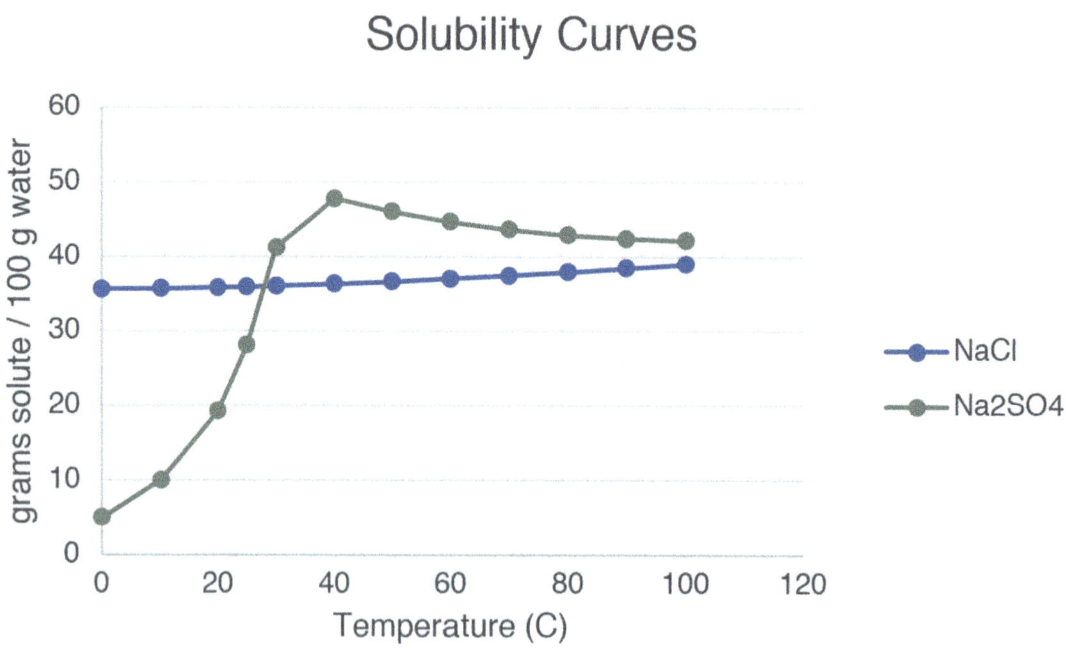

FIGURE 6.14 Solubility limits of sodium chloride and sodium sulfate as a function of temperature.

6.0 MINIMUM LIQUID DISCHARGE TECHNOLOGY APPLICATIONS—AUTOMOTIVE

6.1 Background on the Automotive Industry

The automotive industrial sector consists of companies that design and manufacture cars, vans, and trucks. The text herein focuses on company-owned production plants. However, it should be noted that there is a large supply chain that supports the industry through a variety of parts, materials, and services, including supply of rolled steel and aluminum used to form the car body, windows, materials used to make dashboards and seats, tires, electronics, chemical supplies (e.g., paints, coolants, cleaning agents), and supply of machines/robots/equipment used to build the automobiles.

Company-owned production plants fall into four general categories: Casting plants, engine plants, stamping plants, and assembly plants. Within these operations, significant water use can occur during parts machining, factory temperature control, and painting of newly manufactured vehicles (Moore & Buzby, 2017).

6.1.1 Casting Plants

The production of automobile engines begins with casting of the engine block within a foundry, often referred to as a "casting plant" within the automotive industry. (Casting is also used for production of other automobile components, such as transmission cases.) Two of the most common casting methods are sand casting and die casting. In the sand molding process, a mold is first made from a compacted mixture of sand, clay, and water (called "green sand"). Molten metal is then poured into the mold, and once the metal has cooled, the mold is opened and the sand mixture is shaken off. Die casting entails forcing molten metal into a reusable steel mold (called a "die") under high pressure. Within a die-casting plant, water is used for cooling, quenching, and for wetting to enable release of the engine/part from the die and associated with production of steam.

6.1.2 Engine Plants

Engine blocks received from the casting plant serve as the focal point from which engines are built within the engine plant. High-precision machining and cutting are important production elements within an engine plant. These operations entail the use of coolants (also called "cutting" or "machining fluids"), which are continuously passed across the engine or part being machined. The coolant solutions used within manufacturing operations are prepared at the site and will typically be made up of 5% to 15% pure coolant, with the balance being water. These fluids play multiple important roles, including the following: (a) carrying metal chips or particulate away from the item being machined, (b) providing lubrication, and (c) drawing heat away from the item being cut or machined. Coolant purchase represents a sizable operating expense within an engine plant. As such, efforts are made to extend the life of the coolants as long as possible before disposal. Other aspects of water use within engine plants are often related to cooling, steam production, parts washing, wash booths, general cleaning, and fire suppression.

6.1.3 Stamping Plants

Stamping plants produce the metal shell or body of automobiles. Stamping is not always done in a dedicated plant; stamping is sometimes colocated at an assembly plant. Metal is delivered to the stamping plant in large rolls. The metal is first unrolled, then cut to the correct sizes needed to make each part of the body. These cut sheets of metal are then sent to presses that form the metal sheets into the final shapes needed for the automobile body. Several parts may then be welded together before parts are shipped to the body shop for final assembly. The significant water use within stamping operations is for cooling.

6.1.4 Assembly Plants

Final assembly/production of the automobile occurs within the assembly plant. Assembly plants use—by a wide margin—more water than any of the other automobile plants previously discussed. Assembly plants typically include three significant departments: (a) body shop, (b) paint shop, and (c) general assembly. In the body shop, the sheet metal parts (from stamping) are first fastened together loosely by human workers, after which the car body moves through a series of spot-welding operations, adding more parts to permanently assemble the body (Groover & Kolchin, 1997). The body shop is not a significant water-using operation within the typical assembly plant. Cooling represents the main use of water in the body shop. Some body shops also include body washing, although body washing has been phased out at most facilities to conserve water.

The next significant department in the assembly plant is the paint shop, which is by far the largest water-using department within an automobile assembly plant. Within the paint shop, the car body passes through phosphating and E-coat processes, which essentially consist of a series of dip tanks and rinsing stages. The phosphating process serves to clean the car body, removing any oils or particulate, after which the metal surface of the car body is conditioned by submerging the body in a phosphate dip that etches the metal surface so that primer coat will adhere strongly to the surface. Rinsing plays an important role through the phosphating process and represents significant water consumption for the site.

The car body next enters the first stage of the E-coat process, where it is submerged into a tank containing a paint emulsion through which a high voltage is applied, causing the paint emulsion to deposit on the car body. The body then passes through a series of rinse stages to remove any residual emulsion from the surface. After the body exits E-coat, the surface coating of paint is applied. Air humidification must be controlled within a relatively tight range during the painting process, and this correspondingly represents another water use within the paint shop. Other water users in the paint shop include cooling, skid wash, pump seals, and makeup to the sludge pit.

After exiting the paint shop, the body then proceeds to general assembly, where final assembly of the automobile takes place. Significant water users include fluid fills (e.g., wiper fluid, coolant), cooling, and leak test.

6.2 Wastewater Treatment Within Automotive Plants

The type of wastewater treatment process used at an automotive plant will vary based on a number of circumstances, including the following: (a) discharge permit requirements, (b) where effluent wastewater is discharged to (e.g., public wastewater facility, direct discharge), (c) types of processes generating wastewater at the site, (d) whether or not sanitary wastewater is treated onsite, and (e) water management and efficiency practices within the plant.

Wastewater generated by engine plants will tend to be characterized by high O&G concentrations as well as high COD because of the various machining fluids and chemical agents used in plant operations. Wastewater composition variability can also be an important issue for wastewater from engine plants.

There is a fairly wide variety in wastewater treatment operations found at engine plants. Having equalization at the front of the treatment facility can be valuable because it will help to dampen concentration swings/spikes as well as provide an opportunity for some degree of initial oil removal, although considerations must be made for removal of floated oil. Oil removal is an important element of wastewater treatment at an engine plant, which is often addressed through processes such as an oil-water separator, DAF, or membranes.

Following oil removal, the wastewater will still be characterized by high COD. In some cases, a site may be able to discharge wastewater (following oil removal) to a WRRF that is willing to accept high COD wastewaters. When additional COD removal is needed before discharge, it will be achieved through biological treatment. A variety of biological treatment processes have been used at automotive engine plants; activated sludge and MBR are perhaps the most common amongst new installations. It should be noted that, in some cases, plant operations and/or discharge permit requirements will necessitate the removal of target heavy metals. In these cases, chemical precipitation will be used before biological treatment.

Wastewater assembly plants are characterized by lower concentrations of O&G and COD than engine plants. In most cases, wastewater treatment at assembly plants will consist of chemical precipitation only (for metals removal), followed by discharge to a WRRF.

6.3 Brief Background on Minimum Liquid Discharge Solutions

A company's environmental conservation and sustainable business practices continue to increasingly influence consumer buying decisions. This is especially true for the automotive sector. Consequently, the automotive sector tends to highly value sustainability, with companies each striving to be perceived as a green brand by consumers.

Water use/efficiency are almost always a part of an automotive company's sustainability goals. Water use will typically be tracked and reported within a company's sustainability report in terms of water usage per vehicle, or per "unit" (e.g., m^3/unit or gal/unit), and also sometimes as volume used (e.g., m^3/yr or mil. gal/yr). Water (sustainability) goals within the automotive sector are (almost) always focused on reducing the use of fresh water, rather than focusing on reducing the volume of wastewater produced. However, as a company works to reduce water use, there will be a corresponding reduction in the volume of wastewater produced and discharged. As such, water conservation and MLD can be considered to go hand-in-hand; any reference to "water conservation" herein should be taken to correspondingly provide MLD benefits.

To reduce CapEx required and/or improve financial return, many sites have worked to improve water use efficiency within their manufacturing operations and utilities, often through projects that are low cost and easy to implement. As an example, most automotive sites strive to maximize the number of cycles that their cooling towers are run. In fact, some automotive companies have even made a requirement for their sites to all run cooling towers at or above 4 or 5 COCs. Other water conservation efforts may focus on reducing the volume of makeup water used in certain stages of the phosphate or E-coat processes or to maximize countercurrent flow between stages in these processes. These types of process and utility water efficiency projects will often individually provide incremental water savings; however, when multiple projects are implemented at a site, the overall water savings achieved can be sizable.

Water reuse plays a significant role (upstream) for many of the manufacturing processes within automotive plants. This can include countercurrent flow of water within a paint line at an assembly plant, reuse (and life extension) of machining fluids at an engine plant, as well as many other examples. However, many automotive sites have also implemented larger "end-of-pipe" reuse projects that align with what many people might envision when thinking of reuse. For sites that treat wastewater via biological treatment (particularly for MBR-treated wastewater), it is not uncommon for sites to reuse treated effluent for irrigation, within toilets, or as cooling tower makeup. These types of reuse projects are most common at sites located in water-stressed regions of the world.

Many automotive companies have manufacturing operations within water-stressed regions, such as Mexico and India. In these stressed regions, water availability challenges will frequently necessitate wastewater recycle, particularly for assembly plants. Because of the need for consistent and high-quality water, membranes (nearly always reverse osmosis) play an important role in wastewater recycling facilities found within the automobile manufacturing industry. This often means that wastewater recycle systems will need to include some type of pretreatment in front of the reverse osmosis to ensure influent water quality requirements are met for the reverse osmosis system. Such an example is described for the automotive sector later in this section.

Stormwater recycle has also been implemented at some facilities within the automotive sector. The viability/applicability of stormwater recycle will often be dependent on local circumstances (e.g., rainfall, water availability, the presence of stormwater surcharge fees). One such case study for stormwater reuse is described within *A Framework for the Successful Implementation of Onsite Industrial Water Reuse* (Moore & Buzby, 2017). In that example, high stormwater surcharge fees and water needs for a neighboring industry created an ideal opportunity for stormwater recycle at a General Motors plant.

6.4 Case Study: Advanced Wastewater Recycle in a Water-Stressed Region

6.4.1 Background

A significant automotive manufacturer produces engines, transmissions, and full-size trucks at a large manufacturing complex in Silao, Guanajuato, Mexico. The site directly employs approximately 6000 personnel and there are also approximately 4000 contracted employees onsite on any given day.

Following a period of dramatic growth since 1995, the site made a decision to add nearly 3 million square feet of new production facilities. This would increase the water needs for the site and was the driver for the wastewater recycle project covered in this case study.

The Silao-Romita aquifer, which provides water to the Silao site, has been identified as stressed, and the region is expected to have continued growth in population, agriculture, and industry (Foster et al., 2004). As a leading environmental steward and good community neighbor, this automotive manufacturer embarked on a journey to reduce its water footprint in Silao, while at the same time expanding production at the site and bringing new jobs to the region.

The Silao site relies on six groundwater wells ranging from depths of 300 to 500 m for meeting all freshwater needs. Because of regional water stress, high regional water demands, and slow recharge rates for the aquifer, individual wells at the Silao site cannot be continuously run. As such, the site must frequently rotate the wells in service to obtain the water needed for manufacturing operations. Before the production increase, site groundwater withdrawals averaged slightly less than 2000 m^3/d. Although the site has groundwater rights for approximately 1 000 000 m^3/yr (~2700 m^3/d), the aquifer is overdrawn and has a low recharge rate. Consequently, it is not believed that groundwater yields could meet future needs following the production increase (up to 3100 m^3/d).

Before starting this project, it was noted that the Silao site had already been practicing some extent of water reuse, whereby a portion of the activated sludge effluent was used for irrigation, cooling tower makeup, and restroom toilets. This represents the maximum extent of reuse of biologically treated effluent that was possible for the site as governed by the quality needs of other water users around the site. Further treatment/purification of the wastewater would be required to allow further reuse of wastewater.

Numerous concepts were developed for the wastewater recycle system, and they were rated/compared versus one another relative to key project criteria: CapEx, OpEx, robustness, operability, and percent of wastewater recycled. The concept selected consisted of conversion of the existing CAS over to an MBR, followed by treatment through a multistage reverse osmosis.

6.4.2 Water Balance

A critical step in this project was to determine current water volumetric and quality demands for operations across the site, in addition to developing an accurate projection for future water needs/requirements following the production increase. The team first developed a water balance for current

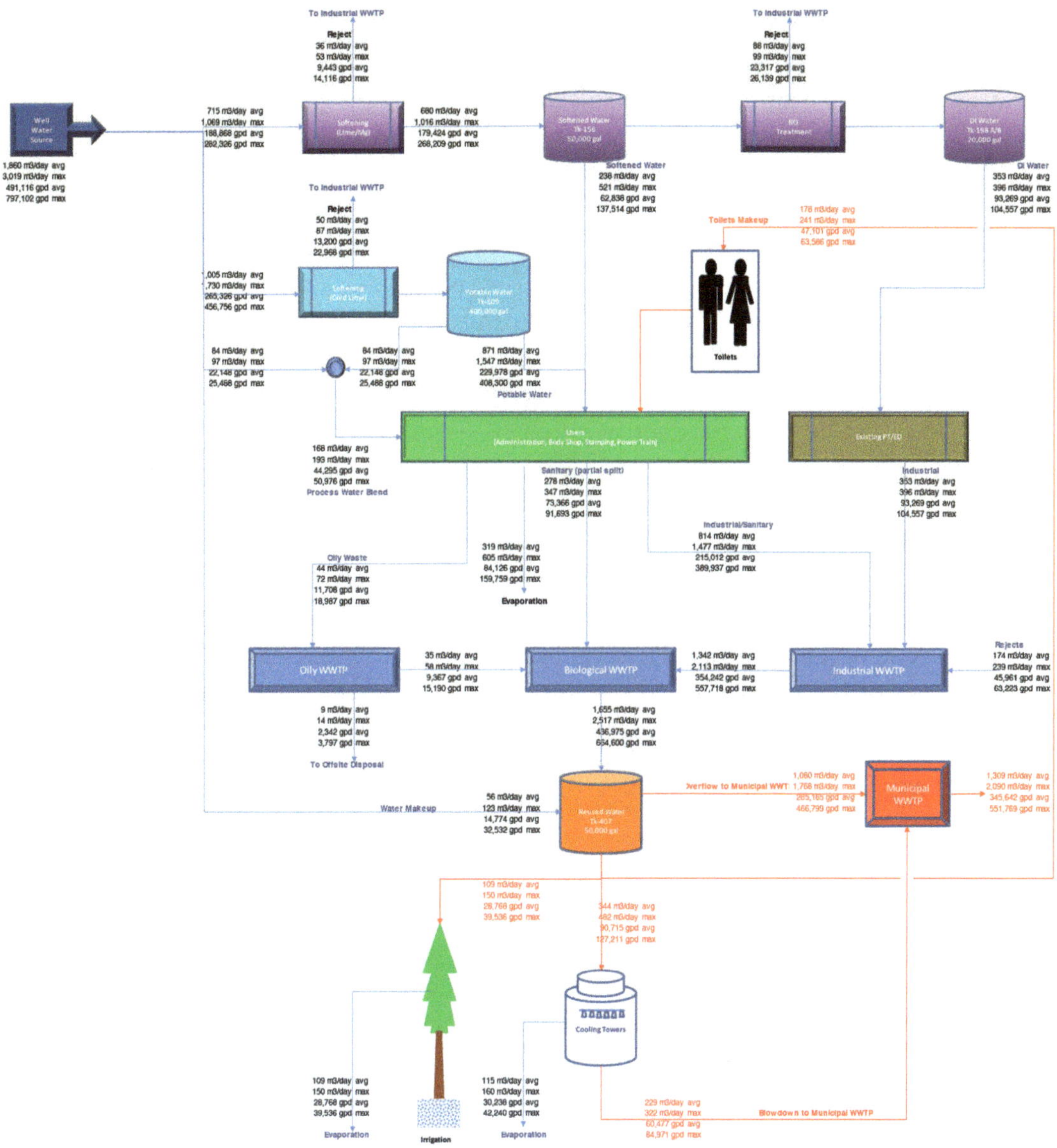

FIGURE 6.15 Simplified water balance depicting conditions before Silao, Mexico, facility expansion.

facility operations (simplified version shown in Figure 6.15). This "current case" balance was expanded to include water quality characteristics using existing (water and wastewater) analytical data from the site and supplementing with new water and wastewater sampling and analyses where needed (Olson et al., 2017).

The team next conducted in-depth interviews with key engineers, area managers, facilities personnel, operators, and contracted personnel to determine the following for each water using operation at the site: (a) present-day water quality needs, (b) projected water quality needs after the site expansion, and (c) projected water volume needs after the expansion. Information from these interviews was used to create a second version of the water balance that reflected projected future water use and wastewater flows.

To ensure correct design and sizing of a recycling facility, it is critical that both typical/average conditions as well as peak/extreme conditions are considered. For this project, the "projected future" balance included average as well as peak flow values.

Information obtained during these interviews was used along with the water balances to identify potential users of recycle water, including water quality specifications and volume needs. It was observed that the total water volume needs for the potential end users was greater than the volume of wastewater that was available for treatment in the recycling facility. This yielded an opportunity to shorten the list of potential users to reduce CapEx. From a CapEx perspective, piping installation costs can add up quickly and can be an important factor in deciding which processes receive recycle water. Potential users that were remotely located and characterized by sufficiently low water volume needs were removed from consideration.

As the site expands, and new automotive manufacturing processes (and water demands) are added, the site will need to recycle as much wastewater as possible to supply the projected water needs of the manufacturing processes (Olson et al., 2017). This means that an increasingly large percentage of site water needs will be getting met by the use of recycle water (versus fresh water). Such a situation can lead to the cycling up of constituent concentrations in the water. Compounds such as silica and total hardness are of particular concern because their presence can be detrimental during certain manufacturing operations within an automotive plant. As such, there needed to be a mechanism for removing these and other constituents from the water.

6.4.3 Wastewater Sources and Water Resource Recovery Facility Loading

Wastewater generated throughout the site is discharged to the WRRF at the Central Utilities Complex (CUC). Wastewater has been organized into three categories based on source and characteristics: oily wastewater, industrial wastewater, and sanitary.

The oily wastewater stream is generated in the engine and transmission plants from grinding, cutting, and machining operations taking place therein, and which cause this stream to have high O&G and COD concentrations. As such, oily wastewater is pretreated for oil removal through screens, oil water separation, and emulsion breaking batch reactors before being combined with the "industrial wastewater" stream (i.e., wastewater produced from all other manufacturing operations). The oily wastewater represents only 2% of the overall hydraulic loading to the WRRF, but it accounts for 30% to 40% of the total COD loading to the WRRF.

The industrial wastewater comprises 81% of the total wastewater flowing to the CUC. Historically, the industrial wastewater has been combined with the pretreated oily wastewater, then flowed to the industrial WRRF. The industrial WRRF uses equalization, pH adjustment with lime, coagulation, and gravity separation with lamella plate settlers, and is then sent to biological treatment. Sanitary wastewater is also treated within biological treatment and comprises approximately 17% of the total wastewater flow being treated within the CUC.

Before this recycle project, biological treatment consisted of anaerobic treatment followed by aerobic treatment. Anaerobic treatment was performed within two alternating batch treatment units, serving as an anaerobic selector. Aerobic treatment was performed within a CAS process. The CAS overflow was treated through sand filtration (1655 m^3/d average). Most of the sand-filtered effluent was sent to the municipality (1080 m^3/d). The remainder of this treated wastewater (631 m^3/d) was being reused for irrigation (109 m^3/d), CUC cooling tower makeup (344 m^3/d), and for toilets at the site (178 m^3/d), as illustrated in Figure 6.15. All water reused for toilets eventually ends up back at the WRRF. In contrast, water reused for irrigation and the CUC cooling tower (i.e., 453 m^3/d) ended up either lost to the environment (into the ground or evaporated) or discharged to the municipal wastewater facility.

6.4.4 Oily Wastewater Loading Reductions and Treatability Evaluation

When the Silao site first opened, the production rate was much lower than present day. At that time, wastewater flows were correspondingly also much lower, and the site's biological treatment system was much smaller (400 m^3/d design capacity). As the site grew, WRRF capacity was added, including addition of an adjacent (2000 m^3/d) biological treatment system. Before this project, the 2000 m^3/d system treated all industrial wastewater and oily wastewater as well as a portion of the sanitary wastewater; the smaller system treated only sanitary wastewater.

The concept developed for this project entails segregation of the oily wastewater and treatment within the smaller biological system, from which effluent will discharge to the municipality (rather than going to the recycling facility). This is beneficial to the operation of the recycle system because the oily wastewater represents a significant COD load with highly variable composition.

It was important to ensure that effective treatment of the oily wastewater could be achieved within the dedicated biological treatment process. The oily wastewater flowrate (35 to 70 m^3/d) is well below the design capacity for the smaller biological system; however, the oily wastewater is also characterized by high COD concentrations (typically 4000 to 12 000 mg/L).

6.4.4.1 Modeling Oily Wastewater Treatment

The smaller biological system consists of four sequential chambers, so it would be possible to operate the first one or two stages anaerobically with the remaining chambers operated aerobically if that provides superior treatment versus operating all aerobically. In addition, if treatment of oily wastewater by itself proves challenging, it could be diluted with sanitary wastewater before treatment in the dedicated biological treatment system. A study was conducted to evaluate these two aspects and determine how best to operate the smaller biological system.

Five-day composite samples were collected for oily wastewater and also for sanitary wastewater. A biomass "seed" sample was also collected from the Silao CAS sludge return. These samples were sent to an outside laboratory specializing in respirometry testing. Tests were conducted to assess biodegradation characteristics and treatability of the oily wastewater with and without dilution with sanitary wastewater, as well as under "aerobic only" versus "anaerobic + aerobic" conditions.

Using results from the testing, the biological treatment system was modeled within GPS-X™. A total of 25 different model scenarios were run to evaluate the effect of influent feed, recycle, and wastage rates on the system performance. Results of the study confirmed the oily wastewater is treatable without need for dilution, with best performance occurring for the configuration of anaerobic followed by aerobic.

6.4.4.2 Oily Waste: Upstream Loading Assessment

A separate study was conducted to identify opportunities for reducing the overall loading of COD to oily wastewater sewer. To understand the sources of oil and high COD materials contributing to the oily wastewater and how/where they are entering the sewer, general operating and housekeeping practices were observed related to coolant and detergent systems in the site's engine and transmission plants. In addition, safety data sheets of the coolant, detergent, and oil materials used in these areas were reviewed. A sampling plan was created and implemented to develop a better understanding of the loading and biodegradability effects from some of the significant upstream oily wastewater generators. This process enabled the development of an array of opportunities for improved fluids management and correspondingly reducing the overall COD and O&G concentrations of the oily wastewater.

6.4.5 Wastewater Recycle Technical Approach

When considering the various potential end users of recycle water and their corresponding volume needs, it was noted that the paint shop constituted the vast majority of water demand (~70%) for this group. However, the paint shop also has the highest water quality requirements of any potential end user of recycle water at the site; paint shop water quality requirements dictated the technical approach (and effluent quality requirements) for the recycling facility. It was thus noted that the recycle process would need to include reverse osmosis treatment to ensure the TDS, silica, and hardness requirements of the paint shop are met.

Next, to meet the COD and TSS feedwater quality needs for the reverse osmosis process, it was decided to convert the CAS system over to an MBR process. In addition, modifications were made to feedwater purification and industrial wastewater treatment to improve water quality and enable greater recovery rates for the reverse osmosis system. A more thorough discussion is provided in the following subsection.

6.4.5.1 Improving the Quality of Feedwater to the Recycling Facility

To minimize CapEx, the team looked beyond the WRRF when conceiving the overall approach for recycling wastewater and maximizing the volume recycled. Reverse osmosis is a critical element to many recycle systems based on the ability of reverse osmosis to produce consistent high-quality water suitable for direct reuse in many industrial applications. However, care must be taken to ensure the water being treated by the reverse osmosis is of acceptable quality to not damage or affect operability of the reverse osmosis. Upon evaluation of historical data as well as sampling performed during this project, it was noted that effluent from the WRRF's CAS process would be poorly suited to reverse osmosis treatment (because of high TSS, silica, hardness, phosphate, COD, and O&G concentrations).

One approach to improve water quality before reverse osmosis treatment was to include additional treatment steps in the recycle system before reverse osmosis treatment. Such an approach can indeed be technically viable; however, the inclusion of additional unit operations also increases the overall project cost.

An alternate approach that can be considered is to consider the facility more holistically. Such an approach entails looking upstream to understand the root causes and evaluate if easier solutions can be used at the source to address (or partly address) the problems. In the Silao example, such an approach was found to be advantageous and enabled the adoption of a simpler recycling facility concept than otherwise would have been possible.

6.4.5.1.1 Chemical Oxygen Demand

When investigating historical performance data for the WRRF, it was noted that effluent COD concentrations were higher than one might expect and often variable (because of COD variability of the influent wastewater). While the effluent COD was sufficiently low to consistently comply with the site's discharge permit, the team was concerned that the COD levels (and COD variability) could be problematic to operation of a recycle system. The following three opportunities were identified and implemented by the team to significantly reduce these issues:

- **Segregation of the oily wastewater:** Although it represents only 2% of hydraulic loading, the oily wastewater stream represents 30% to 40% of the overall COD loading to biological treatment. It was further identified that the oily wastewater was the predominant cause for the concentration spikes/variability observed for the influent to biological treatment. As a final point, it was also found that some of the organics present in the oily wastewater are somewhat recalcitrant or slow to biodegrade (particularly when compared to the COD present in the sanitary and industrial wastewater). It was thus decided to segregate the oily wastewater and treat this stream within a dedicated biological treatment system discharging to the municipal WRRF.
- **Fixing and repairing existing equipment:** It was found that a number of items at the WRRF were in need of rehabilitation or replacement (such as mixers, lamella plates, air diffusers, and expanding sludge management). Fixing or replacing these items was considered to be of high importance, enabling the WRRF to operate as designed and to provide higher quality effluent.
- **Switch the anaerobic treatment process from batch to continuous stirred tank reactor (CSTR):** The anaerobic section of the biological treatment facility consists of two long tanks. Anaerobic treatment had originally been run in a batch mode, such that treatment was only carried out in one tank at a time. As anaerobic treatment was being performed in one tank (Tank A), the other tank (Tank B) was being emptied (to the CAS process) and then refilled. Once Tank B had been refilled, the tanks would switch, with anaerobic treatment occurring in Tank B and Tank A being drained and refilled. This mode of operation limited the overall effective volume for treatment and created a situation where some of the solids would settle as a tank was being

emptied and refilled, leading to deposition on the tank bottoms and corners that reduced working volume. The operation of the anaerobic system was changed from batch mode to a CSTR mode of operation, whereby treatment is continuously carried out in both tanks.

6.4.5.1.2 Total Suspended Solids

A significant amount of floating and poorly settleable fine solids were observed within the Silao biological system. The sanitary wastewater collection system was noted to use an in-channel grinder system before the wet well to the main lift station that transfers solids to the biological WRRF. This grinder system was generating small solids that were passing through the screening devices and ending up in the biological treatment system. As part of the project, the grinder was removed and replaced with a vertical mechanical bar screen to keep large solids from entering the wet well and protecting the lift station pumps.

6.4.5.1.3 Silica

A key operating challenge/risk for reverse osmosis membranes is the opportunity for fouling of the membrane surface. Inorganic mineral scales, organic macromolecules, and the growth of microorganisms all present a risk for fouling of a reverse osmosis membrane (Bonné et al., 2000). Silica fouling of reverse osmosis membranes occurs by the deposition of silica and the subsequent formation of a silica film on the membrane surface (Kim, 2017). In the case of hardness scale or biological fouling, the membrane performance can be relatively readily recovered through low or high pH cleanings, respectively. However, membrane performance degraded by silica fouling is not easily recovered by simple water flushing or cleaning agents, but requires membrane replacements (Koo et al., 2001).

Based on reactive silica concentrations measured in the CAS effluent, as well as projections following the site expansion, it was anticipated that reverse osmosis performance could be limited by silica. The team evaluated potential sources for silica throughout facility operations and found that nearly all silica found in the facility's wastewater had actually come from the well water used by the site, which is characterized by high silica concentrations. Two lime softening systems are used to soften well water at the facility: one softener produces process water for the site, whereas the other softener produces potable water for the site. The softener producing process water is run at pH 11.2 and includes addition of magnesium oxide (MgO). Under this mode of operation, silica will bind to magnesium hydroxide floc that are produced and precipitated, reducing the silica concentration down to $<$20 mg/L.

In contrast, potable water is produced through traditional cold lime softening that is run at a pH of 10.4 with no MgO addition. This mode of operation will not remove silica; the majority of silica found in the wastewater was coming from the potable water. As part of this recycle project, the potable water softening system is being modified to include dosing of MgO. Thereafter, the potable water softener will be run identical to the process water softener, reducing wastewater silica concentrations significantly.

6.4.5.1.4 Phosphate

Precipitation of calcium phosphate scale on reverse osmosis membranes continues to be a significant challenge for wastewater recycling facilities (Malki & Abbas, 2013). Before the site expansion, the wastewater phosphate concentration was typically 20 to 40 mg/L, which raised concern over potential fouling challenges for a reverse osmosis. However, it was identified that a significant portion of the phosphate was coming from chemicals used within the Silao facility's paint shop.

As part of the production expansion, the site was replacing their paint shop. Fortuitously, the new paint shop would be using a different process and would thus be producing effluent with a much lower phosphate concentration. It had been projected that this change would lower wastewater phosphate concentrations to 5 to 7 mg/L. It was believed that biological treatment would further reduce this concentration, producing an acceptable reverse osmosis feed concentration. In the event that this hypothesis was found to be incorrect following startup and acclimation of the MBR system, the team had planned to add ferric chloride dosing during the chemical precipitation process of industrial wastewater treatment to significantly lower wastewater phosphate levels.

6.4.5.2 Biological Treatment

Using (a) the water and constituent mass balances developed in this project (for current and futures cases), along with (b) historical analytical data collected by the site, and (c) new analytical data collected during this project, influent design criteria were developed for biological treatment (Table 6.13). These design criteria exclude loading from the oily wastewater, which will be segregated as discussed earlier.

At the start of this project (and before the expansion), the biological treatment system was receiving an average of 1655 m^3/d of wastewater (2500 m^3/d maximum), as compared to the 2000 m^3/d design capacity of the biological treatment system. The wastewater flow was projected to increase significantly following the site expansion, which included increased water use from a new paint shop pretreatment/electrodeposition systems.

TABLE 6.13 Influent design criteria for biological treatment.

Parameter	Value
Influent temperature, °C	10–30
COD, mg/L	2000
TKN, mg/L	140
Phosphate, as P, mg/L	40
TSS, mg/L	250
TDS, mg/L	1500
O&G, mg/L	10
Alkalinity, mg/L as $CaCO_3$	170
Hardness, mg/L as $CaCO_3$	60
pH	6.0–9.0

Based on planning-level flow estimates developed before the expansion, it was projected that the wastewater flow would increase considerably (2300 m^3/d average; 3100 m^3/d maximum flow).

Biological treatment plays an important role for the recycle process selected for Silao. Based on historical performance data and operations, as well as the increased wastewater flows projected following the expansion, it was noted that the existing Silao biological treatment process would not be able to meet the feedwater quality needs for a reverse osmosis process. Space constraints within the CUC area made it difficult or impossible to expand the size of the CAS process.

To accommodate future wastewater flowrates and improve effluent quality (compared to CAS) without increasing the overall footprint of biological treatment, it was proposed to retrofit the CAS, converting it to an MBR process. The performance for a CAS system is strongly influenced by settleability of the biomass, which limits the MLSS concentration for CAS, typically to 1500 to 4500 mg/L. By replacing the (gravity based) solids separation of a clarifier (for CAS) with a membrane (for MBR), the MLSS limitations of biological treatment become largely decoupled from settleability, allowing for higher MLSS levels, which, in turn, allows for a smaller footprint of the aeration basin.

The recommended range for MLSS is typically listed as 4000 to 10 000 mg/L; however, MBRs are often designed for operation in the 6000 to 10 000 mg/L range to help offset the added equipment cost for an MBR (compared to CAS) by reducing the aeration basin size. Aside from enhancing biological treatment and reducing the COD to a lower, more consistent level, the MBR retrofit will significantly reduce the TSS of the biological treatment effluent, which is important for the feed to the proposed water reuse reverse osmosis system (Olson et al., 2017).

Table 6.14 summarizes design effluent quality requirements for the MBR (i.e., reverse osmosis feed). The MBR has been designed to treat up to 2500 m^3/d based on the upper wastewater flow projections following the expansion. (Although the maximum flow of wastewater to the WRRF was projected to reach as high as 3100 m^3/d, the flows delivered to the MBR could be moderated through use of the WRRF equalization tanks.)

6.4.5.3 Reverse Osmosis Process

The main customer for recycle water is the new paint shop, which requires water with low chloride and silica. To meet the quality requirements, a reverse osmosis system (2000 m^3/d design capacity) has been installed for treatment of MBR permeate flow. To maximize overall extent of wastewater recycle and minimize freshwater demand for the facility, it was important to maximize the overall recovery for the reverse osmosis process.

A multistage reverse osmosis process was specified with a target overall recovery of 80%, producing a reverse osmosis permeate of 1600 m^3/d. Design feedwater quality to the reverse osmosis (Table 6.15) was developed based on samples collected from the CAS effluent, as well as projections for future waste streams and segregation of the oily wastewater. Reverse osmosis permeate quality specifications (Table 6.16) were developed based on the water quality needs for the end users of recycle wastewater.

TABLE 6.14 Design effluent requirements for the MBR treatment system.

Parameter	Value
BOD_5, mg/L	<5
COD, mg/L	<50
TN, mg/L	<8
O&G, mg/L	<0.1
pH, s.u.	6.0–8.0
TSS, mg/L	<0.1
Turbidity, NTU	<0.1

TABLE 6.15 Design feedwater quality to the reverse osmosis system.

Parameter	Units	Concentration
Alkalinity	mg/L, as $CaCO_3$	250
Aluminum, dissolved	mg/L	<0.03
Ammonia	mg/L, as N	< 0.20
Barium, dissolved	mg/L	<0.05
Calcium	mg/L	40
Magnesium	mg/L	25
Ca/Mg hardness	mg/L, as $CaCO_3$	203
Carbonate	mg/L, as CO_3	<5
Chloride	mg/L	250
Fluoride	mg/L	20
Iron, dissolved	mg/L	1.0
Manganese, dissolved	mg/L	<0.2
Nitrate	mg/L, as N	200
Nitrite	mg/L, as N	<0.1
Nitrate + Nitrite	mg/L, as N	200
Organic Carbon, dissolved	mg/L	10
Organic Carbon, total	mg/L	10
Orthophosphate	mg/L, as P	2
Potassium	mg/L	65
Silica, dissolved	mg/L	45
Sodium	mg/L	300
Sulfate	mg/L	300
pH	s.u.	6–8
TDS	mg/L	1500
TSS	mg/L	<0.1
Temperature	°C	10–40

TABLE 6.16 Reverse osmosis permeate design requirements.

Parameter	Units	Concentration
TOC	mg/L	<0.5
Chloride	mg/L	<25
Nitrate	mg/L, as N	<20
Sulfate	mg/L	<10
Total hardness	mg/L as $CaCO_3$	<5
TDS	mg/L	<250

To meet the specified project requirements, a three-stage reverse osmosis system was selected and installed. Figure 6.16 presents a depiction of this arrangement, along with projected water quality and flowrates. In such a configuration, the MBR permeate flows to the first-stage reverse osmosis. Permeate from the first-stage reverse osmosis is collected for reuse within the facility, while reject from the first stage is sent to the second-stage reverse osmosis as feed. Permeate from the second-stage reverse osmosis is combined with Stage 1 permeate and reused within the facility, and reject from

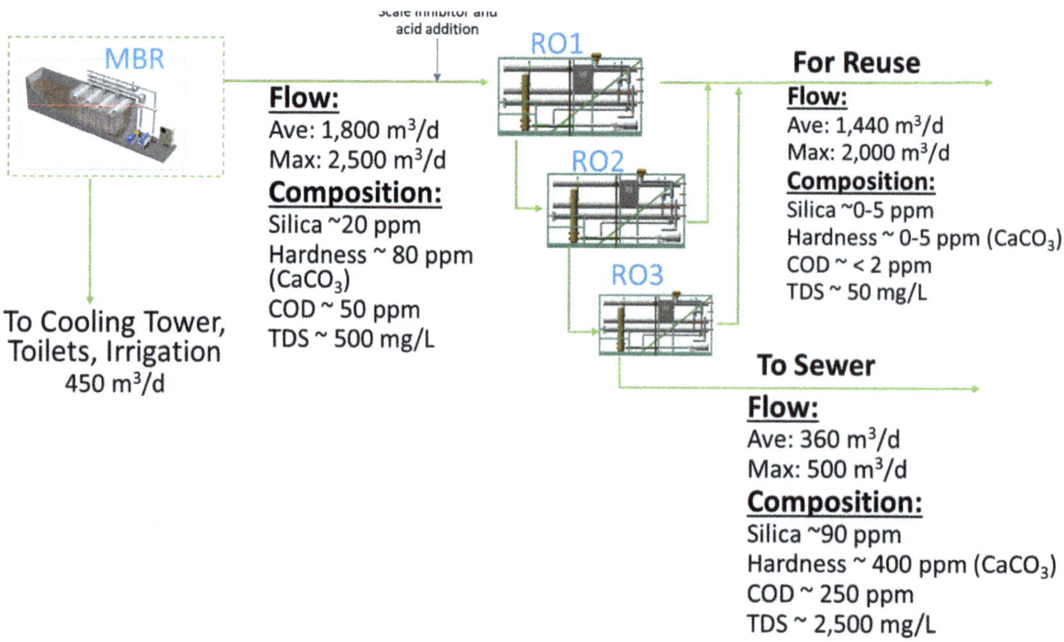

FIGURE 6.16 Recycling facility simplified process flow diagrams, including projected flowrates and water quality characteristics.

the second-stage reverse osmosis goes to the third-stage reverse osmosis. Permeate from the third-stage reverse osmosis is combined with permeate from Stages 1 and 2 and reused on-site; reject from the third-stage reverse osmosis is sent to the municipal sewer.

6.4.6 Startup and Recycling Facility Operations

At the time this case study was being written (April 2019), the construction phase was wrapping up and the project was in the early stages of commissioning. Relative to the construction side, the first MBR train had been running since January 2019, and the second MBR train was in the process of being brought online. However, installation of MgO dosing to the potable water (as described in Section 6.4.5.1.3, "Silica") had not yet been completed.

Table 6.17 summarizes data covering the period from March 21, 2019, through April 22, 2019. As noted earlier in this case study, the sanitary and industrial wastewater streams are combined and processed through biological treatment. It can be seen that the COD levels are being reduced by 92% to 93% through biological treatment (anaerobic and aerobic treatment). Chemical oxygen demand reduction is anticipated to increase slightly as the second MBR train comes online. Since the time the first MBR train was first brought online, the MBR permeate TSS has consistently been measured at 0 mg/L (across 22 sample events), with the exception of a single sample reported at 16 mg/L TSS. It is unclear if that reading represented an artifact of system startup/acclimation or if the reading was erroneous. With the exception of that single sample reading, TSS for the reverse osmosis feed appears excellent.

As described in section 6.4.5.1.4, changes in manufacturing were expected to result in decreased concentrations of phosphate in the industrial wastewater. In practice, the wastewater phosphate concentrations have, indeed, been observed to be lower than they had been before the facility expansion; however, the levels are not as low as had been projected following the expansion and are higher than desired for reverse osmosis feedwater. To address this issue, ferric chloride addition is being incorporated to the coagulation process (to precipitate out phosphate).

The reverse osmosis feed hardness concentration has averaged 151 mg/L, which is less than the value listed in the reverse osmosis feedwater specification (203 mg/L; see Table 6.15). Hardness levels observed to date have been higher than what had been projected, 80 mg/L (Figure 6.16). However, as the second MBR is brought online, the recycling facility will be able to produce larger volumes of (near-zero ppm hardness) water for reuse within facility operations, which will reduce hardness levels for the influent wastewater

TABLE 6.17 Average water quality characteristics over the period of March 21, 2019, to April 22, 2019, during recycling facility commissioning/startup. (Cells containing a dash symbol denote parameters that were not analyzed.)

	Sanitary Wastewater	Industrial Wastewater	Reverse Osmosis Feed	Reverse Osmosis Permeate
Temperature (°C)	—	—	23	—
BOD (mg/L)	589	400	—	—
COD (mg/L)	978	1042	77	—
TSS (mg/L)	501	118	0.7	—
O&G (mg/L)	102	23	—	—
pH	7.9	—	—	5.9
NO_3-N (mg/L)	217	56	6.6	0.1
Phosphate, as P (mg/L)	26	7.9	2.2	0.1
SiO_2 (mg/L)	—	—	65	0.17
Alkalinity (mg/L)	—	—	432	−65
Total Hardness (mg/L as $CaCO_3$)	—	—	151	0
Conductivity (uS/cm)	—	—	-	20
TDS (mg/L)	—	—	-	9.8
Cl^- (mg/L)	—	—	-	1.5
SO_4 (mg/L)	—	—	-	0.4

closer to the level that had been projected. It is also noted that silica concentrations for the reverse osmosis feed are higher than desired (65 mg/L average). However, silica concentration is expected to drop sharply after the potable water softener MgO dosing project is completed, enabling removal of silica from the well water during potable water production. As with hardness, silica levels will also decrease with the increasing extent of wastewater recycle (i.e., once the second MBR is brought online and stabilized).

It was expected that the alkalinity target from Table 6.15 would be difficult to meet, and results to date support this expectation. Alkalinity can increase the formation of carbonate scale upon reverse osmosis membranes, particularly above pH 8.3 (where carbonate exists to a significant extent). To address scaling risk, the reverse osmosis skids have been equipped with scale inhibitor and acid feeds.

Reverse osmosis permeate quality has been good thus far (e.g., compare Table 6.16 and Table 6.15). To date, the reverse osmosis has been operated at 77% to 80% recovery, as compared with the design target of 80%. Following commencement of MgO dosing at the potable water softener, wastewater silica concentrations will decrease markedly, allowing reverse osmosis recovery rates to be increased to consistently meet the design recovery target. Overall, the recycle system is operating favorably.

6.4.7 Future Project Phase: Moving Toward Zero Liquid Discharge

A Phase 2 concept has been developed to include onsite treatment of the brine generated in the reverse osmosis systems, which is presently being sent to the public sewer. Several possible brine management approaches have been developed for consideration. The approaches developed involve various combinations of the following technologies: (a) softening followed by reverse osmosis (to reduce overall brine volume), (b) solar (evaporative) pond (with and without evaporation assistance), (c) thermal evaporator, and (d) crystallizer. The various options have been cross-compared relative to estimated CapEx, OpEx, land requirements, and volume of additional water recovered, among other considerations. Most of the approaches developed for this Phase 2 will enable the site to increase overall recovery of water and move closer to a ZLD treatment process, potentially achieving over 95% recovery.

7.0 REFERENCES

American Petroleum Institute. (n.d.). *Water conservation through power cogeneration*. www.api.org/oil-and-natural-gas/environment/clean-water/surface-water-quality/water-conservation

American Society of Mechanical Engineers. (1994). *Consensus on operating practices for the control of feedwater and boiler water chemistry in modern industrial boilers* (CTRD Vol 34). American Society of Mechanical Engineers.

Bank, T. L. (2011). *Trace metal geochemistry and mobility in the Marcellus Shale*. U.S. EPA Hydrofracturing (HF) Study Technical Workshop 1, February 2011.

Barbera, M., & Gurnari, G. (2018). *Wastewater treatment and reuse in the food industry*. Springer.

Basu, S., & Monroe, S. (2012). *Efficacy of electrocoagulation (EC) as a treatment technology to treat flowback and produced water for reuse in hydrofracturing operation* [Paper presentation]. Texas Reuse Symposium, Austin, Texas, United States.

Blauch, M. E., Myers, R. R., Moore, T. R., Lipinski, B. A., & Houston, N. A. (2009, September 23–25). *Marcellus Shale post-frac flowback waters —Where is all the salt coming from and what are the implications?* [Paper presentation, 125740]. Society of Petroleum Engineers Eastern Regional Meeting, Charleston, West Virginia, United States.

Bluefield Research. (2016). *Managing water inside the fence: Drivers and market opportunities for industrial reuse*. Bluefield Research.

Bonné, P. A. C., Hofman, J. A. M. H., & van Der Hoek, J. P. (2000). Scaling control of RO membranes and direct treatment of surface water. *Desalination, 132*, 109–119.

Boschee, P. (2012). *Handling produced water from hydraulic fracturing, oil and gas facilities.* https://www.spe.org/en/ogf-article-detail/?art=364

Brewers Association. (2013). *Water and wastewater: Treatment/volume reduction manual.* https://www.brewersassociation.org/attachments/0001/1517/Sustainability_-_Water_Wastewater.pdf

Camarillo, M. K., & Stringfellow, W. T. (2018). Biological treatment of oil and gas produced water: A review and meta-analysis. *Lawrence Berkeley National Laboratory, Clean Water Technologies and Environmental Policy, 20*(6).

Cambrian Innovation. (2019). *Wastewater treatment as a service: Water-energy purchase agreement.* https://cambrianinnovation.com/wepa/

Chakraborty, J. N. (2010). *Fundamentals and practices in colouration of textiles New Delhi, India.* Woodhead Publishing India Pvt. Ltd.

CHS Inc. (2019). *Environmental sustainability, smart decisions for a cleaner future.* https://www.chsinc.com/our-company/environmental-sustainability

Ciabatti, I., Tognotti, F., & Lombardi, L. (2010). Treatment and reuse of dyeing effluents by potassium ferrate. *Desalination, 250*(1), 222–228.

Colorado School of Mines. (2009). *Technical assessment of produced water treatment technologies* (1st ed.; RPSEA Project No. 07122-12). Colorado School of Mines.

Eble, K. S., & Feathers, J. (1992a). Process water reuse-1 characterization of streams first step in reuse scheme. *Oil Gas Journal, 90*(38), 86, 88–92.

Eckenfelder, W. W., Ford, D. L., & Englande, A. J. (2009). *Industrial water quality* (4th ed.). McGraw-Hill.

EFX Systems, Inc. (2000). *Field testing and support for application of the GAC-FBR process for gas industry wastes: Biological treatment of produced water.* Final Report Prepared for Gas Research Institute (Contract no. 5095-250-3296).

Elliott, J. (2018). *Canadian refinery recycles all its wastewater.* Chemical Processing, Putman Media®.

Ellis, M., Dillich, S., & Margolis, N. (2001). *Industrial water use and its energy implications*. Prepared by Energetics Incorporated for the U.S. Department of Energy, Office of Energy. Efficiency and Renewable Energy, Office of Industrial Technologies.

Establishment grounds and facilities, 9 C.F.R. § 416.2 (2016).

ExxonMobil. (n.d.). *Singapore integrated manufacturing complex*. https://www.exxonmobil.com.sg/-/media/Singapore/Files/Who-we-are/ExxonMobil-Singapore-Integrated-Manufacturing-Complex-brochure.pdf

Fort, D. J., Meyers, J. D., Page, M. W., & Hercyk, N. L. (2013). Water conservation—Whole effluent toxicity paradox. *Water Environment Research*, *85*, 483–494.

Foster, S., Garduno, H., & Kemper, K. (2004). *The Cotas – progress with stakeholder participation in groundwater management in Guanajuato-Mexico*. World Bank LCR.

Francis, W. R., Mazur, J. J., & Rao, N. M. (1996). *Strategic planning for and implementation of reclaimed municipal waste water as make-up to a refinery cooling system* (International, Paper No. 582). National Association of Corrosion Engineers.

Freedman, D. E., Riley, S. M., Jones, Z. L., Rosenblum, J. S., Sharp, J. O., Spear, J. R., & Cath, T. Y. (2017). Biologically active filtration for fracturing flowback and produced water treatment. *Journal of Water Process Engineering, 18*, 29–40.

Gary, J. H., Handwerk, G. E., & Kaiser, M. J. (2007). *Petroleum refining technology and economics* (5th ed.). CRC Press.

Good manufacturing practice in manufacturing, packing, or holding human food, 21 C.F.R. § 110.80 (2018).

Groover, M. P., & Kolchin, M. G. (1997). *Case study: Automobile final assembly plant*. http://www.mhi.org/downloads/learning/cicmhe/resources/cs_automobile_plant.pdf

Guerra, K., Dahm, K., & Dunfdorf, S. (2011). *Oil and gas produced water management and beneficial use in the Western United States*. U.S. Department of Interior Bureau of Reclamation, Science and Technology Program Report No. 157, p. 71.

Guo, T., Englehardt, J., & Wu, T. (2014). Review of cost versus scale: Water and wastewater treatment and reuse processes. *Water Science & Technology, 69*(2), 223–234.

Hilts, J., Eastuke B., Robertson, E., Knight, M., Edwards, D., Elliott, B., & Arntsen, B. (2018). A review of the co-op refinery complex's wastewater improvement project. *In Proceedings of the 91st Water Environment Federation Technical Exhibition and Conference*, New Orleans, Louisiana, United States.

Iglesias, R., Simón, P., Moragas, L., Arce, A., & Rodriguez-Roda, I. (2017). Cost comparison of full-scale water reclamation technologies with an emphasis on membrane bioreactors. *Water Science and Technology, 75*(11–12), 2562–2570.

India Brand Equity Foundation. (2018). *Textile industry & market growth in India.* India Brand Equity Foundation.

India Brand Equity Foundation. (2019). *Indian textiles and apparel industry analysis.* India Brand Equity Foundation.

Industrial Water World. (2013). *Sustainable water treatment solution reduces costs, conserves water at Ohio refinery.* https://www.watertechonline.com/process-water/article/16212348/sustainable-water-treatment-solution-reduces-costs-conserves-water-at-ohio-refineryInternational HACCP Alliance. (n.d.). Home page. http://www.haccpalliance.org/sub/index.html

International Petroleum Industry Environmental Conservation. (2010). *Petroleum refining water/wastewater use and management.* London, UK.

International Petroleum Industry Environmental Conservation. (2015). *Water case study water minimisation in refining.* BP, Australia, Kwinana Refinery.

Khandegar, V., & Saroha, A. K. (2013). Electrocoagulation for the treatment of textile industry effluent—A review. *Journal of Environmental Management, 128*(15), 949–963.

Kim, K. (2017). *Silica fouling in reverse osmosis systems: The role of chemical interaction with the membrane surface* [Doctoral dissertation, The University of Texas at Austin].

Koo, T., Lee, Y. J., & Sheikholeslami, R. (2001). Silica fouling and cleaning of reverse osmosis membranes. *Desalination, 139*(1–3), 43–56.

KPMG. (2010). *Water sector in India: Overview and focus areas for the future.* KPMG.

Langer, B. S. (1983). *Wastewater reuse and recycle in petroleum refineries.* CEP.

Majcen Le Marechal, A., Križanec, B., Vajnhandl, S., & Volmajer, J. (2012). *Textile finishing industry as an important source of organic pollutants.* InTech. https://doi.org:10.5772/32272

Malki, M., & Abbas, V. (2013). Relationship between phosphate scales and silica fouling in wastewater RO membrane system. *IDA Journal of Desalination and Water Reuse, 5*, 15–24.

Meyers, J. D., Page, M. W., Gantz, R. G., & Hercyk, N. L. (2008). *Water conservation, total dissolved solids, and whole-effluent toxicity testing* (NPRA Paper No. ENV-11-08).

Moore, B., & Buzby, M. (2017). *A framework for the successful implementation of on-site industrial water reuse.* Water Environment & Reuse Foundation, Alexandria, VA (WRRF-14-04).

Narayanan, S. (2012). *Tirupur dyeing units told to close: Madras High Court orders shut down till zero discharge.* DownToEarth. https://www.downtoearth.org.in/news/tirupur-dyeing-units-told-to-close-33025

National Association of Corrosion Engineers International. (2009). *Crude distillation unit-distillation tower overhead system corrosion* (Report Publication No. 34109). National Association of Corrosion Engineers International.

Nichols, E. R. (2018). *Evaluation of a sequencing batch reactor followed by media filtration for organic and nutrient removal from produced water* [Doctoral dissertation, Colorado School of Mines. Arthur Lakes Library].

Olson, D., Moore, B., Rondon, D., & Williams, T. (2017). Maximizing water recycling in automotive manufacturing in a water-stressed region. In *Proceedings of the 90th Annual Water Environment Federation Technical Exhibition and Conference*, Chicago, Illinois, United States.

Ostenberg, R. M. (2009). *Optimization of water usage in petroleum refineries.* Presentation to the Water/Energy Sustainability, Ground Water Protection Council, Salt Lake City, Utah, September 15, 2009. Symposium 34109; Houston, Texas, United States.

Otts, L. E., Jr. (1963). *Water Requirements of the Petroleum Industry*; Geological Survey Water-Supply Paper 1330-G.; U.S. Department of the Interior.

Phillips 66. (n.d.). *Ponca City refinery.* https://www.phillips66.com/refining/ponca-city-refinery

PUB. (2018). *Five clinch top honours in water excellence.* https://www.pub.gov.sg/news/pressreleases/fiveclinchtophonoursinwaterexcellence

Puckorius, P. R., Loretitsch, G. A., & Tvedt, T. J., Jr. (1998). *Reuse water quality for chemical plants, refineries, utilities and air conditioning* (Paper No. 573). National Association of Corrosion Engineers International.

Rajakumar, C. (2019). *World Water Day: India faces massive water crisis in next 10 years.* Greenubuntu Bureau.

Safe Drinking Water Act, 42 U.S. Code § 300f. (1974).

Schutte & Koerting. (n.d.). *Barometric condensers* (Bulletin 5AA). https://www.s-k.com/pdf/5AA_barometric_condensers_brochure.pdf

Sun, P., Elgowainy, A., Wang, M., Han, J., & Henderson, R. J. (2018). Estimation of U.S. refinery water consumption and allocation to refinery products. *Fuel, 221*, 542–557.

The Hartman Group, Inc. (2017). *2017 Sustainability.* The Hartman Group, Inc.

Tomson, M. B., Kan, A. T., Fu, G., & Al-Thubaiti, M. (2003, November 11–14). NORM scale formation, control and relation to gas hydrate control. *Conference Proceedings, Presentation at the 10th International Petroleum Environmental Conference*, Houston, Texas, United States.

Towey, A., & Baldwin, L. S. (2015). *Downstream oil and gas reuse.* Presentation to the WaterReuse.org Webcast.

U.S. Department of Energy. (2009). *Modern shale gas development in the United States: A primer.* U.S. Department of Energy.

U.S. Department of Energy. (2016). *Potential vulnerability of U.S. petroleum refineries to increasing water temperature and/or reduced water availability.* Jacobs Consultancy Executive Summary of Final Report.

U.S. Energy Information Administration. (2011). *The geology of natural gas resources.* Today in Energy. https://www.eia.gov/todayinenergy/detail.php?id=110

U.S. Energy Information Administration. (2016). *Shale gas and oil plays, Lower 48 States (6/30/2016).* Maps: Oil and Gas Exploration, Resources, and Production. https://www.eia.gov/maps/images/shale_gas_lower48.jpg

U.S. Environmental Protection Agency. (1993). *Oil and gas extraction point source category: Offshore subcategory effluent limitations guidelines and new source performance standards—final rule.* U.S. Environmental Protection Agency.

U.S. Environmental Protection Agency. (1996). *Preliminary data summary for the petroleum refining category* (EPA-821-R-96-015). U.S. Environmental Protection Agency.

U.S. Environmental Protection Agency. (2004). *Technical support document for the 2004 effluent guidelines program plan* (EPA-821-R-04-014). U.S. Environmental Protection Agency.

U.S. Environmental Protection Agency. (2015). *Steam Electric Power Generating Effluent Guidelines—2015 Final Rule.* https://www.epa.gov/eg/steam-electric-power-generating-effluent-guidelines-2015-final-rule#reg-requirements

U.S. Environmental Protection Agency. (2019). *Petroleum refining effluent guidelines.* www.epa.gov/eg/petroleum-refining-effluent-guidelines#whatis

U.S. Geological Survey. (2015). Estimated use of water in the United States in 2015, Circular 1441. U.S. Geological Survey.

Venkatesan, V. V. (2012). *Optimize energy costs in refineries, Part II. Chemical processing.* Putman Media®.

Water Tech Online. (2013). *Sustainable water treatment solution reduces costs, conserves water at Ohio refinery.* https://www.watertechonline.com/process-water/article/16212348/sustainable-water-treatment-solution-reduces-costs-conserves-water-at-ohio-refinery

Wu, M. & Xu, H. (2018). *Consumptive water use in the production of ethanol and petroleum gasoline—2018 update.* Argonne National Laboratory, ANL/ESD/09-1 Rev. 2; Argonne, IL.

Young, T., Smoot, S., Peeters, J., Cote, P. (2013). When does building an MBR make sense? How variations of local construction and operating cost parameters impact overall project economics. *Proceedings of the 86th Annual Water Environment Federation Technical Exhibition and Conference*, 8, 6354–6365.

8.0 SUGGESTED READINGS

Eble, K. S., & Feathers, J. (1992b). *Water reuse within a refinery* (AM-92-08). Presented at the National Petroleum Refiners Association Annual Meeting, New Orleans, Louisiana, United States.

Marrs, D. R. (2016, September 26). What refineries should consider when considering reclaimed water for cooling towers. In *Proceedings of the 89th Water Environment Federation Technical Exhibition and Conference*, New Orleans, Louisiana, United States.

U.S. Food and Drug Administration. (n.d.). Hazard Analysis Critical Control Point (HACCP). https://www.fda.gov/Food/GuidanceRegulation/HACCP/default.htm

7

Technology Development Pilots

Al Goodman; Jimmy Yu, PhD; Karen Marie Dietze; Graham Juby, PhD, PE; & Dustin Klempel

1.0 INTRODUCTION

This chapter provides information on applicability of using pilot treatment systems for determining treatment process design criteria, performance, and potential scale-up factors for full-scale treatment systems to recycle and reuse industrial process waters. Pilot treatment systems can be defined as nonpermanent water or wastewater treatment processes that provide sufficient operation to determine the efficiency of the treatment technology for potential use on the actual water/wastewater characteristics and flows expected for full-scale application.

Pilot systems can range in size from bench-top scale systems operated in milliliters per minute to pilot-scale systems operated up to 750 L/m (200 gal/min) or larger in some projects. In general, larger pilot systems are often preferred, as they are expected to yield results that are more representative of the full-scale system. Pilot systems can be operated from several days (typically physical treatment, such as screening) to more than a year. The duration of pilot testing should be selected based on known variations in water/wastewater characteristics, specific treatment process being evaluated, and level of certainty needed for decision-making purposes. It should be noted that pilot testing is different than "jar-testing" or batch testing, because pilot testing typically involves using actual flow-through systems with in-facility sourced flows. Pilots can be done on shipped containers of wastewater; however, continuous flows result in much more accurate testing of the technology and its performance. The purpose of the pilot is to determine what a full-scale treatment system could achieve using the same technology and equipment, such as specific membranes.

2.0 ADVANTAGES OF PILOT TESTING

The main advantages of conducting pilot testing include the following:

- determine efficiency of removal for target contaminants
- assess if piloted technology can achieve required effluent water quality for reuse/recycle
- determine if interferences occur in operation of the pilot technology
- establish design criteria for selected technology
- identify energy requirements (e.g., pumping pressures required for operation)
- evaluate cleaning frequency and extent of fouling of membranes
- determine chemical dosages to identify full-scale requirements
- estimate space allocation required for full-scale, including building footprints
- determine other appurtenances required such as chemical storage, feed systems, and options
- evaluate performance over changes in water chemistry that occur in industrial processes
- validate performance using third-party credibility rather than in-facility evaluations
- convince management of technology and viability of project application to stakeholders (e.g., management, operations)
- confirm capital and operating expense estimates
- generate requirements for performance guarantees from technology vendors

- provide research information for development of technology improvements by manufacturers
- use support from and for universities and research organizations to lower costs
- predict treatment efficiency for future changes in production or process discharges for reuse "(e.g., incorporation of other process/waste flows or stress-testing for future extreme water quality conditions)

An additional advantage is that some pilot systems are available prepackaged and ready for connections on a rental basis.

3.0 DISADVANTAGES OF PILOT TESTING

Although there are more advantages to performing pilot testing, there are some disadvantages that need to be considered as well. These include the following:

- Pilot testing can add considerable time for overall project completion and startup.
- Pilot testing adds costs to project designs for equipment purchase/rental, operations, laboratory tests, and perhaps consulting services for the pilot system.
- Pilot testing may not accurately simulate treatment efficiencies if the water/wastewater stream does not exist (future).
- There is potential for having to release confidential data to regulatory agencies or third parties (vendors).
- Pilot testing could publicize corporate management's intent for production or facility changes.

4.0 SELECTING AND DESIGNING PILOT TREATMENT SYSTEMS

The implementation of pilot testing can determine the actual viability of a treatment technology for the intended application. This then verifies the expected performance, while providing more accurate design details to use in scaling up final design based on the pilot tests. Therefore, when considering designing a pilot testing project, the variability of the influent supply to be treated and future production changes all need to be considered, along with the required chemical parameters and water quality conditions for reuse (e.g., constituent concentrations, pH, temperature).

Recycle water quality required is often difficult for industries to define because the complete chemical matrix of water quality parameters may not be known, defined, or even available. It is often determined that the reuse water quality must meet certain established water quality standards, such as "must meet drinking water quality and regulations." Therefore, a pilot testing program would typically include analysis of common drinking water parameters or some targeted parameters instead of the entire list of U.S. Environmental Protection Agency (U.S. EPA) primary and secondary drinking water standards. (However, defining water quality needs in a such a manner is often not the best way to represent the "actual" water quality needs for processes in the facility.) Depending on the reuse purpose, the pilot system should be flexible to provide or add technology for various additional constituents in the recycle process.

The duration of the pilot test should cover the variations in the normal production processes, allowing sufficient time for the pilot technology to be assessed across these conditions. In physical–chemical pilot tests, the required pilot period may be quite short, whereas in biological treatment pilot tests, effects on performance would be longer as a result of diurnal flow patterns, water quality variations from industrial loads and seasonal changes, time required to establish microbial communities, and so forth. Biological pilot treatment systems also likely may involve more equalization of the influent in industrial systems to minimize the effect on the biology viability, especially regarding pH swings, batch discharges, temperature changes, and effects from regular or infrequent cleaning/disinfection chemicals. As a generalization, physical/chemical pilot tests can run weeks, whereas biological pilot tests require months to capture valuable operating and design data.

5.0 PILOT TREATMENT SETUP CONSIDERATIONS

5.1 Permitting

Pilot treatment systems are often temporary or portable facilities and, therefore, typically do not require special permits for installation or setup; however, that is not always the case. Some regulatory agencies, building commissions, and environmental agencies do require permit applications and approval for pilot systems. Factors that may trigger these requirements include the size of the pilot system (physical space required), if it is considered "permanent" or "temporary," the power required, and the physical location and proximity to facility site boundaries. If the pilot system has a full-time operator, it may be considered as an occupied facility, which could require additional features such as comfort space (office and/or restroom).

An example of permitting required for a pilot system was one installed in California that involved a 9-month trial and used a 6.5-m (20-ft) height upflow anaerobic sludge blanket (UASB) "tower." Even though the system was designed for only 18.9 L/min (5 gal/min), the local permitting agencies required a geotech survey, seismic structural protection for the UASB tower, and anchorage for all treatment tanks. This added time and costs for these designs, studies, and construction of seismic controls and concrete slabs.

5.2 Pretreatment Before a Pilot

Some pretreatment of industrial effluents is often required before initiating a pilot treatment system. This may include screening of large solids that would plug small pilot system piping; pH adjustment; and removal of heavy fats, oils, and grease that might also interfere with small pilot piping and pumping. Equalization for both concentration variations and hydraulic variations needs to be considered, if this is not currently used. Temperature control might also be required, especially for membrane pilot systems. The effects of temperature on membrane performance (increased membrane feed pressure at low temperatures versus increased salt passage at high temperatures)—low water temperatures affect the life-cycle cost of the full-scale system (cost of feed pump, energy consumption), whereas the high water temperatures affect the finished water quality and design criteria of the membrane system. Caution should be taken for pilot flows that tend to foam because some defoamers may have compounds that foul or damage membranes.

5.3 Pilot Physical Setups

When selecting pilot treatment processes, many factors affect the installation. These include the significant considerations outlined in the following subsections.

5.3.1 Location

There needs to be sufficient space within an existing pretreatment facility and/or building available for both installation of the pilot and operational access. Most systems need drainage and electrical connections, so the proximity of these is important. If the location requires pumping to or from the site, this adds costs and complexity to the operation.

One of the main considerations is exterior versus interior locations of the pilot system. Besides the permitting considerations previously noted, exterior locations may also require freeze and wind protection, rain cover and/or containment, fencing, traffic protection, and similar requirements.

5.3.2 Prepackaged Pilots Versus Fabrication

Equipment vendors and manufacturers can often provide prepackaged pilot systems, available for rent or purchase. These prepackaged systems will most often only need piping connections for influent, effluent, and possibly solids removal, along with electrical connection for power. These systems are convenient, and providers of these systems will likely have qualified operator(s) available for startup, operation and training.

Advantages of prepackaged pilot treatment systems include the following:

- These systems are much faster to install/set up than custom-built pilot systems.
- These systems are designed by the full-scale equipment vendor to closely simulate their actual systems.
- These systems can be skid-mounted, installed in shipping containers, or self-contained trailers.
- These systems are available for rent, requiring less capital cost.
- Pilot equipment rental cost may be applied to purchase of full-scale system.
- All controls are integrated to one panel that does not require instrumentation specialists.
- Multiple pilot systems can be "connected" to evaluate different technologies in series.
- Multiple pilots can be compared side-by-side to evaluate performance and operability.

Disadvantages of using prepackaged pilot systems include the following:

- They are limited to a small range of flows and loadings, depending on the pilot system.
- Limited modifications can be performed on piping and prefabricated systems.
- There is limited availability, with typically "first-come, first-committed" approach by vendors.
- Vendors may have only one or two pilot systems available at the time desired, which could add considerable time tothese systems.

- These systems may require crane service for unloading and setup.
- Contracts from vendors may require liquidated damages for delays in pilot time schedule.
- Pilots must be thoroughly cleaned before returning to the vendor, including disposal of unused consumables, which may be considered hazardous (e.g., spent granular activated carbon [GAC]).
- Rental pilot condition can be variable depending on how many times it has been used previously and how thoroughly the vendor rehabilitates it between uses.
- Pilot design and build quality varies among vendors, especially with new technologies or vendors in the market.
- Damages to rental pilots could be incurred.
- Pilot vendors likely will require sharing performance data, which may be considered confidential to industry.

Custom designed and built-onsite pilot systems should also be evaluated. Considerations include designing systems and choosing treatment technologies likely suitable to the intended reuse treatment qualities required. These pilot systems can be designed "in-house" using the industry's environmental and engineering staff, or they can be designed by environmental engineering consultants. Equipment vendors can also design pilot treatment systems; however, these vendor systems may use their technology only and not competing equipment suppliers, and possibly not multiple technologies that they do not provide. There are design-build consultants available that can and do provide these custom-built pilot systems and use recommended technologies for a complete system.

Advantages of custom-built pilot treatment systems include the following:

- Custom systems can use various equipment vendor components (e.g., membranes) allowing for change-out in equipment during the testing period.
- Custom systems can be designed and built for various flows and loadings.
- Custom pilots can be less costly for long-term installations and long pilot durations.
- Custom pilots allow for modifications during the testing period, as may be dictated by performance.
- There is the ability to install sampling locations as needed and where needed.

- Custom pilot systems can be operated more confidentially than rental units (important to reuse evaluations).
- Custom-built system designers and/or vendors can adapt layout and configuration to fit in available space.
- Components on completion of the pilot remain in owner's possession onsite.
- These custom-built systems allow for future pilot evaluation of product changes on future treatment system performance or perhaps at similar existing industrial locations.

Disadvantages of custom-built pilot systems include the following:

- These custom-built systems require higher level of expertise in design and building pilot system.
- An industrial facility may not have a qualified operator familiar with treatment pilot to start up and train personnel.
- These custom systems will require instrumentation and controls to be designed, adding to expertise requirements.
- A facility may build the pilot system and then determine that the chosen technology does not work.
- Custom systems would add time to design, procure, and install compared to rental equipment.
- Custom-built construction costs may not be recovered in the final treatment system like rental equipment cost from a vendor

5.4 Duration of Pilot Testing

One of the key decisions in establishing a pilot-scale treatment system is the length of time that the pilot needs to be operated. This will primarily depend on the variability of the influent, including "normal operating cycles" in production. For example, some food and beverage industries may do product changes several times per day, which require some clean-in-place (CIP) procedures, each of which would produce changes to the wastewater quality that would go through the pilot.

1. A significant system CIP may be completed once per weekend; therefore, the pilot should be operated during at least one full 7-day week period to assess performance under "normal" operating conditions. Other factors affecting piloting duration include laboratory turnaround time for making pilot-scale treatment process control changes

and assessing performance, as well as catching batch discharges or routine spills that may occur infrequently but could have significant effects on the piloted treatment technologies. The duration of pilot testing also depends on the reason for conducting the pilot study in the first place. If the purpose of pilot testing is to screen viable technologies, the pilot study could be relatively short. If, on the other hand, the pilot study is being conducted to optimize operating conditions or refine design criteria, then a longer duration would likely be required.

The duration of pilot testing may also depend on the required level of design/cost certainty. Industries that are less risk averse may opt for longer pilot studies (up to 1 year) to ensure appropriate technology selection and narrow in on key design parameters that affect capital and operating costs.

5.5 Safety During Pilot Testing

As in any industry or treatment facility, safety is of paramount importance. Safety in pilot-scale systems will require assessing effects from chemicals, electrical equipment, motive equipment, temperatures, gases, and general safety requirements. Even small polymer chemical spills can present hazards for falls and slippery conditions.

Depending on the chemicals used, especially acids and caustics, safety showers and eyewashes may be required by local and Occupational Safety and Health Act (OSHA) regulations. Safe segregation and storage of chemicals used in pilot systems must also be implemented. Anaerobic processes can and will generate volatile and toxic gases (such as methane and hydrogen sulfide), which may require intrinsically safe electrical components and additional personnel safety. Emission of gases from anaerobic processes must also be considered relative to odors and offsite wind directions. Other common hazards:

- Tight/cramped areas
- Poor lighting
- Operator fatigue from long days and frequent troubleshooting
- Adverse weather conditions

Additionally, it is common for pilot systems to include a project-specific health and safety plan to identify potential risks, preventive measures, and mitigation strategies.

Safety data sheets for any chemicals being used during pilot operations should be readily available and reviewed by pilot system operators for direction on personal protective equipment, storage requirements, and cleanup recommendations in the event of a spill/release.

Proper training of pilot system operators is important to ensure safe operation and prevention of accidents. Each industry's health and safety officer should review the pilot treatment system plan and implement the necessary safety precautions and personal protective equipment.

5.6 Spill Controls and Security

Because pilot treatment systems may be located in facilities that do not typically have treatment, safe handling of spills and tank overflows need to be considered in the pilot system design and operations. All pilot system effluent, drain, and overflow piping should be coordinated with and routed to appropriate discharge or solids handling locations. Temporary concrete curbs or spill control systems (e.g., tote and drum spill containment and perhaps multiple containment areas) should be considered. This is especially important in the event of a significant system overflow or pipe failure in undrained areas. These can occur if, for example, effluent pumping systems fail.

When pilot systems are set up, evaluation of security is important. Access by unauthorized personnel may damage equipment, disrupt the process, negate the entire pilot testing, and be potentially harmful to personnel. Security measures such as fencing, limited access to buildings, trailers, or pilot buildings and controls are recommended.

5.7 Confidentiality

Some industries may desire that the pilot testing process be confidential, especially where either product changes, production upsizing, or downsizing could occur, or for other reasons. As previously noted, some pilot system rental agreements include requirements to share analytical performance data with the pilot system vendors. These requirements may need to be evaluated by the industry's legal department.

6.0 PILOT TREATMENT OPERATION AND MAINTENANCE

When implementing a pilot project, the operation and maintenance (O&M) of the system is of paramount importance to obtain reliable information. Operation requires routine adjustments of pilot treatment processes, collecting samples, performing onsite analytical testing, calibrating probes and instrumentation, stopping and starting electrical gear, keeping detailed records/log notes, and recognizing needed maintenance. Preparing reports is also often required for proper operation. The most difficult operational requirement is understanding and implementing the process control strategies to produce the best pilot treatment performance possible with the technology available.

Production facility maintenance departments are often charged with maintaining pilot systems when they are implemented. This often causes concerns among maintenance staff who are not familiar with the equipment or technology used and, therefore, are reluctant to react. To reduce these factors, maintenance managers should be brought into the pilot implementation process early on, before the equipment arrives. These staff should also attend training if provided by a vendor for rented equipment.

Critical spare parts should also be procured to maintain continuous longer-term operations of the pilot if this is planned. These may include spare pH probes, dissolved oxygen probes, standby chemical feed pumps and tubing, critical electrical components, and so on.

The operation of the facility may require process monitoring controls to also be available on the pilot system. These include, for example, flow meters, pH probes/monitors, temperature probes, dissolved oxygen probes, oxidation/reduction probes (ORP), turbidimeters, pressure gauges, gas or airflow meters, and so on.

7.0 PILOT STAFFING REQUIREMENTS

There are no set number of operators required for a given pilot treatment system. The staffing required will depend on the technology and equipment requirements, length of the pilot testing, and, most importantly, on the qualifications of the operator(s). Certification of the pilot system operator is typically not required by regulatory agencies; however, this should be reviewed before implementing operation of a pilot system. Certified water and wastewater operators bring a higher level of knowledge to the project. Operations are critical for a pilot system. Because significant capital investment can depend on the outcome of the pilot testing, it is important to find the right professional operator to verify pilot performance.

Operating pilot treatment systems is a great way to train facility staff on challenges and process control strategies before implementing a full-scale treatment facility. This gives the O&M staff insights to what will be required and encountered on a daily basis when full-scale operations begin using similar treatment technologies. This also helps the operator determine where sampling equipment can be installed, what parameters are of significant importance to monitor, and many other O&M facets of the facility. If operators are not available, consultants can often be used to operate the pilot systems.

8.0 LABORATORY TESTING OF PILOT TREATMENT

During any pilot testing project, establishing testing and monitoring is critical to evaluation of the pilot treatment process. The data not only provide performance verification, but also effective scale-up of the technology, allowing for flows and loadings to be properly sized in the future full-scale system.

8.1 Selecting Analytical Testing Parameters

Before selecting analytical tests to be performed during a pilot study, knowledge of the technologies, target contaminants, and analytical methods is required.

8.1.1 Treatment Technology

Selection of analytical testing parameters and frequency of sample collection will depend on the type(s) of treatment technology being evaluated in the pilot study. It is important to understand the potential effects a given technology has on effluent water quality, recovery rates, residuals production, and specific constituent concentrations. For example, if the pilot test is evaluating media filtration for solids removal, then removal of dissolved solids is not likely to occur and would not need to be included in the water quality monitoring program. Conversely, treatment by ultrafiltration or microfiltration membrane filtration will remove colloidal and suspended solids but may also remove large dissolved compounds. Similarly, treatment with polymers for removal of colloidal and suspended solids (typically done in jar-testing procedures) would not likely involve removal of (small or medium molecular weight) dissolved biochemical oxygen demand (BOD) or chemical oxygen demand (COD).

If there are technology vendor(s) providing the pilot testing systems, they will most often recommend analytical testing parameters for verification of treatment efficiencies. If multiple technologies are piloted, such as anaerobic followed by aerobic MBR and reverse osmosis, the individual treatment processes will also likely need influent and effluent monitoring. Testing only the overall treatment facility influent and effluent does not provide sufficient performance data to determine unit process efficiencies and operating conditions, which is important for full-scale design. Testing of individual treatment systems might prove that one of the treatment technology processes is not warranted based on its limited effect on the removal efficiencies.

8.1.2 Selection of Target Analytical Testing

Rather than running an entire list of analytical parameters at every sample point upstream and after each treatment process in (multitechnology) pilot, one could measure only those parameters expected to change through a specific treatment steam. Selection of which water quality parameters to monitor at various sampling locations in the pilot system is driven by the technology being evaluated and its expected effect on water quality will reduce analytical costs, while still achieving some level of confidence for other parameters not tested as frequently. This also can provide faster turnaround time for laboratory results. In some cases, it may be desirable to use surrogate water quality parameters for analytical methods that are either costly or have long turnaround times. Examples of surrogate parameters include conductivity for total dissolved solids (TDS), turbidity for total suspended solids (TSS), UVA-254 for total organic carbon. In cases where surrogate water quality parameters are used to evaluate performance, the surrogate parameter can often be monitored with online instrumentation, allowing for less frequent monitoring of the actual parameters (TDS, TSS, total organic carbon [TOC], etc.). In cases where surrogate parameters are used, it is often desirable to establish a correlation between the actual parameter and the surrogate parameter for analysis of water quality trends.

The best example often used for this is the chemical oxygen demand (COD) test versus the biochemical oxygen demand (BOD) test. If a correlation of the BOD to COD is established, then running the COD test (typically approximately 3 hours) can give a fast indication of the expected BOD that takes 5 days to incubate. Testing for COD could be done onsite, whereas BOD testing is most often done in outside contract laboratories (or municipal laboratories). However, the reader is cautioned that the BOD-to-COD correlation will change as wastewater is processed through many technologies (e.g., biological treatment, GAC, advanced oxidation processes).

8.1.3 Analytical Turnaround Time

Analytical turnaround time refers to the time for providing the laboratory the sample(s) to be tested and the time for getting the results. Some analytical parameters can take over a week to yield results, whereas others may take a day to several days. Establishing the expected turnaround time before initiating pilot testing affects costs at contract laboratories by defining weekend or after-hours workloads. Often, a laboratory may perform the test fairly quickly; however, generation of the written report adds days to a week to the process. Requesting that the laboratory email or phone in results can expedite turnaround time for the information needed to adjust the pilot process. Also, consider whether certain parameters can be analyzed onsite by the operators and the frequency of operational changes and the need to see laboratory results before making further changes.

8.2 Sampling Pilot Systems

Obtaining accurate results for water characteristics is important in the evaluation of treatment technologies.Sometimes, because of small flows and sampling techniques, this affects the representative sample data being used for scale-up of the full treatment system. Collection of flow-composited samples is typically not done in small-scale pilot systems unless there is continuous flow with sufficient volume to install a flowmeter to pace the autosampler.

Time-composited samples are more representative of treatment processes and variations than grab samples. Grab samples should only be used for parameters requiring grab, such as pH, temperature, dissolved oxygen, oxidation–reduction potential (ORP) probes, and so on, or use on-line monitors. Of course, pilot testing batch treatment processes may use grab samples for each batch as being representative. Time-composite samples allow for variations in loading and pilot performance resulting in more accurate average treatment than a grab sample might indicate.

It is best practice to prepare a written sampling plan or matrix indicating where and how samples will be collected, type of samples, frequency, and parameters to test for coordination with the laboratory. Other important factors to consider in sampling include:

- Sample preservation—Multiple samples may be needed, or aliquots of samples with preservation required for some parameters such as metals, BOD, etc. Refer to the latest edition of *Standard Methods for the Examination of Water and Wastewater* (APHA et al., 2017) or U.S. EPA methods for preservation requirements for specific analytical tests. Keeping samples cool is often difficult upon collection in field-pilot systems if there is no cooler or refrigerator available. Cooling of samples should be considered as a required preservation requirement in effective pilot treatment monitoring.
- Training in sampling techniques—Pilot system operators need to be trained on and know how to collect accurate samples. This may include flushing of piping/tubing before collection of samples, not rinsing out preservatives in sample containers provided by laboratories, onsite preservation requirements, delivery to laboratory procedures, etc.

8.3 Contract Laboratory Versus In-House Testing

Many industries do not have in-house laboratories capable of performing the specific analytical tests needed for evaluating pilot system performance; therefore, a contract laboratory may be used. As previously noted, a written sampling plan can help the laboratory prepare for the volume and type of analyses needed, as well as prepare a budget cost for the laboratory work.

Consideration should also be given to the value of a third-party outside laboratory doing the testing to ensure the performance of the pilot is not under total determination by onsite personnel and onsite testing. Many contract laboratories also can provide, set up, and maintain autosamplers or collect samples onsite. This helps reduce the potential for problems in sample preservation that can affect the analytical results. The laboratory can field-filter samples needed for testing metals, sulfides, or other parameters too.

9.0 INTERPRETING DATA

One of the most critical steps in performing a pilot treatment evaluation is interpreting the data collected and applying it to full-scale designs. Although the purpose of the pilot is to verify the technology used is effective for achieving desired water quality, this all depends on the interpretation of data.

Quickly evaluating data as they become available is crucial to optimizing pilot system performance. It is important to note trends in the data, which is easy to do if it is charted or recorded using online instrumentation. Responses of the pilot test operations to changing production procedures, power failure(s) at the pilot, lack of sludge removal, and so on should be recorded and related to the data.

In choosing treatment/recycle technologies using pilot data, all data should be critically evaluated. A cause-and-effect evaluation can relate batch discharges, spills, equipment failures, and similar unplanned events to the responses of the pilot treatment equipment. No effect might be the conclusion, but failure of treatment because of one of these reasons may warrant exclusion of that technology from further application. It could indicate that equalization, or control of specific chemicals (such as CIP chemicals) used in some industries, is required.

10.0 CASE STUDIES OF PILOT TREATMENT PROJECTS

The following actual projects describe sources of industrial wastewaters that were tested using long-term pilot treatment processes. Pilot testing was executed to assess technology and associated operational boundaries based on actual operating conditions. The intended use of the technology assessment results, based on actual operational conditions, was to inform the design of a full-scale treatment process to recycle effluent for reuse; more accurate predictions regarding water reuse could also be achieved. Each pilot case study discussed below provides conclusions from the pilot tests as well as technical information and physical descriptions of equipment used.

10.1 Beverage Process Pilot Treatment for Reuse

10.1.1 Introduction

Water is a critical resource in a beverage production supply chain because it is used in agricultural production, as an ingredient, and during finished goods processing. Sustainable water resource management is necessary to mitigate industrial risk factors such as constraints on growth, operational and supply chain disruptions, conflicts with other stakeholders, loss of social license to operate and grow, and increasing production costs. Additionally, expected rapid population growth will put stress on water resources, thereby increasing the need for sustainable water resource management (Organization for Economic Cooperation and Development, 2008).

PepsiCo is committed to apply sustainable water use in its direct operations and to increase efficiency by 25% by 2025 as part of its overall "Winning with Purpose" water goals. Applying near-zero liquid discharge technology to process water discharge streams and reusing the water for nonfood contact processes may be one of the viable solutions to increase (beverage) facility efficiency. The minimum liquid discharge (MLD) case study outlined below presents the findings of a 12-month pilot study at a beverage manufacturing site using a series of treatment systems designed to treat beverage wastewater to drinking water quality for potential reuse with high recovery efficiency.

10.1.2 Methods

Between March 2016 and February 2017, PepsiCo piloted an MLD process at its Riverside, California, beverage facility. The pilot was designed to treat an average flow of 19 L/min (5 gpm) on a continuous flow basis for approximately 7 months. The influent was taken directly from the bottling operations process wastewater. The target effluent composition was based on drinking water quality. More details pertaining to the influent and effluent characteristics are provided in Sections 10.1.3 and 10.1.4.

Facility personnel were trained on the operation of the treatment components by the pilot system vendors and the engineering consultant. The Riverside facility staff operated the pilot system and collected samples for onsite testing and for outside contract laboratory analyses for the duration of the testing period. Descriptions of the characteristics and the performance of the significant unit process used are provided in Section 10.1.5.

10.1.3 Influent Characteristics

The influent to this pilot exhibited characteristics of wastewater from two points for contribution seen at a bottling facility. The first contributing point for wastewater stems from that of typical bottling facility operations while the second originates from returned goods (i.e., expired product) that are processed onsite before liquid product waste (i.e., "crush") is sent to drain. The processing of returned bottles and cans of product does not occur at all sites; however, this operation does pose unique considerations (i.e., fermentation) when addressing potential characteristics for wastewater from a bottling facility.

In this pilot, influent was passed through biological treatment before membrane filtration, GAC, and then reverse osmosis. Table 7.1 describes the typical influent characteristics into the biological systems.

10.1.4 Effluent Requirements

The success of the pilot study was measured based on the system's ability to meet reuse water quality requirements as defined in 21 C.F.R. 110, *Current Good Manufacturing Practice in Manufacturing, Packing, or Holding Human Food*. The specific requirement referenced to base this case study's success is found under 21 C.F.R. 110.37(a), which reads, "Water supply shall be sufficient for the operations intended and shall be derived from an adequate source. Any water that contacts food or food-contact surfaces shall be safe and of adequate sanitary quality." The final water quality of this facility was intended to meet both the standard set by the 21 C.F.R. 110 regulation as well as the National Primary and Secondary Drinking Water Standards, Title 40-Protection of the Environment, Chapter I-EPA, Parts 141 and 143—National Primary and Secondary Drinking Water Regulations. The pilot treatment system was designed to meet limits for the 13 key parameters listed in Table 7.2.

10.1.5 Pilot Treatment Process

The overall pilot treatment consisted of biological processes (both anaerobic and aerobic) followed by solids separation (ultrafiltration), GAC, and advanced filtrations (reverse osmosis). Figure 7.1 presents the process flow diagram of the pilot, and Figure 7.2 showcases the pilot train pictorially. The following sections provide an in-depth description of each process in the treatment train.

10.1.5.1 Anaerobic System

The anaerobic system used for the pilot was the expanded granular sludge bed technology provided by Veolia Water Technologies, which is similar to a

TABLE 7.1 Pilot influent.

Parameter	Unit	Design Average
General		
Total hardness, as $CaCO_3$	mg/L	100
Organics		
Soluble BOD[a]	mg/L	1100
Total BOD	mg/L	1400
Dissolved Organic Carbon	mg/L	1200
Total Organic Carbon	mg/L	1400
Total COD[b]	mg/L	3500
Solids		
Total Solids	mg/L	2000
Total Volatile Solids	mg/L	1800
Total Dissolved Solids	mg/L	1900
Total Suspended Solids	mg/L	ND[c]
Nutrients		
Total Phosphorus	mg/L	5.5
Ammonia (NH_3-N)	mg/L	ND
Cations		
Sodium	mg/L	41
Potassium	mg/L	10
Calcium	mg/L	30
Barium	mg/L	0.015
Iron	mg/L	0.63
Zinc	mg/L	77
Magnesium	mg/L	6
Total Cations	meq/L	4.03

(continued)

TABLE 7.1 Pilot influent. (*Continued*)

Parameter	Unit	Design Average
	Anions	
Alkalinity, as $CaCO_3$	mg/L	28
Chloride	mg/L	39
Sulfate	mg/L	23
Nitrate (NO_3-N)	mg/L	1.2
Citrate	mg/L	146
Total Anions	meq/L	4.03
	Others	
Silica	mg/L	1.8

[a]BOD refers to biochemical oxygen demand
[b]COD refers to chemical oxygen demand
[c]ND refers to not detected

TABLE 7.2 Process return water requirements for system design.

Parameter	Maximum Contaminant Level	Primary or Secondary Standard
Aluminum	0.05–0.2 mg/L	Secondary
Barium	2 mg/L	Primary
Fluoride	4 mg/L (2 mg/L)	Primary (Secondary)
Nitrate as N	10 mg/L	Primary
Nitrite as N	1 mg/L	Primary
Turbidity	1 NTU	Primary
Chloride	250 mg/L	Secondary
Color	15 units	Secondary
Corrosivity	Noncorrosive (LSI[*] ≥ 1.0)	Secondary
Iron	0.3 mg/L	Secondary
pH	6.5–8.5	Secondary
Sulfate	250 mg/L	Secondary
Total dissolved solids	500 mg/L	Secondary

[*]LSI refers to the Langelier Saturation Index

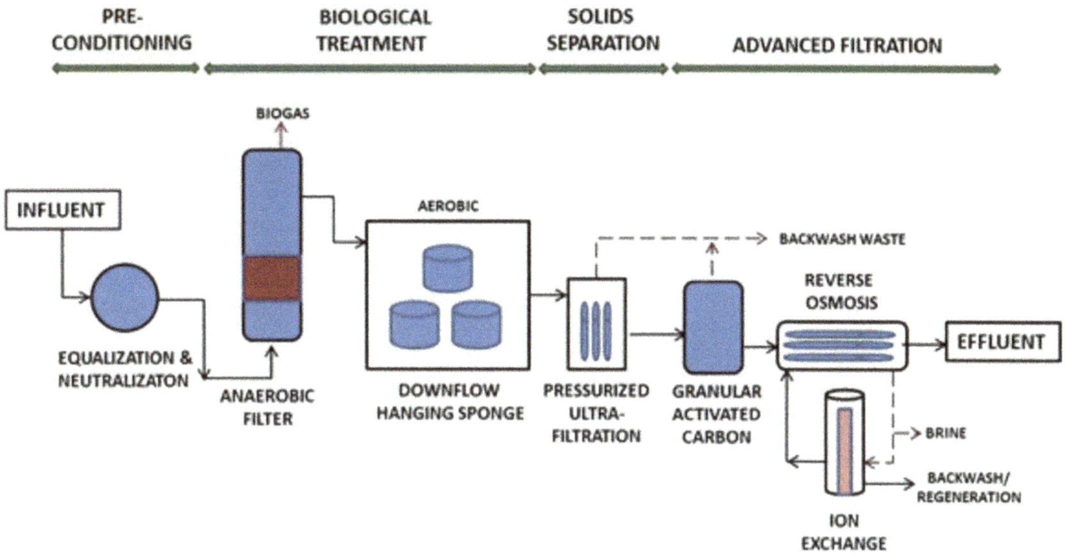

FIGURE 7.1 Process flow diagram of pilot.

high-rate UASB process. The 13-m^3 anaerobic reactor had a height of 9 m (30 ft) and was seeded with active anaerobic granular biomass from PepsiCo's Hayward, California, facility. The reactor pH and temperature were automatically controlled to maintain relatively constant values of pH 6.8 to 7.6 and 35 °C to 37 °C (95 °F to 98.6 °F), respectively. Anaerobic system performance results are summarized in Figure 7.3; the COD removal design basis

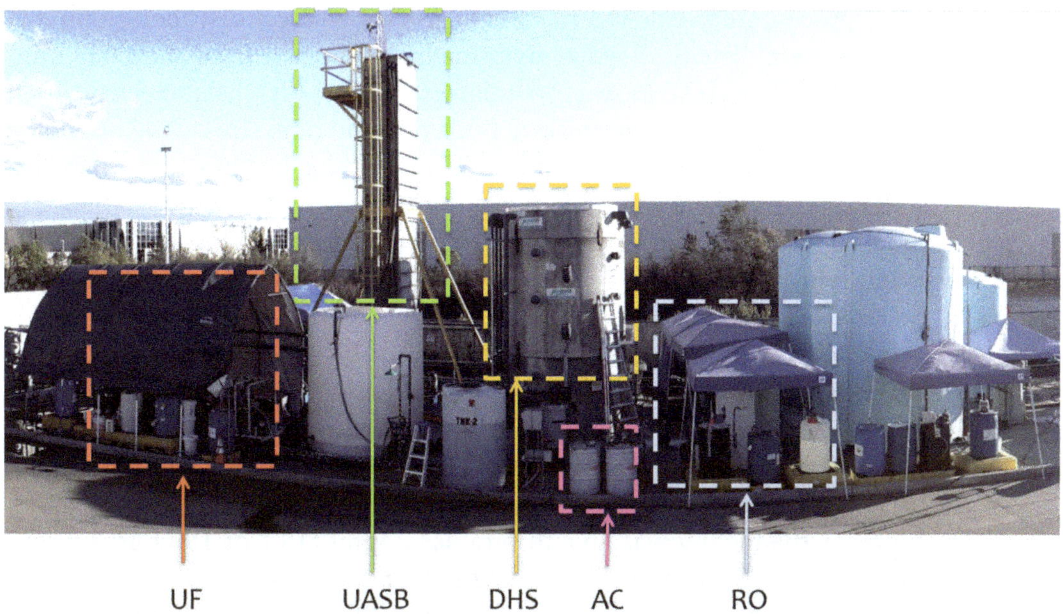

FIGURE 7.2 Pictorial view of the pilot.

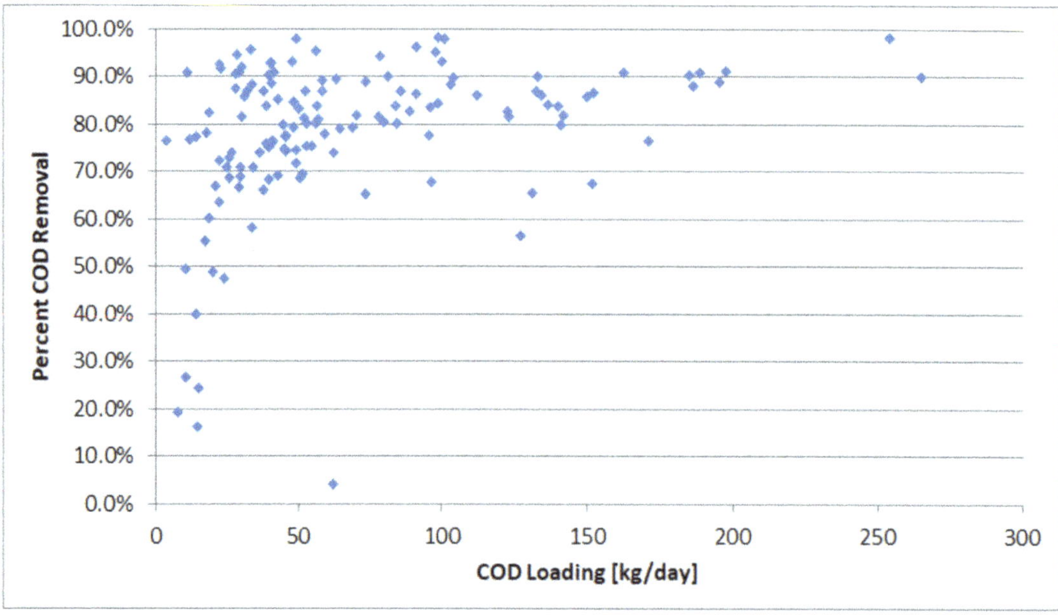

FIGURE 7.3 Anaerobic system performance.

was 75%. Chemical oxygen demand removal was consistent with expected values (74% for all days without crush addition vs. 75% expected).

10.1.5.2 Aerobic System

A downflow hanging sponge (DHS) system was used for the aerobic treatment process step. This is a fixed-film process where a distribution piping/spray system "trickles" water over media from the top of the DHS unit. For this pilot, two DHS reactors (diameter: 2 m [7 ft]; height: 5 m [17 ft]) were in series to minimize construction costs and tank sizes while providing sufficient hydraulic retention time (HRT). Three media "lifts" were installed in each reactor (height per lift = 0.8 m [31.5 in.]).

The media used in the DHS is a random-placed "sponge" inserted in a plastic shroud. This allows aerobic bacteria to grow on the surface while anaerobic bacteria grow inside the porous sponge. The sponge used for this pilot, G3.2, is manufactured by Sekisui Plastics in Japan. Figure 7.4 presents a photograph of one piece of media; a pen has been included, for comparison purposes, to visually indicate how small one piece of media is.

The total sponge volume was 17.2 m^3 (606 cu ft), corresponding to a 50% sponge fill of the reactor volume. The HRT for one DHS was 9 hours; however, the facility was able to produce a total HRT of 18 hours by running both DHS in series. Downflow hanging sponge system performance results are summarized in Figure 7.5; the COD removal design basis was 85%. Chemical oxygen demand removal was less than expected at 61% (vs. 85% expected).

FIGURE 7.4 Photograph of downflow hanging sponge (DHS) media (pen added for scale).

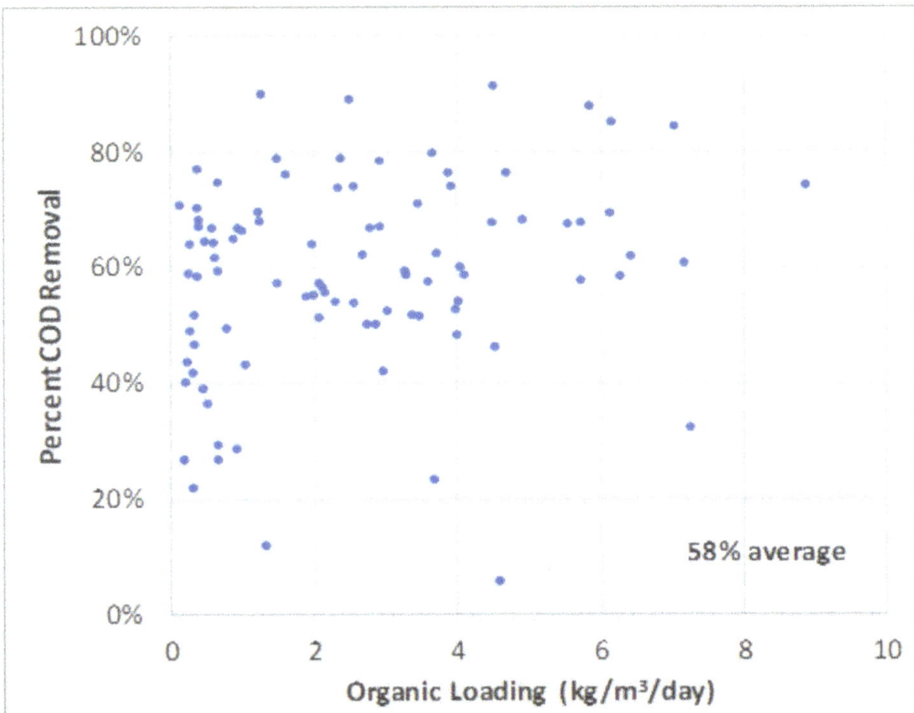

FIGURE 7.5 Aerobic system performance.

10.1.5.3 Ultrafiltration and Granular Activated Carbon

A single pressurized ultrafiltration membrane skid from GE Water & Process Technologies (now part of Suez) was installed, followed by two GAC adsorber units to remove residual BOD in the DHS effluent and to prevent potential fouling of the reverse osmosis membranes. Typically, this system provided approximately a 30% reduction in COD. The ultrafiltration and GAC were primarily used for polishing purposes and as a pretreatment before reverse osmosis operations. To clean the ultrafiltration system, chemicals used included sodium hypochlorite, sodium hydroxide, and citric acid.

10.1.5.4 Reverse Osmosis

Following the GAC adsorbers, the pilot reverse osmosis system operated in "batch mode," which allowed for it to be designed and run as both the primary and secondary reverse osmosis unit. While operating in primary mode, the two-stage reverse osmosis unit operated continuously until there was sufficient reverse osmosis concentrate (i.e., reject stream), approximately 22 710 L (6000 gal), to process through the reverse osmosis system a second time.

However, before recovering more water from the reverse osmosis concentrate, this stream was sent to an ion exchange system to remove sparingly soluble cations. The ion exchange effluent was then collected and treated through the reverse osmosis unit a second time (i.e., "secondary" mode) to recover up to an additional 70% to 80% of the influent.

The reverse osmosis permeate results (see Figure 7.6) reveal that the concentration of TDS, after both stages, fell consistently near 250 mg/L. All other significant parameters, such as disinfection byproducts, metals, and herbicides, were consistently below the U.S. EPA drinking water maximum contaminant levels. Although the reverse osmosis recovery started near 70%, it increased to 94%, which was acceptable for the pilot goals.

10.1.6 Conclusion

Operationally, minimal fouling or scaling of ultrafiltration and reverse osmosis systems was observed. This can be attributed to the injection of sodium hypochlorite and ammonia as a way to maintain monochloramine. Maintenance of monochloramine residual in water being fed to these systems is often necessary to prevent biogrowth. Clean-in-place (i.e., CIP) of the system restored the reverse osmosis performance to original values, which indicated the absence of irreversible fouling/scaling. A weekly caustic CIP was performed out of precaution whereas a water flush for reverse osmosis occurred daily.

The pilot treatment train was able to produce water quality consistent with World Health Organization/U.S. EPA standards from beverage

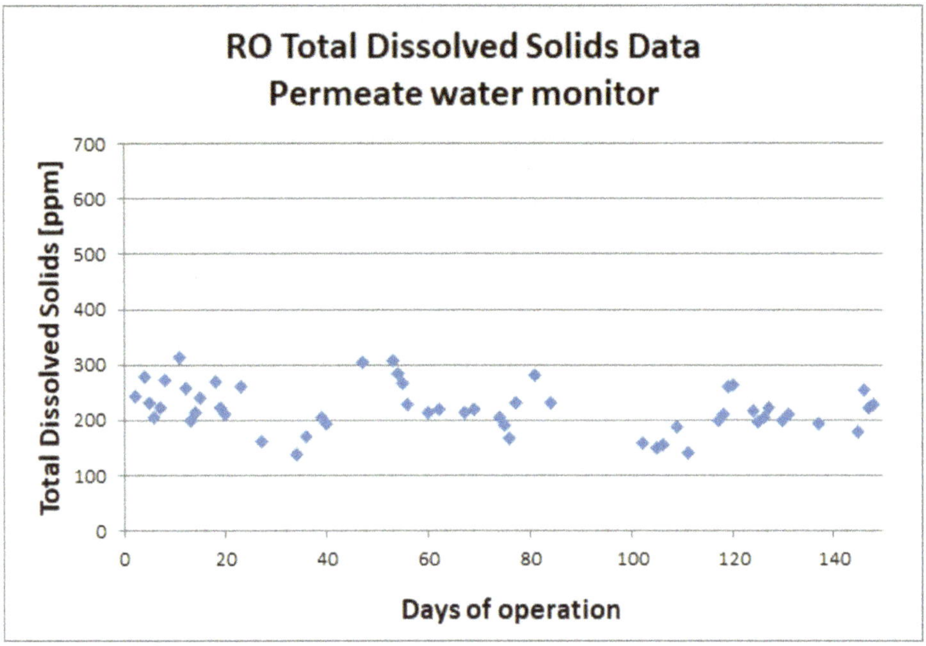

FIGURE 7.6 Reverse osmosis permeate results.

facility wastewater, which wasn't a requirement, but a good result. As a result, water quality data of this stream can be leveraged during discussions with corporate Food Safety and Quality teams to determine where safe and appropriate reuse in the facility can be made possible and if further treatment is needed to do so.

With results from this pilot, design parameters were determined to create conceptual designs for the treatment of beverage process wastewater and served as a basis to inform Opinion of Probable Construction Costs for a full-scale treatment facility. The intent of these designs was to address COD loading differences that may be experienced between varying bottling operations as well as account for different localities with varying limits for wastewater discharge that are included within the PepsiCo network.

These conceptual designs for treatment of beverage process wastewater will continue to inform plans for full-scale implementation throughout the PepsiCo network. Based on this pilot, they will serve an understanding of expected water quality and flow given a selected treatment option, associated costs for equipment and maintenance, operational conditions to maintain systems, as well as any watch-outs to ensure safety compliance. All considerations will help a facility navigate how to best approach integrating sustainable water management practices.

10.2 PETROLEUM PROCESS PRODUCED-WATER PILOT TREATMENT

10.2.1 Overview

A water treatment pilot facility was proposed to evaluate the technology performance and feasibility of a proposed full-scale water treatment process for produced waters from the Moneta Divide Field in Wyoming, United States. The existing facilities consist of oil and gas production units in which the hydrocarbons and formation waters are first separated at the surface using "gun barrel" technology followed by oil water separators and storage. Currently, the Moneta Divide Field produces up to approximately 65 000 barrels of water per day (BWPD), or 10 334 m^3/d that is discharged to a permitted outfall. Before discharge, the water is passed through a series of separators to remove the oil and natural gas, and then the water is cooled to less than 38 °C (100 °F) for discharge.

This section presents the results of the pilot test for the treatment of produced water associated with oil and gas operations for a confidential petroleum industry client, including

- basis of pilot system design,
- description of the pilot treatment process,
- pilot test results of primary treatment processes,
- biological jar test results for methanol removal, and
- conclusions.

The following section summarizes the design criteria of the Moneta Divide Water Treatment Pilot Plant.

10.2.2 Design Flow

The design flow of raw produced water to the pilot treatment system was 0.66 L/s (10.5 gpm). The feed water was cooled to between 37 °C and 49 °C (100 °F and 120 °F) before entering the pilot process feed tank.

10.2.3 Feed Water Quality and Effluent Targets

Table 7.3 summarizes the design raw feed water quality and target treatment limits for the pilot test. Target limits represent standards that would need to be met in the future at the end of pipe for full-scale discharge to surface water under a revised discharge permit, and the current existing outfall limits. The feed water quality is based on a number of samples collected during field operations and pilot testing and is representative of produced water from the Moneta Divide Field.

TABLE 7.3 Feed water quality and effluent targets.

Description	Raw Feed	Future Treatment Limits	Existing Outfall Limits
	Commons		
pH	7.56	6.5–9.0	6.5–9.0
Temperature, °C	37	N/A	N/A
Total alkalinity (mg/L)	2200	N/A	N/A
Nitrogen, ammonia (mg/L)	4.2	<0.1	N/A
Sulfide (mg/L)	0.1	0.01	N/A
Silica	84.0	N/A	N/A
Boron (mg/L)	12.0	0.058	N/A
Phosphorus (mg/L)	0.5	0.10	N/A
	Anions		
Bicarbonate	2626	N/A	N/A
Carbonate	19	N/A	N/A
Chloride (mg/L)	3000	N/A	2000
Sulfate (mg/L)	16	N/A	3000
	Cations		
Calcium (mg/L)	15	N/A	N/A
Magnesium (mg/L)	15	N/A	N/A
Potassium (mg/L)	20	N/A	N/A
Sodium (mg/L)	2896	N/A	N/A
Selenium, recoverable (μg/L)	<0.5	5	5
Iron (+2) (mg/L)	1.10	<0.1	N/A
	Other		
Total suspended solids (mg/L)	20.0	5	N/A
Oil and grease (mg/L)	6.9	10	10
TDS (at 180 °C) (mg/L)	7774	370	N/A
Conductivity (μS/cm)	7500	7500	7500
Total hardness (mg/L as $CaCO_3$)	106.2	N/A	N/A

N/A = not available, not applicable

10.2.4 Process Description

Figure 7.7 illustrates the pilot block flow diagram, which was divided across the following significant components:

- **Water treatment trailer No. 1:** A 16-m (53-ft) long semi-tractor trailer housed all the reverse osmosis pretreatment equipment and reverse osmosis effluent polishing systems. The trailer contained all electrical distribution equipment for power supply and instrumentation and controls. The trailer was heated and insulated.
- **Water treatment trailer No. 2:** This 7-m (24-ft) long semi tractor trailer housed the low-pressure reverse osmosis (LPRO) and high-pressure reverse osmosis (HPRO) systems and HPRO feed tank. The trailer contained all electrical distribution equipment for power supply and instrumentation and controls. The trailer was heated and insulated.
- **Field trailer:** A 6-m (20-ft) long office trailer was located onsite to serve as the office and field laboratory.
- **Outdoor tanks:** The raw water tank, backwash reclaim tank, product water tank, brine concentrate tank, and a pump house were located outside next to the treatment trailers.

For purposes of the pilot test, treated effluent from the pilot facility was combined with the pilot facility brine concentrate and routed back into the existing discharge pipeline downstream of the inlet to the pilot treatment system. All interconnecting piping was with flexible hoses; all hoses were heat-traced and insulated.

10.2.5 Pilot Process Steps

The following process steps were used for this pilot as illustrated in Figure 7.7:

- **Pretreatment of raw water:** The water to be treated was cooled to below 37 °C and pretreated before entering the raw water tank.

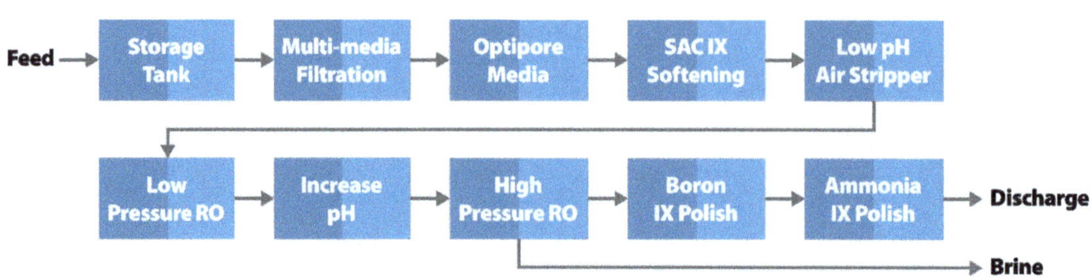

FIGURE 7.7 Pilot treatment system block flow diagram.

Pretreatment included addition of sodium hypochlorite to oxidize/remove sulfide and iron; addition of sulfuric acid to precipitate barium as barium sulfate and to remove some of the alkalinity; and addition of a coagulant to promote settling of the precipitated solids in the tank.

- **Filtration:** Pretreated water from the raw water tank was then pumped to a mixed-media filter containing anthracite, garnet, and sand to remove TSS. Sodium hypochlorite was dosed into the process water stream before the filters to further oxidize any remaining iron and minimize bacterial growth on the filter media.
- **Hydrocarbon removal:** After filtration, dissolved hydrocarbons were removed by adsorption using OPTIPORE™, a hydrocarbon adsorption media capable of high hydrocarbon loading rates. The adsorption media was regenerated onsite by steam stripping and then condensed in a water-cooled heat exchanger to produce a mixture of hydrocarbons and water. The hydrocarbons were skimmed off the top of the condensate tank, which, in the full scale, could be recovered as a saleable byproduct. The condensed water was then either pumped back into the OPTIPORE™ media system or pumped back to the raw water tank for retreatment.
- **Softening:** In the next treatment step, calcium, magnesium and other polyvalent cations (such as iron) were removed using a strong acid cation (SAC) ion-exchange resin. The ion-exchange media was regenerated using a brine solution, and the regeneration waste was directed to the brine concentrate tank. The softened water was directed to the air stripper.
- **Air stripping:** Sulfuric acid was added to lower the pH of the softened effluent to approximately 4.3, and then the water was routed through a tray-type air stripper to remove carbon dioxide. After air stripping, the water was directed to the LPRO feed tank.
- **Low-pressure reverse osmosis:** From the LPRO feed tank, the water was pumped through a LPRO (<400 psig) unit to remove dissolved solids and ammonia. An antiscalant was added before the LPRO unit to inhibit scale formation on the reverse osmosis membranes. Acid was injected ahead of the reverse osmosis system to maintain a pH typically between 3.5 and 4.5. The treated water (permeate) was directed to a permeate tank before final treatment. The concentrated brine was directed to the HPRO unit for additional treatment.
- **High-pressure reverse osmosis:** Before entering the HPRO system, the LPRO reject stream was pH adjusted to between 10.5 and 11.5 with

caustic in a mix tank. An additional antiscalant was added to minimize formation of calcium and barium minerals in the HPRO system. Water from the mix tank was then pumped through a 1-μm cartridge filter and into the HPRO system (<1200 psig) for additional TDS removal. The HPRO permeate flowed to the permeate tank where it was combined with the LPRO permeate before final treatment. The HPRO reject stream (concentrated brine) flowed to the brine concentrate tank.

- **Boron removal:** The mixed reverse osmosis permeate (LPRO and HPRO permeate) was pumped from the permeate tanks to a boron removal unit, where residual boron was then removed to less than 0.050 mg/L using a boron-selective ion-exchange resin. The resin was regenerated using dilute solutions of sulfuric acid and caustic. The resulting concentrated brine regeneration streams were routed to the brine concentrate tank.
- **Ammonia removal:** In the final treatment step, the mixed reverse osmosis permeate passed from the boron removal system through a SAC ion-exchange resin to remove residual ammonia to less than 0.1 mg/L before discharge. The ion-exchange media was regenerated using a sodium chloride salt solution. The resulting strong regenerate brine stream was directed to the brine concentrate tank. After ammonia removal, the water flowed to the product water tank.

The pilot-scale treatment system was capable operating at a feed rate of 0.66 L/s (10.5 gpm) of water supplied from the produced water pipeline (currently carrying 6677 m^3/d or 42 000 BWPD of water), producing a final concentrated brine flow of approximately 0.066 L/s (1.05 gpm). Because there was limited availability to dispose of the concentrated brine stream, the product and brine concentrate streams were stored in separate tanks and then recombined and reinjected back into the pipeline downstream of the takeoff point. The volume of effluent 0.66 L/s (10.5 gpm) is small compared to the current discharge stream (approximately 77 L/s or 1225 gpm). Consequently, the overall effluent quality was essentially the same as the current discharge quality.

The pilot system was started up sequentially over the course of approximately 1 week. Each unit, beginning with the raw feed tank, was started up and tested to confirm proper function and performance before starting up additional treatment units. When completed, the pilot test commenced to continuous operations. During the continuous operations phase, operators collected field data (e.g., pressures, flowrates, field water quality analyses), performed system adjustments, and collected analytical samples.

10.2.6 Overview of Process Performance

This section provides a brief description of each unit operation of the pilot-scale treatment process and a summary of its associated performance during the test.

10.2.6.1 Pretreatment of Raw Water

Raw produced water was fed to the pilot system raw feed tank (48 m^3 or 300 bbl) as a slip stream from a force main. An automatic valve opened and closed to maintain a desired liquid level in the tank. Sodium hypochlorite was metered into the raw water to oxidize/remove sulfide and iron. Sulfuric acid was added to precipitate barium as barium sulfate and to help coalesce oil and grease. Coagulant was added to promote agglomeration of the precipitated solids in the tank.

Analytical and field data showed that the feed tank performed as expected. Barium and iron concentrations were significantly reduced through the tank. Dissolved barium concentrations were reduced from approximately 3 mg/L in the raw feed to <0.1 mg/L in the water exiting the tank. Iron concentrations were below laboratory detection in the facility feed water exiting the tank. A layer of oil and grease developed at the surface of the tank, which was not present in the facility feed water, indicating that oil and grease were accumulating in the tank at the liquid level surface. Although it could not be visually verified, solids were likely collecting at the bottom of the tank. This was evidenced by a noticeable increase in solids in the facility feed water when the automatic valve was in the open position and a decrease in solids in the facility feed shortly after the automatic valve closed.

10.2.6.2 Filtration

Pilot facility feed water was pumped from near the bottom of the raw feed tank through a bank of multimedia filters containing anthracite and sand to remove TSS as well as some diesel-range organics (DRO) and gasoline-range organics (GRO) and free oil/grease. Sodium hypochlorite was added ahead of the multimedia filters to further oxidize any remaining iron and minimize bacterial growth on the filter media. In addition, the chlorine reacted with ammonia present in the water to produce monochloramine, which is a mild disinfectant safe for use in the reverse osmosis membrane system. A coagulant was also added to aid in the filtration of suspended solids, including iron hydroxide and barium sulfate precipitates. The multimedia filters were backwashed as needed, approximately twice per day using filtrate water. Backwash water was collected in the tank and then metered back to the raw water tank.

The media filtration system typically performed as expected. Analytical data indicated that as much as 80% of the DRO/GRO was removed by the multimedia filters. The filtrate turbidity was typically less than approximately 1.0 NTU, and often below 0.1 NTU. Periodically, the turbidity did range well above these values for short durations, typically when the raw feed tank was filling. Suspended solids were reduced to 4 mg/L in the filtrate.

10.2.6.3 Hydrocarbon Removal

After filtration, dissolved hydrocarbons were removed by adsorption using OPTIPORE™ media, an engineered adsorption media capable of relatively high hydrocarbon loading rates. The adsorption media was manually regenerated onsite by steam stripping. The regenerant was then condensed in a water-cooled heat exchanger to produce a mixture of hydrocarbons and water. This mixture was allowed to separate in a tank. The hydrocarbon layer was then decanted, and the condensed water phase was pumped back to the raw water tank for retreatment. Figure 7.8 depicts a sample of the regenerant from the OPTIPORE™ system after it condensed and separated.

FIGURE 7.8 Recovered product from OPTIPORE™ regeneration.

Throughout the pilot test, the OPTIPORE™ system performed well. Analytical data showed that DRO/GRO was reduced by more than 90%. Effluent from the OPTIPORE™ system contained 2.9 mg/L of DRO and 0.13 mg/L of GRO. The effluent was clear and contained no hydrocarbon odor. Total organic carbon in the process water was not significantly reduced; however, a significant component of the TOC was suspected to be present as guar gum or polyacrylamide, which manifested itself in the HPRO system, and is discussed further in the following section.

10.2.6.4 Softening

Calcium, magnesium, and other polyvalent cations (such as barium, aluminum, and iron) remaining in the water were removed using a SAC ion-exchange resin. The SAC system was a packed-bed design used to reduce the volume of regeneration solution. The ion-exchange media was periodically regenerated using a sodium chloride brine solution. Spent regenerant was sent to the concentrate tank.

During the pilot test, the SAC system did remove hardness ions, but did not perform to the level expected or required. The SAC system effluent target was <1 mg/L of hardness as $CaCO_3$; however, hardness concentrations in the effluent ranged from 3 to 8 mg/L as $CaCO_3$. Additional follow-up studies were performed on the SAC resin to account for this lack of performance. From these tests, it was determined that the resin had fouled in operation during the pilot test; fouling was attributed to organics, potentially both the guar gum and/or polyacrylamide previously discussed as well as residual hydrocarbons. The organics were readily removed from the resin through acid cleaning and the ion-exchange capacity of the resin was restored. As such, it was determined that a weak acid cation (WAC) resin would be well suited to the application because WAC resins are regenerated with acid. Figure 7.9 depicts the SAC resin before and after cleaning. The white layer was determined to be organic.

The softened effluent was then pH adjusted using sulfuric acid to a pH ranging from 3.5 to 4.0. This range was chosen because below a pH of 4.3, bicarbonate alkalinity is converted to carbon dioxide gas. The low pH water then flowed to the stacked tray air stripper. The carbon dioxide and any remaining hydrocarbons were stripped from the process water. With the alkalinity removed, the water has little buffering capacity, reducing the quantity of caustic required to adjust pH in downstream operations. Additionally, the lack of alkalinity reduces the potential for carbonate scale buildup in the HPRO system. Effluent from the air stripper was pumped to the LPRO feed tank.

FIGURE 7.9 Strong acid cation resin before and after cleaning.

Field results indicated that the air stripper did not adequately remove carbon dioxide from the process water because of short-circuiting as a result of the system not being level. This condition would be easily rectified in full-scale operations with proper design and installation of a packed tower system of sufficient height. As a result, larger quantities of sodium hydroxide were needed in the pilot system to attain the desired pH adjustment in the HPRO feed tank.

10.2.6.5 Low-Pressure Reverse Osmosis

From the LPRO feed tank, the low pH water was pumped through a set of 1-µm cartridge filters to remove any remaining suspended solids and then to the LPRO (<450 psig) system to remove TDS, boron, and ammonia. Permeate (treated water) from the LPRO flowed to the permeate tank, while reject water from the LPRO flowed to the HPRO feed tank. An antiscalant was metered into the LPRO feed water to prevent scale formation in the LPRO. The LPRO unit was operated at approximately 50% recovery throughout the pilot test.

The LPRO unit showed stable operation throughout the test. Permeate water quality was excellent, and no significant increase in pressure or reduction in permeate flux was observed for the duration of the test.

10.2.6.6 High-Pressure Reverse Osmosis

The LPRO reject flowed to the HPRO feed tank, where it was pH-adjusted to approximately 11 to 11.5. Low dosages of antiscalant and ethylene diaminetetraacetic acid were also added to the HPRO feed tank as safeguards to prevent scaling from barium sulfate and hardness scale, respectively. From the tank, the HPRO feed water was pumped through a set of 1-µm cartridge filters before entering the HPRO unit for TDS and boron removal. Final brine concentrate was collected in the concentrate tank and discharged back to the outfall pipeline as needed. Brine streams entering this tank included the reverse osmosis reject, spent ion-exchange regeneration solutions, and spent reverse osmosis cleaning solutions.

The recovery of the HPRO system was incrementally increased throughout the pilot test. Initially, the unit was operated to achieve an overall reverse osmosis system recovery of 75% (i.e., LPRO + HPRO). After successful operation, it was increased to achieve an overall reverse osmosis recovery of 80%. After several days, the recovery was increased to 85%. Operating at this higher recovery of 85%, a significant operational issue was encountered. The HPRO cartridge filters began to blind at a rapid rate (up to two times per day). Removal and inspection of these filters (see Figure 7.10) showed a whitish, gelatinous residue that was easily removed by rinsing with reverse osmosis permeate.

Continued operation of the HPRO at this recovery led to the eventual fouling of the reverse osmosis membranes from this same foulant, as evidenced by relatively stable HPRO operation followed by a sudden rapid increase in HPRO pressures. Short-duration cleaning (less than 30 minutes) of the HPRO with low-pH water quickly removed the foulant and restored the HPRO operation to baseline conditions. The HPRO required cleaning approximately every 48 hours.

It was determined that the foulant was most likely guar gum; although, a definitive analysis could not be made. To prevent the organic material from coming out of solution in the HPRO system, two solutions were tested. First, a gel-breaking enzyme was added to the raw feed tank in an effort to destroy the organic material before it reached the HPRO feed tank. The gel-breaker was metered into the raw water using a chemical metering pump. After the addition of the gel breaker, longer runtimes were noted in the HPRO, but the organic precipitation problem persisted.

Activated carbon was also used in an attempt to remove the organic material from the LPRO concentrate. A GAC canister was plumbed into the LPRO concentrate line. After implementation of this, the foulant was still collected on the cartridge filters along with GAC fines, which were entrained in the LPRO concentrate from the GAC vessel. As such, this approach was not used further. Pilot operations continued with frequent,

FIGURE 7.10 Organic precipitants collected at high pH on HPRO cartridge filter.

short duration HRPO flushes, as described above. Implications for a full-scale system include provisions for routine flushing and improved filtration upstream of the HPRO system; although additional testing is likely warranted on this issue.

Recovery of the HPRO system was increased again to achieve an overall system recovery of 90% and then 92%. Reverse osmosis system recoveries were based on field flow measurements but were also confirmed through mass balance calculation using analytical data. As shown in Figure 7.11, system pressures increased and permeate fluxes decreased during operation, but both parameters were restored to near baseline conditions with periodic low-pH cleanings, as seen at the lower recoveries. As such, the source of the fouling was attributed to the organic material and not inorganic precipitation, such as silica.

10.2.6.7 Boron Removal

Reverse osmosis permeate was pumped from the permeate tank to a boron removal system where residual boron was removed to less than 1 mg/L using a boron-selective ion-exchange resin. Analytical results showed that the boron removal system achieved boron removal requirements throughout the pilot test.

Performance of the boron removal system was excellent. Analytical data indicated that boron in the combined permeate was approximately 2 ppm and was reduced to <0.1 ppm in the final product water.

10.2.6.8 Ammonia Removal

After the boron removal system, the reverse osmosis permeate passed through a SAC ion-exchange system to remove residual ammonia to less than 1.0 mg/L. The ion-exchange media was regenerated using sodium chloride brine and in conjunction with the upstream softeners. The regenerant brine was sent to the brine concentrate tank. Permeate water leaving the

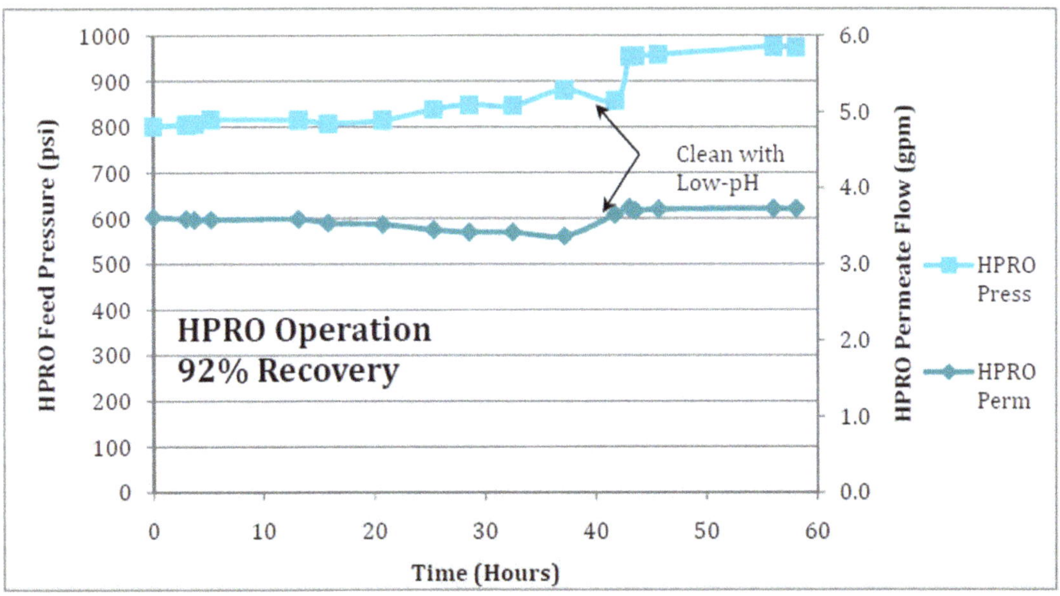

FIGURE 7.11 High-pressure reverse osmosis performance recovery and flows.

ammonia removal system was stored in the product tank. This served as a supply of makeup water for regeneration of the ion-exchange systems and for reverse osmosis cleaning solutions. Excess water was discharged back to the outfall pipeline as needed.

Analytical data showed that ammonia was removed to below discharge requirements throughout the pilot test. Ammonia in the combined reverse osmosis permeate feeding the ammonia SAC ion-exchange system ranged from 7.5 to 16 ppm. Ammonia concentrations in the final product water after the ammonia removal system were <0.5 ppm.

10.2.7 Water Quality Results

Table 7.4 summarizes the feed, effluent, and brine concentrate water qualities when operating the reverse osmosis system at 92% overall recovery.

10.2.8 Biological Methanol Treatment

The end user identified the need to remove methanol late in the pilot process. To remove soluble organics, biological pilot-scale treatment was not performed; however, the initial data from the pilot testing, as described, was used to model the performance of the expected biological treatment processes.

A number of options for removing methanol from the produced water were investigated. Under typical operations, methanol is expected to be present at approximately 150 mg/L, but could range as high as 300 mg/L. The produced water also includes 1200 mg/L of COD, with peaks as high as 2400 mg/L. The treatment requirements for methanol and COD are 1 mg/L and 25 mg/L, respectively.

Biodegradation testing was performed to determine the potential effectiveness of biological removal of COD and methanol. The testing demonstrated that biological treatment was effective for methanol and COD removal. Complications during the test did not allow for test conditions to be optimized. The primary issue was that the activated sludge used to seed the test samples was more dilute than anticipated, which resulted in volatile suspended solids concentrations lower than desired. Even so, methanol and COD removal was significant.

The following three potential options for treating the methanol waste stream were identified from the testing results:

1. Installation of sequencing batch reactor (SBR) equipment in one of the existing onsite influent tanks for full COD removal
2. Installation of a high-rate activated sludge facility following the influent tanks for methanol removal only

TABLE 7.4 Pilot water quality performance results at 92% reverse osmosis recovery.

Sample Date	Raw Influent	Reverse Osmosis Reject	Final Effluent
Commons			
pH (s.u.)	7.6	11.8	6.7
Conductivity (us/cm)	11 500	62 500	302
TDS (mg/L)	7380	41 000	186
Alkalinity (mg/L as $CaCO_3$)	2620	5600	58
Ammonia as N (mg/L)	4.78	17	0.12
Silica (dissolved) (mg/L)	83.9	N/A*	0.4
Boron (dissolved) (mg/L)	12.3	N/A	<0.05
Anions			
Chloride (mg/L)	2400	16 000	49
Sulfate (mg/L)	10	17 000	2
Cations			
Calcium (mg/L)	21	<0.02	<1
Magnesium (mg/L)	4	5.9	<1
Potassium (mg/L)	27	<0.04	<1
Sodium (mg/L)	2770	7100	65
Other Metals			
Aluminum (dissolved) (mg/L)	0.99	N/A	<0.03
Cadmium (dissolved) (mg/L)	<0.001	N/A	<0.001
Chromium (dissolved) (mg/L)	<0.005	N/A	<0.005
Copper (dissolved) (mg/L)	<0.005	N/A	<0.005
Iron (mg/L)	2.08	<0.01	<0.30
Lead (dissolved) (mg/L)	0.002	N/A	<0.001
Selenium (dissolved) (mg/L)	0.005	N/A	<0.001
Zinc (dissolved) (mg/L)	0.07	N/A	0.01
Organics			
GRO (mg/L)	70	1.9	<0.40
DRO (mg/L)	15	13	<0.30
Total petroleum hydrocarbons (mg/L)	74	15	0.022
TOC (mg/L)	292	N/A	52.5

*N/A = not available

3. Installation of a conventional activated sludge facility following the influent tanks for full COD removal

The following would occur under the SBR option:

- Methanol would be removed to less than 1 mg/L.
- The treatment process would remove all of the ammonia. In fact, additional ammonia nitrogen is required to maintain biological activity. By removing the ammonia in the influent tank, the ammonia polishing system could be removed from the tail end of the treatment process.
- The treatment process would be capable of decomposing most of the organics in the water, including free oil and grease and DRO/GRO. Depending on the efficiency of removal, this may eliminate the need for the OPTIPORE™ system.

Because of the lower overall cost and ability to remove all the COD and methanol from the feed water, it was recommended that further investigation of the feasibility of option 1 (SBR) through bench-scale and, if necessary, pilot-scale testing be performed.

Because the treatment process consisted of novel treatment approaches for this application, a primary objective was to provide proof of concept of the treatment process; analytical and field data provided demonstrated the process was effective at achieving treated water for discharge as well as reuse. Additionally, the pilot generated data for future full-scale facility design, operations, and costs. Ultimately, a full-scale facility was constructed by the end user based on data and technology generated from this pilot.

10.3 Snack Foods Pilot for Water Reuse for Food Contact

10.3.1 Overview

A confidential snack foods production facility in the southwestern United States was the site of two pilot tests to evaluate pretreatment technologies to prevent fouling of a newly installed LPRO system that was intended to produce recycled water. This important project was the first food industry to reuse treated process water back in contact with food, so the recycle water quality had to meet drinking water standards. Pilot testing was used onsite when the newly constructed full-scale process water treatment and recycling facility experienced initial fouling of the LPRO system when it was placed online during facility startup. The upstream treatment processes before the LPRO included an oil and grease separator, screening, pH adjustment, primary clarifiers, membrane bioreactors with ultrafilters for membranes, and ultraviolet disinfection.

An initial small-scale pilot test evaluated multiple treatment technologies that might remove the foulant(s) and allow reuse of reverse osmosis permeate in the food processing facility. This initial testing determined that GAC would effectively remove the foulant(s), so a full-scale temporary GAC treatment was installed for long-term pilot testing.

10.3.2 Pilot Description

The pilot was to test various membranes that might be used instead of the installed LPRO membranes that fouled and to determine if GAC could remove the foulant. The first small-scale pilot included a custom-designed reverse osmosis membrane unit to mimic the full-scale LPRO system. A feed tank was batch filled with flow from the existing process water treatment facility and served as a reservoir to feed the pilot pretreatment processes upstream of the LPRO.

The pilot was from a feed tank supplied with flow from the existing process wastewater treatment facility. Subsequently, this initial pilot was run as a continuous-feed process with a slip-stream of pretreated influent continuously flowing through the feed tank. Reject and process effluent were discharged to the process sewer for retreatment through the facility.

10.3.3 Approach to Pilot Setup

Because the process treatment facility was in operation except for the the final treatment process of LPRO for reuse of water, the unexpected fouling needed to be resolved quickly to allow recycle of treated water as designed. Reviews were performed of all safety data sheets to determine the source(s) of chemicals that might cause the LPRO fouling. Several thousand chemicals were reviewed.

Concurrent to the pilot, samples of the LPRO influent were analyzed by five different laboratories to determine potential foulants. This included mass spectrophotometry and chemical testing procedures. Concurrent to these tests, membrane autopsies and scanning-electron microscopy were performed by LPRO vendors using the setup of the initial pilot testing unit.

10.3.4 Pilot Results

The first tests were run on the pilot, three with different LPRO membranes. For each test, the transmembrane pressures were measured as the flowrate was held constant. After each test run, a different LPRO membrane was used, including the same membrane type as designed in the full-scale system The LPRO membranes and the nanofiltration membrane fouled within 15 minutes of initial flow.

As part of the pilot study, one test included running the flow through a GAC canister and then through the pilot membrane. The test through the carbon canister showed no pressure rise through the first hour of operation, indicating removal of some or all of the foulant on the GAC. As verification, the carbon canister was removed, and then the testing was restarted. The pressure curve then showed an immediate rising trend similar to the other test curves without carbon indicating fouling.

Test 2 was performed using a nanofiltration membrane instead of a reverse osmosis membrane to see if the fouling prevailed on the nanofilter. This showed a similar response as the reverse osmosis membranes with a short pressure rise also indicating fouling.

A GAC cartridge filter was placed upstream of the LPRO membrane in the pilot testing apparatus to determine if it would remove the still unidentified foulant chemical(s). The GAC prefilter extended the LPRO run-time to more than a week without appreciable increase in LPRO trans-membrane pressure (TMP). Because considerable costs were being incurred to identify the chemical(s) causing the fouling, the owner and the design-build agreed to add GAC before the LPRO process.

10.3.5 Pilot Large-Scale Testing

To provide some water for reuse in production as desired by the owner, and to verify long-term protection of the LPRO membranes with the GAC, a larger, near-full-scale "pilot" treatment system was designed and installed. This consisted of portable fiberglass pressure vessels filled with GAC configured for parallel operation and temporary polyvinyl chloride (PVC) piping to allow for flow in parallel through these units. Piping also allowed for backwashing of the GAC. Figure 7.12 presents a photograph of this temporary pilot installation.

10.3.6 Pilot Conclusions

Because the larger GAC units protected the reverse osmosis membranes from fouling as verified in both pilot tests and achieved the desired water quality for reuse, the owner decided to install full-scale GAC. The pilot tests convinced upper management of the need for the additional GAC process. Two GAC units were permanently designed and installed with cartridge filters downstream to catch GAC carbon fines before the LPRO membranes. An additional advantage to the recycled water quality was that the GAC also removed a slight amount of residual color. The full-scale treatment and recycling facility has been operating for more than 8 years with 75% of the process wastewater treated and reused for contact with food in snack food production. The GAC is changed approximately two times per year.

FIGURE 7.12 Near full-scale pilot for GAC.

Production facility staff prefer using the recycled water because its water quality for TDS is lower than the local municipal water supply.

10.4 BIOPHARMACEUTICAL PILOT

10.4.1 Introduction

Water is essential to the biopharmaceutical manufacturing process. With the recent 2011 to 2015 California drought and increasing concerns about water security, a California-based biopharmaceutical company recognized the importance of water conservation and water independence. The company set a goal to reduce water use by 20% from 2010 consumptions levels (or reduction by 303 000 m^3/yr [80 mil. gal/yr]) by 2020.

One approach for achieving significant savings would be to use recycled industrial water rather than potable water for on-campus nonpotable uses. To remain independent of outside sources, it was decided to capture, treat, and reuse water from one of the significant production process areas. Because this water source had not been tested before and there was limited analytical data, pilot testing of the proposed treatment train was recommended.

Piloting efforts from November 2017 through April 2018 evaluated microfiltration and reverse osmosis as the recommended technologies for treating biopharmaceutical wastewater in a future water reuse facility. The source water for the pilot was from a waste neutralization cascade (WNC) process, a treatment system that combines production waste streams (production broths and cleaning solutions, but no sanitary waste) from three nearby facilities and adjusts the pH to between 6 and 8 before discharge to the sanitary sewer. The pilot was designed and operated by an engineering consulting firm and constructed onsite by a local contractor.

10.4.2 Pilot Goals

The goals of the pilot were as follows:

- Assess if the proposed treatment train (microfiltration/reverse osmosis) was capable of producing water suitable for use in cooling towers, boilers, cage wash, toilets, and irrigation systems. The most stringent water quality requirements from this group of users were established as the treatment target for the recycle system. In the case where a given water quality parameter did not have a clear target, the goal was to meet or exceed current city water quality.
- Evaluate the reverse osmosis rejection efficiency of organic contaminants, present in the WNC effluent.
- Determine the variability of the WNC effluent in terms of temperature and various water quality parameters.
- Determine the variability of the concentrate and waste streams and their discharge compliance in a potential full-scale 6.3 L/s (100-gpm) facility.
- Evaluate the viability and efficiency of generating electrical energy from the COD-rich reverse osmosis concentrate.

10.4.3 Pilot Setup

Figure 7.13 shows the pilot setup, and Figures 7.14 and 7.15 show the pilot equipment. A heat exchanger maintained influent temperatures below 95 °F (35 °C) to protect the microfiltration membranes. Monitoring of the WNC effluent temperature showed variability to approximately 106 °F (41 °C). The microfiltration and reverse osmosis were leased from reputable providers of full-scale equipment. The microfiltration was a 20-gpm (1.26-L/s) skid with two full-scale polyvinylidene fluoride (PVDF) hollow-fiber membrane modules. The reverse osmosis unit was a 10-gpm (0.63-L/s) closed-circuit reverse osmosis (CCRO) skid with three Dow Filmtec® reverse osmosis membranes. This unit was contained within a shipping container. The

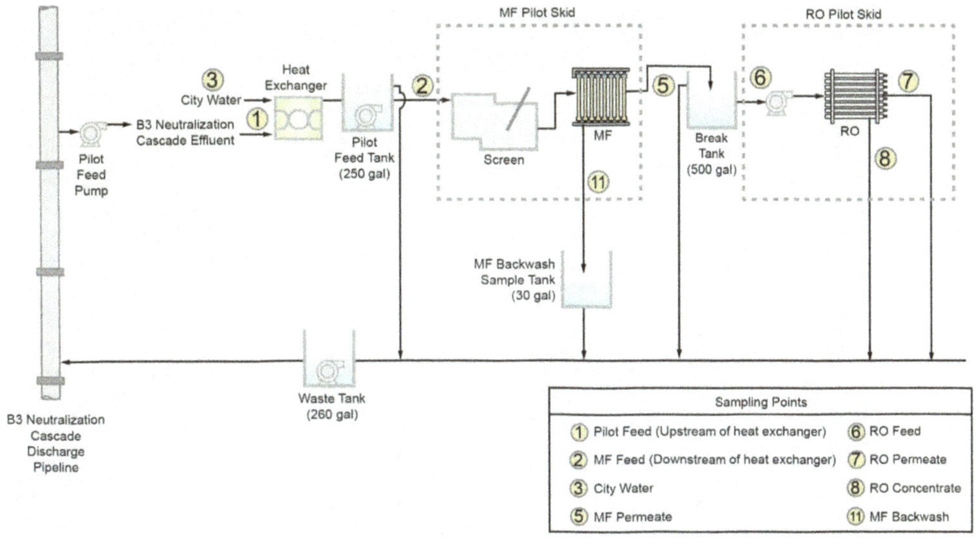

FIGURE 7.13 Pilot process flow diagram and sampling points.

FIGURE 7.14 Microfiltration equipment.

FIGURE 7.15 Closed-circuit reverse osmosis pilot equipment.

CCRO skid was chosen to allow for variable recovery corresponding with highly variable influent TDS levels.

A process flow diagram for the pilot is shown in Figure 7.13, including locations of sampling points throughout the pilot system. Physical parameters were monitored onsite daily, and a wide suite of water quality parameters were laboratory tested weekly.

10.4.4 Influent Characteristics

Table 7.5 presents a summary of the influent water quality during the pilot. The influent water showed a wide variability in COD, TDS, and TOC, among other parameters. The water was low in metals and hardness.

10.4.5 Target Water Qualities

Target water qualities were developed through discussions with operators of the boilers, cooling towers, and other facilities. Irrigation guidelines were taken from the University of California Committee of Consultants (UCCC, 1974) and Water Quality for Agriculture (Ayers & Westcot, 1985). Where specific parameters were not measured or developed, the target water quality was set to meet or exceed city water quality. Because all water users would draw from a common water distribution system, the user with the highest water quality needs dictated the treatment targets for the pilot system.

TABLE 7.5 Feed water characteristics.

Parameters	Units	Reporting Limit[c]	Microfiltration Feed			
			Average	Minimum	Maximum	Count
Physical Parameters						
pH[a]	s.u.	1	6.9	5.1	8.5	76
Electrical conductivity[a]	µS/cm	1	1564	216	8040	77
Temperature[a]	°F	0	74	52	93	74
Turbidity[a]	NTU	0.02	47	4.1	800	75
TSS	mg/L	25	99	5	690	17
Organic Compounds						
TOC	mg/L	0.5	802	22	7900	16
BOD	mg/L	5	1085	9	10 000	15
COD	mg/L	50	1155	170	5000	14
Soluble COD	mg/L	50	949	160	3800	14
Inorganic Compounds[b]						
TDS	mg/L	10	965	330	3370	15
Calcium	mg/L	0.5	3.9	1.2	14.0	15
Magnesium	mg/L	0.5	1.3	0.5	4.6	16
Barium	mg/L	0.005	0.007	0.005	0.016	16
Iron	mg/L	0.1	0.3	0.1	0.6	15
Manganese	mg/L	0.005	0.0	0.0	0.0	14
Copper	mg/L	0.005	0.034	0.011	0.140	16
Lead	mg/L	0.005	0.052	ND	0.750	16
Alkalinity, total as $CaCO_3$	mg/L	10	163	57	470	15
Boron	mg/L	100	0.108	0.100	0.140	14
Ammonia-N	mg/L	0.1	1.3	0.1	4.9	13

(*continued*)

TABLE 7.5 Feed water characteristics. (*Continued*)

Parameters	Units	Reporting Limit[c]	Microfiltration Feed Average	Minimum	Maximum	Count
Microbiological Constituents						
Heterotrophic plate count	CFU/mL	2	Above Range > 7380			11
Total coliform	Present/Absent		Present in 11 out of 15 Samples			15
Escherichia coli	Present/Absent		Present in 10 out of 15 Samples			14

[a]Measured onsite with handheld instruments calibrated daily.
[b]These constituents were sampled at the influent to the pilot reverse osmosis skid, which was assumed to be representative of raw water quality because dissolved constituents were not removed within the microfiltration skid.
[c]Reporting limit was the value at or above which the laboratory reports a result. The method detection limit was lower than the reporting limit.
ND = not detected; below detection limit

Because all water demands are met with city water, the target water quality ultimately became city water quality, or better.

10.4.6 Results

Table 7.6 presents a summary of the product water quality from the pilot facility. This water quality is discussed below.

The primary results of the study were as follows:

- The WNC effluent water quality (Table 7.5) was highly variable in both physical and chemical parameters, with wide ranging turbidities, suspended solids concentrations, COD concentrations, and dissolved solids concentrations.
- There were certain times when the water was not treatable by the microfiltration/reverse osmosis train. Slug loads of organic material would occasionally be sent to the pilot and instantly cause an increase in the microfiltration membrane trans-membrane pressure (TMP), which would shut down the unit. These slugs coincided with high turbidities. Therefore, the full-scale facility could and should be monitored through upstream turbidity analyzers to avoid slug load impacts.
- Figure 7.16 presents water quality samples across the treatment train to show the constituent removal in a visual manner. The color of the microfiltration feed varied depending on the activity in the processing facility.

TABLE 7.6 Summary of water quality produced by pilot system for selected parameters.

Parameters	Units	Reporting Limit[e]	City Water[a]	Reverse Osmosis Permeate			
				Average	Minimum	Maximum	Count
			Physical Parameters				
Temperature[b]	F	0		77.8	61.8	94.3	66
pH[b]	s.u.	1	8.42	6.21	5.2	9.41	67
Specific conductance[b]	µS/cm	1	86	62.0	9.9	520	67
TSS	mg/L	5	5.6	ND	ND	ND	12
Turbidity[b]	NTU	0.02	0.39	0.2	0.07	0.71	66
			Organic Compounds				
TOC[c]	mg/L	0.5	2.22	14.2	3	78	14
Dissolved organic carbon	mg/L			23.7	2.8	78[b]	4
BOD	mg/L	10	ND	26.8	5.9	130	9
COD	mg/L	50	66.8	127.3	13	650	13
			Volatile Organics				
Total trihalomethanes	µg/L	0.3		6.5	ND	29.9	16
Chloroform	µg/L	0.5	ND	6.3	1	18	15
			Inorganic Compounds				
TDS	mg/L	10	26	21	ND	58	13
Calcium	mg/L	0.5	9.5	ND	ND	ND	15
Magnesium	mg/L	0.5	1.5	ND	ND	ND	15
Alkalinity, total as $CaCO_3$	mg/L	10	25	ND	ND	ND	16
Ammonia-N[d]	mg/L	0.1	0.4	0.2	0.1	0.5	13
Sulfate	mg/L	50	6.3	ND	ND	ND	15

(continued)

TABLE 7.6 Summary of water quality produced by pilot system for selected parameters. (*Continued*)

Parameters	Units	Reporting Limit[e]	City Water[a]	Reverse Osmosis Permeate			
				Average	Minimum	Maximum	Count
Microbiological Constituents							
Heterotrophic plate count	MPN/mL	2	2655	3277	738	7380	10
Total coliform	Present/Absent		Present 1 times out of 5	Present 10 times out of 14			
Escherichia coli	Present/Absent		Present 0 times out of 5	Present 4 times out of 14			

[a]Average of five sampling events.
[b]Reverse osmosis permeate values taken onsite with handheld instrumentation. City values were measured at the laboratory.
[c]Value of 78 mg/L of TOC was an outlier. Average without this outlier was 8.8 mg/L of TOC. The maximum TOC without the outlier was 28 mg/L.
[d]Ammonia values do not include ammonia values after the addition of chloramines. Because of the chloramine addition nature at the pilot, ammonium sulfate was overdosed to ensure full chloramine formation. Therefore, the ammonia levels were much higher than they would be, even with normal chloramine addition.
[e]Reporting limit was the value at or above which the laboratory reports a result. The method detection limit was lower than the reporting limit.
ND = not detected; below detection limit

- Two thousand hours of operation were achieved on the microfiltration pilot unit. The microfiltration skid performed as well as expected at a flux of 29.7 L/m^2·h (17.5 gpd/sq ft) with daily minichemical cleans, backwashes every 30 minutes, and a CIP performed every 30 days. There appeared to be no irreversible fouling on the membranes. Chloramine addition was recommended to improve process performance further and reduce buildup in equipment caused by biological growth.
- One thousand hours of operation were achieved on the CCRO pilot unit. The CCRO skid average recovery was 86% at a flux of 17.0 L/m^2·h (10 gpd/sq ft) and a maximum recovery of 95%. A CIP was performed at the 30-day mark and the membranes experienced full recovery, indicating no permanent or irreversible fouling. A membrane autopsy was performed to determine fouling characteristics and ideal cleaning procedures for the full-scale system. The flexibility of the

FIGURE 7.16 Water samples across microfiltration/reverse osmosis treatment train.

CCRO process worked well to maximize the operating recovery of the reverse osmosis system to match the variable influent quality.

- It is well known that reverse osmosis does not remove all inorganic and organic compounds. When it comes to organic compounds it is typically accepted in the industry that molecular weight and hydrophobicity, among other factors, play a role. Typically reverse osmosis is efficient at removing compounds with a molecular weight greater than 200 g/mol. This study showed higher than expected TOC in the reverse osmosis permeate, indicating a significant amount of low molecular weight compounds were present in the feed water. This level of TOC in the finished water was higher than desired, but by using a downstream GAC treatment step it was shown that approximately 80% of the TOC could be removed, reducing TOC levels to equal or below those of city water.
- The waste streams from the treatment train (i.e., microfiltration backwash and reverse osmosis concentrate) were analyzed and shown to not exceed any existing local limits for sewer discharge. Additional acute toxicology tests were performed on the waste streams individually for the *Pimephales promelas* (fathead minnow) and were found to be nontoxic.
- Water quality analysis showed that the permeate from the reverse osmosis process was aggressive with respect to the Langelier Saturation Index. This could be addressed through the addition of calcium

carbonate to restore calcium and alkalinity and adjust the pH. However, soft water is ideal for boilers, so it was recommended to have a dedicated sidestream feed the boilers directly.

- The microfiltration/reverse osmosis system was shown to reliably treat the source water to a quality that meets or exceeds the quality of city water (Table 7.6) and can be used for cooling tower makeup, irrigation, boiler makeup, boiler quench water, cage washing, and toilet flushing. Stabilization of the water is necessary to prevent corrosion within the building plumbing systems and improve the sodium adsorption ratio for irrigation.
- The organic-rich reverse osmosis concentrate was sent to an offsite laboratory to evaluate potential energy generation through microbial fuel cells. The results showed a potential energy production of up to 1.2 kW for a 6.5-L/s (100-gpm) full-scale system. However, because microbial fuel cells are still in development, the return on investment is not yet justifiable and such an energy recovery system would not be included in the full-scale system.

Pilot testing was a valuable and necessary part of this project that provided four important functions. Firstly, it provided an opportunity to obtain more information about the chemical constituents of the raw feed stream (WNC effluent) and its variability in terms of flow and quality, both critical variables for design. Secondly, it confirmed the technical feasibility of the microfiltration/reverse osmosis treatment train as a robust approach to treat the WNC effluent to produce a high-quality product that met or exceeded the quality of city water, making it suitable for replacing city water used for cooling tower makeup, irrigation, boiler makeup, boiler quench water, cage washing, and toilet flushing. The pilot testing also identified additional treatment steps that would be needed to achieve the desired finished water quality, such as the GAC system, for example. Without pilot testing, this addition process step would not have been identified. Thirdly, it provided management valuable information about the operating requirements for a full-scale system so they could prepare for how long term operation would be accomplished. Fourthly, it increased the level of confidence in the company management that a full-scale system could be successfully implemented to treat and recover valuable water from the WNC effluent.

The proposed full-scale treatment train comprises a heat exchanger, chloramine addition, microfiltration, reverse osmosis, GAC, and ultraviolet disinfection, followed by a remineralization with calcium carbonate and final pH adjustment. The number of treatment processes creates a robust train, but also increases the cost and operational complexity.

11.0 CHAPTER SUMMARY

Pilot-scale testing can be effective in determining treatment technology effectiveness, predicting O&M costs, and verifying capital improvement budgets that a treatment facility can achieve MLD and required recycle water quality. If the schedule allows it, most projects will benefit greatly from the insight gained from the time and money invested in pilot testing a proposed treatment system.

12.0 REFERENCES

American Public Health Association; American Water Works Association; & Water Environment Federation. (2017). *Standard methods for the examination of water and wastewater treatment* (23rd ed.). American Public Health Association.

Ayers, R. S., & Westcot, D. W. (1985). Water quality for agriculture. Food and Agriculture Organization of the United Nations Irrigation and Drainage Paper 29 Rev. 1. Food and Agriculture Organization of the United Nations.

Current Good Manufacturing Practice in Manufacturing, Packing, or Holding Human Food, 21 C.F.R. 110.

National Primary and Secondary Drinking Water Standards, Title 40-Protection of the Environment, Chapter I-EPA, Parts 141 and 143.

Organization for Economic Cooperation and Development. (2008). *OECD environmental outlook to 2030.* http://www.oecd.org/env/indicators-modelling-outlooks/40200582.pdf

University of California Committee of Consultants. (1974). *Guidelines for interpretations of water quality for irrigation.* Technical Bulletin. University of California Committee of Consultants.

Index

CPSIA information can be obtained
at www.ICGtesting.com
Printed in the USA
BVHW090221040322
630600BV00007B/34

9 781572 783607